# INSTRUCTOR'S SOLUTIONS

*to accompany*

# APPLIED CALCULUS

**Deborah Hughes-Hallett**
*Harvard University*

**Andrew M. Gleason**
*Harvard University*

**Patti Frazer Lock**
*St. Lawrence University*

**Daniel E. Flath**
*University of South Alabama*

et al.

Prepared by

Elliot Marks
Kyle Niedzwiecki
Radoslav Mladineo
Srdjan Divac
Rebecca Rapoport
Alex Mallozzi
Adrian Iovita
Halip Saifi
Ted Pyne
Mary Prisco
Laura Piscitelli
Nicola Viegi

John Wiley & Sons, Inc.
New York • Chichester • Weinheim • Brisbane • Singapore • Toronto

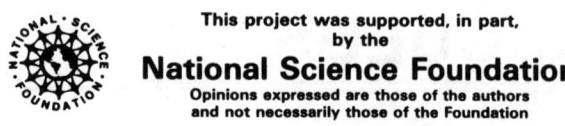
This project was supported, in part, by the
**National Science Foundation**
Opinions expressed are those of the authors and not necessarily those of the Foundation

COVER PHOTO: Tom Bean/Tony Stone Images

Copyright © 1999 by John Wiley & Sons, Inc.

Excerpts from this work may be reproduced by instructors for distribution on a not-for-profit basis for testing or instructional purposes only to students enrolled in courses for which the textbook has been adopted. *Any other reproduction or translation of this work beyond that permitted by Sections 107 or 108 of the 1976 United States Copyright Act without the permission of the copyright owner is unlawful. Requests for permission or further information should be addressed to the Permissions Department, John Wiley & Sons, Inc., 605 Third Avenue, New York, NY 10158-0012.*

ISBN 0-471-17352-5

Printed in the United States of America

10 9 8 7 6 5 4 3 2

Printed and bound by Malloy Lithographing, Inc.

# CONTENTS

CHAPTER 1 .................................. 1

CHAPTER 2 ................................103

CHAPTER 3 ................................139

CHAPTER 4 ................................173

CHAPTER 5 ................................203

CHAPTER 6 ................................259

CHAPTER 7 ................................315

CHAPTER 8 ................................367

APPENDIX ................................415

# CHAPTER ONE

## Solutions for Section 1.1

1. Between 1950 and 1995, we have

$$\text{Average rate of change} = \frac{\text{Change in marine catch}}{\text{Change in years}}$$
$$= \frac{91 - 17}{1995 - 1950}$$
$$= \frac{74}{45}$$
$$= 1.64 \text{ million tons/year}$$

   Between 1950 and 1995, marine catch increased at an average rate of 1.64 million tons each year.

2. (a) The change between 1950 and 1990

$$= \text{Production in 1990} - \text{Production in 1950}$$
$$= 90 - 11 \quad \text{(in millions)}$$
$$= 79 \text{ million bicycles.}$$

   (b) The average rate of change $R$ is the change in amounts (from part (a)) divided by the change in time.

$$R = \frac{90 - 11}{1990 - 1950}$$
$$= \frac{79}{40}$$
$$= 1.975 \quad \text{million bicycles per year.}$$

   This means that production of bicycles has increased on average by 1.975 million bicycles per year between 1950 and 1990.

3. (a)
$$\text{Change between 1991 and 1993} = \text{Sales in 1993} - \text{Sales in 1991}$$
$$= 25021 - 19608$$
$$= 5413 \text{ million dollars.}$$

   (b)
$$\begin{aligned}\text{Average rate of change} \\ \text{between 1991 and 1993}\end{aligned} = \frac{\text{Change in sales}}{\text{Change in time}}$$
$$= \frac{\text{Sales in 1993} - \text{Sales in 1991}}{1993 - 1991}$$
$$= \frac{25021 - 19608}{1993 - 1991}$$
$$= \frac{5413}{2}$$
$$= 2706.5 \text{ million dollars per year.}$$

   This means that Pepsico's sales increased on average by 2706.5 million dollars per year between 1991 and 1993.

**2** CHAPTER ONE /SOLUTIONS

4. (a)
$$\text{Change between 1993 and 1996} = \text{Net profit in 1996} - \text{Net profit in 1993}$$
$$= 452.9 - 258.4$$
$$= 194.5 \text{ million dollars.}$$

(b)
$$\begin{aligned}\text{Average rate of change} \\ \text{between 1993 and 1996}\end{aligned} &= \frac{\text{Change in net profit}}{\text{Change in time}}\\
&= \frac{\text{Net profit in 1996} - \text{Net profit in 1993}}{1996 - 1993}\\
&= \frac{452.9 - 258.4}{1996 - 1993}\\
&= \frac{194.5}{3}\\
&= 64.833 \text{ million dollars per year.}$$

This means that the Gap's net profits increased on average by 64.83 million dollars per year between 1993 and 1996.

(c) The average rate of change was negative, $-19.2$ million dollars per year, between 1991 and 1992.

5. (a) The total change in the public debt during this 13-year period was $4351.2 - 907.7 = 3443.5$ billion dollars.
(b) The average rate of change is given by

$$\begin{aligned}\text{Average rate of change} \\ \text{of the public debt} \\ \text{between 1980 and 1993}\end{aligned} &= \frac{\text{Change in public debt}}{\text{Change in time}}\\
&= \frac{4351.2 - 907.7}{1993 - 1980}\\
&= \frac{3443.5 \text{ billion dollars}}{13 \text{ years}}\\
&= 264.88 \text{ billion dollars per year.}$$

The units for the average rate of change are units of public debt over units of time, or billions of dollars per year. Between 1980 and 1993, the public debt of the United States increased at an average rate of 264.88 billion dollars per year. (This represents an increase of $725,700,000, over 700 million dollars, every day!)

6. Between 1980 and 1985
$$\text{Average rate of change } = \frac{\Delta D}{\Delta t} = \frac{1823.1 - 907.7}{1985 - 1980} = \frac{915.4}{5} = 183.1 \text{ billion dollars per year.}$$

Between 1985 and 1993
$$\text{Average rate of change } = \frac{\Delta D}{\Delta t} = \frac{4351.2 - 1823.1}{1993 - 1985} = \frac{2528.1}{8} = 316.0 \text{ billion dollars per year.}$$

7. (a) The value of exports is higher in 1990 than in 1960. The 1960 figure looks like 600 and the 1990 figure looks like 3300, making the 1990 figure 2700 billion dollars higher than the 1960 figure.
(b) The average rate of change $R$ is the change in values divided by the change in time.
$$R = \frac{3300 - 600}{1990 - 1960}$$
$$= \frac{2700}{30}$$
$$= 90 \text{ billion dollars per year.}$$

Between 1960 and 1990, the value of world exports has increased by an average of 90 billion dollars per year.

8. (a) The average rate of change, $R$, is the change in number of games divided by the change in time.

$$\text{Average from 1983 to 1989} = \frac{8677 - 7957}{1989 - 1983} = \frac{720}{6}$$
$$= 120 \text{ games per year}$$

(b) For each of the years from 1983–1989, the average rate of change is:

$$1983 - 84 : 8029 - 7957 = 72$$
$$1984 - 85 : 240$$
$$1985 - 86 : 91$$
$$1986 - 87 : 220$$
$$1987 - 88 : 7$$
$$1988 - 89 : 90$$

(c)
$$\text{Average of the six figures in part (b)} = \frac{72 + 240 + 91 + 220 + 7 + 90}{6}$$
$$= \frac{720}{6} = 120, \text{ which is the same as part (a)}.$$

9. (a) Negative. Rain forests are being continually destroyed to make way for housing, industry and other uses.
   (b) Positive. Virtually every country has a population which is increasing, so the world's population must also be increasing overall.
   (c) Negative. Since a vaccine for polio was found, the number of cases has dropped every year to almost zero today.
   (d) Negative. As time passes and more sand is eroded, the height of the sand dune decreases.
   (e) Positive. As time passes the price of just about everything tends to increase.

10. (a) The average rate of change $R$ is the difference in amounts divided by the change in time.

$$R = \frac{50.5 - 35.6}{1993 - 1987}$$
$$= \frac{14.9}{6}$$
$$\approx 2.5 \text{ billion dollars/yr}$$

This means that in the years between 1987 and 1993, the amount of money spent on tobacco increased at a rate of approximately $2,500,000,000 per year.

(b) To have a negative rate of change, the amount spent has to decrease during one of these 1 year intervals. Looking at the data, one can see that between 1992 and 1993, the amount spent on tobacco products decreased by 0.4 billion dollars. Thus, the average rate of change is negative between 1992 and 1993.

11. Between 1930 and 1990:

$$\text{Average rate of change} = \frac{103,905 - 29,424}{60}$$
$$= \frac{74,481}{60}$$
$$\approx 1,241 \text{ thousand people/year}$$

This means that, between 1930 and 1990, the labor force increased by an average of 1,241,000 workers per year. Between 1930 and 1950:

$$\text{Average rate of change} = \frac{45,222 - 29,424}{20}$$
$$= \frac{15,798}{20}$$
$$= 789.9 \text{ thousand people/year}$$

**4** CHAPTER ONE /SOLUTIONS

This means that, between 1930 and 1950, the labor force increased by an average of 789,900 workers per year. Between 1950 and 1970:

$$\text{Average rate of change} = \frac{70,920 - 45,222}{20}$$
$$= \frac{25,698}{20}$$
$$\approx 1,285 \quad \text{thousand people/year}$$

This means that, between 1950 and 1970, the labor force increased by an average of about 1,285,000 workers per year.

12. (a)
$$\text{Change between 1991 and 1995} = \text{Sales in 1995} - \text{Sales in 1991}$$
$$= 16202.0 - 4778.6$$
$$= 11423.4 \text{ million dollars.}$$

(b)
$$\begin{aligned}\text{Average rate of change} \\ \text{between 1991 and 1995}\end{aligned} = \frac{\text{Change in sales}}{\text{Change in time}}$$
$$= \frac{\text{Sales in 1995} - \text{Sales in 1991}}{1995 - 1991}$$
$$= \frac{16202.0 - 4778.6}{1995 - 1991}$$
$$= \frac{11423.4}{4}$$
$$= 2855.95 \text{ million dollars per year.}$$

This means that Intel's sales increased on average by 2855.85 million dollars per year between 1991 and 1995.

(c) The average rate of change of sales between 1995 and 1997 is

$$\frac{25070 - 16202}{1997 - 1995} = \frac{8868}{2} = 4434 \text{ million dollars per year.}$$

If sales continue to increase at 4434 million dollars per year, then

$$\begin{aligned}\text{Sales in 1998} &= \text{Sales in 1997} + 4434 = 29504 \text{ million dollars per year.} \\ \text{Sales in 1999} &= \text{Sales in 1998} + 4434 = 33938 \text{ million dollars per year.} \\ \text{Sales in 2000} &= \text{Sales in 1999} + 4434 = 38372 \text{ million dollars per year.} \\ \text{Sales in 2001} &= \text{Sales in 2000} + 4434 = 42806 \text{ million dollars per year.}\end{aligned}$$

If Intel's sales continue to increase at the rate of 4434 million dollars per year, then sales will first reach 40000 million dollars in the year 2001.

13. (a) The average rate of change $R$ of the sperm count is

$$R = \frac{66 - 113}{1990 - 1940} = -0.94 \text{ million sperm per milliliter per year.}$$

(b) We want to find how long it will take 66 million to drop to 20 million, given that annual rate of change is $-0.94$. We write

$$66 + n(-0.94) = 20$$
$$n(-0.94) = -46.$$

Solving for $n$ gives

$$n \approx 49 \text{ years.}$$

The average sperm count would go below 20 million in 2039.

14. (a)
$$\text{Change between 1989 and 1997} = \text{Revenues in 1997} - \text{Revenues in 1989}$$
$$= 172000 - 123212$$
$$= 48788 \text{ million dollars.}$$

(b)
$$\begin{aligned}\text{Average rate of change} \\ \text{between 1989 and 1997}\end{aligned} = \frac{\text{Change in revenues}}{\text{Change in time}}$$
$$= \frac{\text{Revenues in 1997} - \text{Revenues in 1989}}{1997 - 1989}$$
$$= \frac{172000 - 123212}{1997 - 1989}$$
$$= \frac{48788}{8}$$
$$= 6098.5 \text{ million dollars per year.}$$

This means that General Motors' revenues increased on average by 6098.5 million dollars per year between 1989 and 1997.

(c) From 1987 to 1997 there were two one-year time intervals during which the average rate of change in revenues was negative: $-1191$ million dollars per year between 1989 and 1990, and $-4760$ million dollars per year between 1995 and 1996.

## Solutions for Section 1.2

1. (a) The story in (a) matches Graph (IV), in which the person forgot her books and had to return home.
   (b) The story in (b) matches Graph (II), the flat tire story. Note the long period of time during which the distance from home did not change (the horizontal part).
   (c) The story in (c) matches Graph (III), in which the person started calmly but sped up later.
   
   The first graph (I) does not match any of the given stories. In this picture, the person keeps going away from home, but his speed decreases as time passes. So a story for this might be: *I started walking to school at a good pace, but since I stayed up all night studying calculus, I got more and more tired the farther I walked.*

2. (a) The statement $f(1000) = 3500$ means that when $a = 1000$, we have $S = 3500$. In other words, when \$1000 is spent on advertising, the number of sales per month is 3500.
   (b) Graph I, because we expect that as advertising expenditures go up, sales will go up (not down).
   (c) The vertical intercept represents the value of $S$ when $a = 0$, or the sales per month if no money is spent on advertising.

3.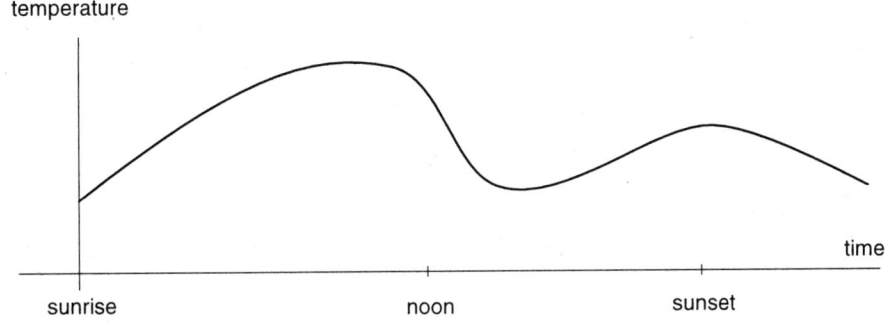

**4.**

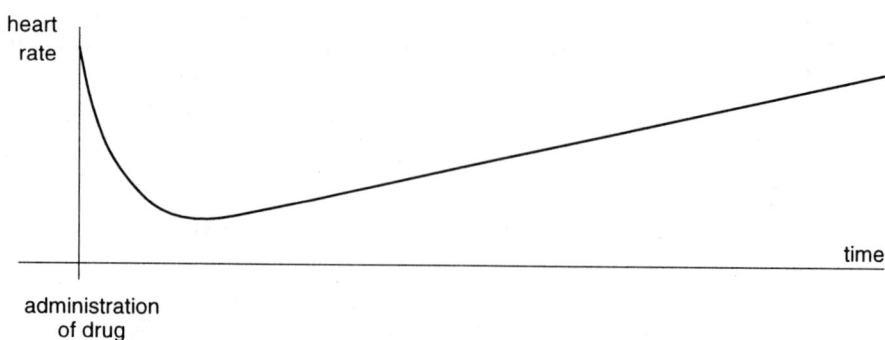

**5.**

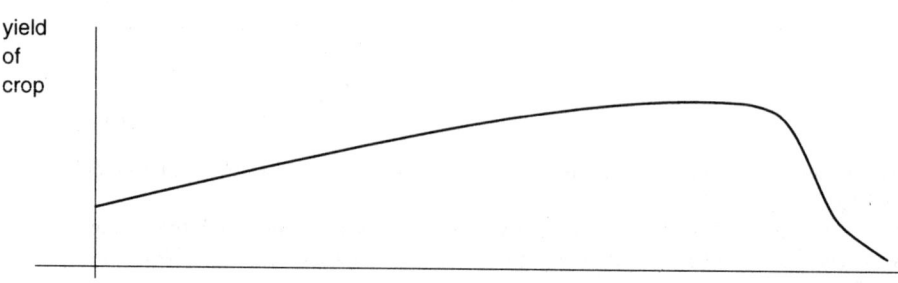

6. At first, as the number of workers increases, productivity also increases. As a result, the curve goes up initially. At a certain point the curve reaches its highest level, after which it goes downward; in other words, as the number of workers increases beyond that point, productivity decreases. This might, for example, be due either to the inefficiency inherent in large organizations or simply to workers getting in each other's way as too many are crammed on the same line. Many other reasons are possible.

7. (a) From the graph, we see $f(3) = 0.14$. This means that after 3 hours, the level of nicotine is 0.14 mg.
   (b) About 4 hours.
   (c) The vertical intercept is 0.4. It represents the level of nicotine in the blood right after the cigarette is smoked.
   (d) A horizontal intercept would represent the value of $t$ when $N = 0$, or the number of hours until all nicotine is gone from the body.
   (e) Over the first three hours, we have

$$\begin{aligned}\text{Average rate of change} &= \frac{f(3) - f(0)}{3 - 0} = \frac{0.14 - 0.4}{3 - 0} \\ &= \frac{-0.26}{3} = -0.087.\end{aligned}$$

   The average rate of change is $-0.087$ mg/hour. In other words, nicotine is leaving the body at the rate of 0.087 mg per hour.
   (f) Negative, because the level of nicotine is going down (not up).

**8.**

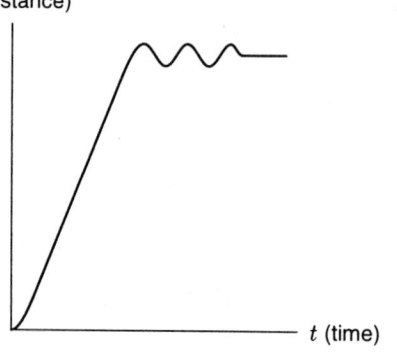

**9.**

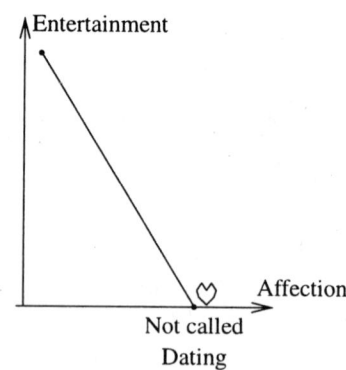

10. (a)
$$f(5) = 2(5) + 3 = 10 + 3 = 13.$$
(b)
$$f(5) = 10(5) - (5)^2 = 50 - 25 = 25.$$
(c) We want the $y$-coordinate of the graph at the point where its $x$-coordinate is 5. Looking at the graph, we see that the $y$-coordinate of this point is 3. Thus
$$f(5) = 3.$$
(d) Looking at the graph, we see that the point on the graph with an $x$-coordinate of 5 has a $y$-coordinate of 2. Thus
$$f(5) = 2.$$
(e) In the table, we must find the value of $f(x)$ when $x = 5$. Looking at the table, we see that when $x = 5$ we have
$$f(5) = 4.1$$

11. (a) At $p = 0$, we see $r = 8$. At $p = 3$, we see $r = 7$.
(b) When $p = 2$, we see $r = 10$. Thus, $f(2) = 10$.

12. (a) We are asked for the value of $y$ when $x$ is zero. That is, we are asked for $f(0)$. Plugging in we get
$$f(0) = (0)^2 + 2 = 0 + 2 = 2.$$
(b) Substituting we get
$$f(3) = (3)^2 + 2 = 9 + 2 = 11.$$
(c)

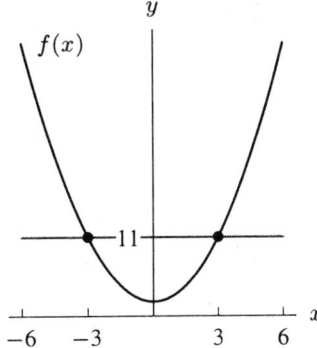

*Figure 1.1*

Looking at Figure 1.1, we see that the graph of $f(x)$ intersects the line $y = 11$ at $x = 3$ and $x = -3$. Thus, when $x$ equals 3 or $x$ equals $-3$ we have $f(x) = 11$.

We can also solve this problem with algebra. Asking what values of $x$ give a $y$-value of 11 is the same as solving
$$y = 11 = x^2 + 2$$
$$x^2 = 9$$
$$x = \pm\sqrt{9} = \pm 3$$

This method also gives the answer: $x$ equals 3 or $-3$.

(d) No. No matter what, $x^2$ is greater than or equal to 0, so $y = x^2 + 2$ is greater than or equal to 2.

13. (a) Substituting $x = 1$ gives $f(1) = 3(1) - 5 = 3 - 5 = -2$.
(b) We substitute $x = 5$:
$$y = 3(5) - 5 = 15 - 5 = 10.$$
(c) We substitute $y = 4$ and solve for $x$:
$$4 = 3x - 5$$
$$9 = 3x$$
$$x = 3$$
(d) Average rate of change $= \dfrac{f(4) - f(2)}{4 - 2} = \dfrac{7 - 1}{2} = \dfrac{6}{2} = 3.$

14. (a) When the car is 5 years old, it is worth $6000.
    (b) Since the value of the car decreases as the car gets older, a possible graph is shown in Figure 1.2.

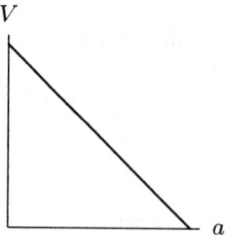

**Figure 1.2**

(c) The vertical intercept is the value of V when $a = 0$, or the value of the car when it is new. The horizontal intercept is the value of $a$ when $V = 0$, or the age of the car when it is worth nothing.

15. (a) Recall that the velocity is the ratio of the distance traveled to the time spent traveling. Thus, between $t = 0$ and $t = 15$, we get average velocity $v$:

$$v = \frac{s(15) - s(0)}{15 - 0}$$
$$= \frac{105 - 0}{15}$$
$$= 7 \text{ ft/sec}$$

Between $t = 10$ and $t = 30$, the car has the average velocity $v$:

$$v = \frac{s(30) - s(10)}{30 - 10}$$
$$= \frac{410 - 55}{20}$$
$$= 17.75 \text{ ft/sec}$$

(b) Between $t = 10$ and $t = 30$, the distance traveled is

$$d = s(30) - s(10)$$
$$= 410 - 55$$
$$= 355 \text{ ft}$$

16. Between $x = 2$ and $x = 10$, the average rate of change

$$= \frac{f(10) - f(2)}{10 - 2} = \frac{1 - 5}{8} = -\frac{4}{8}$$
$$= -0.5$$

It is negative because the curve is sloping downward.

17. The average rate of change $R$ between $x = 1$ and $x = 3$ is

$$R = \frac{f(3) - f(1)}{3 - 1}$$
$$= \frac{18 - 2}{2}$$
$$= \frac{16}{2}$$
$$= 8.$$

## Solutions for Section 1.3

1.  (a) is (V), because slope is positive, vertical intercept is negative
    (b) is (IV), because slope is negative, vertical intercept is positive
    (c) is (I), because slope is 0, vertical intercept is positive
    (d) is (VI), because slope and vertical intercept are both negative
    (e) is (II), because slope and vertical intercept are both positive
    (f) is (III), because slope is positive, vertical intercept is 0

2.  (a) is (V), because slope is negative, vertical intercept is 0
    (b) is (VI), because slope and vertical intercept are both positive
    (c) is (I), because slope is negative, vertical intercept is positive
    (d) is (IV), because slope is positive, vertical intercept is negative
    (e) is (III), because slope and vertical intercept are both negative
    (f) is (II), because slope is positive, vertical intercept is 0

3.  Rewriting the equation as $y = -\frac{5}{2}x + 4$ shows that the slope is $-\frac{5}{2}$ and the vertical intercept is 4.

4.  Slope = $\frac{6-0}{2-(-1)} = 2$ so the equation is $y - 6 = 2(x - 2)$ or $y = 2x + 2$.

5.  $y - c = m(x - a)$

6.  Looking at the graph in the problem, we see that when $x = 0$, the function $f$ takes on the approximate value of 5. Thus
    $$f(0) \approx 5.$$
    When $x = 1$, the function $f$ takes on the approximate value of 7.5. Thus
    $$f(1) \approx 7.5.$$
    When $x = 3$, the function $f$ takes on the approximate value of 12.5. Thus
    $$f(3) \approx 12.5.$$

7.  The intercepts appear to be (0,3) and (7.5,0), giving
    $$\text{Slope} = \frac{-3}{7.5} = -\frac{6}{15} = -\frac{2}{5}.$$
    The $y$-intercept is at (0,3), so
    $$y = -\frac{2}{5}x + 3$$
    is a possible equation for the line (answers may vary).

8.  (a) On the interval from 0 to 1 the value of $y$ decreases by 2. On the interval from 1 to 2 the value of $y$ decreases by 2. And on the interval from 2 to 3 the value of $y$ decreases by 2. Thus, the function has a constant rate of change and it could therefore be linear.
    (b) On the interval from 15 to 20 the value of $s$ increases by 10. On the interval from 20 to 25 the value of $s$ increases by 10. And on the interval from 25 to 30 the value of $s$ increases by 10. Thus, the function has a constant rate of change and could be linear.
    (c) On the interval from 1 to 2 the value of $w$ increases by 5. On the interval from 2 to 3 the value of $w$ increases by 8. Thus, we see that the slope of the function is not constant and so the function is not linear.

9.  For the function given by table (a), we know that the slope is
    $$\text{slope} = \frac{27 - 25}{0 - 1} = -2.$$
    We also know that at $x = 0$ we have $y = 27$. Thus we know that the vertical intercept is 27. The formula for the function is
    $$y = -2x + 27.$$

**10** CHAPTER ONE /SOLUTIONS

For the function in table (b), we know that the slope is

$$\text{slope} = \frac{72-62}{20-15} = \frac{10}{5} = 2.$$

Thus, we know that the function will take on the form

$$s = 2t + b.$$

Substituting in the coordinates $(15, 62)$ we get

$$s = 2t + b$$
$$62 = 2(15) + b$$
$$= 30 + b$$
$$32 = b$$

Thus, a formula for the function would be

$$s = 2t + 32.$$

10. (a) Finding slope $(-50)$ and intercept gives $q = 1000 - 50p$.
    (b) Solving for $p$ gives $p = 20 - 0.02q$.

11. Given that the equation is linear, choose any two points, e.g. $(5.2, 27.8)$ and $(5.3, 29.2)$. Then

$$\text{Slope} = \frac{29.2 - 27.8}{5.3 - 5.2} = \frac{1.4}{0.1} = 14$$

Using the point-slope formula, with the point $(5.2, 27.8)$, we get the equation

$$y - 27.8 = 14(x - 5.2)$$

which is equivalent to

$$y = 14x - 45.$$

12. (a) We know that when $x = 0$, we have

$$4y = -12$$
$$y = -3$$

Thus the $y$-intercept is $-3$. When $y = 0$ we have

$$3x = -12$$
$$x = -4$$

Thus the $x$-intercept is $-4$.

(b) For the line $3x + 4y = -12$, the $x$-intercept is at $-4$ and the $y$-intercept is at $-3$. The distance between these two points is

$$d = \sqrt{(-4-0)^2 + (0-(-3))^2} = \sqrt{16+9} = \sqrt{25} = 5.$$

13. (a) We know that the function for $q$ in terms of $p$ will take on the form

$$q = mp + b.$$

We know that the slope will represent the change in $q$ over the corresponding change in $p$. Thus

$$m = \text{slope} = \frac{4-3}{12-15} = \frac{1}{-3} = -\frac{1}{3}.$$

Thus, the function will take on the form

$$q = -\frac{1}{3}p + b.$$

Substituting the values $q = 3, p = 15$, we get

$$3 = -\frac{1}{3}(15) + b$$
$$3 = -5 + b$$
$$b = 8.$$

Thus, the formula for $q$ in terms of $p$ is

$$q = -\frac{1}{3}p + 8.$$

(b) We know that the function for $p$ in terms of $q$ will take on the form

$$p = mq + b.$$

We know that the slope will represent the change in $p$ over the corresponding change in $q$. Thus

$$m = \text{slope} = \frac{12 - 15}{4 - 3} = -3.$$

Thus, the function will take on the form

$$p = -3q + b.$$

Substituting the values $q = 3, p = 15$ again, we get

$$15 = (-3)(3) + b$$
$$15 = -9 + b$$
$$b = 24.$$

Thus, a formula for $p$ in terms of $q$ is

$$p = -3q + 24.$$

14. (a) We know that the function will take on the form

$$f(t) = mt + b.$$

We know that the slope will be

$$\text{slope} = \frac{18.48 - 19.72}{1 - 0} = -1.24.$$

We also know that when $t = 0$, we have $f(t) = 19.72$. Thus, the vertical intercept is

$$b = 19.72.$$

Hence, we get

$$f(t) = -1.24t + 19.72.$$

(b)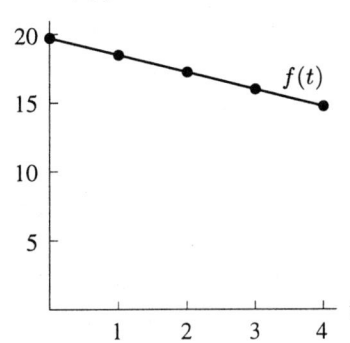

**12** CHAPTER ONE /SOLUTIONS

15. We know that the linear approximation of this function must take on the form
$$Q = mt + b$$
where $t$ is measured in years since 1986. We know that the total change in the gold reserve from the years 1986 ($t = 0$) to 1990 ($t = 4$) is $14.76 - 19.72 = -4.96$. Thus, the slope for the function would be
$$\text{slope} = \frac{-4.96}{4 - 0} = -1.24.$$
We also know that the function takes on a value of 19.72 at the year 1986 (i.e., at $t = 0$). Thus, if our function is to give the amount of gold in the reserve in years after 1986, it will take on the form
$$Q = -1.24t + 19.72 \quad \text{(answer may vary)}.$$

16. We know that the function approximating the data in the graph will take on the form
$$w = mh + b.$$
We know that on the interval from 68 inches to 75 inches the average weight changes from 167 pounds to 200 pounds. Thus, we have
$$\text{slope} = \frac{200 - 167}{75 - 68} = \frac{33}{7} \approx 4.7.$$
Thus, the function will look like
$$w = 4.7h + b.$$
Plugging in the point (68, 167) we get
$$167 = 4.7(68) + b$$
$$= 319.6 + b$$
$$b = -152.6.$$
Thus, a formula for weight in pounds as a function of height in inches is
$$w = 4.7h - 152.6 \quad \text{(answer may vary)}.$$
Note that our approximation holds only for a limited domain since it makes no sense to say that if someone is 0 inches tall, they will weigh $-152.6$ pounds.

The slope of the line is 4.7 pounds per inch. This means that as height goes up by 1 inch, weight increases by about 4.7 pounds.

17. By joining consecutive points we get a line whose slope is the average rate of change. The steeper this line, the greater the average rate of change. See Figure 1.3.

(a)  $C$ and $D$. Steepest slope.
     $B$ and $C$. Slope closest to 0.
(b)  $A$ and $B$, and $C$ and $D$ gives the 2 slopes closest to each other.

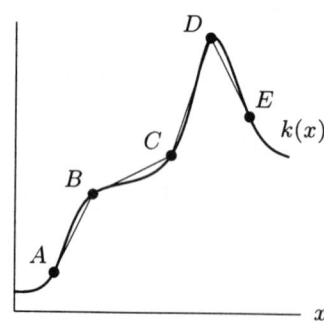

*Figure 1.3*

18. (a)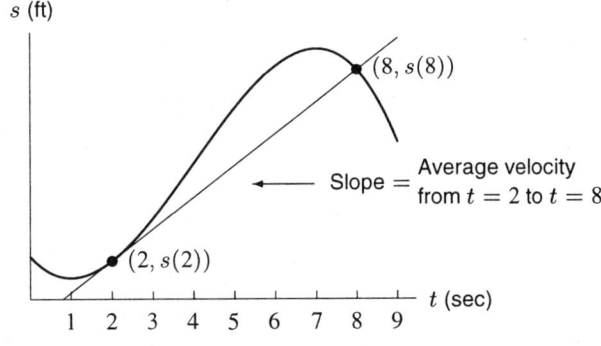

Figure 1.4

(b)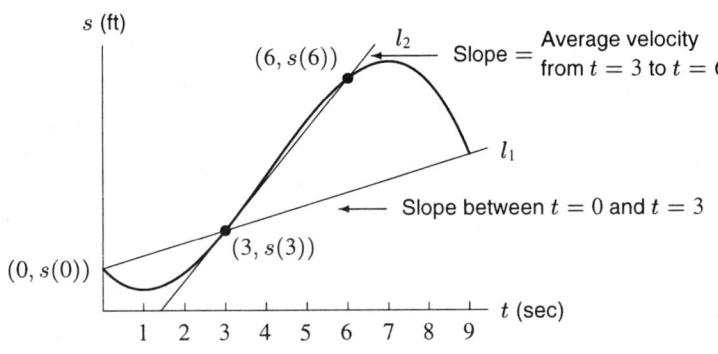

Figure 1.5

The average velocity between two times can be represented by the slope of the line joining these points on the position curve. In Figure 1.5, the average velocity between $t = 0$ and $t = 3$ is equal to the slope of $l_1$, while the average velocity between $t = 3$ and $t = 6$ is equal to the slope of $l_2$. We can see that the slope of $l_2$ is greater than the slope of $l_1$ and so the average velocity between $t = 3$ and $t = 6$ is greater.

(c) Since the average velocity can be represented as the slope of the line between the two values given, we just have to look at the slope of the line passing through $(6, s(6))$ and $(9, s(9))$. We notice that the distance traveled at $t = 6$ is greater than the distance traveled at $t = 9$. Thus, the line between those two points will have negative slope, and thus, the average velocity will be negative.

19.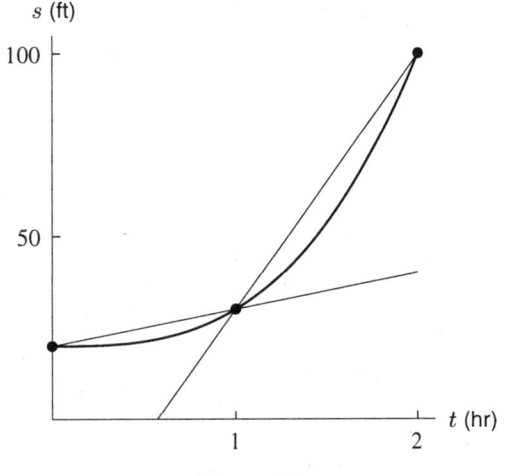

Figure 1.6

The graph in Figure 1.6 satisfies the properties. The graph is always increasing so has positive slope for all lines between any two of its points, making the average velocity positive for all intervals. Furthermore, the line corresponding to the interval from $t = 1$ to $t = 2$ has a greater slope than the interval from $t = 0$ to $t = 1$, making the average velocity of the first half less than that of the second half of the trip.

**14** CHAPTER ONE /SOLUTIONS

20. Each of these questions can also be answered by considering the slope of the line joining the 2 relevant points.
    (a) The average rate of change is positive if the weekly sales are increasing with time and negative if weekly sales are decreasing.
        (i) Since weekly sales are rising from 500 to 1000 from $t = 0$ to $t = 5$, the average rate of change is positive.
        (ii) We can see that the weekly sales at $t = 10$ are greater than the weekly sales at $t = 0$. Thus, the average rate of change is positive.
        (iii) We can see that the weekly sales at $t = 15$ are lower than the weekly sales at $t = 0$. Thus, the average rate of change is negative.
        (iv) We can see that the weekly sales at $t = 20$ are greater than the weekly sales at $t = 0$. Thus, the average rate of change is positive.
    (b) (i) The secant line between $t = 0$ and $t = 5$ is steeper than the secant line between $t = 0$ and $t = 10$, so the slope of the secant line is greater on $0 \leq t \leq 5$. Since average rate of change is represented graphically by the slope of a secant line, the rate of change in the interval $0 \leq t \leq 5$ is greater than that in the interval $0 \leq t \leq 10$.
        (ii) The slope of the secant line between $t = 0$ and $t = 20$ is greater than the slope of the secant line between $t = 0$ and $t = 10$, so the rate of change is larger for $0 \leq t \leq 20$.
    (c) The average rate of change in the interval $0 \leq t \leq 10$ is about
    $$\frac{750 - 500}{10} = \frac{250}{10} = 25 \text{ sales/week/week}$$
    This tells us that for the first ten weeks, the number of weekly sales is growing at an average rate of 25 sales per week per week.

21. The average rate of change $R$ from $x = -2$ to $x = 1$ is:
    $$R = \frac{f(1) - f(-2)}{1 - (-2)} = \frac{3(1)^2 + 4 - (3(-2)^2 + 4)}{1 + 2} = \frac{7 - 16}{3} = -3$$

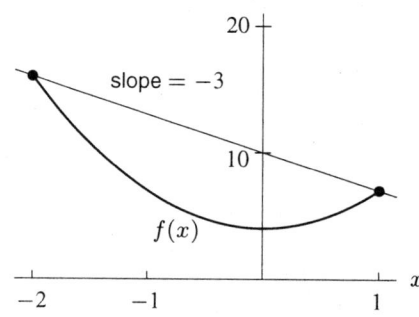

22. (a) The first company's price for a day's rental with $m$ miles on it is $C_1(m) = 40 + 0.15m$. Its competitor's price for a day's rental with $m$ miles on it is $C_2(m) = 50 + 0.10m$.
    (b)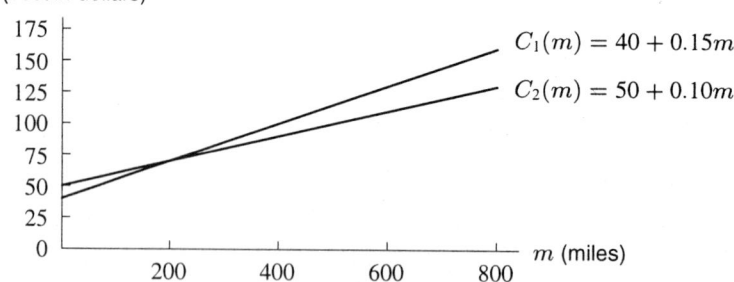
    (c) To find which company is cheaper, we need to determine where the two lines intersect. We let $C_1 = C_2$, and thus
    $$40 + 0.15m = 50 + 0.10m$$
    $$0.05m = 10$$
    $$m = 200.$$
    If you are going more than 200 miles a day, the competitor is cheaper. If you are going less than 200 miles a day, the first company is cheaper.

23. (a) Given the two points $(0, 32)$ and $(100, 212)$, and assuming the graph in Figure 1.7 is a line,
$$\text{Slope} = \frac{212 - 32}{100} = \frac{180}{100} = 1.8.$$

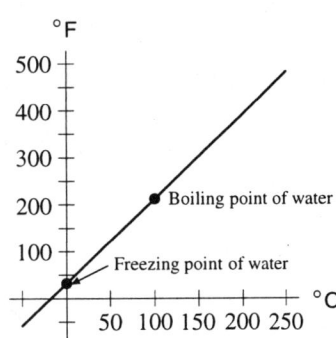

Figure 1.7

(b) The °F-intercept is $(0, 32)$, so
$$°F = 1.8(°C) + 32.$$

(c) If the temperature is 20°Celsius, then
$$°F = 1.8(20) + 32 = 68°F.$$

(d) If $°F = °C$, then
$$°C = 1.8°C + 32$$
$$-32 = 0.8°C$$
$$°C = -40° = °F.$$

24.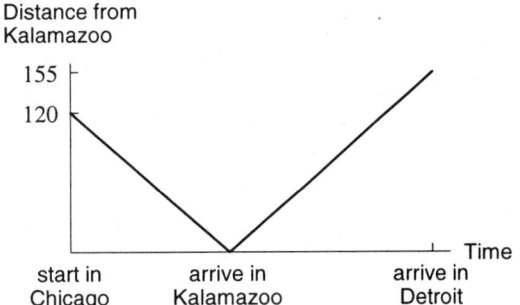

## Solutions for Section 1.4

1. (a) We know that regardless of the number of rides one takes, one must pay $7 to get in. After that, for each ride you must pay another $1.50, thus the function $R(n)$ is
$$R(n) = 7 + 1.5n.$$

(b) Substituting in the values $n = 2$ and $n = 8$ into our formula for $R(n)$ we get
$$R(2) = 7 + 1.5(2) = 7 + 3 = \$10.$$
This means that admission and 2 rides costs $10.
$$R(8) = 7 + 1.5(8) = 7 + 12 = \$19.$$
This means that admission and 8 rides costs $19.

**16** CHAPTER ONE /SOLUTIONS

2. (a) The fixed cost is the cost that would have to be paid even if nothing was produced. That is, the fixed cost is
$$C(0) = 4000 + 2(0) = \$4000.$$

(b) We know that for every additional item produced, the company must pay another \$2. Thus \$2 is the variable cost.

(c) We know that the revenue takes on the form
$$R(q) = p \cdot q$$
where $q$ is the quantity produced and $p$ is the price the company is charging for an item. Thus in our case,
$$p = \frac{R(q)}{q} = \frac{10q}{q} = \$10.$$
So the company is charging \$10 per item.

(d)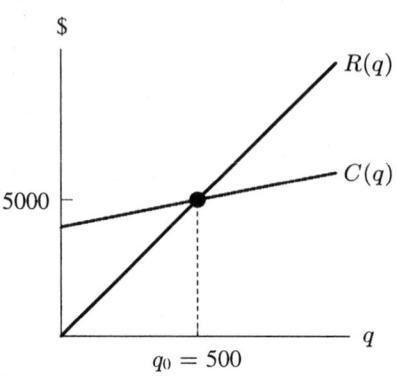

**Figure 1.8**

We know that the company will make a profit for $q > q_0$ since the line $R(q)$ lies above the line $C(q)$ in that region, so revenues are greater than costs.

(e) Looking at Figure 1.8 we see that the two graphs intersect at the point where
$$q_0 = 500.$$
Thus, the company will break even if it produces 500 units. Algebraically, we know that the company will break even for $q_0$ such that the cost function is equal to the revenue function at $q_0$. That is, when
$$C(q_0) = R(q_0).$$
Solving we get
$$C(q_0) = R(q_0)$$
$$4000 + 2q_0 = 10q_0$$
$$8q_0 = 4000$$
$$q_0 = 500.$$
And
$$C(500) = R(500) = \$5,000.$$

3. We know that the fixed cost is the cost that the company would have to pay if no items were produced. Thus
$$\text{Fixed cost} = C(0) = \$5000.$$
We know that the cost function is linear and we know that the slope of the function is exactly the variable cost. Thus
$$\text{Slope} = \text{Variable cost} = \frac{5020 - 5000}{5 - 0} = \frac{20}{5} = 4 \text{ dollars per unit produced.}$$
Thus, the marginal cost is 4 dollars per unit produced. We know that since $C(q)$ is linear
$$C(q) = m \cdot q + b,$$
where $m$ is the slope and $b$ is the value of $C(0)$, the vertical intercept. Or in other words, $m$ is equal to the variable cost and $b$ is equal to the fixed cost. Thus
$$C(q) = 4q + 5000.$$

4. (a) If we think of $q$ as a linear function of $p$, then $q$ is the dependent variable, $p$ is the independent variable, and the slope $m = \Delta q/\Delta p$. We can use any two points to find the slope. If we use the first two points, we get

$$\text{Slope} = m = \frac{\Delta q}{\Delta p} = \frac{460-500}{18-16} = \frac{-40}{2} = -20.$$

The units are the units of $q$ over the units of $p$, or tons per dollar. The slope tells us that, for every dollar increase in price, the number of tons sold every month will decrease by 20.

To write $q$ as a linear function of $p$, we need to find the vertical intercept, $b$. Since $q$ is a linear function of $p$, we have $q = b + mp$. We know that $m = -20$ and we can use any of the points in the table, such as $p = 16$, $q = 500$, to find $b$. Substituting gives

$$q = b + mp$$
$$500 = b + (-20)(16)$$
$$500 = b - 320$$
$$820 = b.$$

Therefore, the vertical intercept is 820 and the equation of the line is

$$q = 820 - 20p.$$

(b) If we now consider $p$ as a linear function of $q$, we have

$$\text{Slope} = m = \frac{\Delta p}{\Delta q} = \frac{18-16}{460-500} = \frac{2}{-40} = -\frac{1}{20} = -0.05.$$

The units of the slope are dollars per ton. The slope tells us that, if we want to sell one more ton of the product every month, we should reduce the price by $0.05.

Since $p$ is a linear function of $q$, we have $p = b + mq$ and $m = -0.05$. To find $b$, we substitute any point from the table such as $p = 16$, $q = 500$ into this equation:

$$p = b + mq$$
$$16 = b + (-0.05)(500)$$
$$16 = b - 25$$
$$41 = b.$$

The equation of the line is

$$p = 41 - 0.05q.$$

Alternatively, notice that we could have taken our answer to part (a), that is $q = 820 - 20p$, and solved for $p$.

5. (a) The cost of producing 500 units is

$$C(500) = 6000 + 10(500) = 6000 + 5000 = \$11{,}000.$$

The revenue the company makes by selling 500 units is

$$R(500) = 12(500) = \$6000.$$

Thus, the cost of making 500 units is greater than the money the company will make by selling the 500 units, so the company does not make a profit.

The cost of producing 5000 units is

$$C(5000) = 6000 + 10(5000) = 6000 + 50000 = \$56{,}000.$$

The revenue the company makes by selling 5000 units is

$$R(5000) = 12(5000) = \$60{,}000.$$

Thus, the cost of making 5000 units is less than the money the company will make by selling the 5000 units, so the company does make a profit.

(b) The break-even point is the number of units that the company has to produce so that in selling those units, it makes as much as it spent on producing them. That is, we are looking for $q$ such that

$$C(q) = R(q).$$

Solving we get

$$C(q) = R(q)$$
$$6000 + 10q = 12q$$
$$2q = 6000$$
$$q = 3000.$$

Thus, if the company produces and sells 3000 units, it will break even.

Graphically, the break-even point, which occurs at (3000, $36,000), is the point at which the graphs of the cost and the revenue functions intersect. (See Figure 1.9.)

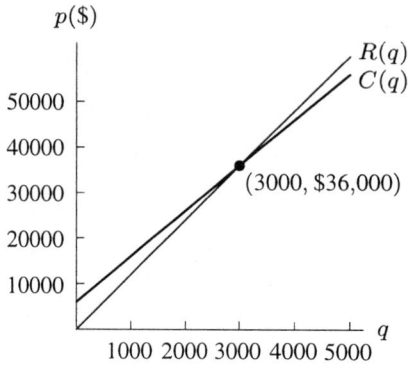

**Figure 1.9**

6. (a) We have $C(q) = 6000 + 2q$ and $R(q) = 5q$ and so

$$\pi(q) = R(q) - C(q) = 5q - (6000 + 2q) = -6000 + 3q.$$

(b) See Figure 1.10. We find the break-even point, $q_0$, by setting the revenue equal to the cost and solving for $q$:

$$\text{Revenue} = \text{Cost}$$
$$5q = 6000 + 2q$$
$$3q = 6000$$
$$q = 2000.$$

The break-even point is $q_0 = 2000$ puzzles. Notice that this is the same answer we get if we set the profit function equal to zero.

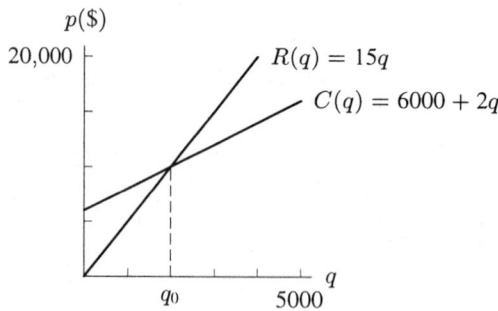

**Figure 1.10**: Cost and revenue functions for the jigsaw puzzle company

7. (a) We know that the cost function will be of the form

$$C(q) = b + m \cdot q$$

where $m$ is the slope of the graph and $b$ is the vertical intercept. We also know that the fixed cost is the vertical intercept and the variable cost is the slope. Thus, we have

$$C(q) = 5000 + 30q.$$

We know that the revenue function will take on the form

$$R(q) = pq$$

where $p$ is the price charged per unit. In our case the company sells the chairs at $50 a piece so

$$R(q) = 50q.$$

(b) Marginal cost is $30 per chair. Marginal revenue is $50 per chair.

(c)
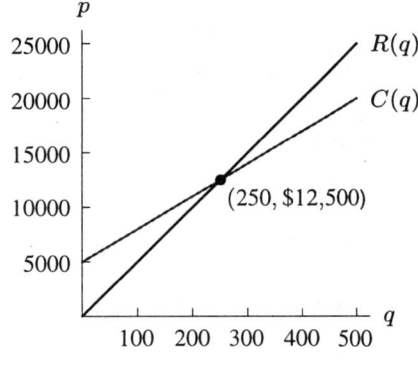

**Figure 1.11**

(d) We know that the break-even point is the number of chairs that the company has to sell so that the revenue will equal the cost of producing these chairs. In other words, we are looking for $q$ such that

$$C(q) = R(q).$$

Solving we get

$$C(q) = R(q)$$
$$5000 + 30q = 50q$$
$$20q = 5000$$
$$q = 250.$$

Thus, the break-even point is 250 chairs and $12,500. Graphically, it is the point in Figure 1.11 where the cost function intersects the revenue function.

8. We know that the cost function will take on the form

$$C(q) = b + m \cdot q$$

where $m$ is the variable cost and $b$ is the fixed cost. We know that the company has a fixed cost of $350,000 and that it costs the company $400 to feed a student. That is, $400 is the variable cost. Thus

$$C(q) = 350,000 + 400q.$$

We know that the revenue function will be of the form

$$R(q) = pq$$

where $p$ is the price that the company charges a student. Since the company intends to charge $800 per student we have

$$R(q) = 800q.$$

We know that the profit is simply the difference between the revenue and the cost. Thus

$$\pi(q) = 800q - (350{,}000 + 400q) = 800q - 350{,}000 - 400q = 400q - 350{,}000.$$

We are asked to find the number of students that must sign up for the plan in order for the company to make money. That is, we are asked to find the number of students $q$ such that

$$\pi(q) > 0.$$

Solving we get

$$400q - 350{,}000 > 0$$
$$400q > 350{,}000.$$
$$q > 875$$

Thus, if more than 875 students sign up, the company will make a profit.

9. (a) The cost function is of the form

$$C(q) = b + m \cdot q$$

where $m$ is the variable cost and $b$ is the fixed cost. Since the variable cost is $20 and the fixed cost is $650,000, we get

$$C(q) = 650{,}000 + 20q.$$

The revenue function is of the form

$$R(q) = pq$$

where $p$ is the price that the company is charging the buyer for one pair. In our case the company charges $70 a pair so we get

$$R(q) = 70q.$$

The profit function is the difference between revenue and cost, so

$$\pi(q) = R(q) - C(q) = 70q - (650{,}000 + 20q) = 70q - 650{,}000 - 20q = 50q - 650{,}000.$$

(b) Marginal cost is $20 per pair. Marginal revenue is $70 per pair. Marginal profit is $50 per pair.

(c) We are asked for the number of pairs of shoes that need to be produced and sold so that the profit is larger than zero. That is, we are trying to find $q$ such that

$$\pi(q) > 0.$$

Solving we get

$$\pi(q) > 0$$
$$50q - 650{,}000 > 0$$
$$50q > 650{,}000$$
$$q > 13{,}000.$$

Thus, if the company produces and sells more than 13,000 pairs of shoes, it will make a profit.

10. (a) The fixed costs are the price of producing zero units, or $C(0)$, which is the vertical intercept. Thus, the fixed costs are roughly $75. The variable cost is the slope of the line. We know that

$$C(0) = 75$$

and looking at the graph we can also tell that

$$C(30) = 300.$$

Thus, the slope or the variable cost is

$$\text{Variable cost} = \frac{300 - 75}{30 - 0} = \frac{225}{30} = 7.50 \text{ dollars per unit.}$$

(b) Looking at the graph it seems that
$$C(10) \approx 150.$$
Alternatively, using what we know from parts (a) and (b) we know that the cost function is
$$C(q) = 7.5q + 75.$$
Thus,
$$C(10) = 7.5(10) + 75 = 75 + 75 = \$150.$$
The total cost of producing 10 items is $150.

11. (a) The company makes a profit when the revenue is greater than the cost of production. Looking at the figure in the problem, we see that this occurs whenever more than roughly 335 items are produced and sold.
    (b) If $q = 600$, revenue $\approx 2400$ and cost $\approx 1750$ so profit is about $650.

12. (a) We know that the fixed cost for this company is the amount of money it takes to produce zero units, or simply the vertical intercept of the graph. Thus, the
$$\text{fixed cost} = \$1000.$$
We know that the variable cost is the price the company has to pay for each additional unit, or in other words, the slope of the graph. We know that
$$C(0) = 1000$$
and looking at the graph we also see that
$$C(200) = 4000.$$
Thus the slope of the line, or the variable cost, is
$$\text{variable cost} = \frac{4000 - 1000}{200 - 0} = \frac{3000}{200} = \$15 \text{ per unit.}$$
(b) $C(q)$ gives the price that the company will have to pay for the production of $q$ units. Thus if
$$C(100) = 2500$$
we know that it will cost the company $2500 to produce 100 items.

13. See Figure 1.12

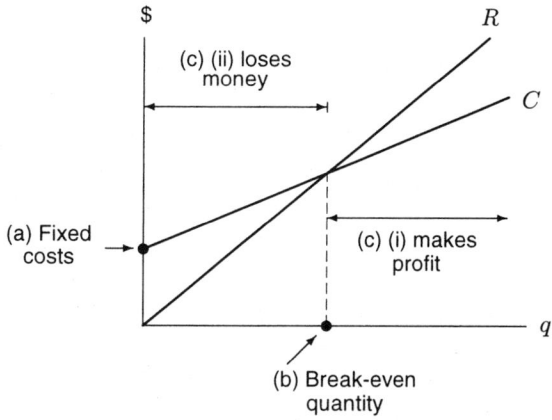

**Figure 1.12**

14. (a) A company with little or no fixed costs would be one that does not need much start-up capital and whose costs are mainly on a per unit basis. An example of such a company is a consulting company, whose major expense is the time of its consultants. Such a company would have little fixed costs to worry about.
    (b) A company with little or no variable costs would be one that can produce a product with little or no additional costs per unit. An example is a computer software company. The major expense of such a company is software development, a fixed cost. Additional copies of its software can be very easily made. Thus, its variable costs are rather small.

**22** CHAPTER ONE /SOLUTIONS

15. (a) We know that the fixed cost of the first price list is $100 and the variable cost is $0.03. Thus, the cost of making $q$ copies under the first option is
$$C_1(q) = 100 + 0.03q.$$
We know that the fixed cost of the second price list is $200 and the variable cost is $0.02. Thus, the cost of making $q$ copies under the second option is
$$C_2(q) = 200 + 0.02q.$$

(b) At 5000 copies, the first price list gives the cost
$$C_1(5000) = 100 + 5000(0.03) = 100 + 150 = \$250.$$
At 5000 copies, the second price list gives the cost
$$C_2(5000) = 200 + 5000(0.02) = 200 + 100 = \$300.$$
Thus, for 5000 copies, the first price list is cheaper.

(c) We are asked to find the point $q$ at which
$$C_1(q) = C_2(q).$$
Solving we get
$$C_1(q) = C_2(q)$$
$$100 + 0.03q = 200 + 0.02q$$
$$100 = 0.01q$$
$$q = 10,000.$$
Thus, if one needs to make ten thousand copies, the cost under both price lists will be the same.

16. (a) We know that the function for the value of the robot at time $t$ will be of the form
$$V(t) = m \cdot t + b.$$
We know that at time $t = 0$ the value of the robot is $15,000. Thus the vertical intercept $b$ is
$$b = 15,000.$$
We know that $m$ is the slope of the line. Also at time $t = 10$ the value is $0. Thus
$$m = \frac{0 - 15,000}{10 - 0} = \frac{-15,000}{10} = -1500.$$
Thus we get
$$V(t) = -1500t + 15,000 \text{ dollars}.$$

(b) The value of the robot in three years is
$$V(3) = -1500(3) + 15,000 = -4500 + 15,000 = \$10,500.$$

17. (a) We know that the function for the value of the tractor will be of the form
$$V(t) = m \cdot t + b$$
where $m$ is the slope and $b$ is the vertical intercept. We know that the vertical intercept is simply the value of the function at time $t = 0$, which is $50,000. Thus
$$b = \$50,000.$$
Since we know the value of the tractor at time $t = 20$ we know that the slope is
$$m = \frac{V(20) - V(0)}{20 - 0} = \frac{10,000 - 50,000}{20} = \frac{-40,000}{20} = -2000.$$
Thus we get
$$V(t) = -2000t + 50,000 \text{ dollars}.$$

(b)

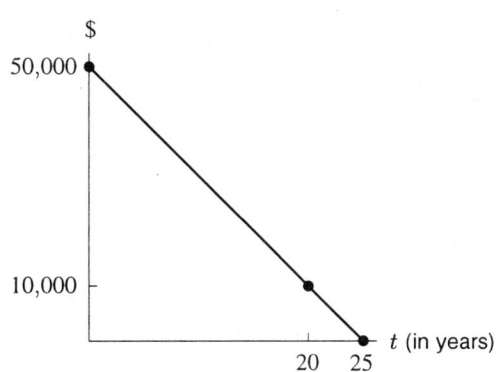

**Figure 1.13**

(c) Looking at Figure 1.13 we see that the vertical intercept occurs at the point $(0, 50{,}000)$ and the horizontal intercept occurs at $(25, 0)$. The vertical intercept tells us the value of the tractor at time $t = 0$, namely, when it was brand new. The horizontal intercept tells us at what time $t$ the value of the tractor will be \$0. Thus the tractor is worth \$50,000 when it is new, and it is worth nothing after 25 years.

18. (a) The amount spent on books will be

$$\text{Amount for books} = \$40 \cdot b$$

where $b$ is the number of books bought. The amount of money spent on outings is

$$\text{money spent on outings} = \$10 \cdot s$$

where $s$ is the number of social outings. Since we want to spend all of the \$1000 budget we end up with

$$40b + 10s = 1000.$$

(b)

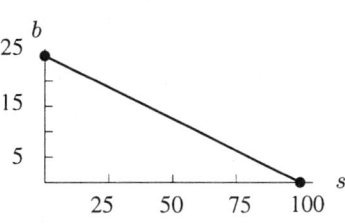

**Figure 1.14**

(c) Looking at Figure 1.14 we see that the vertical intercept occurs at the point

$$(0, 25)$$

and the horizontal intercept occurs at

$$(100, 0).$$

The vertical intercept tells us how many books we would be able to buy if we wanted to spend all of the budget on books. That is, we could buy at most 25 books. The horizontal intercept tells how many social outings we could afford if we wanted to spend all of the budget on outings. That is, we would be able to go on at most 100 outings.

19. (a) We know that the total amount the company will spend on raw materials will be

$$\text{price for raw materials} = \$100 \cdot m$$

where $m$ is the number of units of raw materials the company will buy. We know that the total amount the company will spend on paying employees will be

$$\text{total employee expenditure} = \$25{,}000 \cdot r$$

where $r$ is the number of employees the company will hire. Since the total amount the company spends is \$500,000, we get

$$25{,}000r + 100m = 500{,}000.$$

(b) Solving for $m$ we get

$$25{,}000r + 100m = 500{,}000$$
$$100m = 500{,}000 - 25{,}000r$$
$$m = 5000 - 250r.$$

(c) Solving for $r$ we get

$$25{,}000r + 100m = 500{,}000$$
$$25{,}000r = 500{,}000 - 100m$$
$$r = 20 - \frac{100}{25{,}000}m$$
$$= 20 - \frac{1}{250}m.$$

20. Generally manufacturers will produce more when prices are higher. Therefore, the first curve is a supply curve. Consumers consume less when prices are higher. Therefore, the second curve is a demand curve.

21. (a) Looking at the graph we see that it goes through the point $(20, 17)$, that is, there are 20 items bought when the price is \$17. Looking at the graph we see that it goes through the point $(50, 8)$, that is, there are 50 items bought when the price is \$8.
    (b) Looking at the graph we see that it goes through the point $(30, 13)$, that is, at the price of \$13 per item, 30 items are bought. Looking at the graph we see that it goes through the point $(10, 25)$, that is, at the price of \$25 per item, 10 items are bought.

22. (a) Looking at the graph we see that it goes through the point $(40, 49)$, that is, there are 40 items produced when the price is \$49. Looking at the graph we see that it goes through the point $(60, 77)$, that is, there are 60 items produced when the price is \$77.
    (b) Looking at the graph we see that it goes through the point $(20, 29)$, that is, at the price of \$29 per item, 20 items are produced. Looking at the graph we see that it goes through the point $(50, 62)$, that is, at the price of \$62 per item, 50 items are produced.

23. (a) We know that as the price per unit increases, the quantity supplied increases, while the quantity demanded decreases. So Table 1.20 of the text is the demand curve (since as the price increases the quantity decreases), while Table 1.21 of the text is the supply curve (since as the price increases the quantity increases.)
    (b) Looking at the demand curve data in Table 1.20 we see that a price of \$155 gives a quantity of roughly 14.
    (c) Looking at the supply curve data in Table 1.21 we see that a price of \$155 gives a quantity of roughly 24.
    (d) Since supply exceeds demand at a price of \$155, the shift would be to a lower price.
    (e) Looking at the demand curve data in Table 1.20 we see that if the price is less than or equal to \$143 the consumers would buy at least 20 items.
    (f) Looking at the data for the supply curve (Table 1.21) we see that if the price is greater than or equal to \$110 the supplier will produce at least 20 items.

24.

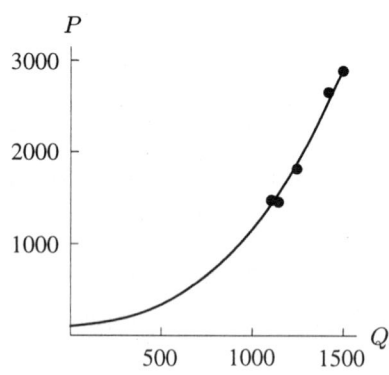

25. (a) We know that the equilibrium point is the point where the supply and demand curves intersect. Looking at the figure in the problem, we see that the price at which they intersect is \$10 per unit and the corresponding quantity is 3000 units.

(b) We know that the supply curve climbs upwards while the demand curve slopes downwards. Thus we see from the figure that at the price of $12 per unit the suppliers will be willing to produce 3500 units while the consumers will be ready to buy 2500 units. Thus we see that when the price is above the equilibrium point, more items would be produced than the consumers will be willing to buy. Thus the producers end up wasting money by producing that which will not be bought, so the producers are better off lowering the price.

(c) Looking at the point on the rising curve where the price is $8 per unit, we see that the suppliers will be willing to produce 2500 units, whereas looking at the point on the downward sloping curve where the price is $8 per unit, we see that the consumers will be willing to buy 3500 units. Thus we see that when the price is less than the equilibrium price, the consumers are willing to buy more products than the suppliers would make and the suppliers can thus make more money by producing more units and raising the price.

26. Since the price $p$ is on the vertical axis and the quantity $q$ is on the horizontal axis we would like a formula which expresses $p$ in terms of $q$. Thus

$$75p + 50q = 300$$
$$75p = 300 - 50q$$
$$p = 4 - \frac{2}{3}q.$$

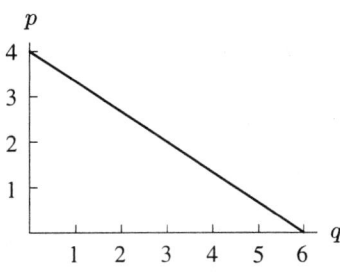

**Figure 1.15**

From the graph in Figure 1.15 we see that the vertical intercept occurs at $p = 4$ dollars and the horizontal intercept occurs at $q = 6$. This tells us that at a price of $4 or more nobody would buy the product. On the other hand even if the product were given out for free, no more than 6 units would be demanded.

27. We know that a formula for passengers versus price will take the form

$$N = mp + b$$

where $N$ is the number of passengers on the boat when the price of a tour is $p$ dollars. We know two points on the line thus we know that the slope is

$$\text{slope} = \frac{650 - 500}{20 - 25} = \frac{150}{-5} = -30.$$

Thus the function will look like

$$N = -30p + b.$$

Plugging in the point $(20, 650)$ we get

$$N = -30p + b$$
$$650 = (-30)(20) + b$$
$$= -600 + b$$
$$b = 1250$$

Thus a formula for the number of passengers as a function of tour price is

$$N = -30p + 1250.$$

28. (a) Let
$$I = \text{number of Indian peppers}$$
$$M = \text{number of Mexican peppers}.$$

Then (from the given information)
$$1{,}200I + 900M = 14{,}000$$

is the Scoville constraint.

(b) Solving for $I$ yields
$$I = \frac{14{,}000 - 900M}{1{,}200}$$
$$= \frac{35}{3} - \frac{3}{4}M.$$

29. (a) $k = p_1 s + p_2 l$ where $s = $ # of liters of soda and $l = $ # of liters of oil.

(b) If $s = 0$, then $l = \frac{k}{p_2}$. Similarly, if $l = 0$, then $s = \frac{k}{p_1}$. These two points give you enough information to draw a line containing the points which satisfy the equation.

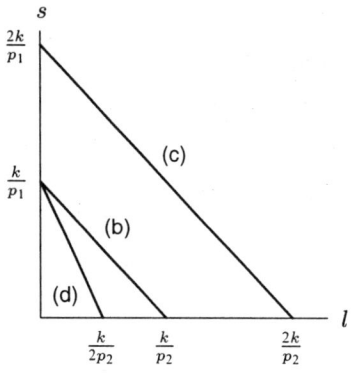

(c) If the budget is doubled, we have the constraint: $2k = p_1 s + p_2 l$. We find the intercepts as before. If $s = 0$, then $l = \frac{2k}{p_2}$; if $l = 0$, then $s = \frac{2k}{p_1}$. The intercepts are both twice what they were before.

(d) If the price of oil doubles, our constraint is $k = p_1 s + 2p_2 l$. Then, calculating the intercepts gives that the $s$ intercept remains the same, but the $l$ intercept gets cut in half. In other words, $s = 0$ means $l = \frac{k}{2p_2} = \frac{1}{2}\frac{k}{p_2}$. Therefore the maximum amount of oil you can buy is half of what it was previously.

30. (a) The equilibrium price and quantity occur when demand equals supply. If we graph these functions on the same axes, we get Figure 1.16.

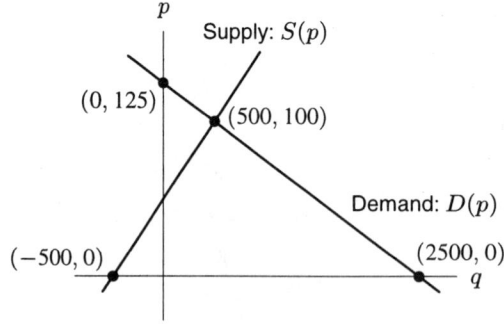

Figure 1.16

We can see from the graph in Figure 1.16 that the supply and demand curves intersect at the point (500, 100). The equilibrium price is $100 and the equilibrium quantity is 500. This answer can also be obtained algebraically, by solving

$$D(p) = S(p)$$
$$2500 - 20p = 10p - 500$$
$$3000 = 30p$$
$$p = 100$$
$$q = D(p) = D(100)$$
$$= 2500 - 20(100)$$
$$= 2500 - 2000$$
$$= 500.$$

(b) When a tax is imposed on a product, the price that the producer receives is less than the price paid by the consumer. In this case, a $6 tax is imposed on the product. If $p$ is the price the consumer pays, the quantity supplied depends on $p - 6$ and is given by:

$$\text{Quantity supplied} = S(p - 6) = 10(p - 6) - 500$$
$$= 10p - 560$$

We can find the new equilibrium price and quantity by graphing the new supply function with the demand function.

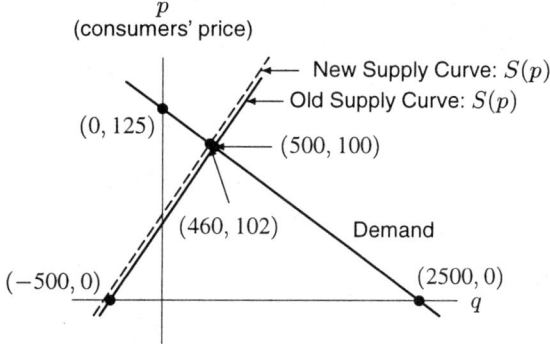

Figure 1.17: Specific tax shifts supply curve

From Figure 1.17, we see that the new equilibrium price (including tax) is $102 and the new equilibrium quantity is 460 units. We can also obtain these results algebraically:

$$\text{Demand} = \text{Supply}$$
$$2500 - 20p = 10p - 560$$
$$3060 = 30p$$
$$p = 102,$$

so $q = 10(102) - 560 = 460$.

(c) The tax paid by the consumer is $2, since the new equilibrium price of $102 is $2 more than the old equilibrium price of $100. Since the tax is $6, the producer pays $4 of the tax and receives $102 - $6 = $96 per item after taxes.

(d) The tax received by the government per unit product is $6. Thus, the total revenue received by the government is equal to the tax per unit times the number of units sold, which is just the equilibrium quantity. Thus,

$$\text{Revenue} = \text{Tax} \cdot \text{Quantity} = 6(460) = \$2760.$$

31. (a) The tax is imposed on the consumer, and so the price that the consumer pays is $p + 0.05p = 1.05p$. The demand is calculated by replacing the price $p$ by the effective price $p + 0.05p$. The new demand function is

$$\text{Quantity demanded} = D(p + 0.05p) = 100 - 2(p + 0.05p) = 100 - 2(1.05p).$$

**28** CHAPTER ONE /SOLUTIONS

The equilibrium price occurs when the new demand function equals the supply function:

$$\text{Demand} = \text{Supply}$$
$$D(1.05p) = S(p)$$
$$100 - 2(1.05p) = 3p - 50$$
$$150 = 5.1p$$
$$p = \$29.41.$$

The equilibrium quantity $q$ is given by

$$q = S(29.41) = 3(29.41) - 50 = 88.23 - 50 = 38.23 \text{ units.}$$

(b) Since the pre-tax price was \$30 and the suppliers' new price is \$29.41 per unit,

$$\text{Tax paid by supplier} = \$30 - \$29.41 = \$0.59.$$

The consumers' new price is $1.05p = 1.05(29.41) = \$30.88$ per unit and the pre-tax price was \$30, so

$$\text{Tax paid by consumer} = 30.88 - 30 = \$0.88.$$

The total tax paid per unit by suppliers and consumers together is $0.59 + 0.88 = \$1.47$.

32. (a) The sales tax affects the supply function by replacing $p$ with its effective price to the producer, that is $p - 0.05p = 0.95p$. Thus,

$$\text{Quantity supplied} = S(p - 0.05p) = 3(0.95p) - 50 = 2.85p - 50.$$

The new equilibrium price occurs when the demand curve intersects the new supply curve. This occurs when we set the supply function equal to the demand function:

$$\text{Demand} = \text{Supply}$$
$$2.85p - 50 = 100 - 2p$$
$$4.85p = 150$$
$$p \approx \$30.93.$$

The equilibrium quantity $q$ is the value of the demand at the equilibrium value:

$$S(0.95p) = D(p) = 100 - 2(30.93)$$
$$= 100 - 61.86$$
$$= 38.14 \text{ units.}$$

(b) Since the pretax price was \$30 and consumers' new price is \$30.93,

$$\text{Tax paid by consumers} = 30.93 - 30 = \$0.93$$

The supplier keeps $0.95p = 0.95(30.93) = \$29.38$ per unit, so

$$\text{Tax paid by suppliers} = 30 - 29.38 = \$0.62.$$

The total tax per unit paid by consumers and suppliers together is $0.93 + 0.62 = \$1.55$.

33. We know that at the point where the price is \$1 per scoop the quantity must be 240. Thus we can fill in the graph as follows:

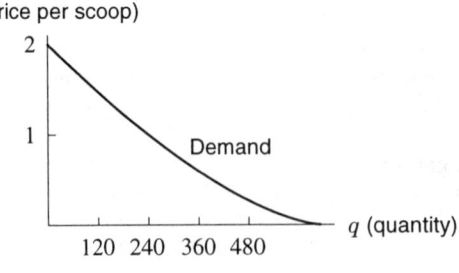

**Figure 1.18**

(a) Looking at Figure 1.18 we see that when the price per scoop is half a dollar, the quantity given by the demand curve is roughly 360 scoops.

(b) Looking at Figure 1.18 we see that when the price per scoop is $1.50, the quantity given by the demand curve is roughly 120 scoops.

34. (a)

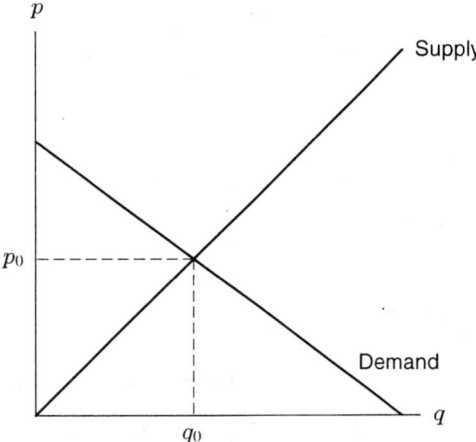

(b) If the slope of the supply curve increases then the supply curve will intersect the demand curve sooner, resulting in a higher equilibrium price $p_1$ and lower equilibrium quantity $q_1$. Intuitively, this makes sense since if the slope of the supply curve increases. The amount produced at a given price decreases.

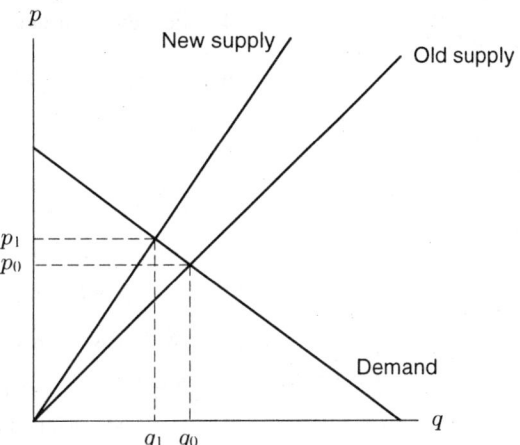

(c) When the slope of the demand curve becomes more negative, the demand function will decrease more rapidly and will intersect the supply curve at a lower value of $q_1$. This will also result in a lower value of $p_1$ and so the equilibrium price $p_1$ and equilibrium quantity $q_1$ will decrease. This follows our intuition, since if demand for a product lessens, the price and quantity purchased of the product will go down.

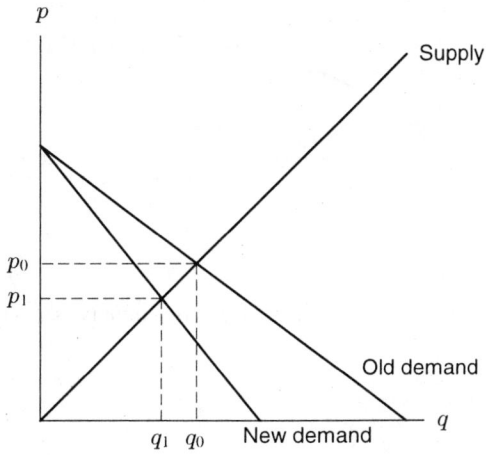

## Solutions for Section 1.5

1. One possible answer is shown in Figure 1.19.

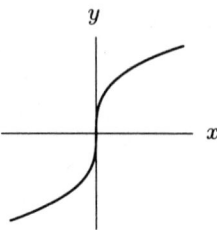

**Figure 1.19**

2. If $p$ is proportional to $t$, then $p = kt$ for some fixed constant $k$. From the values $t = 10, p = 25$, we have $25 = k(10)$, so $k = 2.5$. To see if $p$ is proportional to $t$, we must see if $p = 2.5t$ gives all the values in the table. However, when we check the values $t = 20, p = 60$, we see that $60 \neq 2.5(20)$. Thus, $p$ is not proportional to $t$.

3. For some constant $c$, we have $K = cv^2$.

4. For some constant $k$, we have $F = \dfrac{k}{d^2}$.

5. If distance is $d$, then $v = \dfrac{d}{t}$.

6. We have $V = kr^3$. You may know that $V = \dfrac{4}{3}\pi r^3$.

7. (a) $8^{2/3} = (8^{1/3})^2 = 2^2 = 4$.

   (b) $9^{-(3/2)} = (9^{1/2})^{(-3)} = 3^{(-3)} = \dfrac{1}{3^3} = \dfrac{1}{27}$.

8. $y = 3x^{-2}; k = 3, p = -2$.

9. $y = 5x^{1/2}; k = 5, p = 1/2$.

10. $y = \tfrac{3}{8}x^{-1}; k = \tfrac{3}{8}, p = -1$.

11. Not a power function.

12. $y = \tfrac{5}{2}x^{-1/2}; k = \tfrac{5}{2}, p = -1/2$.

13. $y = 9x^{10}; k = 9, p = 10$.

14. $y = 0.2x^2; k = 0.2, p = 2$

15. Not a power function

16. $y = 5^3 \cdot x^3 = 125x^3; k = 125, p = 3$

17. $y = 8x^{-1}; k = 8, p = -1$

18. $y = (1/5)x; k = 1/5, p = 1$

19. Not a power function

20. $y = x^{2/3}$ is larger as $x \to \infty$.

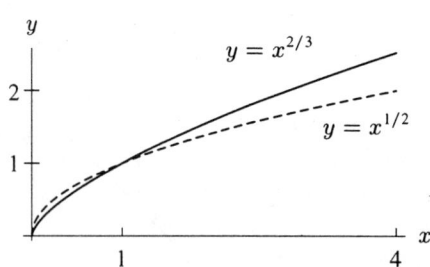

21. $y = x^4$ goes to positive infinity in both cases.

22. $y = -x^7$ goes to negative infinity as $x \to \infty$, and goes to positive infinity as $x \to -\infty$.

23.

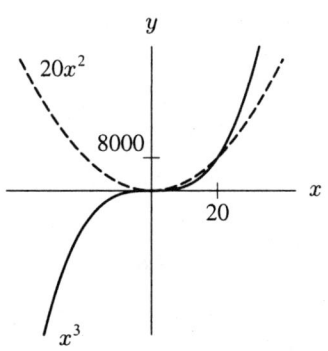

$f(x) = x^3$ is larger as $x \to \infty$.

24. As $x \to \infty$, $f(x) = x^5$ has the largest positive values. As $x \to -\infty$, $g(x) = -x^3$ has the largest positive values.

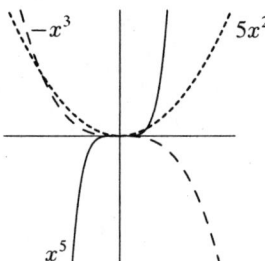

25. For $y = x$, average rate of change $= \frac{10-0}{10-0} = 1$.
For $y = x^2$, average rate of change $= \frac{100-0}{10-0} = 10$.
For $y = x^3$, average rate of change $= \frac{1000-0}{10-0} = 100$.
For $y = x^4$, average rate of change $= \frac{10000-0}{10-0} = 1000$.
So $y = x^4$ has the largest average rate of change. For $y = x$, the line is the same as the original function.

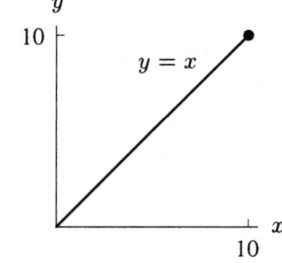
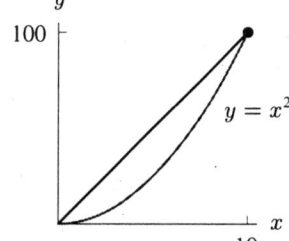
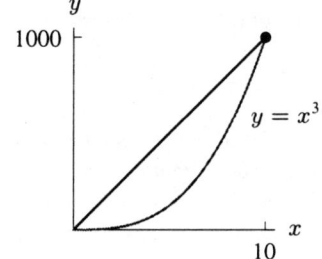
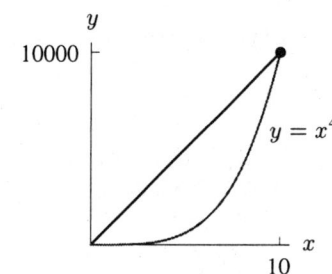

26. Since
$$S = kM^{2/3}$$
we have
$$18{,}600 = k(70^{2/3})$$
and so
$$k = 1095.$$
We have $S = 1095 M^{2/3}$. If $M = 60$, then
$$S = 1095(60^{2/3}) = 16{,}782 \text{ cm}^2.$$

27. Since $N$ is inversely proportional to the square of $L$, we have
$$N = \frac{k}{L^2}.$$
As $L$ increases, $N$ decreases, so there are more species at small lengths.

28. (a) $T$ is proportional to the fourth root of $B$, and so
$$T = k\sqrt[4]{B} = k \cdot B^{1/4}.$$
(b) $148 = k \cdot (5230)^{1/4}$ and so $k = 148/(5230)^{1/4} = 17.4$.
(c) Since $T = 17.4 B^{1/4}$, for a human with $B = 70$ we have
$$T = 17.4(70)^{1/4} = 50.3 \text{ seconds}$$
It takes about 50 seconds for all the blood in the body to circulate and return to the heart.

29. Let $M =$ blood mass and $B =$ body mass. Then $M = k \cdot B$. Using the fact that $M = 150$ when $B = 3000$, we have
$$M = k \cdot B$$
$$150 = k \cdot 3000$$
$$k = 150/3000 = 0.05$$
We have $M = 0.05B$. For a human with $B = 70$, we have $M = 0.05(70) = 3.5$ kilograms of blood.

30. (a) We have $T = k\sqrt{l}$. For the grandfather clock, $T = 1.924$ and $l = 3$ and so
$$T = k\sqrt{l}$$
$$1.924 = k\sqrt{3}$$
$$k = 1.924/\sqrt{3} = 1.111$$
Thus $T = 1.111\sqrt{l}$.
(b) Foucault's pendulum has $l = 197.0$, and so
$$T = 1.111\sqrt{197} = 15.59 \text{ seconds}$$

31. If $y = k \cdot x^3$, then we have $y/x^3 = k$ with $k$ a constant. In other words, the ratio
$$\frac{y}{x^3} = \frac{\text{Weight}}{(\text{Length})^3}$$
should all be approximately equal (and the ratio will be the constant of proportionality $k$). For the first fish, we have
$$\frac{y}{x^3} = \frac{332}{(33.5)^3} = 0.0088.$$
If we check all 11 data points, we get the following values of the ratio $y/x^3$: 0.0088, 0.0088, 0.0087, 0.0086, 0.0086, 0.0088, 0.0087, 0.0086, 0.0087, 0.0088, 0.0088. These numbers are indeed approximately constant with an average of about 0.0087. Thus, $k \approx 0.0087$ and the allometric equation $y = 0.0087 x^3$ fits this data well.

32. Looking at the given data, it seems that Galileo's hypothesis was incorrect. The first table suggests that velocity is not a linear function of distance, since the increases in velocity for each foot of distance are themselves getting smaller. Moreover, the second table suggests that velocity is instead proportional to *time*, since for each second of time, the velocity increases by 32 ft/sec.

33. (a) Note that
$$\frac{17.242}{50} = 0.34484,$$
$$\frac{25.863}{75} = 0.34484,$$
$$\frac{34.484}{100} = 0.34484,$$
$$\frac{51.726}{150} = 0.34484,$$

and finally
$$\frac{68.968}{200} = 0.34484$$

Thus, the proportion of $A$ to $m$ remains constant for all $A$ and corresponding $m$, and so $A$ is proportional to $m$.

(b) Since
$$\frac{31.447}{8} \approx 3.931$$
and
$$\frac{44.084}{12} \approx 3.674$$

we see that the proportion of $r$ to $m$ does not remain constant for different values of $r$ and corresponding values of $m$. Thus, we see that $r$ and $m$ are not proportional.

34. Substituting $w = 65$ and $h = 160$, we have

(a)
$$s = 0.01(65^{0.25})(160^{0.75}) = 1.3 \text{ m}^2.$$

(b) We substitute $s = 1.5$ and $h = 180$ and solve for $w$:
$$1.5 = 0.01 w^{0.25}(180^{0.75}).$$

We have
$$w^{0.25} = \frac{1.5}{0.01(180^{0.75})} = 3.05.$$

Since $w^{0.25} = w^{1/4}$, we take the fourth power of both sides, giving
$$w = 86.8 \text{ kg}.$$

(c) We substitute $w = 70$ and solve for $h$ in terms of $s$:
$$s = 0.01(70^{0.25})h^{0.75},$$

so
$$h^{0.75} = \frac{s}{0.01(70^{0.25})}.$$

Since $h^{0.75} = h^{3/4}$, we take the 4/3 power of each side, giving
$$h = \left(\frac{s}{0.01(70^{0.25})}\right)^{4/3} = \frac{s^{4/3}}{(0.01^{4/3})(70^{1/3})}$$

so
$$h = 112.6 s^{4/3}.$$

35. Let $D(v)$ be the stopping distance required by an Alpha Romeo as a function of its velocity. The assumption that stopping distance is proportional to the square of velocity is equivalent to the equation
$$D(v) = kv^2$$

where $k$ is a constant of proportionality. To determine the value of $k$, we use the fact that $D(70) = 177$.
$$D(70) = k(70)^2 = 177.$$

Thus,
$$k = \frac{177}{70^2} \approx 0.0361.$$
It follows that
$$D(35) = \left(\frac{177}{70^2}\right)(35)^2 = \frac{177}{4} = 44.25 \text{ ft}$$
and
$$D(140) = \left(\frac{177}{70^2}\right)(140)^2 = 708 \text{ ft}.$$
Thus, at half the speed it requires one fourth the distance, whereas at twice the speed it requires four times the distance, as we would expect from the equation. (We could in fact have figured it out that way, without solving for $k$ explicitly.)

36. The curve is obviously not linear – it both decreases and increases. It is $U$-shaped, which rules out, $-x^2$, $x^3$, and $-x^3$. This leaves only $x^2$ as the closest shape.

37. The graph described is shown in Figure 1.20. It most closely resembles the function $x^3$: it cannot be any of the other four functions mentioned.

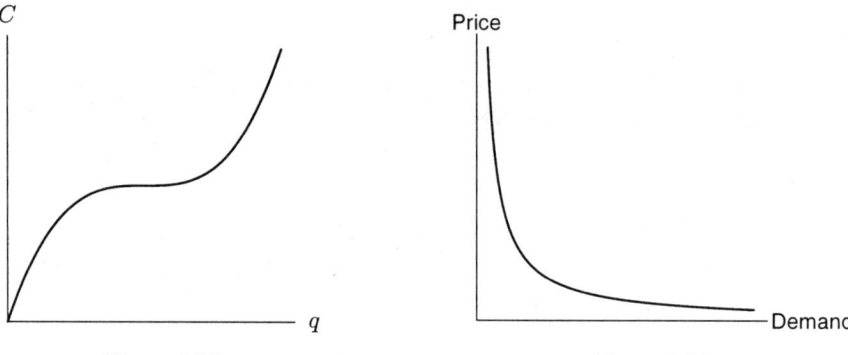

**Figure 1.20**     **Figure 1.21**

38. (a) We know that the function will be of the form
$$D = k \cdot \frac{1}{P}.$$
Thus, the graph of this function will look like Figure 1.21.
(b) Looking at the graph we see that as the price gets closer and closer to zero, the demand for the object is increasing. That is, as the object gets closer and closer to being given away for free, innumerably many more people will want to buy it.
(c) We see that if the price is very large the demand gets closer to zero. That is, as the object gets more and more expensive, fewer and fewer people will want to buy the object, but there will always be some demand for it.

39. As $t$ increases $w$ decreases, so the function is decreasing. The rate at which $w$ is decreasing is itself decreasing: as $t$ goes from 0 to 4, $w$ decreases by 42, but as $t$ goes from 4 to 8, $w$ decreases by 36. Thus, the function is concave up.

40. (a) This is the graph of a linear function, which increases at a constant rate, and thus corresponds to $k(t)$, which increases by 0.3 over each interval of 1.
(b) This graph is concave down, so it corresponds to a function whose increases are getting smaller, as is the case with $h(t)$, whose increases are 10, 9, 8, 7, and 6.
(c) This graph is concave up, so it corresponds to a function whose increases are getting bigger, as is the case with $g(t)$, whose increases are 1, 2, 3, 4, and 5.

41. The values in Table 1.1 suggest that this limit is 0. The graph of $y = 1/x$ in Figure 1.22 suggests that $y \to 0$ as $x \to \infty$ and so supports the conclusion.

**TABLE 1.1**

| $x$ | 100 | 1000 | 1,000,000 |
|---|---|---|---|
| $1/x$ | 0.01 | 0.001 | 0.000001 |

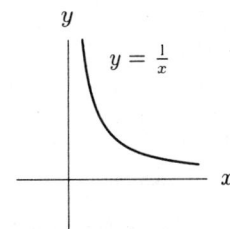

*Figure 1.22*

42. If $f(x) = -x^2$, we have $\lim_{x \to \infty} f(x) = -\infty$ and $\lim_{x \to -\infty} f(x) = -\infty$.
43. Possible graphs are shown. There are many possible answers.

(a)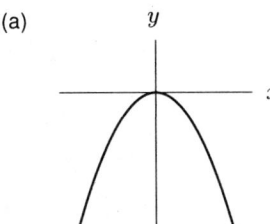

*Figure 1.23*

(b)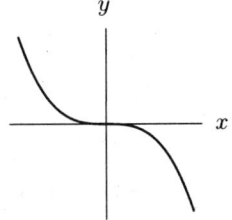

*Figure 1.24*

(c)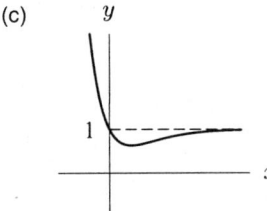

*Figure 1.25*

## Solutions for Section 1.6

1.  (a) Initial amount = 100; exponential growth; growth rate = 7% = 0.07.
    (b) Initial amount = 5.3; exponential growth; growth rate = 5.4% = 0.054.
    (c) Initial amount = 3,500; exponential decay; decay rate = −7% = −0.07.
    (d) Initial amount = 12; exponential decay; decay rate = −12% = −0.12.
2.  (a) Town (i) has the largest percent growth rate, at 12%.
    (b) Town (ii) has the largest initial population, at 1000.
    (c) Yes, town (iv) is decreasing in size, since the decay rate is 0.9, which is less than 1.

3.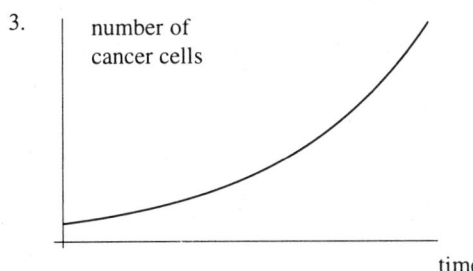

4.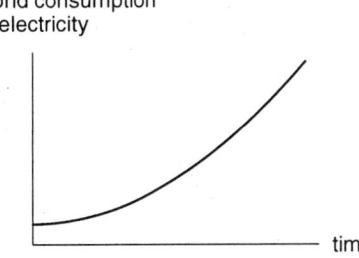

5.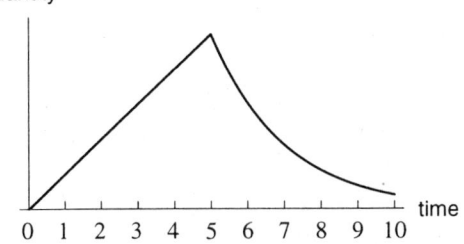

6.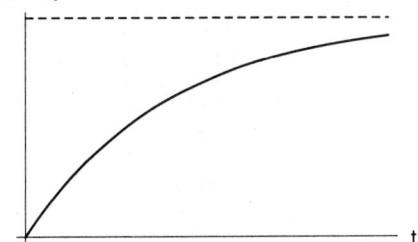

7. (a) This is a linear function, corresponding to $g(x)$, whose rate of decrease is constant, 0.6.
   (b) This graph is concave down, so it corresponds to a function whose rate of decrease is increasing, like $h(x)$. (The rates are $-0.2, -0.3, -0.4, -0.5, -0.6$.)
   (c) This graph is concave up, so it corresponds to a function whose rate of decrease is decreasing, like $f(x)$. (The rates are $-10, -9, -8, -7, -6$.)

8. The graphs of the functions are in Figure 1.26.

   (a) The drug is represented by $g(t)$, as this is the only function which increases and then levels off.
   (b) The radioactive carbon-14 is represented by $h(t)$, as this is the only decreasing function.
   (c) The population is represented by $f(t)$, which is increasing but not leveling off.

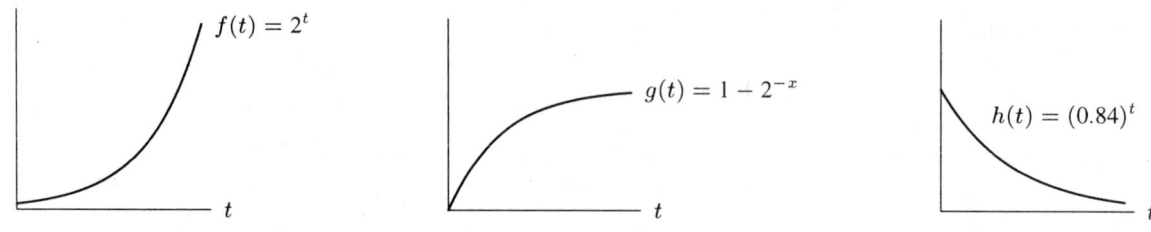

**Figure 1.26**: Graphs of exponential functions

9. Since $2^x$ is always positive, the graph of $y = 2^x$ is above the $x$-axis. The fact that $2^{-x} = 1/2^x$ tells us that $2^{-x}$ is always positive as well, and that $2^{-x}$ is small where $2^x$ is large and vice versa. The graph of $y = 2^{-x}$ is large for negative $x$ and small for positive $x$. The graphs of $y = 2^x$ and $y = 2^{-x}$ are reflections of one another in the $y$-axis. (This happens because $2^{-x}$ is obtained from $2^x$ by replacing $x$ by $-x$.)

10. We see that $\frac{1.09}{1.06} \approx 1.03$, and therefore $h(s) = c(1.03)^s$; $c$ must be 1. Similarly $\frac{2.42}{2.20} = 1.1$, and so $f(s) = a(1.1)^s$; $a = 2$. Lastly, $\frac{3.65}{3.47} \approx 1.05$, so $g(s) = b(1.05)^s$; $b \approx 3$.

11. The values of $f(x)$ given seem to increase by a factor of 1.4 for each increase of 1 in $x$, so we expect an exponential function with base 1.4. To assure that $f(0) = 4.30$, we multiply by the constant, obtaining
$$f(x) = 4.30(1.4)^x.$$

12. Each increase of 1 in $t$ seems to cause $g(t)$ to decrease by a factor of 0.8, so we expect an exponential function with base 0.8. To make our solution agree with the data at $t = 0$, we need a coefficient of 5.50, so our completed equation is
$$g(t) = 5.50(0.8)^t.$$

13. (a) In this case we know that
$$f(1) - f(0) = 12.7 - 10.5 = 2.2$$
while
$$f(2) - f(1) = 18.9 - 12.7 = 6.2.$$
Thus, the function described by these data is not a linear one. Next we check if this function is exponential.
$$\frac{f(1)}{f(0)} = \frac{12.7}{10.5} \approx 1.21$$
while
$$\frac{f(2)}{f(1)} = \frac{18.9}{12.7} \approx 1.49$$
thus $f(x)$ is not an exponential function either.

(b) In this case we know that
$$s(0) - s(-1) = 30.12 - 50.2 = -20.08$$
while
$$s(1) - s(0) = 18.072 - 30.12 = -12.048.$$
Thus, the function described by these data is not a linear one. Next we check if this function is exponential.
$$\frac{s(0)}{s(-1)} = \frac{30.12}{50.2} = 0.6,$$
$$\frac{s(1)}{s(0)} = \frac{18.072}{30.12} = 0.6,$$
and
$$\frac{s(2)}{s(1)} = \frac{10.8432}{18.072} = 0.6.$$
Thus, $s(t)$ is an exponential function. We know that $s(t)$ will be of the form
$$s(t) = P_0 a^t$$
where $P_0$ is the initial value and $a = 0.6$ is the base. We know that
$$P_0 = s(0) = 30.12.$$
Thus,
$$s(t) = 30.12 a^t.$$
Since $a = 0.6$, we have
$$s(t) = 30.12(0.6)^t.$$

(c) In this case we know that
$$\frac{g(2) - g(0)}{2 - 0} = \frac{24 - 27}{2} = \frac{-3}{2} = -1.5,$$
$$\frac{g(4) - g(2)}{4 - 2} = \frac{21 - 24}{2} = \frac{-3}{2} = -1.5,$$
and
$$\frac{g(6) - g(4)}{6 - 4} = \frac{18 - 21}{2} = \frac{-3}{2} = -1.5.$$
Thus, $g(u)$ is a linear function. We know that
$$g(u) = m \cdot u + b$$
where $m$ is the slope and $b$ is the vertical intercept, or the value of the function at zero. So
$$b = g(0) = 27$$
and from the above calculations we know that
$$m = -1.5.$$
Thus,
$$g(u) = -1.5u + 27.$$

14. (a) We must check that the proportional change in the function is approximately constant. We get
$$\frac{428}{349} \approx 1.23,$$
$$\frac{521}{428} \approx 1.22,$$
$$\frac{665}{521} \approx 1.28,$$
$$\frac{822}{665} \approx 1.24,$$
$$\frac{1055}{822} \approx 1.28,$$
and
$$\frac{1348}{1055} \approx 1.28.$$
Thus, we see that the function is approximately exponential with a two-year growth factor of 1.26, or 26%.

(b) Looking at the data we see that in 1974 the yearly expenditure was $521 per capita while in 1980 the yearly expenditure was $1054 per capita. Thus, the expenditure more or less doubled in 6 years. So the doubling time is about six years.

15. (a) Since $P(t)$ is an exponential function, it will be of the form $P(t) = P_0 a^t$, where $P_0$ is the initial population and $a$ is the base. $P_0 = 200$, and a 5% growth rate means $a = 1.05$. Thus, we get
$$P(t) = 200(1.05)^t.$$

(b) The graph is shown in Figure 1.27.

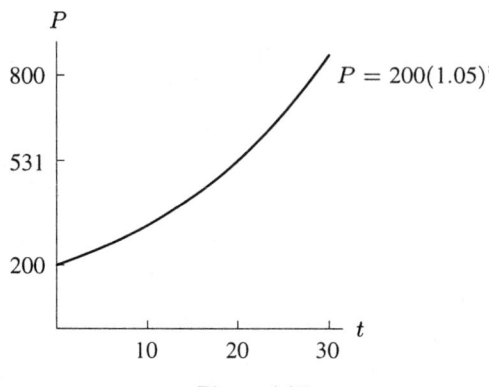

**Figure 1.27**

(c) Evaluating gives that $P(10) = 200(1.05)^{10} \approx 326$.
(d) From the graph we see that the population is 400 at about $t = 15$, so the doubling time appears to be about 15 years.

16.

**TABLE 1.2**

| $t$ (hours)   | 0   | 2   | 4   | 6    | 8     | 10     |
|---------------|-----|-----|-----|------|-------|--------|
| Nicotine (mg) | 0.4 | 0.2 | 0.1 | 0.05 | 0.025 | 0.0125 |

From the table it appears that it will take just over 6 hours for the amount of nicotine to reduce to 0.04 mg.

17. (a) Since $P(t)$ is an exponential function, it will be of the form $P(t) = P_0 a^t$. We have $P_0 = 1$, since 100% is present at time $t = 0$, and $a = 0.975$, because each year 97.5% of the contaminant remains. Thus,
$$P(t) = (0.975)^t.$$

(b) The graph is shown in Figure 1.28.

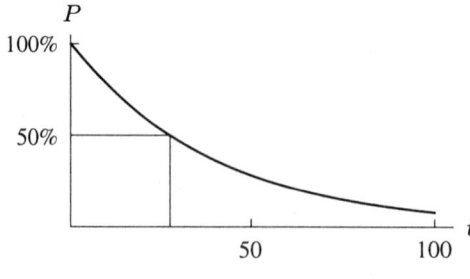

**Figure 1.28**

(c) The half-life is about 27 years, since $(0.975)^{27} \approx 0.5$.
(d) At time $t = 100$ there appears to be about 8% remaining, since $(0.975)^{100} \approx 0.08$.

18. (a) The city's population is increasing at 5% per year. Thus, the rate of growth of the city depends on its size. It therefore has an exponential rate of growth. Furthermore, since the city's population is increasing, the graph of the population must be an increasing exponential. Thus, it must be either $A$ or $B$. We notice that the next problem has a higher rate of growth and thus we choose the exponential curve that grows more slowly, $B$.
    (b) Using the same reason as part (a), we know that the graph of the population is a positive exponential curve. Comparing it to the previous city, it has a faster rate of growth and thus we choose the faster growing exponential curve, $A$.
    (c) The population is growing by a constant amount. Thus, the graph of the population of the city is a line. Since the population is growing, we know that the line must have positive slope. Thus, the city's population is best described by graph $C$.
    (d) The city's population doesn't change. That means that the population is a constant. It is thus a line with 0 slope, or graph $E$.

    The remaining curves are $D$ and $F$. Curve $D$ shows a population decreasing at an exponential rate. Comparison of the slope of curve $D$ at the points where it intersects curves $A$ and $B$ shows that the slope of $D$ is less steep than the slope of $A$ or $B$. From this we know that the city's population is decreasing at a rate less than $A$ or $B$'s is increasing. It is decreasing at a rate of perhaps 4% per year.

    Graph $F$ shows a population decreasing linearly – by the same amount each year. By noticing that graph $F$ is more downward sloping than $C$ is upward sloping, we can determine that the annual population decrease of a city described by $F$ is greater than the annual population increase of a city described by graph $C$. Thus, the population of city $F$ is decreasing by more than 5000 people per year.

19. (a) At $t = 0$, the population is 1000. The population doubles (reaches 2000) at about $t = 4$, so the population doubled in 4 years.
    (b) At $t = 3$, the population is about 1700. The population reaches 3400 at about $t = 7$. The population doubled in 4 years.
    (c) No matter when you start, the population doubles in 4 years.

20. The doubling time $t$ depends only on the growth rate; it is the solution to
$$2 = (1.02)^t,$$
since $1.02^t$ represents the factor by which the population has grown after time $t$. Trial and error shows that $(1.02)^{35} \approx 1.9999$ and $(1.02)^{36} \approx 2.0399$, so that the doubling time is about 35 years.

21. (a) The slope is given by
$$m = \frac{P - P_1}{t - t_1} = \frac{100 - 50}{20 - 0} = \frac{50}{20} = 2.5.$$
We know $P = 50$ when $t = 0$, so
$$P = 2.5t + 50.$$
    (b) Given $P = P_0 a^t$ and $P = 50$ when $t = 0$,
$$50 = P_0 a^0, \text{ so } P_0 = 50.$$
Then, using $P = 100$ when $t = 20$
$$100 = 50a^{20}$$
$$2 = a^{20}$$
$$a = 2^{1/20} \approx 1.035265.$$
And so we have
$$P = 50(1.035265)^t.$$
The completed table is found in Table 1.3.

**TABLE 1.3** *The cost of a home*

| $t$ | (a) Linear Growth (price in $1000 units) | (b) Exponential Growth (price in $1000 units) |
|---|---|---|
| 0  | 50  | 50 |
| 10 | 75  | 70.71 |
| 20 | 100 | 100 |
| 30 | 125 | 141.42 |
| 40 | 150 | 200 |

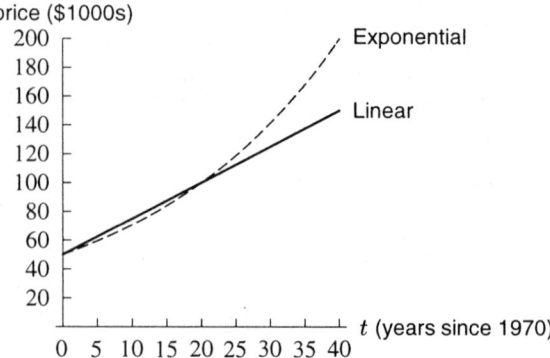

(c)

(d) Since economic growth (inflation, investments) are usually measured in percentage change per year, the exponential model is probably more realistic.

22. (a) The differences in production, $p$, over the 5-year intervals are 0.55, 0.59, 0.56, and 0.59. The average of the increases in $p$ over an interval is 0.5725, and the interval is 5 years long. Thus, if we choose to fit a linear function to this data, we try a slope of

$$m = \frac{0.5725}{5} = 0.1145.$$

Using this and the point (0, 5.35) with the point-slope formula, we get

$$p - p_1 = m(t - t_1)$$
$$p - 5.35 = 0.1145(t - 0)$$
$$p = 0.1145t + 5.35$$

where $t$ is the number of years since 1975.

If we choose to fit an exponential to the production data, we calculate the ratios of successive terms, giving 1.103, 1.100, 1.086, and 1.084. The average of these values is 1.093, so we can use the function $p = 5.35(1.093)^t$.

For the population data, if we try to fit a linear function, we calculate the increases in population $P$ over the 5-year intervals, giving 3.7, 4, 4.3, and 4.5. Since these differences vary very widely, a linear function will not fit the population data well.

To fit an exponential to the population data, we look at the ratios of successive terms, which are 1.070, 1.070, 1.071, 1.069. Since these values are pretty close, the population appears to be growing exponentially by a factor of about 1.07 every 5 years. Thus, we expect to be able to fit the formula

$$P = P_0 a^t.$$

We will let $P_0 = 53.2$, and then use the fact that $P = 56.9$ when $t = 5$ to solve for $a$:

$$56.9 = 53.2a^5.$$

Solving gives

$$a^5 = \frac{56.9}{53.2} = 1.070 \quad \text{so} \quad a = (1.070)^{1/5} = 1.0136.$$

Thus, if $t$ is the number of years since 1975,

$$P = 53.2(1.0136)^t.$$

(b) Since the region had neither surpluses nor shortages in 1975 (it was exactly self-supporting), we can assume that the amount of grain needed per person per year is given by

$$\frac{5.35 \text{ million tons}}{53.2 \text{ million people}} = 0.1005639 \text{ tons/person}.$$

From 1980 to 1995, the ratios were

$$1980: \frac{5.90}{56.9} = 0.1036907 \text{ tons/person}$$

$$1985: \frac{6.49}{60.9} = 0.1065681 \text{ tons/person}$$

$$1990: \frac{7.05}{65.2} = 0.1081288 \text{ tons/person}$$

$$1995: \frac{7.64}{69.7} = 0.1096126 \text{ tons/person}$$

so the region was not only self-supporting, it had an increasing surplus.

(c) If the population continues to grow like an exponential function while production only grows like a linear function, eventually the population will get too big for the production and there will be shortages. This is because exponential functions always overtake linear functions. If population and production both grow exponentially, there will be no shortages, as the growth factor for the population is smaller than the growth factor for the production.

23. The flattest curve must correspond to $y = 100x^2$. The steepest curve must be $y = 3^x$, since exponential curves always overtake polynomial curves. The "middle" curve must be $y = x^5$. Notice that since $3^0 = 1$, the scale along the $y$-axis must be quite large, so the $x$ values shown are large. For values of $x$ less than one, the curve $y = 100x^2$ lies above the curve $y = x^5$. The scale in the graph is too large to show this.

24. The graphs are shown in Figure 1.29.

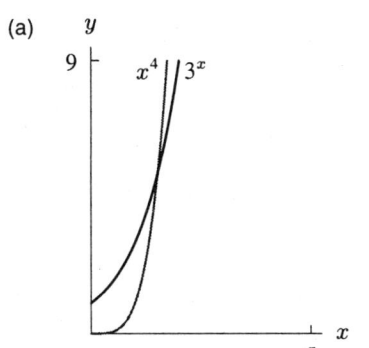
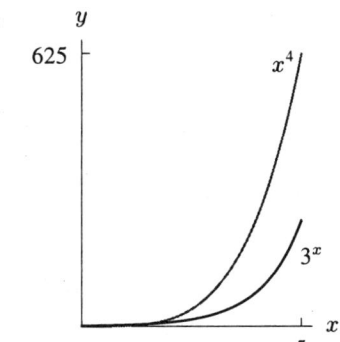
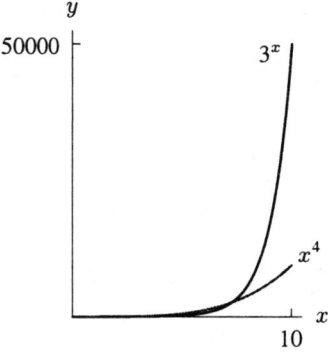

**Figure 1.29**

25. (a) Since the growth is exponential, the equation is of the form $P(t) = P_0 a^t$, where $P_0 = 5.6$, the population at time $t = 0$. Since the growth rate is 1.2%, $a = 1.012$. Thus, $P(t) = 5.6(1.012)^t$, in billions.

(b)
$$\text{Average rate of change} = \frac{P(2000) - P(1994)}{2000 - 1994}$$
$$= \frac{6.015 - 5.6}{6}$$
$$= 0.069 \text{ billion people per year increase.}$$

(c)
$$\text{Average rate of change} = \frac{P(2020) - P(2010)}{2020 - 2010}$$
$$= \frac{7.636 - 6.778}{10}$$
$$= 0.086 \text{ billion people per year increase.}$$

## Solutions for Section 1.7

1. Since $e^{0.08} = 1.0833$, and $e^{-0.3} = 0.741$, we have
$$P = e^{0.08t} = \left(e^{0.08}\right)^t = (1.0833)^t \text{ and}$$
$$Q = e^{-0.3t} = \left(e^{-0.3}\right)^t = (0.741)^t.$$

2. (a) (i) $P = 1000(1.05^t)$; (ii) $P = 1000e^{0.05t}$
   (b) (i) 1629; (ii) 1649

3. (a) Town $D$ is growing fastest, since the rate of growth $k = 0.12$ is the largest.
   (b) Town $C$ is largest now, since $P_0 = 1200$ is the largest.
   (c) Town $B$ is decreasing in size, since $k$ is negative ($k = -0.02$).

4. We use the equation $B = Pe^{rt}$, where $P$ is the initial principal, $t$ is the time in years the deposit is in the account, $r$ is the annual interest rate and $B$ is the balance. After $t = 5$ years, we will have

$$B = (10{,}000)e^{(0.08)5} = (10{,}000)e^{0.4} \approx \$14{,}918.25.$$

5. (a) If the interest is added only once a year (i.e. annually), then at the end of a year we have $1.055x$ where $x$ is the amount we had at the beginning of the year. After two years, we'll have $1.055(1.055x)$ and after eight years, we'll have $(1.055)^8 x$. Since we started with $1000, after eight years we'll have $(1.055)^8 (1000) \approx \$1534.69$.
   (b) If an initial deposit of $\$P_0$ is compounded continuously at interest rate $r$ then the account will have $P = P_0 e^{rt}$ after $t$ years. In this case, $P = P_0 e^{rt} = 1000e^{(0.055)(8)} \approx \$1552.71$.

6. We know that the formula

$$P(t) = P_0 e^{rt}$$

gives the balance after $t$ years in an account compounded continuously, where $P_0$ is the initial deposit and $r$ is the nominal rate. In our case this gives us

$$P(t) = 1000e^{0.06t}$$

as the balance in the account after $t$ years. Table 1.4 gives us the balance in the account for a number of different values of $t$.

**TABLE 1.4**

| $t$ | 1 | 2 | 3 | 4 | 5 | 6 |
|---|---|---|---|---|---|---|
| $P(t)$ | 1061.84 | 1127.50 | 1197.22 | 1271.25 | 1349.86 | 1433.33 |
| $t$ | 7 | 8 | 9 | 10 | 11 | 12 |
| $P(t)$ | 1521.96 | 1616.07 | 1716.01 | 1822.12 | 1934.79 | 2054.43 |

Thus, we see that the account doubles in approximately 11.5 years. Alternatively we could look at the graph of $P(t)$ and see at what value of $t$ the graph takes on a value of $2000. (See Fig 1.30.)

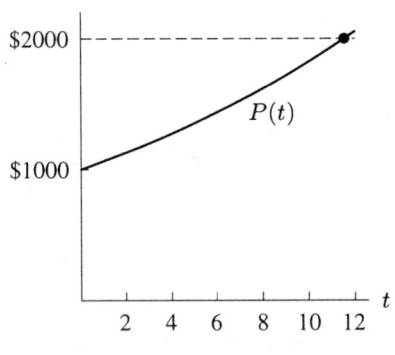

*Figure 1.30*

Again we see that the the balance is $2000 after roughly 11.5 years.

7. (a) (I)
   (b) (IV)
   (c) (II) and (IV)
   (d) (II) and (III)

8. We know that the formula for the balance after $t$ years in an account which is compounded continuously is
$$P(t) = P_0 e^{rt}$$
where $P_0$ is the initial deposit and $r$ is the nominal rate. In our case we are told that the nominal rate is
$$r = 0.08$$
so we have
$$P(t) = P_0 \cdot e^{0.08t}.$$
We are asked to find the value for $P_0$ such that after three years the balance would be $10,000. That is, we are asked for $P_0$ such that
$$10,000 = P_0 \cdot e^{0.08(3)}.$$
Solving we get
$$\begin{aligned} 10,000 &= P_0 \cdot e^{0.08(3)} \\ &= P_0 \cdot e^{0.24} \\ &\approx P_0(1.27125) \\ P_0 &\approx \frac{10,000}{1.27125} \\ &\approx 7866.28 \end{aligned}$$
Thus, $7866.28 should be deposited into this account so that after three years the balance would be $10,000.

9. In both cases the initial deposit was $20. Compounding continuously earns more interest than compounding annually at the same nominal rate. Therefore, curve A corresponds to the account which compounds interest continuously and curve B corresponds to the account which compounds interest annually. We know that this is the case because curve A is higher than curve B over the interval, implying that bank account A is growing faster, and thus is earning more money over the same time period.

10. (a) The pressure $P$ at 6198 meters is given in terms of the pressure $P_0$ at sea level to be
$$\begin{aligned} P &= P_0 e^{-0.00012h} \\ &= P_0 e^{(-0.00012)6198} \\ &= P_0 e^{-0.74376} \\ &\approx 0.4753 P_0 \quad \text{or about 47.5\% of sea level pressure.} \end{aligned}$$

(b) At $h = 12,000$ meters, we have
$$\begin{aligned} P &= P_0 e^{-0.00012h} \\ &= P_0 e^{(-0.00012)12,000} \\ &= P_0 e^{-1.44} \\ &\approx 0.2369 P_0 \quad \text{or about 23.7\% of sea level pressure.} \end{aligned}$$

11. (a) At the end of the year, the landlord's investment amounts to $1000 e^{0.06} \approx \$1061.84$, and so has earned $61.84 in interest. Each year, the landlord pays the tenant $1000(0.05) = \$50$. Therefore, the landlord made $11.84 in interest.

(b) At the end of the year, the landlord's investment amounts to $1000 e^{0.04} = \$1040.81$, and so has earned $40.81 in interest. By paying the tenant $50, the landlord loses $9.19.

12.
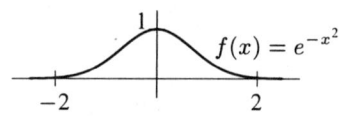

*Figure 1.31*

(a) The function $f$ is increasing for negative $x$. The function $f$ is decreasing for all positive $x$.
(b) The graph of $f$ is concave down at $x = 0$.
(c) As $x \to \infty$, $f(x) \to 0$. As $x \to -\infty$, $f(x) \to 0$.

**44** CHAPTER ONE /SOLUTIONS

13. (a)

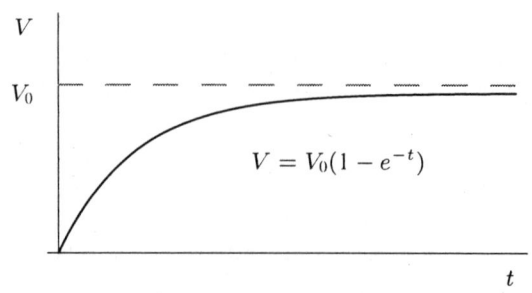

(b) $V_0$ represents the terminal velocity of the raindrop or the maximum speed it can attain as it falls (although a raindrop starting at rest will never quite reach $V_0$ exactly).

## Solutions for Section 1.8

1. $y = \ln e^x$ is a straight line with slope 1, passing through the origin. This is so because $y = \ln e^x = x \ln e = x \cdot 1 = x$. So this function is really $y = x$ in disguise. See Figure 1.32.

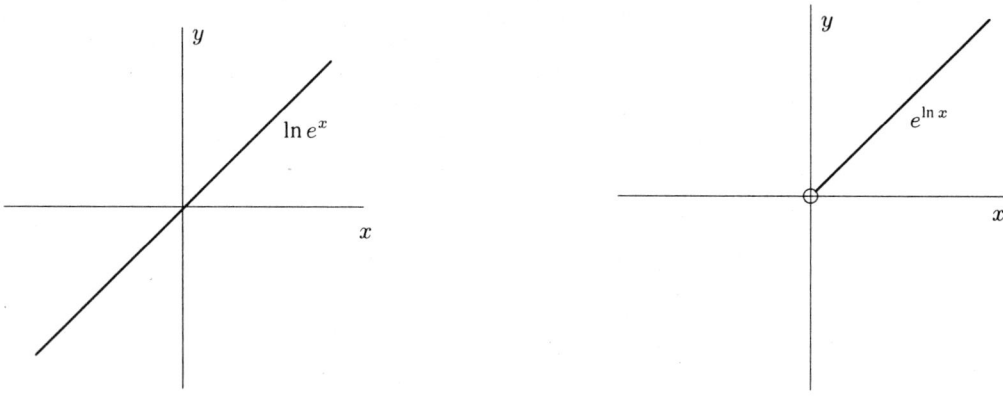

Figure 1.32

Figure 1.33

2. Since $e^{\ln x} = x$ for $x > 0$, this equation is $y = x$ for $x > 0$. Its graph is therefore a straight line, with slope 1, to the right of the origin. See Figure 1.33.

3. Taking natural logs of both sides we get
$$\ln(5^t) = \ln 7.$$
This gives
$$t \ln 5 = \ln 7$$
or in other words
$$t = \frac{\ln 7}{\ln 5} \approx 1.209.$$

4. Taking natural logs of both sides we get
$$\ln 130 = \ln(10^t).$$
This gives
$$t \ln 10 = \ln 130$$
or in other words
$$t = \frac{\ln 130}{\ln 10} \approx 2.1139.$$

5. Taking natural logs of both sides we get
$$\ln 2 = \ln(1.02^t).$$
This gives
$$t \ln 1.02 = \ln 2$$
or in other words
$$t = \frac{\ln 2}{\ln 1.02} \approx 35.003.$$

6. Taking natural logs of both sides we get
$$\ln 10 = \ln(2^t).$$
This gives
$$t \ln 2 = \ln 10$$
or in other words
$$t = \frac{\ln 10}{\ln 2} \approx 3.3219$$

7. Dividing both sides by 25 we get
$$4 = 1.5^t.$$
Taking natural logs of both side gives
$$\ln(1.5^t) = \ln 4.$$
This gives
$$t \ln 1.5 = \ln 4$$
or in other words
$$t = \frac{\ln 4}{\ln 1.5} \approx 3.419.$$

8. Dividing both sides by 10 we get
$$5 = 3^t.$$
Taking natural logs of both sides gives
$$\ln(3^t) = \ln 5.$$
This gives
$$t \ln 3 = \ln 5$$
or in other words
$$t = \frac{\ln 5}{\ln 3} \approx 1.465.$$

9. Taking natural logs of both sides (and assuming $b \neq 1$) gives
$$\ln(b^t) = \ln a.$$
This gives
$$t \ln b = \ln a$$
or in other words
$$t = \frac{\ln a}{\ln b}.$$

10. Taking natural logs of both sides we get
$$\ln(e^t) = \ln 10$$
which gives
$$t = \ln 10 \approx 2.3026.$$

**46** CHAPTER ONE /SOLUTIONS

11. Dividing both sides by 2 we get
$$2.5 = e^t.$$
Taking the natural log of both sides gives
$$\ln(e^t) = \ln 2.5.$$
This gives
$$t = \ln 2.5 \approx 0.9163.$$

12. Taking natural logs of both sides we get
$$\ln(e^{3t}) = \ln 100.$$
This gives
$$3t = \ln 100$$
or in other words
$$t = \frac{\ln 100}{3} \approx 1.535.$$

13. Dividing both sides by 6 gives
$$e^{0.5t} = \frac{10}{6} = \frac{5}{3}.$$
Taking natural logs of both sides we get
$$\ln(e^{0.5t}) = \ln\left(\frac{5}{3}\right).$$
This gives
$$0.5t = \ln\left(\frac{5}{3}\right) = \ln 5 - \ln 3$$
or in other words
$$t = 2(\ln 5 - \ln 3) \approx 1.0217.$$

14. Dividing both sides by $P$ we get
$$\frac{B}{P} = e^{rt}.$$
Taking the natural log of both sides gives
$$\ln(e^{rt}) = \ln\left(\frac{B}{P}\right).$$
This gives
$$rt = \ln\left(\frac{B}{P}\right) = \ln B - \ln P.$$
Dividing by $r$ gives
$$t = \frac{\ln B - \ln P}{r}.$$

15. Dividing both sides by $P$ we get
$$2 = e^{0.3t}.$$
Taking the natural log of both sides gives
$$\ln(e^{0.3t}) = \ln 2.$$
This gives
$$0.3t = \ln 2$$
or in other words
$$t = \frac{\ln 2}{0.3} \approx 2.3105.$$

16. Taking natural logs of both sides we get
$$\ln(7 \cdot 3^t) = \ln(5 \cdot 2^t)$$
which gives
$$\ln 7 + \ln(3^t) = \ln 5 + \ln(2^t)$$
or in other words
$$\ln 7 - \ln 5 = \ln(2^t) - \ln(3^t)$$
This gives
$$\ln 7 - \ln 5 = t \ln 2 - t \ln 3 = t(\ln 2 - \ln 3)$$
Thus, we get
$$t = \frac{\ln 7 - \ln 5}{\ln 2 - \ln 3} \approx -0.8298.$$

17. Taking natural logs of both sides we get
$$\ln(5e^{3t}) = \ln(8e^{2t}).$$
This gives
$$\ln 5 + \ln(e^{3t}) = \ln 8 + \ln(e^{2t})$$
or in other words
$$\ln(e^{3t}) - \ln(e^{2t}) = \ln 8 - \ln 5.$$
This gives
$$3t - 2t = \ln 8 - \ln 5$$
or in other words
$$t = \ln 8 - \ln 5 \approx 0.47.$$

18. $P = P_0(e^{0.2})^t = P_0(1.2214)^t$. Exponential growth because $0.2 > 0$ or $1.2214 > 1$.
19. $P = 10(e^{0.917})^t = 10(2.5018)^t$. Exponential growth because $0.917 > 0$ or $2.5018 > 1$.
20. $P = P_0(e^{-0.73})^t = P_0(0.4819)^t$. Exponential decay since $-0.73 < 0$ or $0.4819 < 1$.
21. $P = 79(e^{-2.5})^t = 79(0.0821)^t$. Exponential decay because $-2.5 < 0$ or $0.0821 < 1$.
22. $P = 7(e^{-\pi})^t = 7(0.0432)^t$. Exponential decay because $-\pi < 0$ or $0.0432 < 1$.
23. $P = 2.91(e^{0.55})^t = 2.91(1.733)^t$. Exponential growth since $0.55 > 0$ or $1.733 > 1$.
24. $P = (5 \cdot 10^{-3})(e^{-1.9 \cdot 10^{-2}})^t = (5 \cdot 10^{-3})(0.9812)^t$. Exponential decay since $-1.9 \cdot 10^{-2} < 0$ or $0.9812 < 1$.
25. We want $2^t = e^{kt}$ so $2 = e^k$ and $k = \ln 2 = 0.693$. Thus $P = P_0 e^{0.693t}$.
26. We want $1.7^t = e^{kt}$ so $1.7 = e^k$ and $k = \ln 1.7 = 0.5306$. Thus $P = 10e^{0.5306t}$.
27. We want $0.9^t = e^{kt}$ so $0.9 = e^k$ and $k = \ln 0.9 = -0.1054$. Thus $P = 174e^{-0.1054t}$.
28. We want $0.2^t = e^{kt}$ so $0.2 = e^k$ and $k = \ln 0.2 = -1.6094$. Thus $P = 5.23e^{-1.6094t}$.
29. To solve for the value of $t$
$$100{,}000 = 50{,}000 e^{0.045t},$$
we divide by 50,000 giving
$$2 = e^{0.045t}.$$
Taking natural logs gives
$$\ln 2 = 0.045t$$
$$t = \frac{\ln 2}{0.045} = 15.4 \text{ years}$$

**48** CHAPTER ONE /SOLUTIONS

30. We solve for $t$

$$0.6 = 2.4e^{-0.004t}$$

by dividing by 2.4, giving

$$\frac{0.6}{2.4} = 0.25 = e^{-0.004t}.$$

Taking natural logs gives

$$\ln 0.25 = -0.004t$$
$$t = \frac{\ln 0.25}{-0.004} = \frac{-1.3863}{-0.004} = 346.6 \text{ years.}$$

31. (a) Substituting $t = 0.5$ in $P(t)$ gives

$$P(0.5) = 1000e^{-0.5(0.5)}$$
$$= 1000e^{-0.25}$$
$$\approx 1000(0.7788)$$
$$= 778.8 \approx 779 \text{ trout}$$

Substituting $t = 1$ in $P(t)$ gives

$$P(1) = 1000e^{-0.5}$$
$$\approx 1000(0.6065)$$
$$= 606.5 \approx 607 \text{ trout}$$

(b) Substituting $t = 3$ in $P(t)$ gives

$$P(3) = 1000e^{-0.5(3)}$$
$$= 1000e^{-1.5}$$
$$\approx 1000(0.2231)$$
$$= 223.1 \approx 223 \text{ trout}$$

This tells us that after 3 years the fish population has gone down to about 22% of the initial population.

(c) We are asked to find the value of $t$ such that $P(t) = 100$. That is, we are asked to find the value of $t$ such that

$$100 = 1000e^{-0.5t}.$$

Solving this gives

$$100 = 1000e^{-0.5t}$$
$$0.1 = e^{-0.5t}$$
$$\ln 0.1 = \ln e^{-0.5t} = -0.5t$$
$$t = \frac{\ln 0.1}{-0.5}$$
$$\approx 4.6$$

Thus after roughly 4.6 years, the trout population will have decreased to 100.

(d) A graph of the population is given in Figure 1.34.

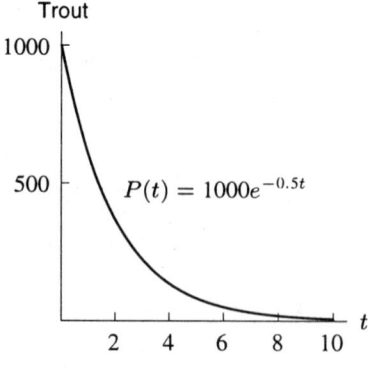

**Figure 1.34**

Looking at the graph, we see that as time goes on there are fewer and fewer trout in the pond. A possible cause for this may be the presence of predators, such as fisherman. The rate at which the fish die decreases due to the fact that there are fewer fish in the pond, which means that there are fewer fish that have to try to survive. Notice that the survival chances of any given fish remains the same, it is just the overall mortality that is decreasing.

32. (a) $P = 3.6(1.034)^t$
    (b) Since $P = 3.6e^{kt} = 3.6(1.034)^t$, we have

$$e^{kt} = (1.034)^t$$
$$kt = t\ln(1.034)$$
$$k = 0.0334$$

Thus, $P = 3.6e^{0.0334t}$.

(c) The annual growth rate is 3.4%, while the continuous growth rate is 3.3%. Thus the growth rates are not equal, though for small growth rates (such as these), they are close. The annual growth rate is larger.

33. We know that the formula for the balance in an account after $t$ years is

$$P(t) = P_0 e^{rt}$$

where $P_0$ is the initial deposit and $r$ is the nominal rate. In our case the initial deposit is $12,000 that is

$$P_0 = 12,000$$

and the nominal rate is

$$r = 0.08.$$

Thus we get

$$P(t) = 12,000 e^{0.08t}.$$

We are asked to find $t$ such that $P(t) = 20,000$, that is we are asked to solve

$$20,000 = 12,000 e^{0.08t}$$
$$e^{0.08t} = \frac{20,000}{12,000} = \frac{5}{3}$$
$$\ln e^{0.08t} = \ln(5/3)$$
$$0.08t = \ln 5 - \ln 3$$
$$t = \frac{\ln 5 - \ln 3}{0.08} \approx 6.39$$

Thus after roughly 6.39 years there will be $20,000 in the account.

34. Since we are assuming continuous compounding we know that the formula for the account balance after $t$ years is

$$P(t) = P_0 e^{rt}$$

where $P_0$ is the initial deposit and $r$ is the nominal rate. We know that the initial deposit was $5000 and we know that

$$P(4) = 8080.$$

Thus, we are asked to solve for $r$ in the equation

$$8080 = P(4) = 5000 e^{4r}.$$

Solving we get

$$8080 = 5000 e^{4r}$$
$$e^{4r} = \frac{8080}{5000} = 1.616$$
$$\ln e^{4r} = \ln 1.616$$
$$4r = \ln 1.616$$
$$r = \frac{\ln 1.616}{4} \approx 0.12.$$

Thus, the continuous rate is roughly 12% which implies that the annual rate of return, $a$ is

$$a = (e^{0.12} - 1) \approx 0.127.$$

That is, the annual rate of return is roughly 12.7%.

35. We know that the world population is an exponential function over time. Thus the function for world population will be of the form
$$P(t) = P_0 e^{rt}$$
where $P_0$ is the initial population, $t$ is measured in years after 1994 and $r$ is the continuous rate of change. We know that the initial population is the population in the year 1994 so
$$P_0 = 5.6 \text{ billion}.$$
We are also told that in the year 2030 the population will be 8.5 billion, that is
$$P(36) = 8.5 \text{ billion}.$$
Solving for $r$ we get
$$8.5 = 5.6 e^{36r}$$
$$e^{36r} = \frac{8.5}{5.6} \approx 1.518$$
$$\ln e^{36r} \approx \ln 1.518$$
$$36r \approx \ln 1.518$$
$$r \approx \frac{\ln 1.518}{36} \approx 0.0115917.$$
Thus, we know that the continuous rate $r$ is $1.15917\%$. Thus the annual growth rate $a$ is
$$a = e^{0.0115917} - 1 \approx 0.01166.$$
Or in other words, the annual growth rate is $1.166\%$.

36. (a) We know that the formula for the account balance at time $t$ in an account compounded continuously is given by the formula
$$P(t) = P_0 e^{rt}$$
where $P_0$ is the initial deposit and $r$ is the annual rate. Thus, in our case the formula would be
$$P(t) = 5000 e^{0.04t}.$$
Substituting the value $t = 8$ we get
$$P(8) = 5000 e^{0.04(8)}$$
$$= 5000 e^{0.32}$$
$$\approx 5000(1.377128)$$
$$= 6885.64.$$
Thus, the balance at the end of eight years would be about $\$6885.64$.

(b) We are asked to solve for the rate, $r$, that would give us an $\$8000$ balance at the end of eight years. In other words we are asked to solve for $r$ in the equation
$$8000 = 5000 e^{8r}.$$
Solving we get
$$8000 = 5000 e^{8r}$$
$$e^{8r} = \frac{8000}{5000} = 1.6$$
$$\ln e^{8r} = \ln 1.6$$
$$8r = \ln 1.6$$
$$r = \frac{\ln 1.6}{8} \approx 0.059.$$
Thus, the rate we would need in order to have a balance of $\$8000$ at the end of eight years is about $5.9\%$.

37. Let $n$ be the infant mortality of Senegal. As a function of time $t$, $n$ is given by
$$n = n_0(0.90)^t.$$
To find when $n = 0.50 n_0$ (so the number of cases has been reduced by 50%), we solve
$$0.50 = (0.90)^t,$$
$$\ln(0.50) = t\ln(0.90),$$
$$t = \frac{\ln(0.50)}{\ln(0.90)} \approx 6.58 \text{ years.}$$

38. Let $t =$ number of years since 1980. Then the number of vehicles, $V$, in millions, at time $t$ is given by
$$V = 170(1.04)^t$$
and the number of people, $P$, in millions, at time $t$ is given by
$$P = 227(1.01)^t.$$
There is an average of one vehicle per person when $\dfrac{V}{P} = 1$, or $V = P$. Thus, we must solve for $t$ the equation:
$$170(1.04)^t = 227(1.01)^t,$$
which implies
$$\left(\frac{1.04}{1.01}\right)^t = \frac{(1.04)^t}{(1.01)^t} = \frac{227}{170}$$
Taking logs on both sides,
$$t \ln \frac{1.04}{1.01} = \ln \frac{227}{170}.$$
Therefore,
$$t = \frac{\ln\left(\frac{227}{170}\right)}{\ln\left(\frac{1.04}{1.01}\right)} \approx 9.9 \text{ years.}$$
So there was, according to this model, about one vehicle per person in 1990.

39. (a) (i) We know that the formula for the account balance at time $t$ is given by
$$P(t) = P_0 e^{rt}$$
(since the account is being compounded continuously), where $P_0$ is the initial deposit and $r$ is the annual rate. In our case the initial deposit is
$$P_0 = 24$$
and the annual rate is
$$r = 0.05.$$
Thus, the formula for the balance is
$$P(t) = 24e^{0.05t}.$$
We are asked for the amount of money in the account in the year 2000, that is we are asked for the amount of money in the account 374 years after the initial deposit. This gives
$$\text{amount of money in the account in the year 2000} = P(374)$$
$$= 24e^{0.05(374)}$$
$$= 24e^{18.7}$$
$$\approx 24(132{,}223{,}000)$$
$$= 3{,}173{,}352{,}000.$$
Thus, in the year 2000, there would be about 3.17 billion dollars in the account.

(ii) Here the annual rate is
$$r = 0.07.$$
Thus, the formula for the balance is
$$P(t) = 24e^{0.07t}.$$
Again, we want $P(374)$. This gives

$$\begin{aligned}
\text{Amount of money in the account in the year 2000} &= P(374) \\
&= 24e^{0.07(374)} \\
&= 24e^{26.18} \\
&\approx 24(234{,}330{,}890{,}000) \\
&= 5{,}623{,}941{,}360{,}000.
\end{aligned}$$

Thus, in the year 2000, there would be about 5.62 trillion dollars in the account.

(b) Here the annual rate is
$$r = 0.06.$$
Thus, the formula for the balance is
$$P(t) = 24e^{0.06t}.$$
We are asked to solve for $t$ such that
$$P(t) = 1{,}000{,}000.$$
Solving we get

$$\begin{aligned}
1{,}000{,}000 = B(t) &= 24e^{0.06t} \\
\frac{1{,}000{,}000}{24} &= e^{0.06t} \\
e^{0.06t} &\approx 41666.7 \\
\ln e^{0.06t} &\approx \ln 41666.7 \\
0.06t &\approx \ln 41666.7 \\
t &\approx \frac{\ln 41666.7}{0.06} \approx 177.3.
\end{aligned}$$

Thus, there would be one million dollars in the account 177.3 years after the initial deposit, that is roughly in 1803.

40. Since the population of koala bears is growing exponentially, we have $P = P_0 e^{kt}$. Since the initial population is 18, we have $P_0 = 18$, so
$$P = 18e^{kt}.$$
We use the fact that $P = 500$ in 1993 (when $t = 70$) to find $k$:
$$5000 = 18e^{k(70)}$$

Divide both sides by 18:
$$\frac{5000}{18} = e^{70k}$$

Take logs of both sides:
$$\ln\left(\frac{5000}{18}\right) = 70k$$

Divide both sides by 70:
$$k = \frac{\ln(5000/18)}{70} = 0.08$$

The koala bear population increased at a rate of 8% per year. The population is given by
$$P = 18e^{0.08t},$$
where $t$ is in years since 1923. In the year 2010, we have $t = 87$, and
$$P = 18e^{0.08(87)} = 18{,}965 \text{ koala bears}.$$

As a result of this rapid growth, local authorities are trying to find new ways to slow down the rate of growth of the koala bears.

41. Marine catch, $M$ (in millions), is increasing exponentially, so
$$M = M_0 e^{kt}.$$
If we let $t$ be the number of years since 1950, we have $M_0 = 17$ and
$$M = 17e^{kt}.$$
Since $M = 91$ when $t = 45$, we can solve for $k$:
$$91 = 17e^{k(45)}$$
$$91/17 = e^{45k}$$
$$\ln(91/17) = 45k$$
$$k = \frac{\ln(91/17)}{45} = 0.037.$$
Marine catch has increased by 3.7% per year. Since
$$M = 17e^{0.037t}$$
in the year 2000, we have $t = 50$ and
$$M = 17e^{0.037(50)} = 108 \text{ million tons}.$$

42. (a) We know that an exponential function for solid waste generated in the US is given by
$$W(t) = W_0 e^{rt}$$
where $W_0$ is the waste at time $t = 0$ and $r$ is the rate at which waste is being produced yearly. If we let 1960 be $t = 0$ we get
$$W_0 = 82.3$$
and our formula becomes
$$W(t) = 82.3e^{rt}.$$
We also know that in the year 1980, that is at time $t = 20$, the amount of solid waste generated in the US was 139.1 million tons. Thus we get
$$W(20) = 139.1.$$
Solving for $r$ gives
$$139.1 = W(20) = 82.3e^{20r}$$
$$e^{20r} = \frac{139.1}{82.3}$$
$$\ln e^{20r} = \ln \frac{139.1}{82.3}$$
$$20r = \ln \frac{139.1}{82.3}$$
$$r = \frac{\ln(139.1/82.3)}{20} \approx 0.0262411$$
Thus the exponential formula for waste generated $t$ years after 1960 is
$$W(t) = 82.3e^{0.0262411t}$$

(b) We get that in the year 2000, or $t = 40$, the amount of solid waste generated is
$$W(40) = 82.3e^{0.0262411(40)}$$
$$= 82.3e^{1.049644}$$
$$\approx 82.3(2.856634)$$
$$\approx 235.1$$

Thus, under the assumption that the growth is exponential, we can expect about 235.1 million tons of solid waste to be produced in the US in the year 2000. This is 39.1 million more tons than we had predicted under the assumption that the growth is linear.

43. (a) We know the decay follows the equation
$$P = P_0 e^{-kt},$$
and that 10% of the pollution is removed after 5 hours (meaning that 90% is left). Therefore,
$$0.90 P_0 = P_0 e^{-5k}$$
$$k = -\frac{1}{5}\ln(0.90).$$

Thus, after 10 hours:
$$P = P_0 e^{-10((-0.2)\ln 0.90)}$$
$$P = P_0 (0.9)^2 = 0.81 P_0$$

so 81% of the original amount is left.

(b) We want to solve for the time when $P = 0.50 P_0$:
$$0.50 P_0 = P_0 e^{t((0.2)\ln 0.90)}$$
$$0.50 = e^{\ln(0.90^{0.2t})}$$
$$0.50 = 0.90^{0.2t}$$
$$t = \frac{5\ln(0.50)}{\ln(0.90)} \approx 32.9 \text{ hours.}$$

(c)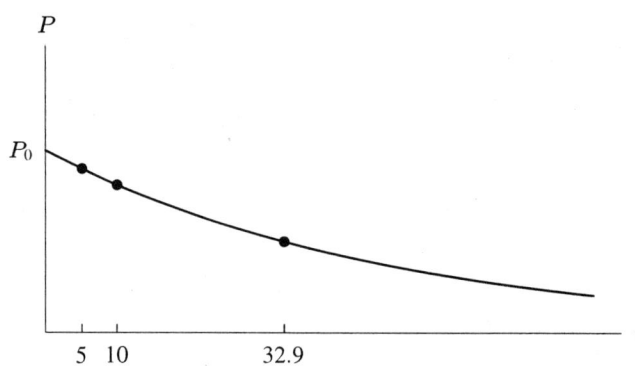

(d) When highly polluted air is filtered, there is more pollutant per liter of air to remove. If a fixed amount of air is cleaned every day, there is a higher amount of pollutant removed earlier in the process.

44.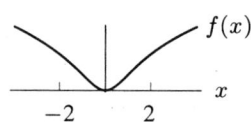

Figure 1.35

(a) The function $f$ is increasing when $x > 0$ and decreasing when $x < 0$.
(b) We can see from the graph that the function $f$ is concave up at $x = 0$.
(c) Even though the logarithmic function grows very slowly, it does keep increasing so as $x \to \infty$, $f(x) \to \infty$. And since we square $x$ in this function, $f(x) \to \infty$ as $x \to -\infty$ too.

## Solutions for Section 1.9

1. We know that the formula for the total process is
$$P(t) = P_0 a^t$$

where $P_0$ is the initial size of the process and $a - 1$ is the annual rate. That is,
$$a = 1 + 0.07 = 1.07.$$
We are asked to find the doubling time. Thus, we are asked to solve
$$2P_0 = P_0(1.07)^t$$
$$1.07^t = 2$$
$$\ln 1.07^t = \ln 2$$
$$t \ln 1.07 = \ln 2$$
$$t = \frac{\ln 2}{\ln 1.07} \approx 10.24.$$
Thus, the doubling time is roughly 10.24 years.

Alternatively we could have used the "rule of 70" to get that the doubling time is
$$\frac{70}{7} = 10.$$

2. We know that a function modeling the quantity of substance at time $t$ will be an exponential decay function satisfying the fact that at $t = 10$ hours, 4% of the substance has decayed, or in other words 96% of the substance is left. Thus, a formula for the quantity of substance left at time $t$ is
$$Q(t) = Q_0 e^{kt},$$
where $Q_0$ is the initial quantity. We solve for $k$:
$$0.96 Q_0 = Q_0 e^{k \cdot 10}$$
$$0.96 = e^{k \cdot 10}$$
$$\ln 0.96 = 10k$$
$$k = \frac{\ln 0.96}{10} \approx -0.004.$$
We are asked to find the time $t$ at which the quantity of material will be half of the initial quantity. Thus, we are asked to solve
$$0.5 Q_0 = Q_0 e^{-0.004t}$$
$$0.5 = e^{-0.004t}$$
$$\ln 0.5 = -0.004t$$
$$t = \frac{\ln 0.5}{-0.004} \approx 173.$$
Thus, the half life of the material is roughly 173 hours.

3. (a) Using the formula for exponential decay, $A = A_0 e^{-kt}$, we substitute $A = 10.32$ when $t = 0$. Since $e^0 = 1$, $A_0 = 10.32$

Since the half-life is 12 days, when $t = 12$, $A = \frac{10.32}{2} = 5.16$. Substituting this into the formula, we have
$$5.16 = 10.32 e^{-12k},$$
$$0.5 = e^{-12k}, \quad \text{and, taking ln of both sides,}$$
$$\ln 0.5 = -12k,$$
$$k = -\frac{\ln(0.5)}{12} \approx 0.057762.$$
The full equation is
$$A = 10.32 e^{-0.057762t}.$$

(b) We want to solve for $t$ when $A = 1$. Plugging into the equation from (a) yields
$$1 = 10.32 e^{-0.057762t},$$
$$\frac{1}{10.32} \approx 0.096899 = e^{-0.057762t},$$
$$\ln 0.096899 = -0.057762t,$$
$$t \approx \frac{-2.33408}{-0.057762} = 40.41 \text{ days}.$$

**56** CHAPTER ONE /SOLUTIONS

4. (a) Using the "rule of 70," we get that the doubling time of the investment is

$$\frac{70}{8} = 8.75.$$

That is, it would take about 8.75 years for the investment to double.

(b) We know that the formula for the balance at the end of $t$ years is

$$B(t) = Pa^t$$

where $P$ is the initial investment and $a$ is 1+(interest per year). We are asked to solve for the doubling time, which amounts to asking for the time $t$ at which

$$B(t) = 2P.$$

Solving, we get

$$2P = P(1 + 0.08)^t$$
$$2 = 1.08^t$$
$$\ln 2 = \ln 1.08^t$$
$$t \ln 1.08 = \ln 2$$
$$t = \frac{\ln 2}{\ln 1.08} \approx 9.01.$$

Thus, the actual doubling time is 9.01 years. And so our estimation by the "rule of 70" was off by a quarter of a year.

5. Since the factor by which the prices have increased after time $t$ is given by $(1.05)^t$, the time after which the prices have doubled solves

$$2 = (1.05)^t$$
$$\ln 2 = \ln(1.05^t) = t \ln(1.05)$$
$$t = \frac{\ln 2}{\ln 1.05} \approx 14.21 \text{ years}.$$

6. The population has increased by a factor of $\frac{56,000,000}{40,000,000} = 1.4$ in 10 years. Thus we have the formula

$$P = 40,000,000(1.4)^{t/10},$$

and $t/10$ gives the number of 10-year periods that have passed since 1980.

In 1980, $t/10 = 0$, so we have $P = 40,000,000$.
In 1990, $t/10 = 1$, so $P = 40,000,000(1.4) = 56,000,000$.
In 2000, $t/10 = 2$, so $P = 40,000,000(1.4)^2 = 78,400,000$.
To find the doubling time, solve $80,000,000 = 40,000,000(1.4)^{t/10}$, to get $t \approx 20.6$ years.

7. We have $P_0 = 500$, so $P = 500e^{kt}$. We can use the fact that $P = 1,500$ when $t = 2$ to find $k$:

$$1,500 = 500e^{kt}$$
$$3 = e^{2k}$$
$$\ln 3 = 2k$$
$$k = \frac{\ln 3}{2} \approx 0.55.$$

The size of the population is given by

$$P = 500e^{0.55t}.$$

At $t = 5$, we have

$$P = 500e^{0.55(5)} \approx 7,821.$$

8. Since the population doubles in 2 hours, we have $P = 2P_0$ when $t = 2$, so

$$2P_0 = P_0 e^{k(2)}$$
$$2 = e^{2k}$$
$$\ln 2 = 2k$$
$$k = \frac{\ln 2}{2} \approx 0.35.$$

So $P = P_0 e^{0.35t}$. We want to find $t$ when $P = 3P_0$:

$$3P_0 = P_0 e^{0.35t}$$
$$3 = e^{0.35t}$$
$$\ln 3 = 0.35t$$
$$t = \frac{\ln 3}{0.35} \approx 3.14 \text{ hours.}$$

The population triples in about 3.14 hours.

9. We use $P = P_0 e^{kt}$. Since 70% remains after 20 hours, we have $P = 0.70 P_0$ when $t = 20$. Solving for $k$ gives:

$$0.70 P_0 = P_0 e^{k(20)}$$
$$0.70 = e^{20k}$$
$$\ln(0.70) = 20k$$
$$k = \frac{\ln(0.70)}{20} \approx -0.018.$$

We have $P = P_0 e^{-0.018t}$, and we find $t$ when $P = 0.5 P_0$:

$$0.5 P_0 = P_0 e^{-0.018t}$$
$$0.5 = e^{-0.018t}$$
$$\ln(0.5) = -0.018t$$
$$t = \frac{\ln(0.5)}{-0.018} \approx 38.5.$$

The half-life is about 38.5 hours.

10. We use $P = P_0 e^{kt}$. Since 200 grams are present initially, we have $P = 200 e^{kt}$. To find $k$, we use the fact that the half-life is 8 years:

$$P = 200 e^{kt}$$
$$100 = 200 e^{k(8)}$$
$$0.5 = e^{8k}$$
$$\ln(0.5) = 8k$$
$$k = \frac{\ln(0.5)}{8} \approx -0.087.$$

We have
$$P = 200 e^{-0.087t}.$$

The amount remaining after 12 years is

$$P = 200 e^{-0.087(12)} = 70.4 \text{ grams.}$$

To find the time until 10% remains, we use $P = 0.10 P_0 = 0.10(200) = 20$:

$$20 = 200 e^{-0.087t}$$
$$0.10 = e^{-0.087t}$$
$$\ln(0.10) = -0.087t$$
$$t = \frac{\ln(0.10)}{-0.087} = 26.5 \text{ years.}$$

**58** CHAPTER ONE /SOLUTIONS

11. We use $P = P_0 e^{kt}$. Since the half-life is 3 days, we can find $k$:

$$P = P_0 e^{kt}$$
$$0.50 P_0 = P_0 e^{k(3)}$$
$$0.50 = e^{3k}$$
$$\ln(0.50) = 3k$$
$$k = \frac{\ln(0.50)}{3} \approx -0.231.$$

After one day,
$$P = P_0 e^{-0.231(1)} = P_0(0.79).$$

About 79% of the original dose is in the body one day later. After one week (at $t = 7$), we have

$$P = P_0 e^{-0.231(7)} = P_0(0.20).$$

After one week, 20% of the dose is still in the body.

12. We have $Q(t) = Q_0 e^{kt}$ with $Q_0 = 100$. We use the half-life to find $k$:

$$0.5 Q_0 = Q_0 e^{k \cdot 4}$$
$$0.5 = e^{4k}$$
$$\ln 0.5 = 4k$$
$$k = \frac{\ln 0.5}{4} \approx -0.173.$$

Thus, $Q(t) = 100 e^{-0.173t}$, and we solve for $t$ when $Q(t) = 5$:

$$5 = 100 e^{-0.173t}$$
$$0.05 = e^{-0.173t}$$
$$\ln(0.05) = -0.173 t$$
$$t = \frac{\ln(0.05)}{-0.173} = 17.3 \text{ hours}.$$

13. (a) We want to find $t$ such that
$$0.15 Q_0 = Q_0 e^{-0.000121 t},$$
so $0.15 = e^{-0.000121 t}$, meaning that $\ln 0.15 = -0.000121 t$, or $t = \dfrac{\ln 0.15}{-0.000121} \approx 15{,}678.7$ years.

(b) Let $T$ be the half-life of carbon-14. Then
$$0.5 Q_0 = Q_0 e^{-0.000121 T},$$
so $0.5 = e^{-0.000121 T}$, or $T = \frac{\ln 0.5}{-0.000121} \approx 5{,}728.5$ years.

14. $e^{-k(5730)} = 0.5$ so $k = 1.21 \cdot 10^{-4}$. Thus, $e^{-1.21 \cdot 10^{-4} t} = 0.995$, so $t = \dfrac{\ln 0.995}{-1.21 \cdot 10^{-4}} = 41.43$ years, so the painting is a fake.

15. You should choose the payment which gives you the highest present value. The immediate lump-sum payment of $2800 obviously has a present value of exactly $2800, since you are getting it now. We can calculate the present value of the installment plan as:

$$\text{PV} = 1000 e^{-0.06(0)} + 1000 e^{-0.06(1)} + 1000 e^{-0.06(2)}$$
$$\approx \$2828.68.$$

Since the installment payments offer a (slightly) higher present value, you should accept this option.

16. (a) We calculate the future values of the two options:

$$FV_1 = 6e^{0.1(3)} + 2e^{0.1(2)} + 2e^{0.1(1)} + 2e^{0.1(0)}$$
$$\approx 8.099 + 2.443 + 2.210 + 2$$
$$= \$14.752 \text{ million.}$$

$$FV_2 = e^{0.1(3)} + 2e^{0.1(2)} + 4e^{0.1(1)} + 6e^{0.1(0)}$$
$$\approx 1.350 + 2.443 + 4.421 + 6$$
$$= \$14.214 \text{ million.}$$

As we can see, the first option gives a higher future value, so he should choose Option 1.

(b) From the future value we can easily derive the present value using the formula $PV = FVe^{-rt}$. So the present value is

$$\text{Option 1: } PV = 14.752e^{0.1(-3)} \approx \$10.929 \text{ million.}$$
$$\text{Option 2: } PV = 14.214e^{0.1(-3)} \approx \$10.530 \text{ million.}$$

17. (a) The total present value for each of the two choices are in the following table. Choice 2 is the preferred choice since it has the larger present value.

| Choice 1 | | | Choice 2 | | |
|---|---|---|---|---|---|
| Year | Payment | Present value | | Payment | Present value |
| 0 | 2000 | 2000 | | 3000 | 3000 |
| 1 | 3000 | $3000/(1.05) = 2857.14$ | | 3000 | $3000/(1.05) = 2857.14$ |
| 2 | 4000 | $4000/(1.05)^2 = 3628.12$ | | 3000 | $3000/(1.05)^2 = 2721.09$ |
| | Total | 8485.26 | | Total | 8578.23 |

(b) The difference between the choices is an extra $1000 now ($3000 in Choice 2 instead of $2000 in Choice 1) versus an extra $1000 in two years ($4000 in Choice 1 instead of $3000 in Choice 2). Thus, Choice 2 will have the larger present value no matter what the interest rate.

18. (a) The total present value for each of the two choices are in the following table. Choice 1 is the preferred choice since it has the larger present value.

| Choice 1 | | | Choice 2 | | |
|---|---|---|---|---|---|
| Year | Payment | Present value | | Payment | Present value |
| 0 | 1500 | 1500 | | 1900 | 1900 |
| 1 | 3000 | $3000/(1.05) = 2857.14$ | | 2500 | $2500/(1.05) = 2380.95$ |
| | Total | 4357.14 | | Total | 4280.95 |

(b) The difference between the choices is an extra $400 now ($1900 in Choice 2 instead of $1500 in Choice 1) versus an extra $500 one year from now ($3000 in Choice 1 instead of $2500 in Choice 2). Since $400 \times 1.25 = 500$, Choice 2 is better at interest rates above 25%.

19. (a) The following table gives the present value of the cash flows. The total present value of the cash flows is $116,224.95.

| Year | Payment | Present value |
|---|---|---|
| 1 | 50000 | $50000/(1.075) = 46,511.63$ |
| 2 | 40000 | $40000/(1.075)^2 = 34,613.30$ |
| 3 | 25000 | $25000/(1.075)^3 = 20,124.01$ |
| 4 | 20000 | $20000/(1.075)^4 = 14,976.01$ |
| | Total | 116,224.95 |

(b) The present value of the cash flows, $116,224.95, is larger than the price of the machine, $97,000, so the recommendation is to buy the machine.

**60** CHAPTER ONE /SOLUTIONS

20. The following table contains the present value of each of the expenses. Since the total present value of the expenses, $352.01, is less than the price of the extended warranty, it is not worth purchasing the extended warranty.

| | Present value of expenses | |
|---|---|---|
| Year | Expense | Present value |
| 2 | 150 | $150/(1.05)^2 = 136.05$ |
| 3 | 250 | $250/(1.05)^3 = 215.96$ |
| | Total | 352.01 |

21. The following table contains the present value of each of the expenses. Since the total present value of the repairs, $255.15, is more than the cost of the service contract, you should buy the service contract.

| | Present value of repairs | |
|---|---|---|
| Year | Repairs | Present Value |
| 1 | 50 | $50/(1.0725) = 46.62$ |
| 2 | 100 | $100/(1.0725)^2 = 86.94$ |
| 3 | 150 | $150/(1.0725)^3 = 121.59$ |
| | Total | 255.15 |

## Solutions for Section 1.10

1.

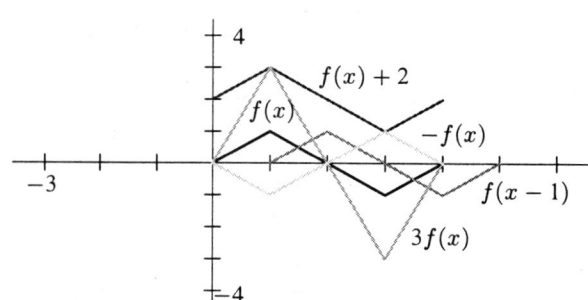

Figure 1.36

2.

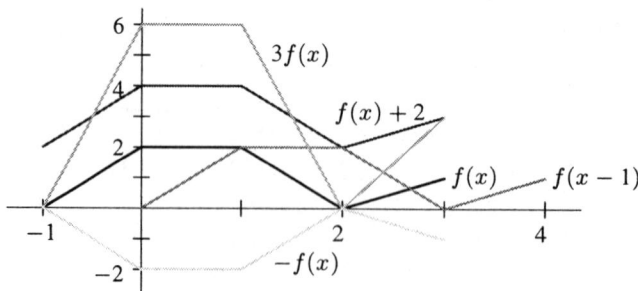

Figure 1.37

3.

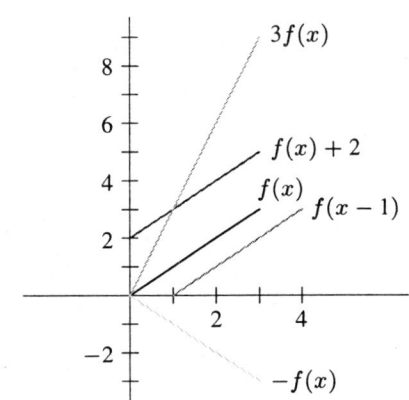

**Figure 1.38**

4.

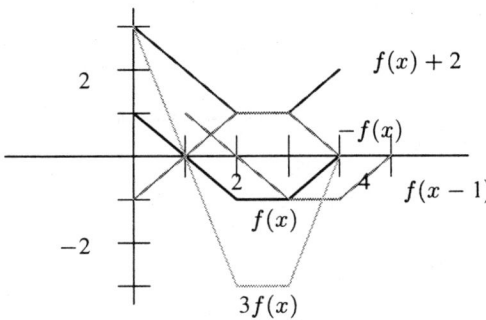

**Figure 1.39**

5. (a) The equation is $y = 2x^2 + 1$. Note that its graph is narrower than the graph of $y = x^2$ which appears in grey. See Figure 1.40.

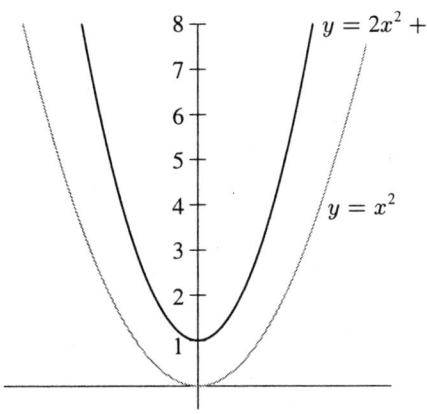

**Figure 1.40**

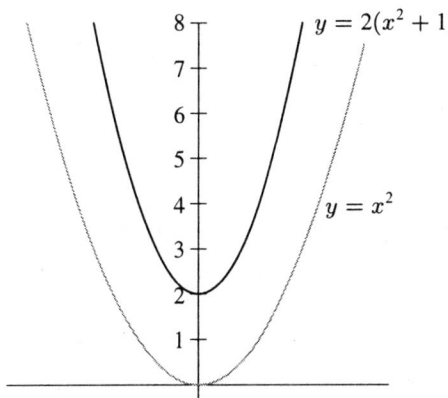

**Figure 1.41**

(b) $y = 2(x^2 + 1)$ moves the graph up one unit and *then* stretches it by a factor of two. See Figure 1.41.

(c) No, the graphs are not the same. Note that stretching vertically leaves any point whose $y$-value is zero in the same place but moves any other point. This is the source of the difference because if you stretch it first, its lowest point stays at the origin. Then you shift it up by one and its lowest point is $(0, 1)$. Alternatively, if you shift it first, its lowest point is $(0, 1)$ which, when stretched by 2, becomes $(0, 2)$.

6.

(a)

| $x$ | $f(x)+3$ |
|---|---|
| 0 | 13 |
| 1 | 9 |
| 2 | 6 |
| 3 | 7 |
| 4 | 10 |
| 5 | 14 |

(b)

| $x$ | $f(x-2)$ |
|---|---|
| 2 | 10 |
| 3 | 6 |
| 4 | 3 |
| 5 | 4 |
| 6 | 7 |
| 7 | 11 |

(c)

| $x$ | $5g(x)$ |
|---|---|
| 0 | 10 |
| 1 | 15 |
| 2 | 25 |
| 3 | 40 |
| 4 | 60 |
| 5 | 75 |

(d)

| $x$ | $-f(x)+2$ |
|---|---|
| 0 | -8 |
| 1 | -4 |
| 2 | -1 |
| 3 | -2 |
| 4 | -5 |
| 5 | -9 |

(e)

| $x$ | $g(x-3)$ |
|---|---|
| 3 | 2 |
| 4 | 3 |
| 5 | 5 |
| 6 | 8 |
| 7 | 12 |
| 8 | 15 |

(f)

| $x$ | $f(x)+g(x)$ |
|---|---|
| 0 | 12 |
| 1 | 9 |
| 2 | 8 |
| 3 | 12 |
| 4 | 19 |
| 5 | 26 |

7.

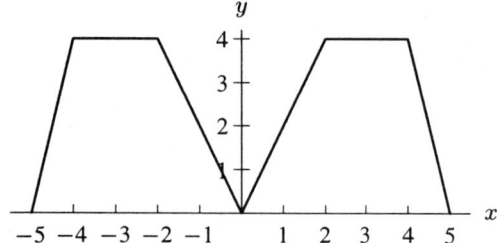

**Figure 1.42**: Graph of $y=2f(x)$

8.

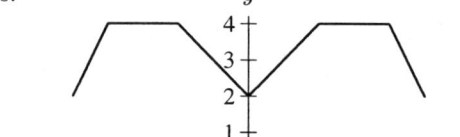

**Figure 1.43**: Graph of $y=f(x)+2$

9.

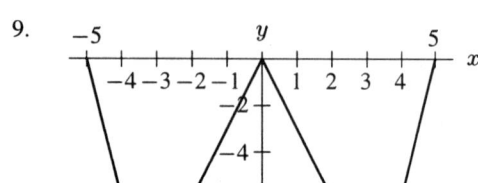

**Figure 1.44**: Graph of $y=-3f(x)$

10.

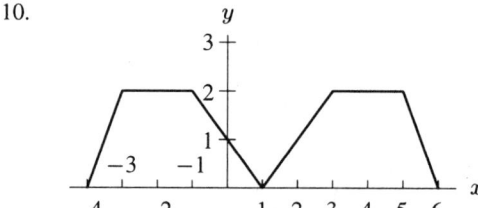

**Figure 1.45**: Graph of $y=f(x-1)$

11.

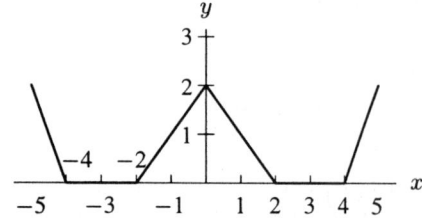

**Figure 1.46**: Graph of $2-f(x)$

12.

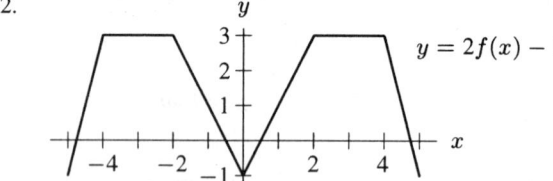

**Figure 1.47**

13. $f(g(1)) = f(2) \approx 0.4$.

14. $g(f(2)) \approx g(0.4) \approx 1.1$.

15. $f(f(1)) \approx f(-0.4) \approx -0.9$.

16. (a) Demand curves are plots of decreasing functions, so we sketch the graph of any decreasing function, as in Figure 1.48.

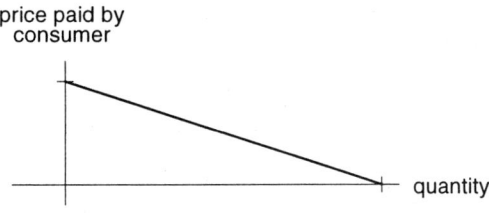

**Figure 1.48**

(b) The new demand is the old demand at $2 more. Thus, if point $(q, p)$ is in the old demand curve, then the point $(q, p - 2)$ is on the new demand curve which includes tax. The new curve is a vertical shift down 2 units. Both curves graphed together look similar to Figure 1.49.

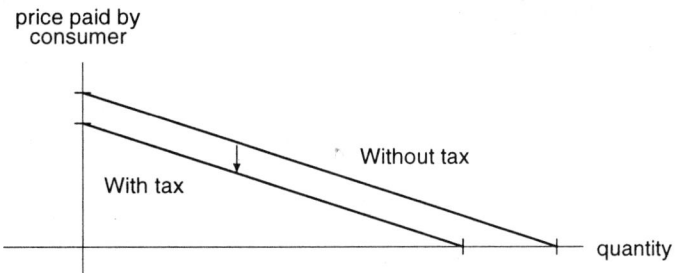

**Figure 1.49**: Note: graph not drawn to scale

17.

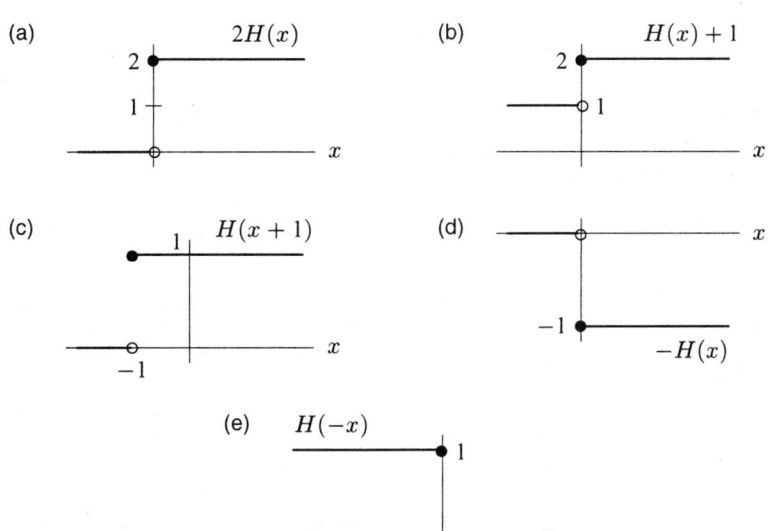

18. (a) Let $u = 5t^2 - 2$. We now have $y = u^6$.
    (b) Let $u = -0.6t$. We now have $P = 12e^u$.
    (c) Let $u = q^3 + 1$. We now have $C = 12 \ln u$.

19. (a) Let $u = 3x - 1$. We now have $y = 2^u$.
    (b) Let $u = 5t^2 + 10$. We now have $P = \sqrt{u}$.
    (c) Let $u = 3r + 4$. We now have $w = 2 \ln u$.

20. (a) $f(g(x)) = f(x+3) = 2(x+3)^2 = 2(x^2 + 6x + 9) = 2x^2 + 12x + 18$.
    (b) $g(f(x)) = g(2x^2) = 2x^2 + 3$.
    (c) $f(f(x)) = f(2x^2) = 2(2x^2)^2 = 8x^4$.

21. Notice that $f(2) = 4$ and $g(2) = 5$.
    (a) $f(2) + g(2) = 4 + 5 = 9$
    (b) $f(2) \cdot g(2) = 4 \cdot 5 = 20$
    (c) $f(g(2)) = f(5) = 5^2 = 25$
    (d) $g(f(2)) = g(4) = 3(4) - 1 = 11$.

22. (a) $g(h(x)) = g(x^3 + 1) = \sqrt{x^3 + 1}$.
    (b) $h(g(x)) = h(\sqrt{x}) = (\sqrt{x})^3 + 1 = x^{3/2} + 1$.
    (c) $h(h(x)) = h(x^3 + 1) = (x^3 + 1)^3 + 1$.
    (d) $g(x) + 1 = \sqrt{x} + 1$.
    (e) $g(x+1) = \sqrt{x+1}$.

23. (a) $f(g(t)) = f\left(\dfrac{1}{t+1}\right) = \left(\dfrac{1}{t+1} + 7\right)^2$.
    (b) $g(f(t)) = g((t+7)^2) = \dfrac{1}{(t+7)^2 + 1}$.
    (c) $f(t^2) = (t^2 + 7)^2$.
    (d) $g(t-1) = \dfrac{1}{(t-1)+1} = \dfrac{1}{t}$.

24. (a) We know that $f(x) = 2x + 3$ and $g(x) = \ln x$. Thus,
    $$g(f(x)) = g(2x + 3) = \ln(2x + 3).$$
    (b) We know that $f(x) = 2x + 3$ and $g(x) = \ln x$. Thus,
    $$f(g(x)) = f(\ln x) = 2\ln x + 3.$$
    (c) We know that $f(x) = 2x + 3$. Thus,
    $$f(f(x)) = f(2x + 3) = 2(2x + 3) + 3 = 4x + 6 + 3 = 4x + 9.$$

25. $\ln(\ln(x))$ means take the ln of the value of the function $\ln x$. (See Figure 1.51.) On the other hand, $\ln^2(x)$ means take the function $\ln x$ and square it. (See Figure 1.52.) For example, consider each of these functions evaluated at $e$. Since $\ln e = 1$, $\ln^2 e = 1^2 = 1$, but $\ln(\ln(e)) = \ln(1) = 0$. See the graphs in Figures 1.50–1.52. (Note that $\ln(\ln(x))$ is only defined for $x > 1$.)

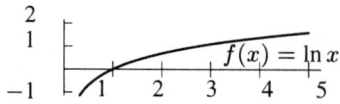

Figure 1.50

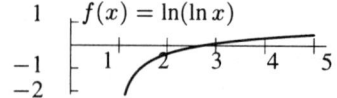

Figure 1.51

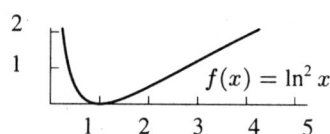

Figure 1.52

26. (a) This is a graph of $y = x^3$ shifted to the right 2 units and up 1 unit. A possible formula is $y = (x-2)^3 + 1$.
    (b) This is a graph of $y = -x^2$ shifted to the left 3 units and down 2 units. A possible formula is $y = -(x+3)^2 - 2$.

# Solutions for Section 1.11

1. (I) Degree $\geq 3$, leading coefficient negative.
   (II) Degree $\geq 4$, leading coefficient positive.
   (III) Degree $\geq 4$, leading coefficient negative.
   (IV) Degree $\geq 5$, leading coefficient negative.
   (V) Degree $\geq 5$, leading coefficient positive.

2. (a) The degree of $x^2 + 10x - 5$ is 2 and its leading coefficient is positive.
   (b) In a large window, the function looks like $x^2$. See Figure 1.53.

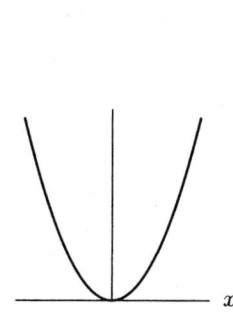

**Figure 1.53**

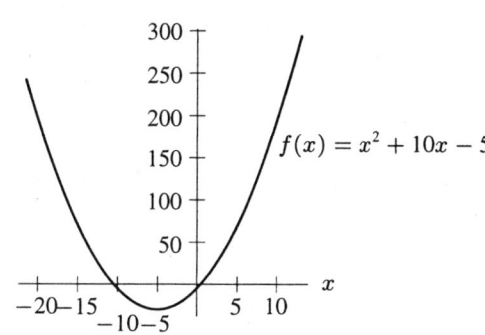

**Figure 1.54**

   (c) As $x \to \infty$, $f(x) \to \infty$. As $x \to -\infty$, $f(x) \to \infty$.
   (d) The function has one turning point which is what we expect of a quadratic. See Figure 1.54.

3. (a) The degree is 3 and the leading coefficient is positive.
   (b) In a large window, the function looks like $5x^3$. See Figure 1.55.

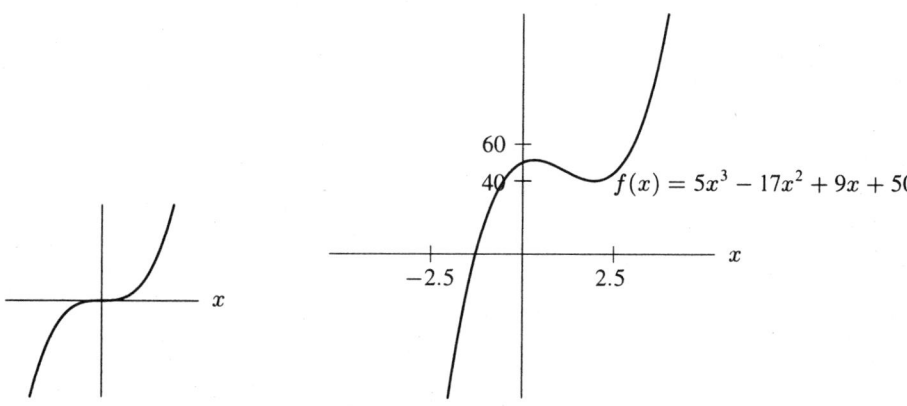

**Figure 1.55**

**Figure 1.56**

   (c) As $x \to -\infty$, $f(x) \to -\infty$. As $x \to \infty$, $f(x) \to \infty$.
   (d) See Figure 1.56. The function has two turning points. Cubics have at most 2 turning points.

4. (a) The degree of $8x - 3x^2$ is 2 and the leading coefficient is negative.
   (b) The function looks like $-3x^2$ in a large window. See Figure 1.57.

**66** CHAPTER ONE /SOLUTIONS

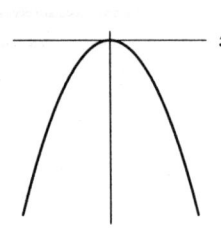

Figure 1.57

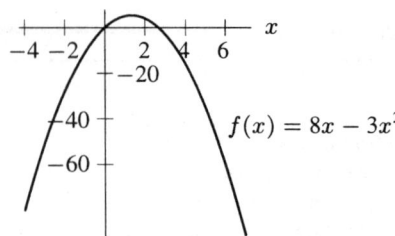

Figure 1.58

(c) As $x \to -\infty$, $f(x) \to -\infty$. As $x \to \infty$, $f(x) \to -\infty$.
(d) See Figure 1.58. The function has 1 turning point as all quadratic functions do.

5. (a) The degree of $17 + 8x - 2x^3$ is 3 and the leading coefficient is negative.
   (b) In a large window, the function looks like $-2x^3$. See Figure 1.59.

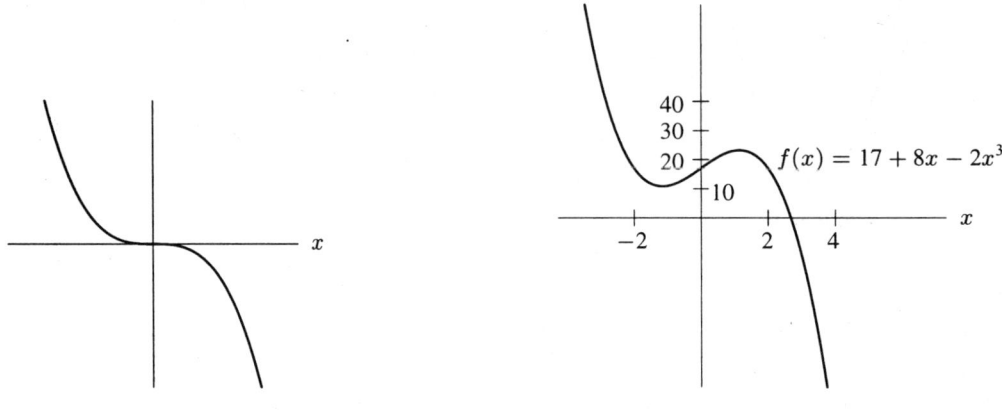

Figure 1.59           Figure 1.60

(c) As $x \to -\infty$, $f(x) \to \infty$. As $x \to \infty$, $f(x) \to -\infty$.
(d) See Figure 1.60. The function has 2 turning points. Cubics all have 0, 1, or 2 turning points.

6. (a) The degree of $-9x^5 + 72x^3 + 12x^2$ is 5 and its leading coefficient is negative.
   (b) In a large window, the graph of $f(x)$ looks like $-9x^5$. See Figure 1.61.

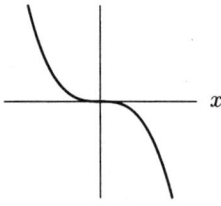

Figure 1.61

(c) As $x \to -\infty$, $f(x) \to \infty$. As $x \to \infty$, $f(x) \to -\infty$.
(d) It is very hard to see what's going on with this function by looking at just one graph. If we look at the graph of $f(x)$ from $x = -4$ to $x = 4$ in Figure 1.62, we see two obvious turning points but can't tell what's going on near the origin. If we look at the graph from $x = -0.5$ to $x = 0.5$ in Figure 1.63, we can see what's going on near the origin but we can no longer see the two turning points we saw in the larger scale plot. But we see that there are two turning points near the origin as well. Thus, there are 4 turning points total. This is one less than the degree of the polynomial, which is the maximum it could be.

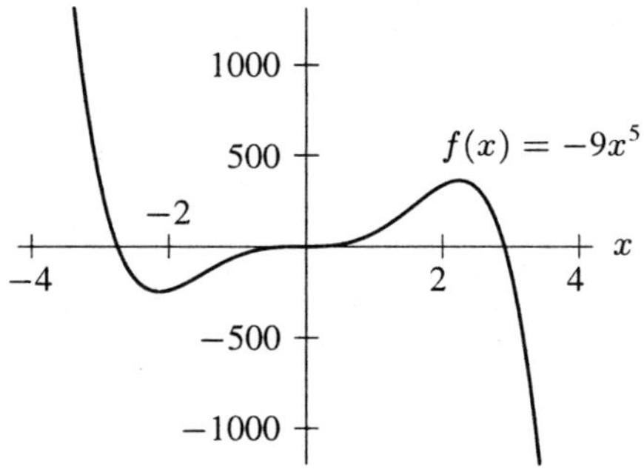

*Figure 1.62* 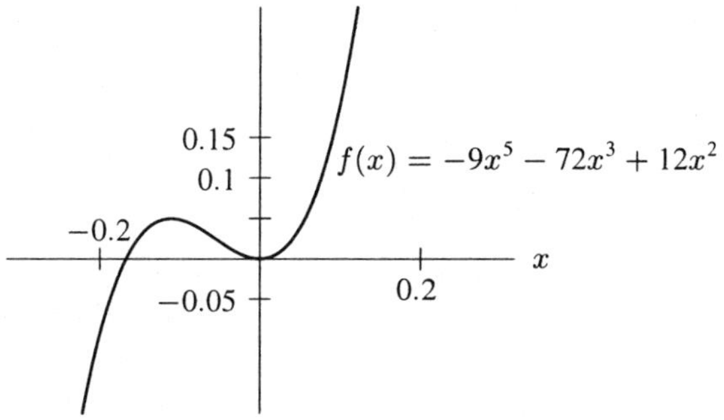            *Figure 1.63*

7. (a) The degree of $f(x) = 0.01x^4 + 2.3x^2 - 7$ is 4 and the leading coefficient is positive.
   (b) In a large window, the graph looks like $0.01x^4$. See Figure 1.64.

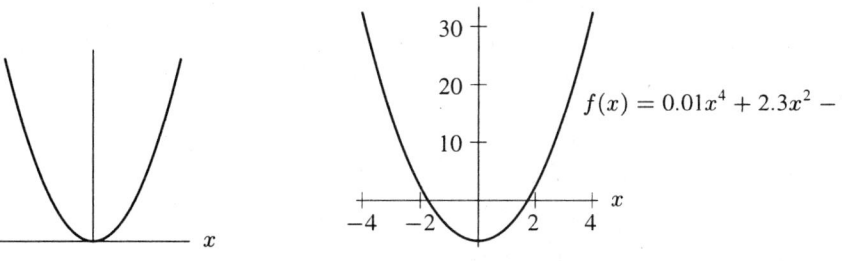

*Figure 1.64*            *Figure 1.65*

   (c) As $x \to -\infty$, $f(x) \to \infty$. As $x \to \infty$, $f(x) \to \infty$.
   (d) See Figure 1.65. The function has 1 turning point which is less than the degree of 4.

8. (a) The function $f(x) = 100 + 5x - 12x^2 + 3x^3 - x^4$ has degree 4 and a negative leading coefficient.
   (b) In a large window, the function looks like $-x^4$. See Figure 1.66.

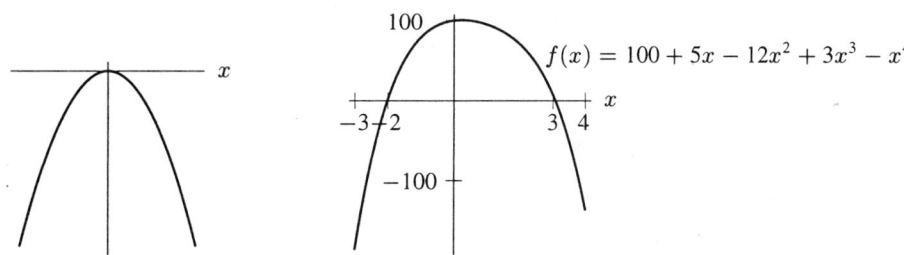

*Figure 1.66*            *Figure 1.67*

   (c) As $x \to \pm\infty$, $f(x) \to -\infty$.
   (d) See Figure 1.67. The function has 1 turning point which is less than the degree of the function.

9. (a) The function $f(x) = 0.2x^7 + 1.5x^4 - 3x^3 + 9x - 15$ has degree 7 and a positive leading coefficient.
   (b) In a large window, the function looks like $0.2x^7$. See Figure 1.68.

**68** CHAPTER ONE /SOLUTIONS

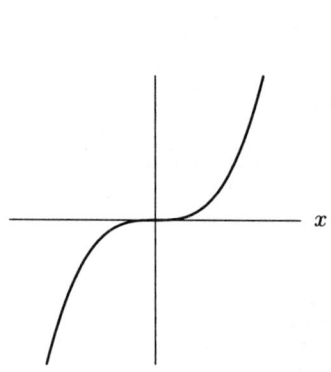

**Figure 1.68**

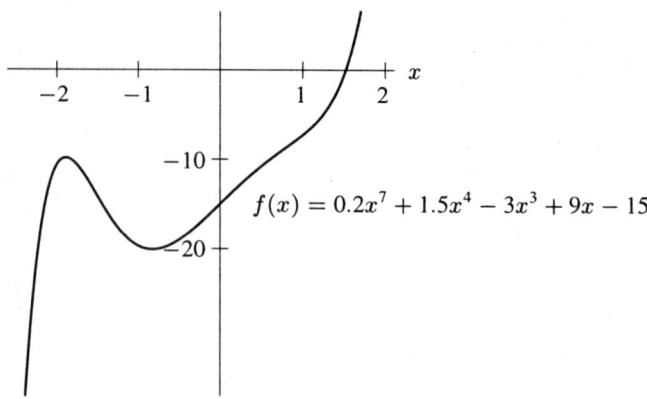

**Figure 1.69**

(c) As $x \to -\infty$, $f(x) \to -\infty$. As $x \to \infty$, $f(x) \to \infty$.
(d) See Figure 1.69. The function has 2 turning points which is less than the degree.

10. The graphs are shown in Figure 1.70. The graphs suggest that (a) is the shape shown in Figure 1.99 in the text, whereas (b) is the shape of $y = x^3$, and (c) is the shape of $y = x^3$ reflected over the axis.

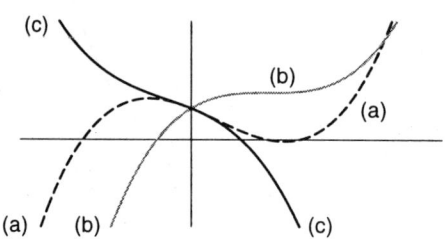

**Figure 1.70**

11. (a) A graph of the profit function is given in Figure 1.71.

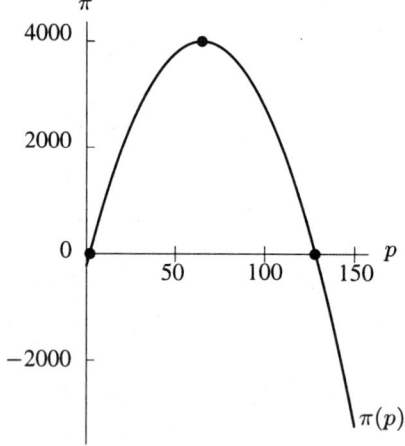

**Figure 1.71**

(b) Looking at the graph in Figure 1.71 we see that the maximum profit occurs when the price of a pound of dog food is 65¢ and the maximum profit itself is $4000.
(c) Looking at the graph we see that the profit is positive (i.e. a profit is actually made) when the price of a pound of dog food is between roughly 2¢ and $1.28.

12. (a) We know that the demand function will be of the form
$$q = mp + b$$
where $m$ is the slope and $b$ is the vertical intercept. We know that the slope is
$$m = \frac{q(30) - q(25)}{30 - 25} = \frac{460 - 500}{5} = \frac{-40}{5} = -8.$$
Thus, we get
$$q = -8p + b.$$
Plugging in the point $(30, 460)$ we get
$$460 = -8(30) + b = -240 + b,$$
so that
$$b = 700.$$
Thus, the demand function is
$$q = -8p + 700.$$

(b) We know that the revenue is given by
$$R = pq$$
where $p$ is the price and $q$ is the demand. Thus, we get
$$R = p(-8p + 700) = -8p^2 + 700p.$$

(c)
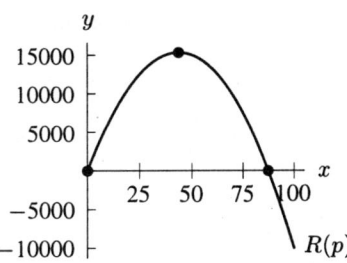

**Figure 1.72**

Looking at the graph, we see that maximal revenue is attained when the price charged for the product is roughly $44. At this price the revenue is roughly $15,300.

13. (a) We know that
$$q = 3000 - 20p,$$
$$C = 10{,}000 + 35q,$$
and
$$R = pq.$$
Thus, we get
$$\begin{aligned} C &= 10{,}000 + 35q \\ &= 10{,}000 + 35(3000 - 20p) \\ &= 10{,}000 + 105{,}000 - 700p \\ &= 115{,}000 - 700p \end{aligned}$$
and
$$\begin{aligned} R &= pq \\ &= p(3000 - 20p) \\ &= 3000p - 20p^2. \end{aligned}$$

(b) The graph of the cost and revenue functions are shown in Figure 1.73.

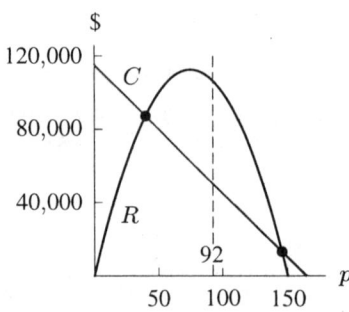

**Figure 1.73**

(c) It makes sense that the revenue function looks like a parabola, since if the price is too low, although a lot of people will buy the product, each unit will sell for a pittance and thus the total revenue will be low. Similarly, if the price is too high, although each sale will generate a lot of money, few units will be sold because consumers will be repelled by the high prices. In this case as well, total revenue will be low. Thus, it makes sense that the graph should rise to a maximum profit level and then drop.

(d) The club makes a profit whenever the revenue curve is above the cost curve in Figure 1.73. Thus, the club makes a profit when it charges roughly between $40 and $145.

(e) We know that the maximal profit occurs when the difference between the revenue and the cost is the greatest. Looking at Figure 1.73, we see that this occurs when the club charges roughly $92.

14. (a) The line given by $(0, 2)$ and $(1, 1)$ has slope $m = \frac{2-1}{-1} = -1$ and $y$-intercept 2, so its equation is
$$y = -x + 2.$$
The points of intersection of this line with the parabola $y = x^2$ are given by
$$x^2 = -x + 2$$
$$x^2 + x - 2 = 0$$
$$(x + 2)(x - 1) = 0.$$
The solution $x = 1$ corresponds to the point we are already given, so the other solution, $x = -2$, gives the $x$-coordinate of $C$. When we substitute back into either equation to get $y$, we get the coordinates for $C$, $(-2, 4)$.

(b) The line given by $(0, b)$ and $(1, 1)$ has slope $m = \frac{b-1}{-1} = 1 - b$, and $y$-intercept at $(0, b)$, so we can write the equation for the line as we did in part (a):
$$y = (1 - b)x + b.$$
We then solve for the points of intersection with $y = x^2$ the same way:
$$x^2 = (1 - b)x + b$$
$$x^2 - (1 - b)x - b = 0$$
$$x^2 + (b - 1)x - b = 0$$
$$(x + b)(x - 1) = 0$$
Again, we have the solution at the given point $(1, 1)$, and a new solution at $x = -b$, corresponding to the other point of intersection $C$. Substituting back into either equation, we can find the $y$-coordinate for $C$ is $b^2$, and thus $C$ is given by $(-b, b^2)$. This result agrees with the particular case of part (a) where $b = 2$.

15. (a) $R(P) = kP(L - P)$, where $k$ is a positive constant.
(b) A possible graph is shown below.

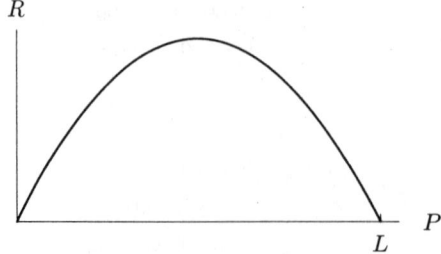

## Solutions for Section 1.12

1. It makes sense that sunscreen sales would be lowest in the winter (say from December through February) and highest in the summer (say from June to August). It also makes sense that sunscreen sales would depend almost completely on time of year, so the function should be periodic with a period of 1 year.

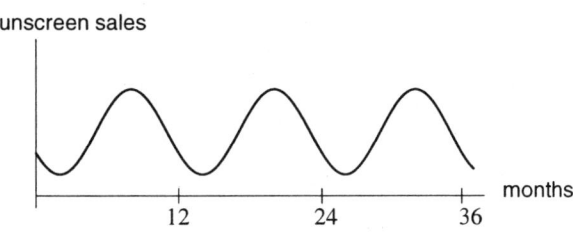

**Figure 1.74**

2. The levels of both hormones certainly look periodic. In each case, the period seems to be about 28 days. Progesterone appears to peak from the 17th through 21st day. Estrogen appears to peak around the 12th day. (Note: the days given in the answer are approximate so your answer may differ slightly.)

3. (a) The period looks to be about 12 months. This means that the number of mumps cases oscillates and repeats itself approximately once a year.

$$\text{Amplitude} = \frac{\max - \min}{2} = \frac{11,000 - 2,000}{2} = 4,500 \text{ cases}$$

This means that the minimum and maximum number of cases of mumps are within $9,000 (= 4,500 \cdot 2)$ cases of each other.

(b) Assuming cyclical behavior, the number of cases in 30 months will be the same as the number of cases in 6 months which is 2,000 cases. (30 months equals 2 years and six months). The number of cases in 45 months equals the number of cases in 3 years and 9 months which assuming cyclical behavior is the same as the number of cases in 9 months. This is about 2,000 as well.

4. The function $y = \cos t$ reaches a high of 1 and a low of $-1$ so the amplitude is $\frac{1-(-1)}{2} = 1$. The period of $\cos t$ is $2\pi$ since it repeats itself after $2\pi$.

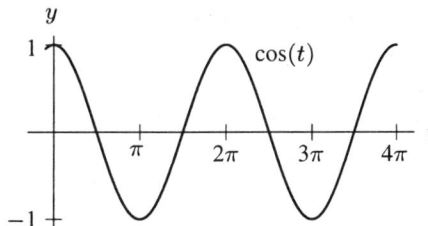

**Figure 1.75**

5.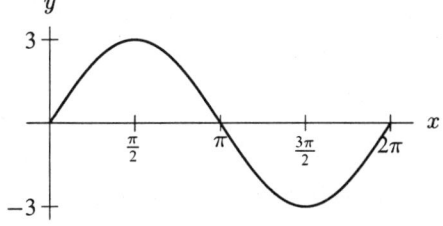

**Figure 1.76**

The amplitude is 3; the period is $2\pi$.

6.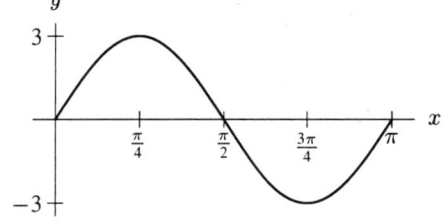

**Figure 1.77**

The amplitude is 3; the period is $\pi$.

7.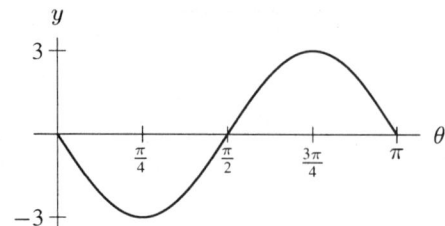

Figure 1.78

The amplitude is 3; the period is $\pi$.

8.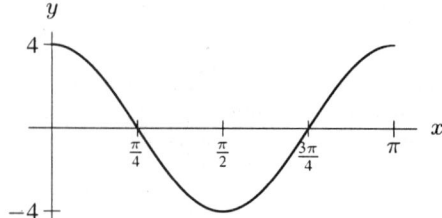

Figure 1.79

The amplitude is 4; the period is $\pi$.

9.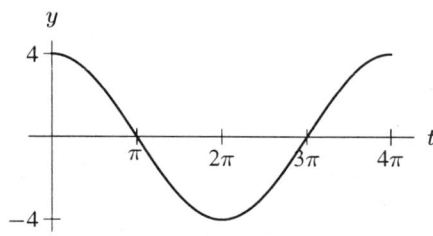

Figure 1.80

The amplitude is 4; the period is $4\pi$.

10.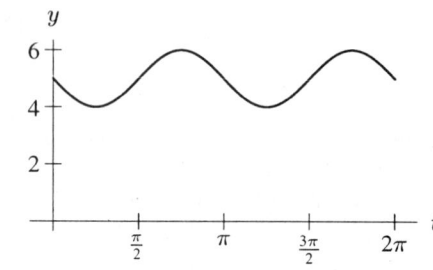

Figure 1.81

The amplitude is 1; the period is $\pi$.

11. This graph is a cosine curve with period $6\pi$ and amplitude 5, so it is given by $f(x) = 5\cos\left(\dfrac{x}{3}\right)$.

12. This graph is a sine curve with period $8\pi$ and amplitude 2, so it is given by $f(x) = 2\sin\left(\dfrac{x}{4}\right)$.

13. This graph is an inverted sine curve with amplitude 4 and period $\pi$, so it is given by $f(x) = -4\sin(2x)$.

14. This graph is the same as in Problem 12 but shifted up by 2, so it is given by $f(x) = 2\sin\left(\dfrac{x}{4}\right) + 2$.

15. This graph is an inverted cosine curve with amplitude 8 and period $20\pi$, so it is given by $f(x) = -8\cos\left(\dfrac{x}{10}\right)$.

16. This graph has period 5, amplitude 1 and no vertical shift or horizontal shift from $\sin x$, so it is given by

$$f(x) = \sin\left(\frac{2\pi}{5}x\right).$$

17. This graph has period 6, amplitude 5 and no vertical or horizontal shift, so it is given by

$$f(x) = 5\sin\left(\frac{2\pi}{6}x\right) = 5\sin\left(\frac{\pi}{3}x\right).$$

18. This can be represented by a sine function of amplitude 3 and period 18. Thus,

$$f(x) = 3\sin\left(\frac{\pi x}{9}\right).$$

19. This graph has period 8, amplitude 3, and a vertical shift of 3 with no horizontal shift. It is given by

$$f(x) = 3 + 3\sin\left(\frac{2\pi}{8}x\right) = 3 + 3\sin\left(\frac{\pi}{4}x\right).$$

20. From the example, we know that $y = 5 + 4.9\cos(\frac{\pi}{6}t)$, where $t$ represents hours after midnight and $y$ represents the height of the water in feet.

$$\text{At 3:00 am, } t = 3 \text{ so } y = 5 + 4.9\cos\left(\frac{\pi}{6} \cdot 3\right) = 5 + 4.9(0) = 5 \text{ feet.}$$

$$\text{At 4:00 am, } t = 4 \text{ so } y = 5 + 4.9\cos\left(\frac{\pi}{6} \cdot 4\right) = 5 + 4.9(-0.5) = 2.55 \text{ feet.}$$

$$\text{At 5:00 pm, } t = 17 \text{ so } y = 5 + 4.9\cos\left(\frac{\pi}{6} \cdot 17\right) \approx 5 + 4.9(-0.866) \approx 0.76 \text{ feet.}$$

21.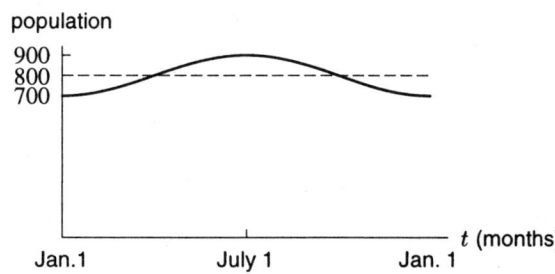

Figure 1.82

22. The period is $2\pi$.

23. (a) $f(g(x)) = f(2x + 1) = 5\sin(2x + 1)$
    (b) $g(f(x)) = g(5\sin x) = 2(5\sin x) + 1 = 10\sin x + 1$
    (c) $g(g(x)) = g(2x + 1) = 2(2x + 1) + 1 = 4x + 2 + 1 = 4x + 3$

24. If we graph the data, it certainly looks like there's a pattern from $t = 20$ to $t = 45$ which begins to repeat itself after $t = 45$. And the fact that $f(20) = f(45)$ helps to reinforce this theory.

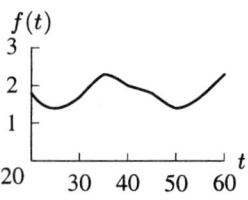

Figure 1.83

Since it looks like the graph repeats after 45, the period is $(45 - 20) = 25$. The amplitude $= \frac{1}{2}(\text{max} - \text{min}) = \frac{1}{2}(2.3 - 1.4) = 0.45$. Assuming periodicity,

$$f(15) = f(15 + 25) = f(40) = 2.0$$
$$f(75) = f(75 - 25) = f(50) = 1.4$$
$$f(135) = f(135 - 25 - 25 - 25) = f(60) = 2.3$$

25. (a) The function appears to vary between 5 and $-5$, and so the amplitude is 5.
    (b) The function begins to repeat itself at $x = 8$, and so the period is 8.
    (c) The function is at its highest point at $x = 0$, so we use a cosine function. It is centered at 0 and has amplitude 5, so we have $f(x) = 5\cos(Bx)$. Since the period is 8, we have $8 = 2\pi/B$ and $B = \pi/4$. The formula is

$$f(x) = 5\cos\left(\frac{\pi}{4}x\right).$$

26. Since the volume of the function varies between 2 and 4 liters, the amplitude is 1 and the graph is centered at 3. Since the period is 3, we have
$$3 = \frac{2\pi}{B} \quad \text{so} \quad B = \frac{2\pi}{3}.$$
The correct formula is (b). This function is graphed in Figure 1.84 and we see that it has the right characteristics.

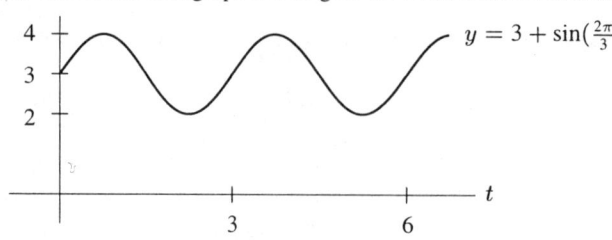

**Figure 1.84**

27. (a) $D =$ the average depth of the water.
    (b) $A =$ the amplitude $= 15/2 = 7.5$.
    (c) Period $= 12.4$ hours. Thus $(B)(12.4) = 2\pi$ so $B = 2\pi/12.4 \approx 0.507$.
    (d) $C$ is the time of a high tide.

28. (a) $f(t) = -0.5 + \sin t$, $g(t) = 1.5 + \sin t$, $h(t) = -1.5 + \sin t$, $k(t) = 0.5 + \sin t$.
    (b) $g(t) = 1 + k(t)$; $g(t) = 1.5 + \sin t = 1 + 0.5 + \sin t = 1 + k(t)$.
    (c) Since $-1 \le \sin t \le 1$, adding 1.5 everywhere we get $0.5 \le 1.5 + \sin t \le 2.5$ and since $1.5 + \sin t = g(t)$, we get $0.5 \le g(t) \le 2.5$. Similarly, $-2.5 \le -1.5 + \sin t = h(t) \le -0.5$.

29. (a) Aside from the fact that the graph *looks* like a periodic function, the graph pretty clearly repeats itself and reaches approximately the same minimum and maximum each year.
    (b) The maximum occurs in the 2nd quarter and the minimum occurs in the 4th quarter. It seems reasonable that people would drink more beer in the summer and less in the winter. Thus, production would be highest the quarter just before summer and lowest the quarter just before winter.
    (c) The period is 4 quarters or 1 year.
    $$\text{Amplitude} = \frac{\text{max} - \text{min}}{2} \approx \frac{55 - 45}{2} = 5 \text{ million barrels}$$

30. (a) It makes sense that temperature would be a function of time of day. Assuming similar weather for all the days of the experiment, time of day would probably be the over-riding factor on temperature. Thus, the temperature pattern should repeat itself each day.
    (b) The maximum seems to occur at 19 hours or 7 pm. It makes sense that the river would be the hottest toward the end of daylight hours since the sun will have been beating down on it all day. The minimum occurs at around 9 am which also makes sense because the sun hasn't been up long enough to warm the river much by then.
    (c) The period is one day. The amplitude $\approx \dfrac{32 - 28}{2} = 2°\text{C}$

## Solutions for Chapter 1 Review

1. (a) The statement $f(12) = 60$ says that when $p = 12$, we have $q = 60$. When the price is \$12, we expect to sell 60 units.
   (b) Decreasing, because as price increases, we expect less to be sold.

2.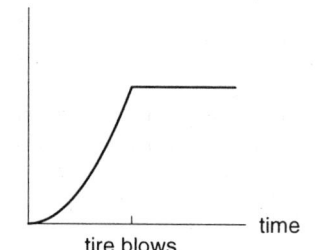

3.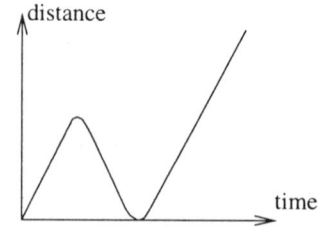

**SOLUTIONS TO REVIEW PROBLEMS FOR CHAPTER ONE** **75**

4. There are three peaks in the curve, each one of which occurs around one of the three meal times. The first peak is about 7 am, the next one right before noon and the third one around 6 pm. From the relative magnitudes of the peaks, more gas is used to cook dinner than lunch, and more gas is used to cook lunch than breakfast.

5. (a) As $x$ gets larger and larger, the value of the function gets closer and closer to 3.
   (b) Many answers are possible, such as the following.

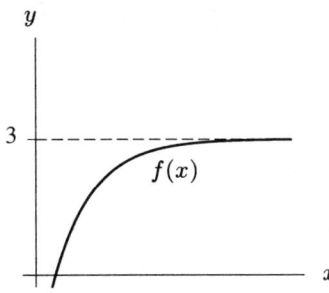

*Figure 1.85*: Answer to (i)

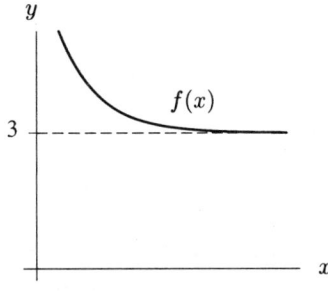

*Figure 1.86*: Answer to (ii)

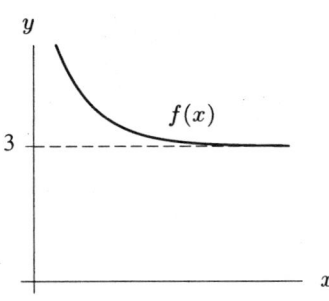

*Figure 1.87*: Answer to (iii)

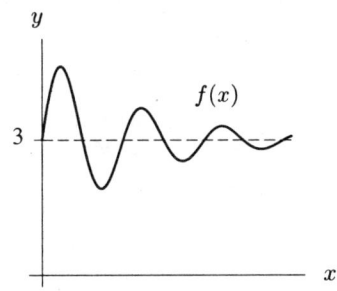

*Figure 1.88*: Answer to (iv)

6. (a) As $x$ gets more and more negative, the value of the function gets closer and closer to 3.
   (b) Many answers are possible, such as the following.

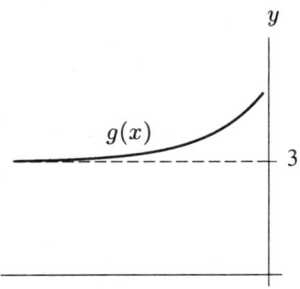

*Figure 1.89*: Answer to (i)

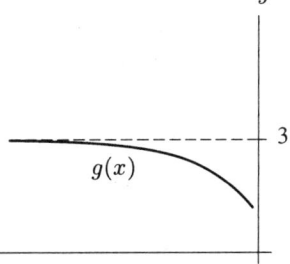

*Figure 1.90*: Answer to (ii)

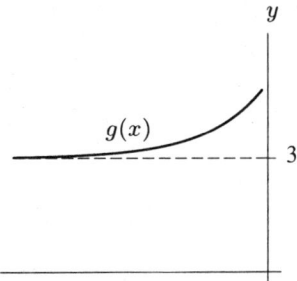

*Figure 1.91*: Answer to (iii)

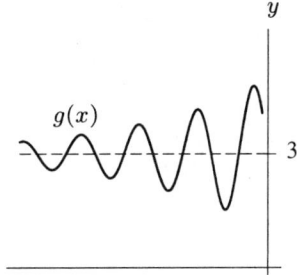

*Figure 1.92*: Answer to (iv)

7. One possibility is $f(x) = -x^3$; see Figure 1.93.

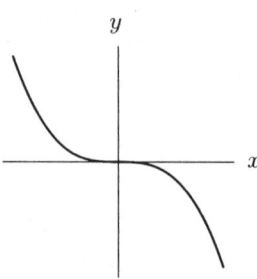

**Figure 1.93**: $f(x) = -x^3$

8. (a) Four zeros, at approximately $x = -4.6, 1.2, 2.7,$ and $4.1$.
   (b) $f(2)$ is the $y$-value corresponding to $x = 2$, so $f(2)$ is about $-1$. Likewise, $f(4)$ is about $0.4$.
   (c) Decreasing near $x = -1$, increasing near $x = 3$.
   (d) Concave up near $x = 2$, concave down near $x = -4$.
   (e) Increasing for $x < -1.5$ and for $2 < x < 3.5$.

9. (a) It was decreasing from March 2 to March 5 and increasing from March 5 to March 9.
   (b) From March 5 to 8, the average temperature increased, but the rate of increase went down, from 12° between March 5 and 6 to 4° between March 6 and 7 to 2° between March 7 and 8.
   From March 7 to 9, the average temperature increased, and the rate of increase went up, from 2° between March 7 and 8 to 9° between March 8 and 9.

10. We are looking for a linear function $y = f(x)$ that, given a time $x$ in years, gives a value $y$ in dollars for the value of the refrigerator. We know that when $x = 0$, that is, when the refrigerator is new, $y = 950$, and when $x = 7$, the refrigerator is worthless, so $y = 0$. Thus $(0, 950)$ and $(7, 0)$ are on the line that we are looking for. The slope is then given by
$$m = \frac{950}{-7}$$
It is negative, indicating that the value decreases as time passes. Having found the slope, we can take the point $(7, 0)$ and use the point-slope formula:
$$y - y_1 = m(x - x_1).$$
So,
$$y - 0 = -\frac{950}{7}(x - 7)$$
$$y = -\frac{950}{7}x + 950.$$

11. We will let
$$T = \text{amount of fuel for take-off,}$$
$$L = \text{amount of fuel for landing,}$$
$$P = \text{amount of fuel per mile in the air,}$$
$$m = \text{the length of the trip in miles.}$$
Then $Q$, the total amount of fuel needed, is given by
$$Q(m) = T + L + Pm.$$

12. (a) We know that the function for the cost of running the theater is of the form
$$C = mq + b$$
where $q$ is the number of customers, $m$ is the variable cost and $b$ is the fixed cost. Thus, the function for the cost is
$$C = 2q + 5000.$$
We know that the revenue function is of the form
$$R = pq$$

where $p$ is the price charged per customer. Thus, the revenue function is

$$R = 7q.$$

The theater makes a profit when the revenue is greater than the cost, that is when

$$R > C.$$

Substituting $R = 7q$ and $C = 2q + 5000$, we get

$$R > C$$
$$7q > 2q + 5000$$
$$5q > 5000$$
$$q > 1000.$$

Thus, the theater makes a profit when it has more than 1000 customers.

(b) The graph of the two functions is shown in Figure 1.94.

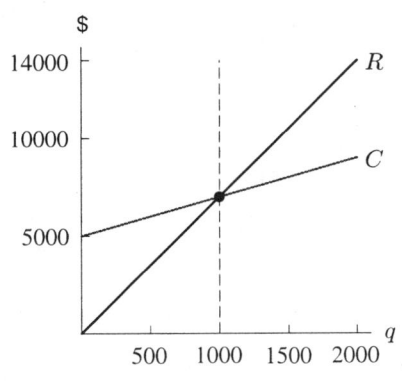

**Figure 1.94**

13. We know that a function is linear if its slope is constant. A function is concave up if the slope increases as $t$ gets larger. And a function is concave down if the slope decreases as $t$ gets larger. Looking at the points $t = 10, t = 20$ and $t = 30$ we see that the slope of $F(t)$ decreases since

$$\frac{F(20) - F(10)}{20 - 10} = \frac{22 - 15}{10} = 0.7$$

while

$$\frac{F(30) - F(20)}{30 - 20} = \frac{28 - 22}{10} = 0.6$$

Looking at the points $t = 10, t = 20$ and $t = 30$ we see that the slope of $G(t)$ is constant

$$\frac{G(20) - G(10)}{20 - 10} = \frac{18 - 15}{10} = 0.3$$

and

$$\frac{G(30) - G(20)}{30 - 20} = \frac{21 - 18}{10} = 0.3$$

Also note that the slope of $G(t)$ is constant everywhere.

Looking at the points $t = 10, t = 20$ and $t = 30$ we see that the slope of $H(t)$ increases since

$$\frac{H(20) - H(10)}{20 - 10} = \frac{17 - 15}{10} = 0.2$$

while

$$\frac{H(30) - H(20)}{30 - 20} = \frac{20 - 17}{10} = 0.3$$

Thus $F(t)$ is concave down, $G(t)$ is linear and $H(t)$ is concave up.

**14.** (a) Advertising is generally cheaper in bulk; spending more money will give better and better marginal results initially, hence concave up. (Spending $5,000 could give you a big newspaper ad reaching 200,000 people; spending $100,000 could give you a series of TV spots reaching 50,000,000 people.) A graph is shown below, left. But after a certain point, you may "saturate" the market, and the increase in revenue slows down – hence concave down.

(b) The temperature of a hot object decreases at a rate proportional to the difference between its temperature and the temperature of the air around it. Thus, the temperature of a very hot object decreases more quickly than a cooler object. The graph is decreasing and concave up. (We are assuming that the coffee is all at the same temperature.)

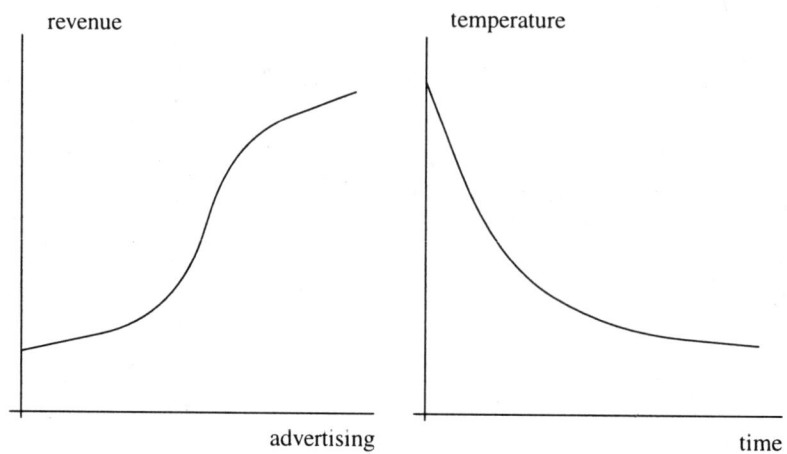

**15.** (a) (I) The incidence of cancer increase with age, but the rate of increase slows down slightly. The graph is nearly linear. This type of cancer is closely related to the aging process.

(II) In this case a peak is reached at about age 55, after which the incidence decreases.

(III) This type of cancer has an increased incidence until the age of about 48, then a slight decrease, followed by a gradual increase.

(IV) In this case the incidence rises steeply until the age of 30, after which it levels out completely.

(V) This type of cancer is relatively frequent in young children, and its incidence increases gradually from about the age of 20.

(VI) This type of cancer is not age-related – all age-groups are equally vulnerable, although the overall incidence is low (assuming each graph has the same vertical scale).

(b) Graph (V) shows a relatively high incidence rate for children. Leukemia behaves in this way.

(c) Graph (III) could represent cancer in women with menopause as a significant factor. Breast cancer is a possibility here.

(d) Graph (I) shows a cancer which might be caused by toxins building up in the body. Lung cancer is a good example of this.

**16.** One possible graph is given in Figure 1.95.

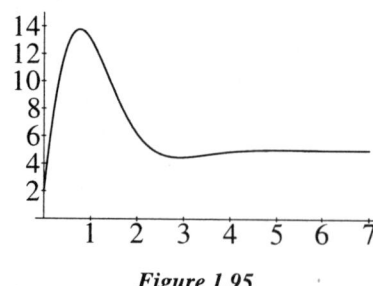

*Figure 1.95*

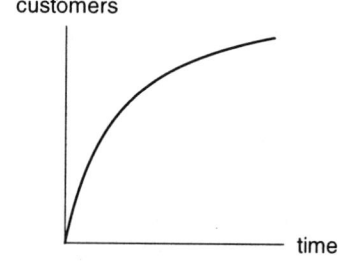

*Figure 1.96*

**17.** (a) See Figure 1.96.

(b) "The rate at which new people try it" is the rate of change of the total number of people who have tried the product. Thus the statement of the problem is telling you that the graph is concave down—the slope is positive but decreasing, as the graph shows.

**18.** (a) The death rate from stomach cancer has decreased fairly steadily. For lung cancer the death rate has increased fairly steadily. Colon/rectum and prostrate cancer deaths increased until about 1945 and then have stayed level. Death rates from cancer of the esophagus have remained steady. Liver cancer deaths have decreased gradually. Deaths from leukemia and cancer of the pancreas have increased slightly.

(b) Lung cancer deaths have had the greatest average rate of change – that line is by far the steepest on the graph.

$$\text{Average rate of change} = \frac{43 - 2.5}{1967 - 1930}$$
$$\approx 1.1 \quad \text{additional deaths per 100,000 per year between 1930 and 1967.}$$

(c) Stomach cancer deaths have the most negative slope on the graph.

$$\text{Average rate of change} = \frac{10 - 29}{1967 - 1930}$$
$$\approx -0.5 \text{ or 0.5 fewer deaths}$$

per 100,000 per year between 1930 and 1967.

**19.** (a) is $g(x)$ since it is linear. (b) is $f(x)$ since it has decreasing slope; the slope starts out about 1 and then decreases to about $\frac{1}{10}$. (c) is $h(x)$ since it has increasing slope; the slope starts out about $\frac{1}{10}$ and then increases to about 1.

**20.** (a) When $l = 20{,}000$ we have $J = 10$, so

$$10 = a(20{,}000)^{0.3}$$
$$a = \frac{10}{20{,}000^{0.3}} \approx 0.5125.$$

(b) When $l = 0.2$, we have $J = 0.5125(0.2)^{0.3} \approx 0.3162$.

(c) Using $a = 0.5125$ with $J = 20$, we have

$$20 = 0.5125 l^{0.3}$$
$$l^{0.3} = \frac{20}{0.5125}$$
$$l = \left(\frac{20}{0.5125}\right)^{10/3} \approx 201{,}583 \text{ dynes/cm}^2.$$

(This corresponds approximately to the loudness of thunder.)

**21.** By plotting $b$ against $G^{-1/3}$ we see a straight line with slope $\approx 42.5$. Alternatively, by calculating $b/G^{-1/3}$ for each of the data points, we find a common value of approximately 42.4. These give an estimate of $a \approx 42.45$. Figure 1.97 shows the function $b = 42.45 G^{-1/3}$, which seems to model well the data provided in the problem.

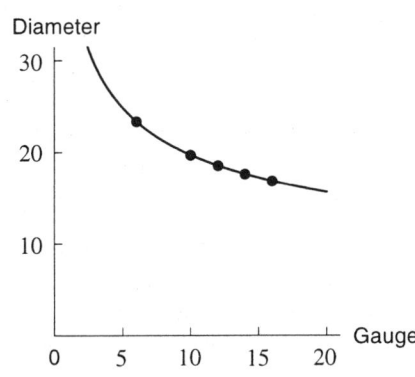

*Figure 1.97:* The gauge and diameter of a shotgun's bore and the function $b = 42.45 G^{-1/3}$

**80** CHAPTER ONE /SOLUTIONS

**22.** We use $Q = Q_0 e^{kt}$ and let $t = 0$ represent 8 am. Then $Q_0 = 100$, so

$$Q = 100 e^{kt}.$$

For the husband, the half-life is four hours. We solve for $k$:

$$50 = 100 e^{k \cdot 4}$$
$$0.5 = e^{4k}$$
$$\ln 0.5 = 4k$$
$$k = \frac{\ln 0.5}{4} \approx -0.173.$$

We have $Q = 100 e^{-0.173 t}$. At 10 pm, we have $t = 14$ so the amount of caffeine for the husband is

$$Q = 100 e^{-0.173(14)} = 8.87 \text{ mg} \approx 9 \text{ mg}.$$

For the pregnant woman, the half-life is 10. Solving for $k$ as above, we obtain

$$k = \frac{\ln 0.5}{10} = -0.069.$$

At 10 pm, the amount of caffeine for the pregnant woman is

$$Q = 100 e^{-0.069(14)} \approx 38 \text{ mg}.$$

**23.** Since $3^x$ is always positive while $x^3$ is negative, we know that $3^x > x^3$ for $x < 0$. Looking at the large-scale graph in Figure 1.98, we see that $3^x$ is also clearly bigger than $x^3$ for $x > 4$. We zoom in on the interval $(0, 4)$ to see the behavior there, shown in Figure 1.99. We see that the graphs are very close to each other near $x = 3$ (where the values are equal), but elsewhere, $3^x > x^3$. To figure out what is going on near $x = 3$, we zoom in again (see Figure 1.100), and notice that $x^3 > 3^x$ on the interval from about 2.5 to 3. Thus $3^x > x^3$ if $x < 2.5$ (approximately) or $x > 3$.

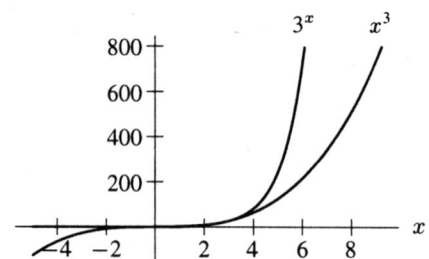

Figure 1.98

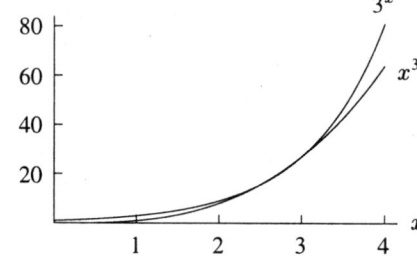

Figure 1.99

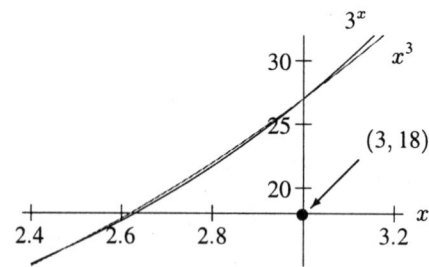

Figure 1.100

Alternatively, graph $f(x) = 3^x - x^3$ for $0 \leq x \leq 3.5$ and $-2 \leq y \leq 4$. This shows $3^x > x^3$ for $x > 3$ and $x < 2.48$.

**24.** (a) We know that the equilibrium point is the point where the supply and demand curves intersect. They appear to intersect at a price of $250 per unit, so the corresponding quantity is 750 units.

(b) We know that the supply curve climbs upwards while the demand curve slopes downwards. At the price of $300 per unit the suppliers will be willing to produce 875 units while the consumers will be ready to buy 625 units. Thus, we see that when the price is above the equilibrium point, more items would be produced than the consumers will be willing to buy. Thus, the producers end up wasting money by producing that which will not be bought, so the producers are better off lowering the price.

(c) Looking at the point on the rising curve where the price is $200 per unit, we see that the suppliers will be willing to produce 625 units, whereas looking at the point on the downward sloping curve where the price is $200 per unit, we see that the consumers will be willing to buy 875 units. Thus, we see that when the price is less than the equilibrium price, the consumers are willing to buy more products than the suppliers would make and the suppliers can thus make more money by producing more units and raising the price.

**25.** (a) A graph of $P$ against $t$ is shown in Figure 1.101.

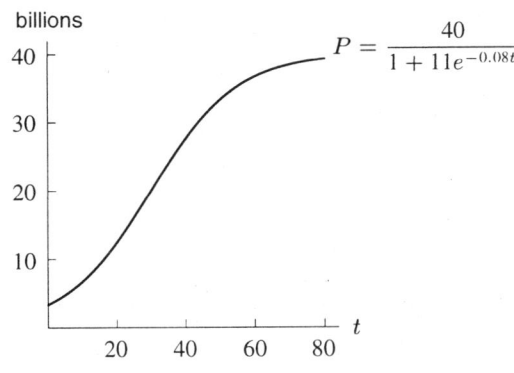

**Figure 1.101**

(b) We are asked to find the time $t$ such that $P(t) = 20$. We can find this either by tracing along the graph, which tells us that $P = 20$ when $t \approx 30$. Alternatively, we can solve analytically:

$$20 = P(t) = \frac{40}{1 + 11e^{-0.08t}}$$

$$1 + 11e^{-0.08t} = \frac{40}{20} = 2$$

$$11e^{-0.08t} = 1$$

$$e^{-0.08t} = \frac{1}{11}$$

$$\ln e^{-0.08t} = \ln \frac{1}{11}$$

$$-0.08t \approx -2.4$$

$$t \approx \frac{-2.4}{-0.08} = 30.$$

Thus, 30 years from 1990 the population of the world should be 20 billion. In other words the population of the world will be 20 billion in the year 2020.

We are asked to find the time $t$ such that $P(t) = 39.9$. By tracing along the curve, we find $P = 39.9$ when $t \approx 105$. Alternatively, we can solve analytically:

$$39.9 = P(t) = \frac{40}{1 + 11e^{-0.08t}}$$

$$1 + 11e^{-0.08t} = \frac{40}{39.9} = 1.00251$$

$$11e^{-0.08t} = 0.00251$$

$$e^{-0.08t} = \frac{0.00251}{11}$$

**82** CHAPTER ONE /SOLUTIONS

$$\ln e^{-0.08t} = \ln \frac{0.00251}{11}$$
$$-0.08t \approx -8.39$$
$$t \approx \frac{-8.39}{-0.08} \approx 105.$$

Thus, 105 years from 1990 the population of the world should be 39.9 billion. In other words the population of the world will be 39.9 billion in the year 2095.

(c) We are asked for the difference in populations at the years 2000 and 1990. That is we are asked for

$$P(10) - P(0).$$

Substituting $t = 0$, we get

$$P(0) = \frac{40}{1 + 11e^{-0.08(0)}}$$
$$= \frac{40}{1 + 11e^0}$$
$$= \frac{40}{1 + 11(1)}$$
$$= \frac{40}{1 + 11}$$
$$= \frac{40}{12}$$
$$\approx 3.33.$$

Thus, in the year 1990 the population would be approximately 3.33 billion. Substituting $t = 10$, we get

$$P(10) = \frac{40}{1 + 11e^{-0.08(10)}}$$
$$= \frac{40}{1 + 11e^{-0.8}}$$
$$= \frac{40}{1 + 11(0.449)}$$
$$= \frac{40}{1 + 4.939}$$
$$= \frac{40}{5.939}$$
$$\approx 6.73.$$

Thus, in the year 2000 the population will be about 6.73 billion. The increase in population between 1990 and 2000 is

$$P(10) - P(0) \approx 6.73 \text{ billion } - 3.33 \text{ billion } = 3.40 \text{ billion}.$$

26. Because the population is growing exponentially, the time it takes to double is the same, regardless of the population levels we are considering. For example, the population is 20,000 at time 3.7, and 40,000 at time 6.0. This represents a doubling of the population in a span of $6.0 - 3.7 = 2.3$ years.

How long does it take the population to double a second time, from 40,000 to 80,000? Looking at the graph once again, we see that the population reaches 80,000 at time $t = 8.3$. This second doubling has taken $8.3 - 6.0 = 2.3$ years, the same amount of time as the first doubling.

Further comparison of any two populations on this graph that differ by a factor of two will show that the time that separates them is 2.3 years. Similarly, during any 2.3 year period, the population will double. Thus, the doubling time is 2.3 years.

Suppose $P = P_0 a^t$ doubles from time $t$ to time $t + d$. We now have $P_0 a^{t+d} = 2P_0 a^t$, so $P_0 a^t a^d = 2P_0 a^t$. Thus, canceling $P_0$ and $a^t$, $d$ must be the number such that $a^d = 2$, no matter what $t$ is.

**27.**

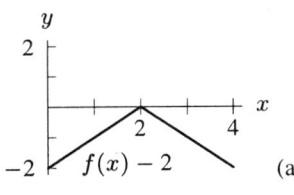
(a)

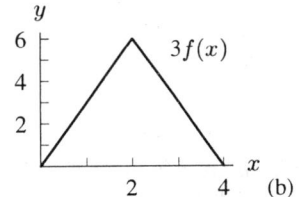
(b)

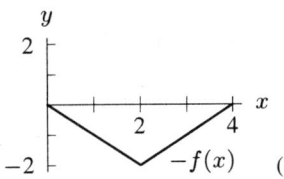
(c)

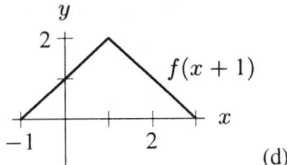
(d)

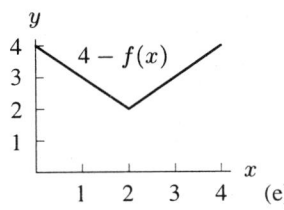
(e)

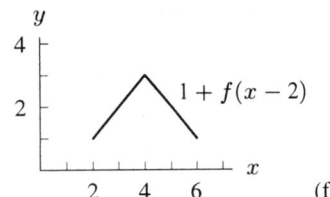
(f)

**Figure 1.102**

**28.** The formula for compound interest is $P_0 e^{rt}$ where $P_0$ is initial investment, $r$ is the interest rate and $t$ is number of years. So after 5 years, \$10,000 at 8% compounded continuously is $\$10{,}000 e^{(0.08)(5)} \approx \$14{,}918.25$. After 10 years, we have $\$10{,}000 e^{(0.08)(10)} \approx \$22{,}255.41$.

**29.** The formula which models compounding interest continuously for $t$ years is $P_0 e^{rt}$ where $P_0$ is the initial deposit and $r$ is the interest rate. So if we want to double our money while getting 6% interest, we want to solve the following for $t$:

$$P_0 e^{0.06t} = 2P_0$$
$$e^{0.06t} = 2$$
$$0.06t = \ln(2)$$
$$t = \frac{\ln 2}{0.06} \approx 11.6 \text{ years}$$

Alternatively, by the Rule of Seventy, we have

$$\text{Double time} = \frac{70}{6} \approx 11.67 \text{ years.}$$

**30.** (a)

$$e^{-0.015t} = \frac{1}{2}$$
$$-0.015t = \ln\left(\frac{1}{2}\right)$$
$$t = \frac{\ln\left(\frac{1}{2}\right)}{-0.015} \approx 46.21 \quad \text{years.}$$

(b)

$$e^{-0.024t} = \frac{1}{2}$$
$$-0.024t = \ln\left(\frac{1}{2}\right)$$
$$t = \frac{\ln\left(\frac{1}{2}\right)}{-0.024} \approx 28.88 \quad \text{years.}$$

**84** CHAPTER ONE /SOLUTIONS

(c) If we find the half-life assuming a decay rate of $-0.02$, we get:

$$e^{-0.02t} = \frac{1}{2}$$

$$t = \frac{\ln\left(\frac{1}{2}\right)}{-0.02} \approx 34.66 \quad \text{years.}$$

This is significantly different from the half-lives we found for decay rates of $-0.015$ and $-0.024$, so round-off error is very significant in this sort of problem.

31. Looking at $y_1$, we see that it is a decreasing function of $x$. Furthermore, we see that the rate of decrease of $y_1$ decreases as $x$ increases. Thus, given the choices, $y_1$ must be inversely proportional to $x$, so we have

$$y_1 = \frac{k}{x}.$$

We know that at $x = 10$ we have $y_1 = 300$, so

$$300 = \frac{k}{10},$$

and we get for the constant of proportionality

$$k = 3000.$$

Hence,

$$y_1 = \frac{3000}{x}.$$

Looking at $y_2$, we see that $y_2$ is an increasing function of $x$. Also, notice that the rate of increase of $y_2$ increases as $x$ increases. Therefore, $y_2$ must be proportional to $x^2$, so we have

$$y_2 = kx^2.$$

We know that at $x = 10$ we have $y_2 = 200$, so

$$200 = k(10)^2 = 100k,$$

and we get for the constant of proportionality

$$k = 2.$$

Thus,

$$y_2 = 2x^2.$$

Looking at $y_3$, we see that the rate of increase of $y_3$ is constant for all given $x$ values. Therefore, $y_3$ must be linearly proportional to $x$, and we can write

$$y_3 = kx.$$

We know that at $x = 10$ we have $y_3 = 2.50$, so

$$2.50 = k(10),$$

and we get for the constant of proportionality

$$k = 0.25.$$

Hence,

$$y_3 = 0.25x.$$

32. We know that if the population increases exponentially the formula for the population at time $t$ is of the form

$$P(t) = P_0 e^{rt}$$

where $P_0$ is the population at time $t = 0$ and $r$ is the rate of growth; we can measure time in years. If we let 1950 be the initial time $t = 0$ we get

$$P_0 = 2.564$$

so

$$P(t) = 2.564 e^{rt}.$$

We also know that in 1980, time $t = 30$, we have

$$P(30) = 4.478.$$

Thus we get

$$4.478 = P(30)$$
$$= 2.564e^{30r}$$
$$e^{30r} = \frac{4.478}{2.564} \approx 1.746$$
$$\ln e^{30r} = \ln 1.746$$
$$30r = \ln 1.746$$
$$r = \frac{\ln 1.746}{30} \approx 0.0186$$

Thus, the formula for the population, assuming exponential growth, is

$$P(t) = 2.564e^{0.0186t} \text{ billion}.$$

Trying our formula for the year 1991, time $t = 41$, we get

$$P(41) = 2.564e^{0.0186(41)}$$
$$= 2.564e^{0.7626}$$
$$\approx 2.564(2.14)$$
$$\approx 5.49 \text{ billion}$$

Thus, we see that we were not too far off the mark when we approximated the population by an exponential function.

33. (a) The graph of the profit is shown in Fig 1.103.

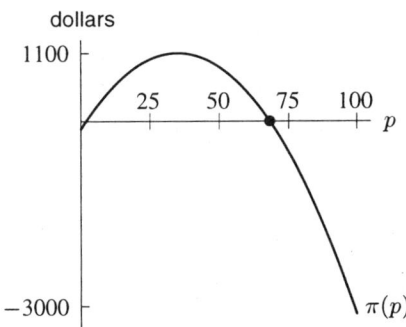

**Figure 1.103**

Looking at the graph, we can see that the value of $p$ that will maximize profits occurs at about $p = 35$ dollars.

(b) We know that the company will make a profit whenever the profit is greater than zero, or in other words, for all prices $p$ such that the graph of $\pi(p)$ is above the $x$-axis. Looking at the graph we see that this occurs approximately when the price per skateboard is between $2 and $68.

34. (a) We know that the cost function will be of the form

$$C = mq + b$$

where $m$ is the variable cost and $b$ is the fixed cost. In this case this gives

$$C = 5q + 7000.$$

We know that the revenue function is of the form

$$R = pq$$

where $p$ is the price per shirt. Thus in this case we have

$$R = 12q.$$

(b) We are given
$$q = 2000 - 40p.$$
We are asked to find the demand when the price is \$12. Plugging in $p = 12$ we get
$$q = 2000 - 40(12) = 2000 - 480 = 1520.$$
Given this demand we know that the cost of producing $q = 1520$ shirts is
$$C = 5(1520) + 7000 = 7600 + 7000 = \$14{,}600.$$
The revenue from selling $q = 1520$ shirts is
$$R = 12(1520) = \$18{,}240.$$
Thus the profit is
$$\pi(12) = R - C$$
or in other words
$$\pi(12) = 18{,}240 - 14{,}600 = \$3640.$$

(c) Since we know that
$$q = 2000 - 40p,$$
$$C = 5q + 7000,$$
and
$$R = pq,$$
we can write
$$C = 5q + 7000 = 5(2000 - 40p) + 7000 = 10{,}000 - 200p + 7000 = 17{,}000 - 200p$$
and
$$R = pq = p(2000 - 40p) = 2000p - 40p^2.$$
We also know that the profit is the difference between the revenue and the cost so
$$\pi(p) = R - C = 2000p - 40p^2 - (17{,}000 - 200p) = -40p^2 + 2200p - 17{,}000.$$

(d) Looking at Figure 1.104 we see that the maximum profit occurs when the company charges about \$27.50 per shirt. At this price, the profit is about \$13,250.

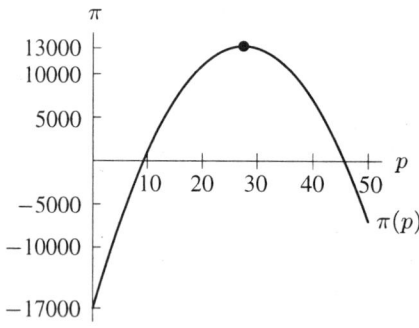

**Figure 1.104**

**35.** (a) $P = 1000 + 50t$
(b) $P = 1000(1.05)^t$

**36.** (a) Since $Q = 25$ at $t = 0$, we have $Q_0 = 25$ (since $e^0 = 1$). We then plug the value at $t = 5$ into the equation
$$Q = 25e^{rt}$$
to find $r$. Doing so, we get
$$391 = 25e^{r(5)}$$
$$\frac{391}{25} = 15.64 = e^{5r}$$
$$\ln 15.64 = 5r$$
$$0.55 \approx r.$$
And so the equation is
$$Q = 25e^{0.55t}.$$

(b) At the time $t$ when the population has doubled,
$$2 = e^{0.55t}$$
$$\ln 2 = 0.55t$$
$$t = \frac{\ln 2}{0.55} \approx 1.3 \text{ months}.$$

(c) At the time $t$ when the population is 1000 rabbits,
$$1000 = 25e^{0.55t}$$
$$40 = e^{0.55t}$$
$$\ln 40 = 0.55t$$
$$t = \frac{\ln 40}{0.55} \approx 6.7 \text{ months}.$$

**37.** We use the equation $B = Pe^{rt}$. We want to have a balance of $B = \$20,000$ in $t = 6$ years, with an annual interest rate of 10%.
$$20,000 = Pe^{(0.1)6}$$
$$P = \frac{20,000}{e^{0.6}}$$
$$\approx \$10,976.23.$$

**38.** To find a half-life, we want to find at what $t$ value $Q = \frac{1}{2}Q_0$. Plugging this into the equation of the decay of plutonium-240, we have
$$\frac{1}{2} = e^{-0.00011t}$$
$$t = \frac{\ln(1/2)}{-0.00011} \approx 6{,}301 \text{ years}.$$

The only difference in the case of plutonium-242 is that the constant $-0.00011$ in the exponent is now $-0.0000018$. Thus, following the same procedure, the solution for $t$ is
$$t = \frac{\ln(1/2)}{-0.0000018} \approx 385{,}081 \text{ years}.$$

**39.** If $C$ is the amount of carbon-14 originally present in the skull, then after 5730 years, $\frac{1}{2}C$ is left. Hence,
$$\frac{1}{2}C = Ce^{-k \cdot 5730}$$
$$\frac{1}{2} = e^{-k \cdot 5730}$$
$$\ln\left(\frac{1}{2}\right) = -k \cdot 5730$$
$$k = \frac{-\ln(1/2)}{5730} = \frac{\ln 2}{5730}.$$

The question asks how long it took for the skull to lose 80% of its carbon-14. In other words, when will the amount of remaining carbon-14 be $\frac{1}{5}C$?

$$\text{Remaining carbon-14} = \frac{1}{5}C = Ce^{-(\ln 2)t/5730}$$

$$\frac{1}{5} = e^{-(\ln 2)t/5730}$$

$$\ln\left(\frac{1}{5}\right) = -\frac{\ln 2}{5730}t.$$

Since $\ln(1/5) = -\ln 5$, we have

$$t = \frac{\ln 5}{\ln 2}(5730) \approx 13{,}300 \text{ years},$$

the approximate age of the skull.

40. (a) Each day prices are multiplied by 1.001, so after 365 days prices are $(1.001)^{365} \approx 1.44$ times what they were originally. Therefore, they increase by about 44% a year.
    (b) Guess about two years, since $1.44^2 \approx (1.001)^{2(365)} = (1.001)^{730} \approx 2.074$, so it's a good guess.

41. In effect, your friend is offering to give you $17,000 now in return for the $19,000 lottery payment one year from now. Since $19000/17000 = 1.11764\cdots$, your friend is charging you 11.7% interest per year, compounded annually. You can expect to get more by taking out a loan as long as the interest rate is less than 11.7%. In particular, if you take out a loan, you have the first lottery check of $19,000 plus the amount you can borrow to be paid back by a single payment of $19000 at the end of the year. At 8.25% interest, compounded annually, the present value of 19,000 one year from now is $19000/(1.0825) = 17551.96$. Therefore the amount you can borrow is the total of the first lottery payment and the loan amount, that is, $19000 + 17551.96 = 36551.96$. So you do better by taking out a one-year loan at 8.25% per year, compounded annually, than by accepting your friend's offer.

42. The following table contains the present value of each of the payments, though it does not take into account the resale value if you buy the machine.

| Buy | | | Lease | | |
|---|---|---|---|---|---|
| Year | Payment | Present Value | | Payment | Present Value |
| 0 | 12000 | 12000 | | 2650 | 2650 |
| 1 | 580 | $580/(1.0775) = 538.28$ | | 2650 | $2650/(1.0775) = 2459.40$ |
| 2 | 464 | $464(1.0775)^2 = 399.65$ | | 2650 | $2650/(1.0775)^2 = 2282.50$ |
| 3 | 290 | $290/(1.0775)^3 = 231.82$ | | 2650 | $2650/(1.0775)^3 = 2118.33$ |
|  | Total | 13,169.75 | | Total | 9,510.23 |

Now we consider the $5000 resale.

$$\text{Present value of resale} = \frac{5000}{(1.0775)^3} = 3996.85.$$

The net present value associated with buying the machine is the present value of the payments minus the present value of the resale price, which is

$$\text{Present value of buying} = 13{,}169.75 - 3996.85 = 9172.90$$

Since the present value of the expenses associated with buying ($9172.90) is smaller than the present value of leasing ($9510.23), you should buy the machine.

43. The pomegranate is at ground level when $f(t) = -16t^2 + 64t = -16t(t-4) = 0$, so when $t = 0$ or $t = 4$. At time $t = 0$ it is thrown, so it must hit the ground at $t = 4$ seconds. The symmetry of its path with respect to time may convince you that it reaches its maximum height after 2 seconds. Alternatively, we can think of the graph of $f(t) = -16t^2 + 64t = -16(t-2)^2 + 64$, which is a downward parabola with vertex (i.e., highest point) at $(2, 64)$. The maximum height is $f(2) = 64$ feet.

**44.** (a) A graph of world CFC production is in Figure 1.105.

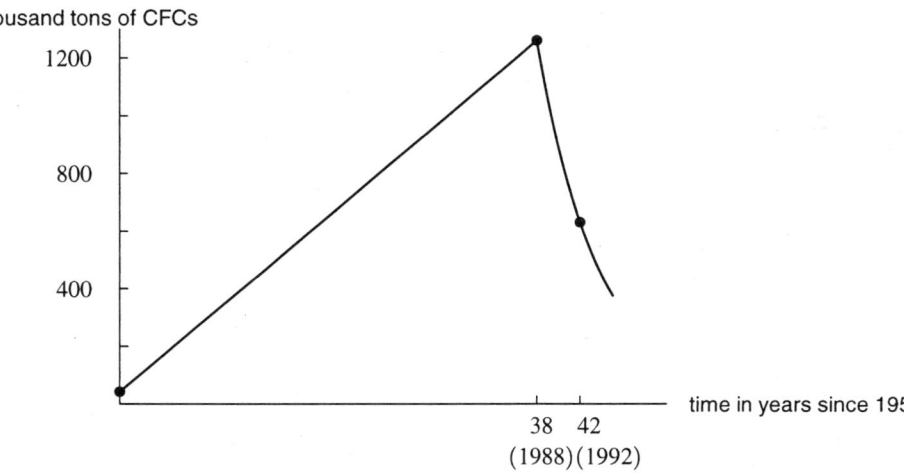

**Figure 1.105**

(b) If we assume that CFC production will eventually drop to zero but that it will take quite a while to do so, a function of the form $P = P_0 e^{-kt}$ could be used to model CFC production. We assume CFC production continues to drop by half every 4 years and write
$$P = P_0 e^{-kt}$$
where $P_0 = 1260$ and $t$ is years since 1988. Substituting $P = 1260/2$ when $t = 1992 - 1988$ gives
$$\frac{1}{2} = e^{-k(4)}$$
Solving for $k$, we get
$$\ln\left(\frac{1}{2}\right) = -4k.$$
So
$$k = -\frac{\ln(1/2)}{4} \approx 0.17.$$
Thus,
$$P = 1260 e^{-0.17t}.$$

**45.** (a) This moves the graph one unit to the left.
(b) A non-constant polynomial tends toward $+\infty$ or $-\infty$ as $x \to \infty$. This polynomial $p$ does not. Therefore, $p$ must be a constant function, i.e. its graph is a horizontal line.

**46.** (a) We are expecting to see a graph of a function which approaches its horizontal asymptote from below. Thus, VIII is the appropriate graph.
(b) We are expecting to see a parabola opening upwards. Thus III is the appropriate graph.
(c) We are expecting to see a trigonometric curve. Notice that the function takes on both positive and negative values and is periodic. Thus, VII is the appropriate graph.
(d) We are expecting to see an upside-down parabola. Thus, I is the appropriate graph.
(e) We are expecting to see a positive exponential curve with a positive horizontal asymptote. Thus, IX is the appropriate graph.
(f) We are expecting to see a function which takes on large negative values when $x$ is a large negative number and which takes on large positive values when $x$ is a positive number. Since this function is a cubic, it may have two "twists". Thus, VI is the appropriate graph.
(g) We expect to see a function which is undefined at zero but is very large next to zero and gets more negative as $x$ gets larger. Thus, II is an appropriate graph.
(h) We expect to see a trigonometric curve that is periodic. We can also observe that all values of the function are greater than or equal to zero since $\cos x \geq -1$. Thus, IV is an appropriate graph.
(i) We are expecting to see a graph which has both a horizontal asymptote at $y = 0$ and a vertical asymptote at $x = 0$. Thus, V is the appropriate graph.

47. (a) Suppose $p(x) = ax^2 + bx + c$. Then
$$p(-x) = ax^2 - bx + c.$$
Since $p(x) = p(-x)$, we have
$$ax^2 + bx + c = ax^2 - bx + c,$$
so $2bx = 0$ for all $x$, so $b = 0$. Therefore, the solution is $p(x) = ax^2 + c$ for any $a$ and any $c$.

(b) Suppose $p(x) = ax^2 + bx + c$, so
$$p(2x) = a(2x)^2 + b(2x) + c = 4ax^2 + 2bx + c.$$
Since $p(2x) = 2p(x)$, we have
$$4ax^2 + 2bx + c = 2(ax^2 + bx + c)$$
$$2ax^2 - c = 0.$$
Since $2ax^2 - c = 0$ for all $x$, we must have $a = c = 0$. Therefore, the solution is $p(x) = bx$ for any $b$.

48. (a) $r(p) = kp(A - p)$, where $k > 0$ is a constant.
(b) $p = A/2$.

49. (a) The rate $R$ is the difference of the rate at which the glucose is being injected, which is given to be constant, and the rate at which the glucose is being broken down, which is given to be proportional to the amount of glucose present. Thus we have the formula
$$R = k - aG$$
where $k$ is the rate that the glucose is being injected, $a$ is the constant relating the rate that it is broken down to the amount present, and $G$ is the amount present.

(b)
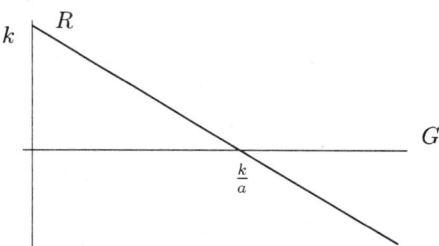

50. If $r$ was the average yearly inflation rate, in decimals, then $\frac{1}{4}(1+r)^3 = 2{,}400{,}000$, so $r = 211.53$, i.e. $r = 21{,}153\%$.

51. (a) The rate of change of the fish population is the difference between the rate they are reproducing, and the rate they are being caught. We are given that they are reproducing at a rate of 5% of $P$ where $P$ is their population, and being caught at a constant rate $Y$. Thus the net rate of change $R$ (in fish per year) is given by
$$R = (0.05)P - Y$$

(b)
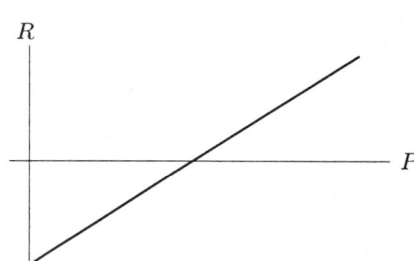

52. From the graph the period appears to be about $\pi$:

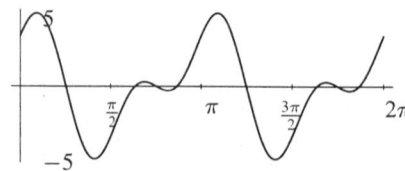

**53.** The graph looks like a sine function of amplitude 7 and period 10. If the equation is of the form

$$y = 7\sin(kt),$$

then $2\pi/k = 10$, so $k = \pi/5$. Therefore, the equation is

$$y = 7\sin\left(\frac{\pi t}{5}\right).$$

**54.** The graph looks like an upside-down sine function with amplitude $(90 - 10)/2 = 40$ and period $\pi$. Since the function is oscillating about the line $x = 50$, the equation is

$$x = 50 - 40\sin(2t).$$

**55.** Depth $= d = 7 + 1.5\sin\left(\dfrac{\pi}{3}t\right)$

**56.** (a)

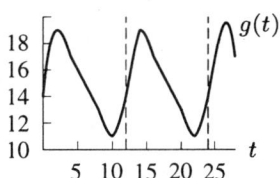

**Figure 1.106**

From the graph in Figure 1.106, the period appears to be about 12, and the table tells us that $g(0) = 14 = g(12) = g(24)$, which supports this guess.

$$\text{Amplitude} \approx \frac{\text{max} - \text{min}}{2} \approx \frac{19 - 11}{2} = 4$$

(b) To estimate $g(34)$, notice that $g(34) = g(24 + 10) = g(10) = 11$ (assuming the function is periodic with period 12). Likewise, $g(60) = g(24 + 12 + 12 + 12) = 14$.

## Solutions to the Projects

**1.** (a) Compounding daily (continuously),

$$\begin{aligned}P &= P_0 e^{rt} \\ &= \$450{,}000 e^{(0.06)(213)} \\ &\approx \$1.5977 \times 10^{11}\end{aligned}$$

This amounts to approximately $160 billion.

(b) Compounding yearly,

$$\begin{aligned}A &= \$450{,}000\,(1 + 0.06)^{213} \\ &= \$450{,}000(1.06)^{213} \approx \$450{,}000(245555.29) \\ &\approx \$1.10499882 \times 10^{11}\end{aligned}$$

This is only about $110.5 billion.

(c) We first wish to find the interest that will accrue during 1990. For 1990, the principal is $\$1.105 \times 10^{11}$. At 6% annual interest, during 1990 the money will earn

$$0.06 \times \$1.105 \times 10^{11} = \$6.63 \times 10^9.$$

**92** CHAPTER ONE /SOLUTIONS

The number of seconds in a year is

$$\left(365 \frac{\text{days}}{\text{year}}\right) \left(24 \frac{\text{hours}}{\text{day}}\right) \left(60 \frac{\text{mins}}{\text{hour}}\right) \left(60 \frac{\text{secs}}{\text{min}}\right) = 31{,}536{,}000 \text{ sec.}$$

Thus, over 1990, interest is accumulating at the rate of

$$\frac{\$6.63 \times 10^9}{31{,}536{,}000 \text{ sec}} \approx \$210.24 /\text{sec.}$$

2. (a) Since the population center is moving west at 50 miles per 10 years, or 5 miles per year, if we start in 1990, when the center is in Steelville, its distance $d$ west of Steelville $t$ years after 1990 is given by

$$d = 5t.$$

(b) It moved over 700 miles over the 200 years from 1790 to 1990, so its average speed was greater than $700/200 = 3.5$ miles/year, somewhat slower than its present rate.

(c) According to the function in (a), after 300 years the population center would be 1500 miles west of Steelville, in California, which seems rather unlikely.

## Solutions to Problems on Fitting Formulas to Data

1. (a) A line seems to fit the data in Figure 1.107 reasonably well.

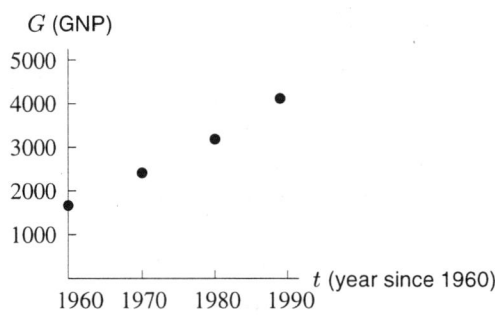

**Figure 1.107**: GNP as a function of year

(b) If we let $G$ represent the GNP and $t$ represent the years that have passed since 1960, we get from a calculator the linear regression equation:

$$G = F(t) = 1623 + 83.65t.$$

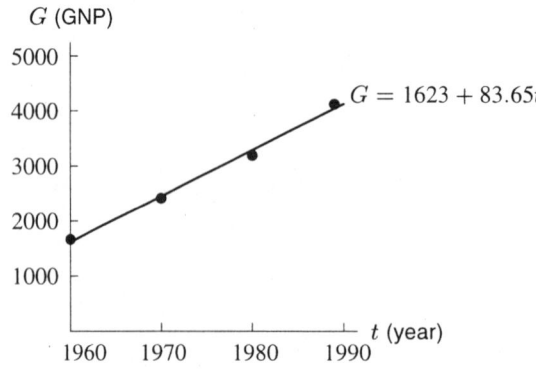

**Figure 1.108**: GNP data and linear regression function

(c) Using our equation, we estimate the GNP in 1985 as

$$F(25) = 1623 + (83.65)(25)$$
$$\approx 1623 + 2091$$
$$= 3714.$$

We can estimate the GNP in 2020 with

$$F(60) = 1623 + (83.65)(60) = 1623 + 5019$$
$$= 6642.$$

Our estimate is likely to work better for the year 1985 than for the year 2020. This is because this sort of estimation works better for years that are closer to the existing data points. Thus, our estimation will be more accurate for 1985, which is within the range of the data points given and hence involves interpolation, than 2020, for which we need extrapolation.

2. (a) We can see from the equation of the regression line that the slope is negative. That means that as time passes during the study, pH is decreasing. Thus, from the data about pH values given in the problem, we can see that the acidity level in the precipitation is getting higher.

(b) The pH at the beginning of the study is the value of $P$ when $t = 0$. Thus,

$$P = 5.43 - 0.0053(0)$$
$$= 5.43.$$

The pH at the end of the study is the value of $P$ when $t = 150$, or

$$P = 5.43 - 0.0053(150)$$
$$\approx 5.43 - 0.80$$
$$= 4.63.$$

(c) The slope of the regression line is the coefficient of $t$ in the equation of the regression line, $-0.0053$. The negative sign tells us that the pH of the rain is decreasing over time and that the precipitation is becoming more acidic. The pH goes down about 0.0053 each week.

3. (a) Using our calculator, we get the stride rate $S$ as a function of speed $v$,

$$S = 0.080v + 1.77$$

(b)
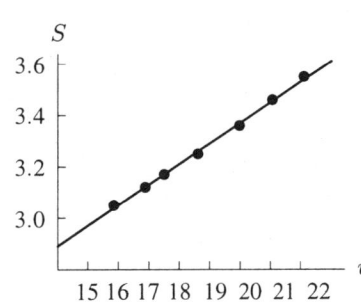

The line seems to fit the data quite well.

(c) Using our formula, we substitute in the given speed of 18 ft/sec into our regression line and get

$$S = (0.080)(18) + 1.77$$
$$= 1.44 + 1.77$$
$$= 3.21,$$

To find the stride rate when the speed is 10 ft/sec, we substitute $v = 10$ and get

$$S = (0.080)(10) + 1.77$$
$$= .8 + 1.77$$
$$= 2.57.$$

We can have more confidence in our estimate for the stride at the speed 18 ft/sec rather than our estimate at 10 ft/sec, because the speed 18 ft/sec lies well within the domain of the data set given, and thus involves interpolation, whereas the speed 10 ft/sec lies well to the left of the plotted data points and is thus a more speculative value, involving extrapolation.

4. (a)
$$\text{The average rate of change between 1960 and 1990} = \frac{354.0 - 316.8}{1990 - 1960}$$
$$= \frac{37.2}{30}$$
$$= 1.24 \text{ additional parts per million of } CO_2 \text{ per year.}$$

The concentration of $CO_2$ has increased, on the average, by about 1.24 parts per million each year.

(b)
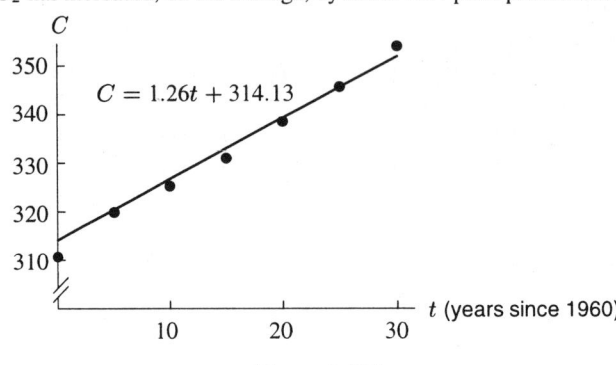

**Figure 1.109**

The slope 1.26 is very close to the average rate of change found in part (a). In the year 2000 ($t = 40$), we have
$$C = 1.26(40) + 314.13$$
$$C = 364.53.$$

5. (a) The annual growth rate of an exponential function of the form $Ar^t$ is just $(r - 1)$, analogous to the rate of interest in interest problems. Thus, in the given function
$$r - 1 = 1.0026 - 1.00$$
$$= 0.0026.$$

This means that the $CO_2$ concentration grows by 0.26% every year.

(b) The $CO_2$ concentration given by the model for 1900, when we substitute $t = 0$ into the function, is
$$C = 272.27(1.0026)^0$$
$$= 272.27(1)$$
$$= 272.27.$$

The $CO_2$ concentration given by the model for 1980 is, substituting $t = 80$
$$C = 272.27(1.0026)^{80}$$
$$\approx 335.1.$$

This is not too far from the real concentration of 338.5.

6. Graphing the data points in Figure 1.110, we get

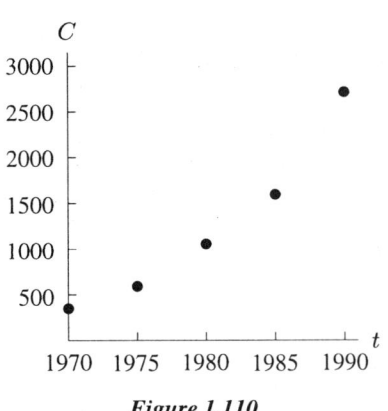

Figure 1.110

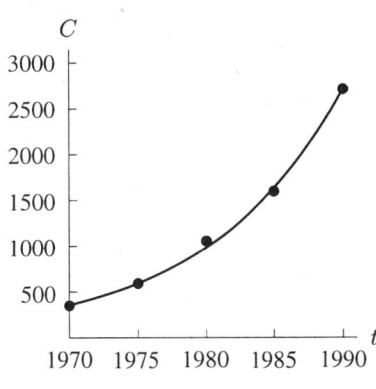

Figure 1.111

From Figure 1.110, it seems that an exponential model will fit these data best.

Letting $C$ represent the per capita health care expenditures and $t$ the years since 1970, we get the exponential regression function

$$C = 357(1.107)^t.$$

Figure 1.111 shows the regression function fits the data quite well.

7. (a) The data is plotted in Figure 1.112.

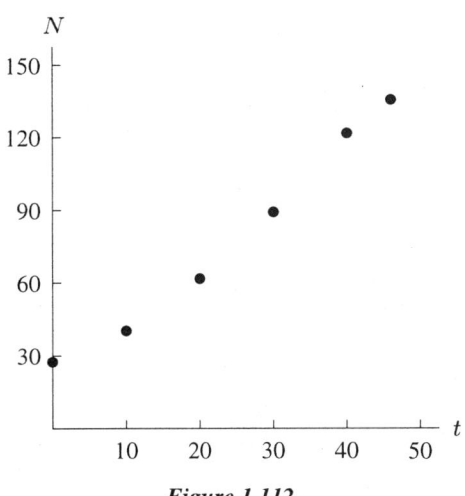

Figure 1.112

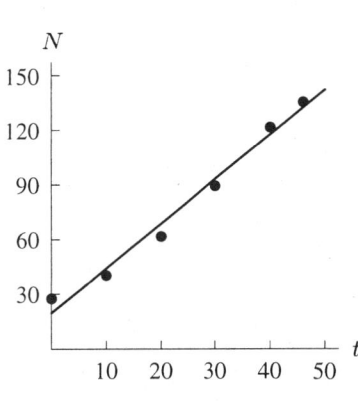

Figure 1.113

(b) It is not obvious which model is best.

(c) The regression line is

$$N = 2.45t + 19.7$$

The line and the data are graphed in Figure 1.113. The estimated number of cars in the year 2000, according to this model is, substituting $t = 60$,

$$N = 2.45(60) + 19.7$$
$$= 147 + 19.7$$
$$= 166.7.$$

(d) The slope represents the fact that every year the number of passenger cars in the US grows by approximately 2,450,000.

(e) The exponential regression function is

$$N = 28.7(1.036)^t.$$

The exponential curve and the data are graphed in Figure 1.114.

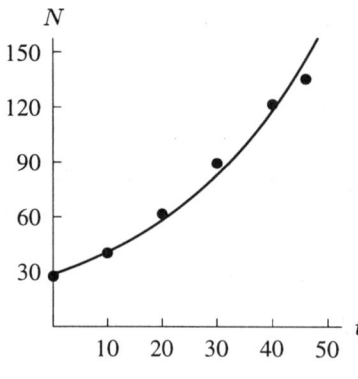

Figure 1.114

The estimated number of cars in the year 2000, according to this model is

$$N = 28.7(1.036)^{60}$$
$$\approx 28.7(8.35)$$
$$\approx 239.6.$$

This is much larger than the prediction obtained from the linear model.

(f) The annual growth rate is as $a - 1$ for the exponential function $N = N_0 a^t$. Thus, the rate is 0.036 so the annual rate of growth in the number of US passenger cars is 3.6% according to our exponential model.

8. (a) See Figure 1.115. Over this time interval (which is fairly short), both a linear and an exponential model to fit the data reasonably well.

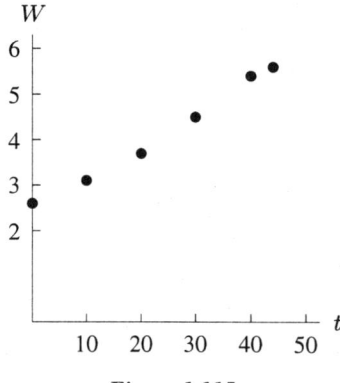

Figure 1.115

(b) The exponential regression curve for this data set, where $W$ is world population and $t$ is the time in years since 1950, is

$$W = 2.6(1.018)^t.$$

(c) The annual rate of growth is

$$1.018 - 1.00 = 0.018,$$

or 1.8%.

(d) For the year 2000, we let $t = 50$ and get

$$W = 2.6(1.018)^{50}$$
$$\approx 2.6(2.44)$$
$$\approx 6.3.$$

For the year 2050, we let $t = 100$ and get
$$W = 2.6(1.018)^{100}$$
$$\approx 15.5.$$

Thus, according to our model, there will be 6.3 billion people on the earth by the year 2000 and 15.5 billion people by the year 2050. Now we note that we are more confident with the year 2000 prediction as the 2050 prediction is far beyond the domain of our data set, thus the predictive value of our model at such a distant time is limited.

9. (a) A linear model seems to provide the best fit. See Figure 1.116. Note that we are looking at a data set with a negative slope.

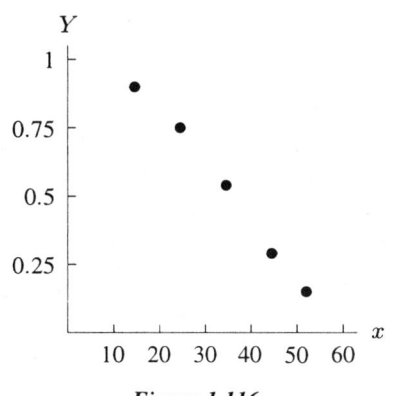

Figure 1.116

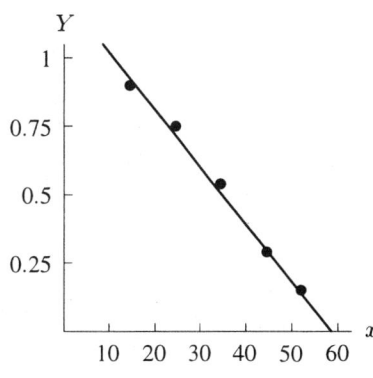

Figure 1.117

(b) The best linear regression function for this data set is
$$Y = -0.021x + 1.23$$

The slope of this function is $-0.021$. The negative sign indicates that the farther the distance from the goal line, the less likely a field goal kick is to succeed. The success rate goes down by about 0.021 for each additional yard from the goal line. See Figure 1.117.

(c) The best exponential regression function is
$$Y = 2.17(0.954)^x,$$

which is plotted in Figure 1.118.

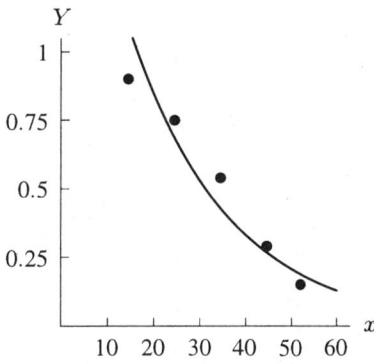

Figure 1.118

We get the success rate predicted by substituting $x = 50$ into the exponential regression function
$$Y = 2.17(0.954)^x$$
$$= 2.17(0.954)^{50}$$
$$\approx 2.17(0.095)$$
$$\approx 0.206.$$

Thus, the predicted success rate is 20.6%.

(d) Of the two graphs, the linear model seems to fit the data best.

10. (a) See Figure 1.119.

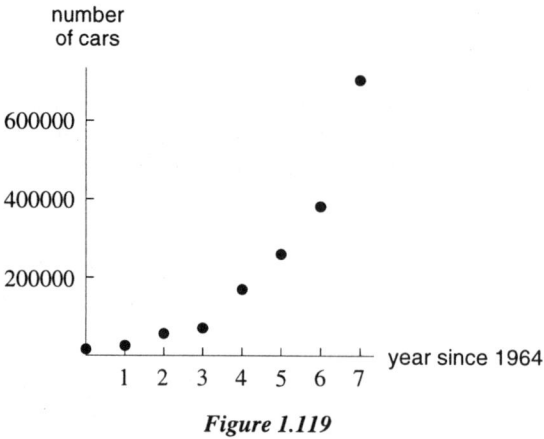

**Figure 1.119**

(b) The data appears to be more exponential.
(c) Using a calculator or computer, we find that the best exponential function is

$$C = 15{,}862.61 \cdot (1.7268)^t,$$

where $C$ is the number of imported Japanese cars and $t$ is in years since 1964. See Figure 1.120.

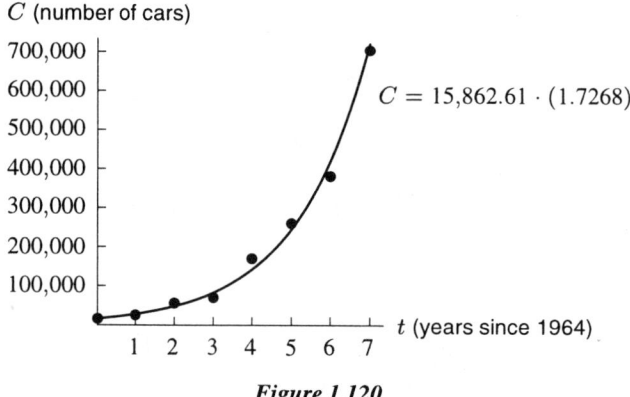

**Figure 1.120**

(d) Since the base is 1.7268, the annual percentage growth rate was about 73%.
(e) No, extrapolation past 1971 (given by $t = 7$) will not be very accurate, since the total number of Japanese cars imported (according to this model) will quickly surpass the total number of cars in the US.

11. There are several possible answers to this questions depending on whether or not you choose to "smooth out" the given curve. In all cases the leading coefficient is positive. If you didn't "smooth out" the function, the graph changes direction 10 times so the degree of the polynomial would be at least 11. If you "smoothed" the function so that it changes direction 4 times, the function would have at least degree 5. Finally, if you "smoothed out" the function so that it only changes direction twice, the degree would be at least 3.

12. In fitting the functions we assume the graphs display all the major tendencies of the data. (For example, we assume in graph (c) that the data does not begin to climb to the right of the region shown).

   (a) This function seems to have a slower growth as $x$ increases, almost levelling off before growing more steeply again. The best function is a cubic polynomial.
   (b) This data set seems to increase to a maximum, level off, then decrease rapidly. It also has a vertical axis of symmetry. It thus looks like the graph of a quadratic function.
   (c) This function seems to level off, having a horizontal asymptote. It thus seems to resemble an exponential function. Notice that this function has the shape of an exponential decay function with a negative growth rate.

(d) This data set clearly resembles a linear function. Notice that the line has negative slope.
(e) This quickly increasing function resembles an exponential function. Notice that it has a positive growth rate.
(f) Although this data set has a slight aberration from linear, it resembles a linear function most strongly. Remember that data can often deviate from a precise line or curve. Often we have to use our best judgement to separate superficial anomalies and actual trends.

13. (a)

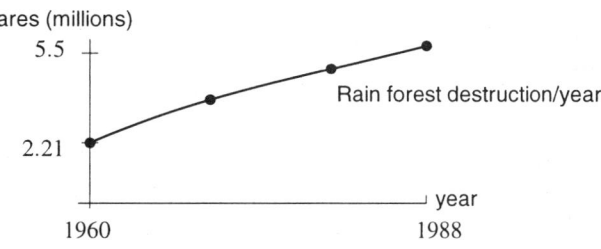

Figure 1.121

(b) The data are increasing and the curve is slightly concave down. The fact that the data are increasing means that more hectares of rain forest are being destroyed every year. The fact that the curve is concave down means the rate of increase of rain forest destruction is decreasing.
(c) Using a graphing calculator, we get $y \approx -1885 + 249 \ln x$.
(d) When $x = 2000$, $y \approx 7.6 \times 10^6$ hectares.

## *Solutions to Problems on Compound Interest and the Number e*

1. For $20 \leq x \leq 100$, $0 \leq y \leq 1.2$, this function looks like a horizontal line at $y \approx 1.0725$ (In fact, the graph approaches this line from below.) Now, $e^{0.07} \approx 1.0725$, which strongly suggests that, as we already know, as $x \to \infty$, $\left(1 + \frac{0.07}{x}\right)^x \to e^{0.07}$.

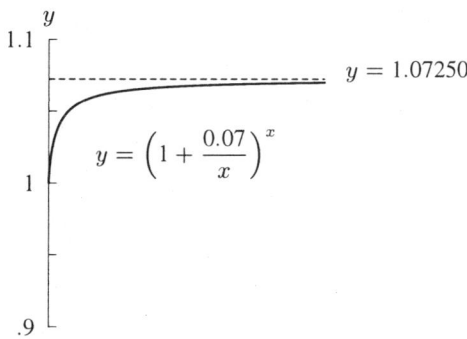

2. In one year, an investment of $P_0$ becomes $P_0 e^{0.06}$. Using a calculator, we see that
$$P_0 e^{0.06} = P_0(1.0618365)$$
So the effective annual yield is about 6.18%.

3. $e^r = 1.05$ so $r = \ln(1.05) = 0.0488$ or 4.88%.
4. $e^{0.08} = 1.0833$ so the effective annual yield= 8.33%.
5. (a) (i)
$$\left(1 + \frac{0.05}{1000}\right)^{1000} = 1.05126978\ldots,$$
so the effective annual yield is $5.126978\ldots\%$.

(ii)
$$\left(1+\frac{0.05}{10000}\right)^{10000} = 1.05127096\ldots,$$
so the effective annual yield is $5.127096\ldots\%$.

(iii)
$$\left(1+\frac{0.05}{100000}\right)^{100000} = 1.05127108\ldots,$$
so the effective annual yield is $5.127108\ldots\%$.

(b) The effective annual rates in part (a) are closing in on $5.127\%$, so this is the effective annual yield for a 5% annual rate compounded continuously.

(c) $e^{0.05} = 1.05127109\ldots$. Since continuous compounding is equivalent to multiplying by $e^{0.05}$, the effective annual yield for continuous compounding is $0.05127109\ldots \approx 5.127\%$.

6. (a)
$$\left(1+\frac{0.04}{10000}\right)^{10000} \approx 1.0408107$$
$$\left(1+\frac{0.04}{100000}\right)^{100000} \approx 1.0408108$$
$$\left(1+\frac{0.04}{1000000}\right)^{1000000} \approx 1.0408108$$

Effective annual yield: $4.08108\%$

(b) $e^{0.04} \approx 1.048108$ as expected.

7. (a) $e^{0.06} = 1.0618365$ which means the bank balance has increased by approximately $6.18\%$.

(b) $e^{0.06t} = 2$, so $t = (\ln 2)/0.06 = 11.55$ years.

(c) $e^{rt} = 2$ so $t = (\ln 2)/r$.

8. (a) Using the formula $A = A_0(1+\frac{r}{n})^{nt}$, we have $A = 10^6(1+\frac{1}{12})^{12} \approx 10^6(2.61303529) \approx 2{,}613{,}035$ zaïre after one year.

(b) Compounding daily, $A = 10^6(1+\frac{1}{365})^{365} \approx 10^6(2.714567) \approx 2{,}714{,}567$ zaïre. Compounding hourly, $A = 10^6(1+\frac{1}{8760})^{8760} \approx 10^6(2.7181267) \approx 2{,}718{,}127$ zaïre. Compounding each minute, $A = 10^6(1+\frac{1}{525600})^{525600} \approx 10^6(2.718280) \approx 2{,}718{,}280$ zaïre

(c) The amount does not seem to be increasing without bound, but rather it seems to level off at a value just over $2{,}718{,}000$ zaïre. A close upper limit might be $2{,}718{,}300$ (amounts may vary). In fact, the limit is $(e \times 10^6)$ zaïre.

9. We know that for a given annual rate, the higher the frequency of compounding, the higher the effective annual yield. So the effective yield of (a) will be greater than that of (c) which is greater than that of (b). Also, the effective annual yield of (e) will be greater than that of (d). Now the effective annual yield of (e) will be less than the effective annual yield of 5.5% annual rate, compounded twice a year, and the latter will be less than the yield from (b). Thus $d < e < b < c < a$. Matching these up with our choices, we get
(d) I, (e) II, (b) III, (c) IV, (a) V.

10. (a) Compounding 33% interest 12 times should be the same as compounding the yearly rate $R$ once, so we get
$$\left(1+\frac{R}{100}\right)^1 = \left(1+\frac{33}{100}\right)^{12}.$$
Solving for $R$, we obtain $R = 2963.51$. The yearly rate, $R$, is $2963.51\%$. Notice how much more this is than $12(33\%) \approx 400\%$.

(b) The monthly rate $r$ satisfies
$$\left(1+\frac{4.6}{100}\right)^1 = \left(1+\frac{r}{100}\right)^{12}$$
$$1.046^{\frac{1}{12}} = 1 + \frac{r}{100}$$
$$r = 100(1.046^{\frac{1}{12}} - 1) = 0.3755.$$

The monthly rate is $0.3755\%$.

11. (a) Its cost in 1989 would be $1000 + \frac{1290}{100} \cdot 1000 = 13900$ cruzados.
    (b) The monthly inflation rate $r$ solves
    $$\left(1 + \frac{1290}{100}\right)^1 = \left(1 + \frac{r}{100}\right)^{12},$$
    since we compound the monthly inflation 12 times to get the yearly inflation. Solving for $r$, we get $r = 24.52\%$. Notice that this is much different than $\frac{1290}{12} = 107.5\%$.

# CHAPTER TWO

## Solutions for Section 2.1

1. distance

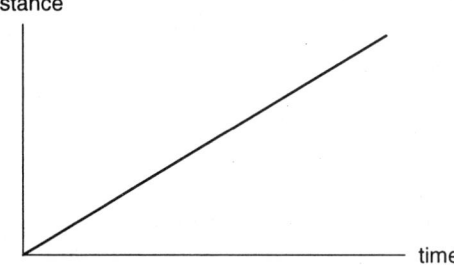

2. distance

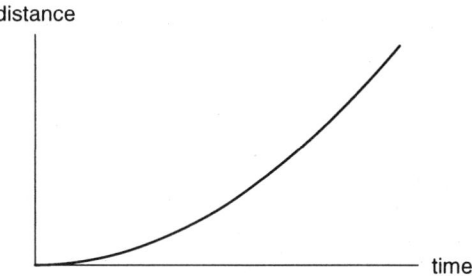

3. distance

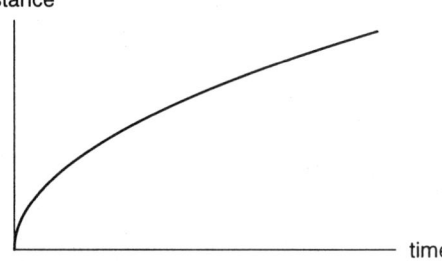

4. distance

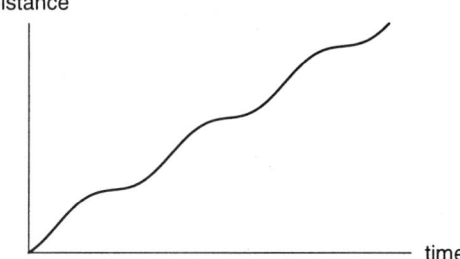

5. (a) The average velocity between $t = 3$ and $t = 5$ is

$$\frac{\text{Distance}}{\text{Time}} = \frac{s(5) - s(3)}{5 - 3} = \frac{25 - 9}{2} = \frac{16}{2} = 8 \text{ ft/sec.}$$

(b) Using an interval of size 0.1, we have

$$\left(\begin{array}{c}\text{Instantaneous velocity}\\ \text{at } t = 3\end{array}\right) \approx \frac{s(3.1) - s(3)}{3.1 - 3} = \frac{9.61 - 9}{0.1} = 6.1.$$

Using an interval of size 0.01, we have

$$\left(\begin{array}{c}\text{Instantaneous velocity}\\ \text{at } t = 3\end{array}\right) \approx \frac{s(3.01) - s(3)}{3.01 - 3} = \frac{9.0601 - 9}{0.01} = 6.01.$$

From this we guess that the instantaneous velocity at $t = 3$ is about 6 ft/sec.

6. (a) The size of the tumor when $t = 0$ months is $S(0) = 2^0 = 1$ cubic millimeter. The size of the tumor when $t = 6$ months is $S(6) = 2^6 = 64$ cubic millimeters. The total change in the size of the tumor is $S(6) - S(0) = 64 - 1 = 63$ mm$^3$.

(b) The average rate of change in the size of the tumor during the first six months is:

$$\text{Average rate of change} = \frac{S(6) - S(0)}{6 - 0} = \frac{64 - 1}{6} = \frac{63}{6} = 10.5 \text{ cubic millimeters/month.}$$

(c) We will consider intervals to the right of $t = 6$:

| $t$ (months) | 6 | 6.001 | 6.01 | 6.1 |
|---|---|---|---|---|
| $S$ (cubic millimeters) | 64 | 64.0444 | 64.4452 | 68.5935 |

$$\text{Average rate of change} = \frac{68.5935 - 64}{6.1 - 6} = \frac{4.5935}{0.1} = 45.935$$

$$\text{Average rate of change} = \frac{64.4452 - 64}{6.01 - 6} = \frac{0.4452}{0.01} = 44.52$$

$$\text{Average rate of change} = \frac{64.0444 - 64}{6.001 - 6} = \frac{0.0444}{0.001} = 44.4$$

We can continue taking smaller intervals but the value of the average rate will not change much. Therefore, we can say that a good estimate of the growing rate of the tumor at $t = 6$ months is 44.4 cubic millimeters/month.

7.

| Slope | −3 | −1 | 0 | 1/2 | 1 | 2 |
|---|---|---|---|---|---|---|
| Point | F | C | E | A | B | D |

8. The slope is positive at $A$ and $D$; negative at $C$ and $F$. The slope is most positive at $A$; most negative at $F$.

9. For the interval $0 \leq t \leq 0.8$, we have

$$\left( \begin{array}{c} \text{Average velocity} \\ 0 \leq t \leq 0.8 \end{array} \right) = \frac{s(0.8) - s(0)}{0.8 - 0} = \frac{6.5}{0.8} = 8.125 \text{ ft/sec.}$$

$$\left( \begin{array}{c} \text{Average velocity} \\ 0 \leq t \leq 0.2 \end{array} \right) = \frac{s(0.2) - s(0)}{0.2 - 0} = \frac{0.5}{0.2} = 2.5 \text{ ft/sec.}$$

$$\left( \begin{array}{c} \text{Average velocity} \\ 0.2 \leq t \leq 0.4 \end{array} \right) = \frac{s(0.4) - s(0.2)}{0.4 - 0.2} = \frac{1.3}{0.2} = 6.5 \text{ ft/sec.}$$

To find the velocity at $t = 0.2$, we find the average velocity to the right of $t = 0.2$ and to the left of $t = 0.2$ and average them. So a reasonable estimate of the velocity at $t = 0.2$ is the average of $\frac{1}{2}(6.5 + 2.5) = 4.5$ ft/sec.

10. (a) The average rate of change of a function over an interval is represented graphically as the slope of the secant line to its graph over the interval. See Figure 2.1. Segment $AB$ is the secant line to the graph in the interval from $x = 0$ to $x = 3$ and segment $BC$ is the secant line to the graph in the interval from $x = 3$ to $x = 5$.

We can easily see that slope of $AB >$ slope of $BC$. Therefore, the average rate of change between $x = 0$ and $x = 3$ is greater than the average rate of change between $x = 3$ and $x = 5$.

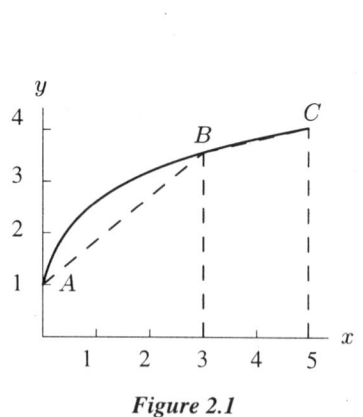

Figure 2.1

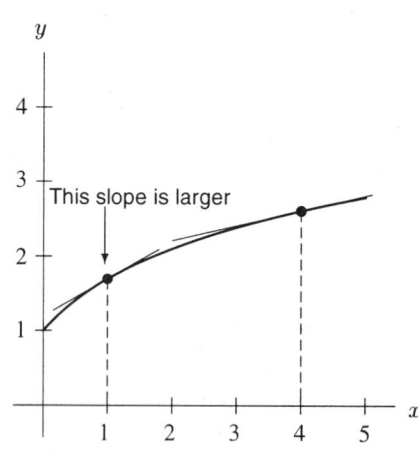

Figure 2.2

(b) We can see from the graph in Figure 2.2 that the function is increasing faster at $x = 1$ than at $x = 4$. Therefore, the instantaneous rate of change at $x = 1$ is greater than the instantaneous rate of change at $x = 4$.

(c) The units of rate of change are obtained by dividing units of cost by units of product: thousands of dollars/kilogram.

11. We want to approximate $P'(0)$ and $P'(2)$. Since for small $h$

$$P'(0) \approx \frac{P(h) - P(0)}{h} \quad \text{and} \quad P'(2) \approx \frac{P(2+h) - P(2)}{h},$$

if we take $h = 0.01$, we get

$$P'(0) \approx \frac{1.15(1.014)^{0.01} - 1.15}{0.01} = 0.01599 \text{ billion/year}$$
$$= 16.0 \text{ million people/year}$$
$$P'(2) \approx \frac{1.15(1.014)^{2.01} - 1.15(1.014)^2}{0.01} = 0.0164 \text{ billion/year}$$
$$= 16.4 \text{ million people/year}$$

12. (a) Since 69.6 percent live in the city in 1960 and 20 percent in 1860, we have

$$\text{Average rate of change} = \frac{69.9 - 20}{1960 - 1860} = \frac{49.9}{100} = 0.499.$$

Thus the average rate of change is 0.499 percent/year.

(b) By looking at the population in 1960 and 1970 we see that

$$\text{Average rate of change} = \frac{73.5 - 69.9}{1970 - 1960} = \frac{3.6}{10} = 0.36$$

which gives an average rate of change of 0.36 percent/year between 1960 and 1970. Alternatively, we look at the population in 1950 and 1960 and see that

$$\text{Average rate of change} = \frac{69.9 - 64}{1960 - 1950} = \frac{5.9}{10} = 0.59$$

giving an average rate of change of 0.59 percent/year between 1950 and 1960. We see that the rate of change in the year 1960 is somewhere between 0.36 and 0.59 percent/year. A good estimate is $(0.36 + 0.59)/2 = 0.475$ percent per year.

(c) By looking at the population in 1830 and 1860 we see that

$$\text{Average rate of change} = \frac{20 - 9}{1860 - 1830} = \frac{11}{30} = 0.37$$

giving an average rate of change of 0.37 percent/year between 1830 and 1860. Alternatively we can look at the population in 1800 and 1830 and see that

$$\text{Average rate of change} = \frac{9 - 6}{1830 - 1800} = \frac{3}{30} = 0.1$$

giving an average rate of change of 0.1 percent/year between 1800 and 1830. We see that the rate of change at the year 1830 is somewhere between 0.1 and 0.37. This tells us that in the year 1830 the percent of the population in urban areas is changing by a rate somewhere between 0.1 percent/year and 0.37 percent/year.

(d) Looking at the data we see that the percent gets larger as time goes by. Thus, the function is always increasing.

13. (a) When $t = 0$, the ball is on the bridge and its height is $f(0) = 36$, so the bridge is 36 feet above the ground.

(b) After 1 second, the ball's height is $f(1) = -16 + 50 + 36 = 70$ feet, so it traveled $70 - 36 = 34$ feet in 1 second, and its average velocity was 34 ft/sec.

(c) At $t = 1.001$, the ball's height is $f(1.001) = 70.017984$ feet, and its velocity about

$$\text{Velocity} = \frac{70.017984 - 70}{1.001 - 1} = 17.984 \approx 18 \text{ ft/sec}.$$

(d) The graph is shown in Figure 2.3. Since the coordinates of the peak are about $(1.6, 75)$ the ball reaches a height of about 75 feet. The velocity of the ball is zero when it is at its peak since the tangent is horizontal there.

(e) The ball reaches its maximum height when $t \approx 1.6$.

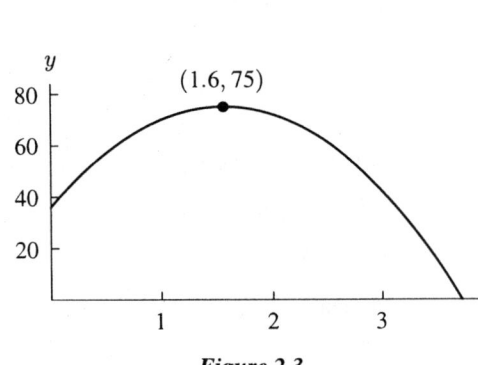

Figure 2.3

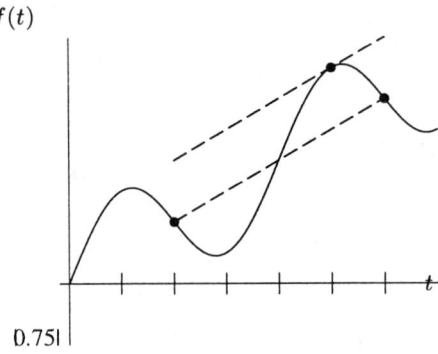

Figure 2.4

14. $0 <$ slope at $C <$ slope at $B <$ slope of $AB < 1 <$ slope at $A$. (Note that the line $y = x$, has slope 1.)
15. One possibility is shown in Figure 2.4.

## Solutions for Section 2.2

1. (a) The average rate of change is the slope of the secant line in Figure 2.5, which shows that this slope is positive.
   (b) The instantaneous rate of change is the slope of the graph at $x = 3$, which we see from Figure 2.6 is negative.

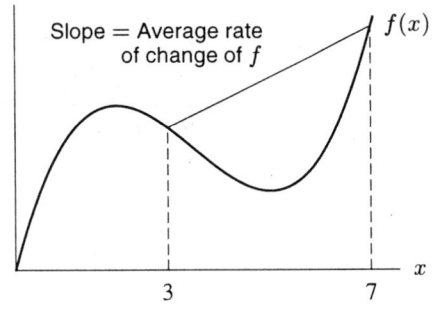

Figure 2.5

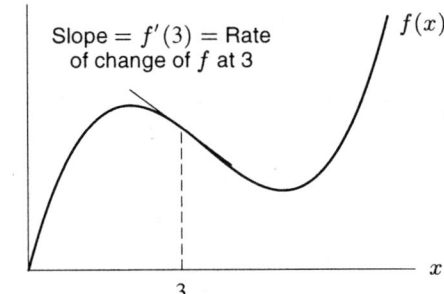

Figure 2.6

2. (a) From Figure 2.7 we can see that for $x = 1$ the value of the function is decreasing. Therefore, the derivative of $f(x)$ at $x = 1$ is negative.
   (b) $f'(1)$ is the derivative of the function at $x = 1$. This is the rate of change of $f(x) = 2 - x^3$ at $x = 1$. We estimate this by computing the average rate of change of $f(x)$ over intervals near $x = 1$.

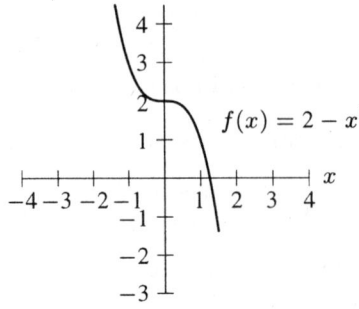

Figure 2.7

Using the intervals $0.999 \le x \le 1$ and $1 \le x \le 1.001$, we see that

$$\left(\begin{array}{c}\text{Average rate of change}\\ \text{on } 0.999 \le x \le 1\end{array}\right) = \frac{[2-1^3]-[2-0.999^3]}{1-0.999} = \frac{1-1.002997}{0.001} = -2.997,$$

$$\left(\begin{array}{c}\text{Average rate of change}\\ \text{on } 1 \le x \le 1.001\end{array}\right) = \frac{[2-1.001^3]-[2-1^3]}{1.001-1} = \frac{0.996997-1}{0.001} = -3.003.$$

It appears that the rate of change of $f(x)$ at $x = 1$ is approximately $-3$, so we estimate $f'(1) = -3$.

3. $P'(0)$ is the derivative of the function $P(t) = 200(1.05)^t$ at $t = 0$. This is the same as the rate of change of $P(t)$ at $t = 0$. We estimate this by computing the average rate of change over intervals near $t = 0$.

   If we use the intervals $-0.001 \le t \le 0$ and $0 \le t \le 0.001$, we see that:

$$\left(\begin{array}{c}\text{Average rate of change}\\ \text{on } -0.001 \le t \le 0\end{array}\right) = \frac{200(1.05)^0 - 200(1.05)^{-0.001}}{0-(-0.001)} = \frac{200-199.990}{0.001} = \frac{0.010}{0.001} = 10,$$

$$\left(\begin{array}{c}\text{Average rate of change}\\ \text{on } 0 \le t \le 0.001\end{array}\right) = \frac{200(1.05)^{0.001} - 200(1.05)^0}{0.001-0} = \frac{200.010-200}{0.001} = \frac{0.010}{0.001} = 10.$$

It appears that the rate of change of $P(t)$ at $t = 0$ is 10, so we estimate $P'(0) = 10$.

4. (a) Looking at Figure 2.8 we see that the slope of the tangent to $f(x)$ at $x = -1$ is near zero or slightly negative while the slopes of the tangents to $f(x)$ at $x = 0$ and $x = 1$ are positive. Thus

$$f'(-1) \le 0, \qquad f'(0) > 0, \qquad \text{and } f'(1) > 0.$$

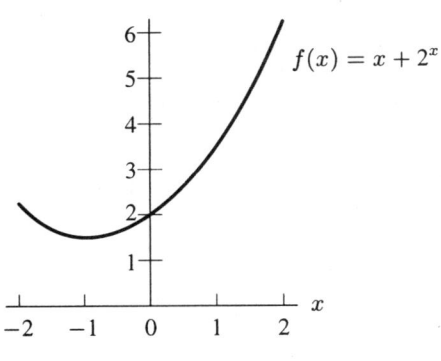

Figure 2.8

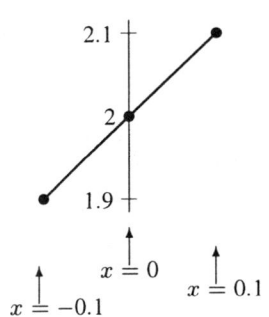

Figure 2.9

(b) When we zoom in, we see the graph in Figure 2.9, so the slope appears to be about 1. We see that $f'(0) \approx 1$.

5. We know that $f'(2)$ is the derivative of $f(x) = x^3 - 2x$ at 2. This is the same as the rate of change of $x^3 - 2x$ at 2. We estimate this by looking at the average rate of change over intervals near 2. If we use the intervals $1.999 \le x \le 2$ and $2 \le x \le 2.001$, we see that

$$\left(\begin{array}{c}\text{Average rate of change}\\ \text{on } 1.999 \le x \le 2\end{array}\right) = \frac{[2^3-2(2)]-[1.999^3-2(1.999)]}{2-1.999} = \frac{4-3.990}{0.001} = 10,$$

$$\left(\begin{array}{c}\text{Average rate of change}\\ \text{on } 2 \le x \le 2.001\end{array}\right) = \frac{[2.001^3-2(2.001)]-[2^3-2(2)]}{2.001-2} = \frac{4.010-4}{0.001} = 10.$$

It appears that the rate of change of $f(x)$ at $x = 2$ is 10, so we estimate $f'(2) = 10$.

6. (a) Figure 2.10 shows that for $t = 2$ the function $g(t) = 0.8^t$ is decreasing. Therefore, $g'(2)$ is negative.

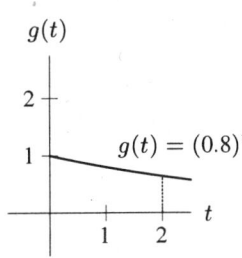

Figure 2.10

(b) To estimate $g'(2)$ we take the small interval between $t = 2$ and $t = 2.001$ to the right of $t = 2$.

$$g'(2) \approx \frac{g(2.001) - g(2)}{2.001 - 2} = \frac{0.8^{2.001} - 0.8^2}{0.001} = \frac{0.6399 - 0.64}{0.001} = \frac{-0.0001}{0.001} = -0.1$$

7. (a) Since the values of $P$ go up as $t$ goes from 4 to 6 to 9, we see that $f'(6)$ appears to be positive. The percent of households with microwave ovens is increasing at $t = 6$.

   (b) We estimate $f'(2)$ using the difference quotient for the interval to the right of $t = 2$, as follows:

$$f'(2) \approx \frac{\Delta P}{\Delta t} = \frac{21 - 14}{4 - 2} = \frac{7}{2}.$$

$f'(2) = 3.5$ is telling us that the percent of households with microwave ovens in the United States was increasing at a rate of 3.5 percentage points per year when $t = 2$ (that means 1980).

Similarly:

$$f'(9) \approx \frac{\Delta P}{\Delta t} = \frac{79 - 61}{12 - 9} = \frac{18}{3} = 6.$$

$f'(9) = 6$ is telling us that the percent of households in the United States with microwave ovens was increasing at a rate of 6 percentage points per year when $t = 9$ (that means 1987).

8. (a) The function $N = f(t)$ is decreasing when $t = 1950$. Therefore, $f'(1950)$ is negative. That means that the number of farms in the United States, in millions, was decreasing in 1950.

   (b) The function $N = f(t)$ is decreasing in 1960 as well as in 1980 but it is decreasing faster in 1960 than in 1980. Therefore, $f'(1960)$ is more negative than $f'(1980)$.

9. (a) At points $A$, $B$, and $D$ the function is increasing. Therefore, the derivative of the function at points $A$, $B$, and $D$ is positive. At points $C$ and $F$, the function is decreasing. Therefore, the derivative of the function at points $C$ and $F$ is negative. At point $E$, the function is neither decreasing nor increasing. Therefore, the derivative of the function at point $E$ is zero.

   (b) The derivative is the most positive where the graph of the function is the steepest and increasing, as at point $D$. The derivative is the most negative where the graph of the function is the steepest and decreasing, as at point $F$.

10. See Figure 2.11

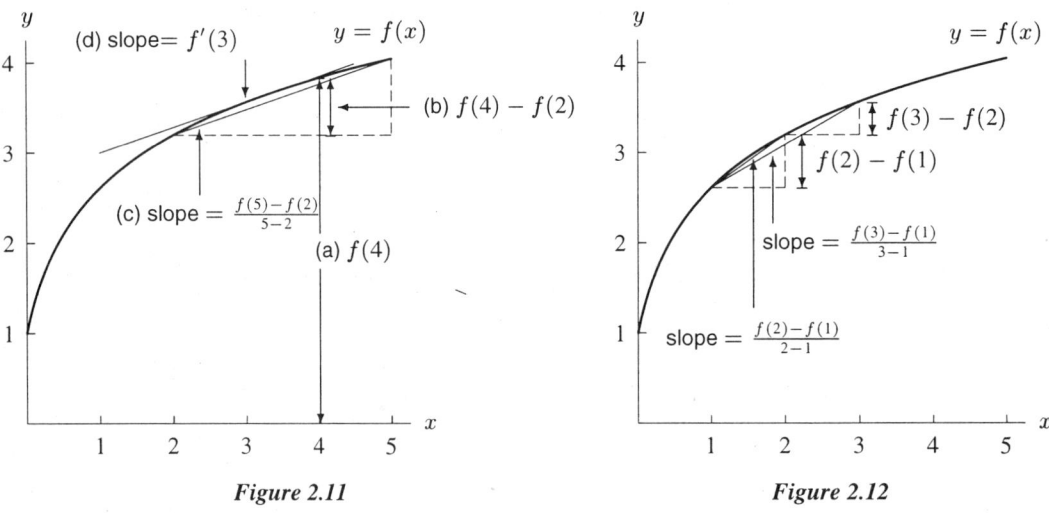

Figure 2.11                    Figure 2.12

11. (a) $f(4) > f(3)$ since $f$ is increasing. See Figure 2.12.
    (b) From Figure 2.12, it appears that $f(2) - f(1) > f(3) - f(2)$.
    (c) $\dfrac{f(2) - f(1)}{2 - 1}$ represents the slope of the secant line connecting the graph at $x = 1$ and $x = 2$. This is greater than the slope of the secant line connecting the graph at $x = 1$ and $x = 3$ which is $\dfrac{f(3) - f(1)}{3 - 1}$.
    (d) The function is steeper at $x = 1$ than at $x = 4$ so $f'(1) > f'(4)$.

12. Figure 2.13 shows the quantities in which we are interested.

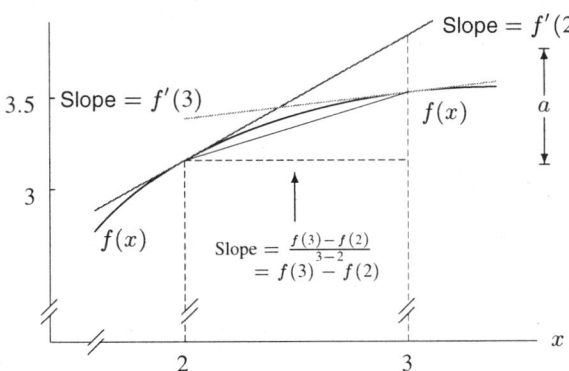

*Figure 2.13*

The quantities $f'(2), f'(3)$ and $f(3) - f(2)$ have the following interpretations:

- $f'(2)$ = slope of the tangent line at $x = 2$
- $f'(3)$ = slope of the tangent line at $x = 3$
- $f(3) - f(2) = \frac{f(3)-f(2)}{3-2}$ = slope of the secant line from $f(2)$ to $f(3)$.

From Figure 2.13, it is clear that $0 < f(3) - f(2) < f'(2)$. By extending the secant line past the point $(3, f(3))$, we can see that it lies above the tangent line at $x = 3$. Thus $0 < f'(3) < f(3) - f(2) < f'(2)$. From the figure, the height $a$ appears less than 1, so $f'(2) = \frac{a}{3-2} = \frac{a}{1} < 1$.

Thus
$$0 < f'(3) < f(3) - f(2) < f'(2) < 1.$$

13.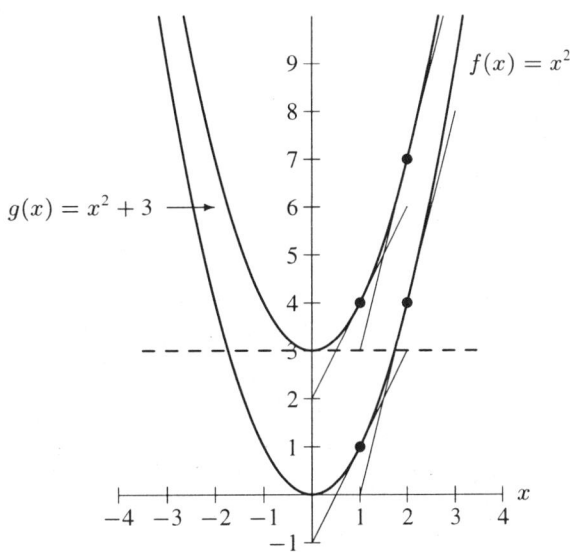

*Figure 2.14*

(a) The tangent line to the graph of $f(x) = x^2$ at $x = 0$ coincides with the $x$-axis and therefore is horizontal (slope = 0). The tangent line to the graph of $g(x) = x^2 + 3$ at $x = 0$ is the dashed line indicated in the figure and it also has a slope equal to zero. Therefore both tangent lines at $x = 0$ are parallel.

We see in Figure 2.14 that the tangent lines at $x = 1$ appear parallel, and the tangent lines at $x = 2$ appear parallel. The slopes of the tangent lines at any value $x = a$ will be equal.

(b) Adding a constant shifts the graph vertically, but does not change the slope of the curve.

14.

**TABLE 2.1**

| $x$ | 2.998 | 2.999 | 3.000 | 3.001 | 3.002 |
|---|---|---|---|---|---|
| $x^3 + 4x$ | 38.938 | 38.969 | 39.000 | 39.031 | 39.062 |

We see that each $x$ increase of 0.001 leads to an increase in $f(x)$ by about 0.031, so $f'(3) \approx \frac{0.031}{0.001} = 31$.

15.

**TABLE 2.2**

| $x$ | 1.998 | 1.999 | 2.000 | 2.001 | 2.002 |
|---|---|---|---|---|---|
| $g(x)$ | 31.8403 | 31.9201 | 32.0000 | 32.0801 | 32.1603 |

**TABLE 2.3**

| $x$ | $-2.002$ | $-2.001$ | $-2.000$ | $-1.999$ | $-1.998$ |
|---|---|---|---|---|---|
| $g(x)$ | $-32.1603$ | $-32.0801$ | $-32.0000$ | $-31.9201$ | $-31.8403$ |

Looking at Table 2.2, we'd estimate $g'(2) = \frac{0.0801}{0.001} = 80.1$. From Table 2.3, we estimate $g'(-2) = \frac{0.0801}{0.001} = 80.1$, so $g'(2) = g'(-2)$. This is what we'd expect from looking at the graph. (The function $g(x)$ is an odd function, with a graph which is symmetric around the origin, so we'd expect tangent lines at symmetric points to have the same slope.)

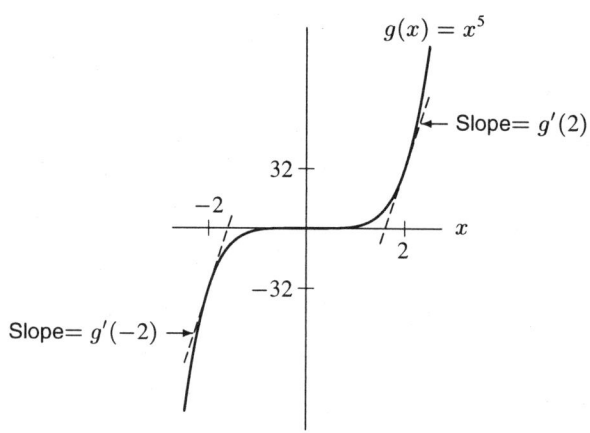

*Figure 2.15*

16. We use the interval $x = 2$ to $x = 2.01$:

$$f'(2) \approx \frac{f(2.01) - f(2)}{2.01 - 2} = \frac{5^{2.01} - 5^2}{0.01} = \frac{25.4056 - 25}{0.01} = 40.56$$

For greater accuracy, we can use the smaller interval $x = 2$ to $x = 2.001$:

$$f'(2) \approx \frac{f(2.001) - f(2)}{2.001 - 2} = \frac{5^{2.001} - 5^2}{0.001} = \frac{25.040268 - 25}{0.001} = 40.268$$

17. (a) A graph of $f(x)$ and the tangent line to the point $(2, \ln 2)$ are shown in Figure 2.16. Looking at the graph we see that as $x$ goes from 2 to 4 the corresponding $y$-increment in the tangent line is approximately of length 1, making the slope of the tangent and the derivative at $x = 2$

$$f'(2) \approx \frac{1}{4 - 2} = 0.5$$

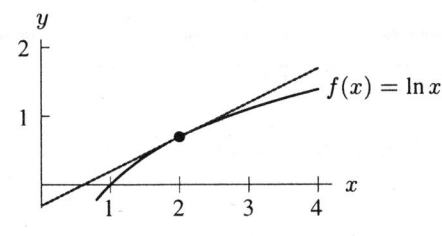

**Figure 2.16**

(b) We know that

$$f'(2) \approx \frac{f(2+h) - f(2)}{h} \quad \text{for small} \quad h$$
$$= \frac{\ln(2+h) - \ln(2)}{h}$$

Substituting small values of $h$ gives Table 2.4.

**TABLE 2.4**

| $h$ | 0.1 | 0.01 | 0.001 |
|---|---|---|---|
| $f'(2)$ | 0.4879 | 0.4988 | 0.4999 |

Thus this approximation of the derivative also gives us

$$f'(2) \approx 0.5$$

18. (a) $f'(t)$ is negative, because the average number of hours worked a week has been *decreasing* over time.
    $g'(t)$ is positive, because hourly wage has been increasing.
    $h'(t)$ is positive, because average weekly earnings has been increasing.
    (b) We use a difference quotient to the right for our estimates.
    (i)
    $$f'(1965) \approx \frac{37.1 - 38.8}{1970 - 1965} = -0.34 \quad \text{hours/year}$$
    $$f'(1985) \approx \frac{34.5 - 34.9}{1990 - 1985} = -0.08 \quad \text{hours/year.}$$
    In 1965, the average number of hours worked by a production worker in a week was decreasing at the rate of 0.34 hours per year. In 1985, it was decreasing at a rate of 0.08 hours per year.
    (ii)
    $$g'(1965) \approx \frac{3.23 - 2.46}{1970 - 1965} = \$0.15 \quad \text{per year}$$
    $$g'(1985) \approx \frac{10.01 - 8.57}{1990 - 1985} = \$0.29 \quad \text{per year}$$
    In 1965, the hourly wage was increasing at a rate of $0.15 per year. In 1985, it was increasing at a rate of $0.29 per year.
    (iii)
    $$h'(1965) \approx \frac{119.83 - 95.45}{1970 - 1965} = \$4.88 \quad \text{per year}$$
    $$h'(1985) \approx \frac{345.35 - 299.09}{1990 - 1985} = \$9.25 \quad \text{per year}$$
    In 1965, average weekly earnings was increasing at a rate of $4.88 a year. In 1985, it was increasing at a rate of $9.25 a year.

**112** CHAPTER TWO /SOLUTIONS

## Solutions for Section 2.3

1. The graph is that of the line $y = -2x + 2$. The slope, and hence the derivative, is $-2$. See Figure 2.17.

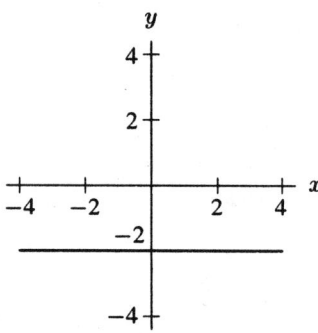

Figure 2.17

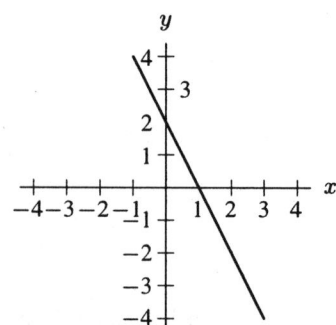

Figure 2.18

2. See Figure 2.18.

3. The slope of this curve is approximately $-1$ at $x = -4$ and at $x = 4$, approximately $0$ at $x = -2.5$ and $x = 1.5$, and approximately $1$ at $x = 0$.

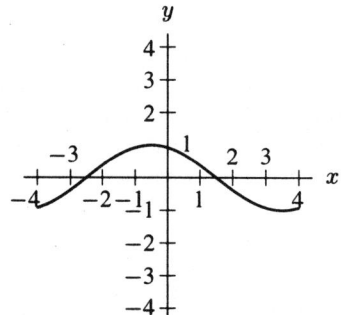

4.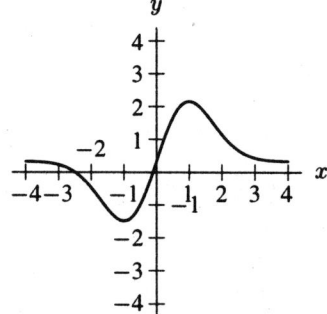

5.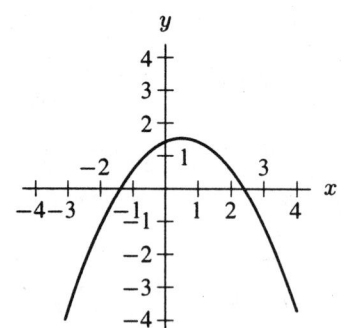

6.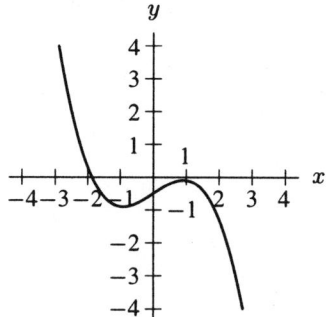

7. On the intervals where the graph of $f(t)$ is horizontal, the slope is zero, and so $f'(t) = 0$. This means the graph of $f'(t)$ lies on the horizontal axis on these intervals. On the small interval where the graph of $f(t)$ is not horizontal, it is increasing, and so $f'(t)$ is positive on this interval. (This means that the graph of $f'(t)$ lies above the horizontal axis.) The function $f(t)$ is increasing quite rapidly on this small interval, and so the derivative $f'(t)$ gets quite large here. A possible graph of $f'(t)$ is shown in Figure 2.19.

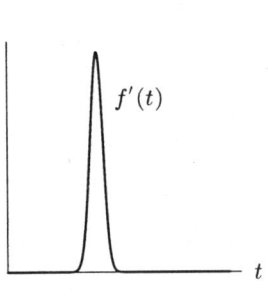

**Figure 2.19**

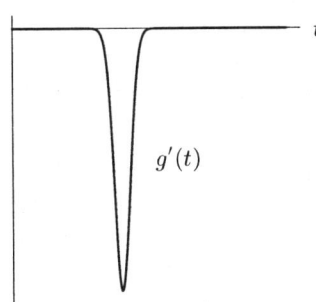

**Figure 2.20**

8. A possible graph for $g'$ is shown in Figure 2.20. On intervals where $g(t)$ is constant (i.e. flat), $g' = 0$. The small interval where $g$ decreases very quickly corresponds to the downwards spike on the graph of $g'$. At the point where $g'$ hits the bottom of its spike, the rate of decrease of $g(t)$ is greatest, i.e., at the mid-point of the step.

9.

(a)
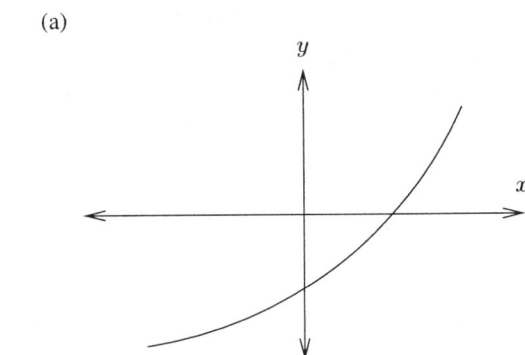

(b)
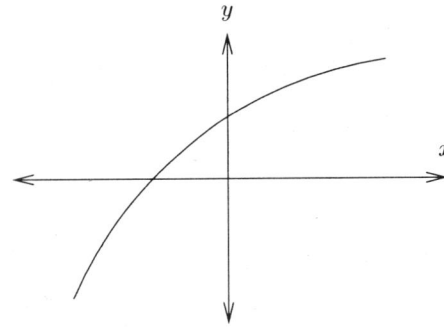

(c)
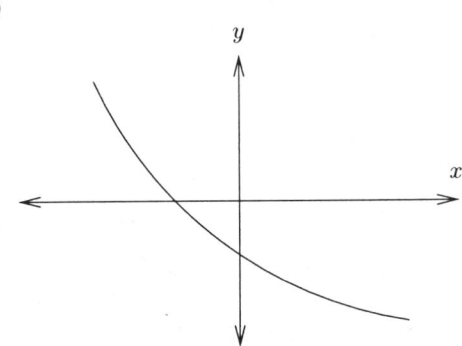

(d)
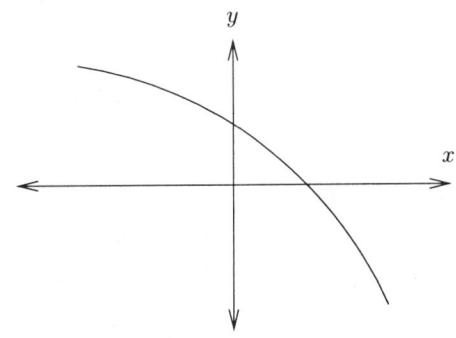

10. From the table we have

$$f'(0) \approx \frac{13 - 18}{1 - 0} = -5.$$

Between 0 and 1, the rate of change is $-5$. Similarly, between 1 and 2, the rate of change is $-3$. Continuing in this way, we obtain the following values for $f'(x)$.

| $x$ | 0 | 1 | 2 | 3 | 4 | 5 | 6 | 7 |
|---|---|---|---|---|---|---|---|---|
| $f'(x)$ | $-5$ | $-3$ | $-1$ | 0 | 2 | 4 | 6 | 9 |

The rate of change of $f(x)$ is positive for $4 \leq x \leq 8$, negative for $0 \leq x \leq 3$. The rate of change appears to be increasing, and so is greatest at about $x = 8$.

**11.**

| $x$ | $x^3$ |
|---|---|
| 0.998 | 0.9940 |
| 0.999 | 0.9970 |
| 1.000 | 1.0000 |
| 1.001 | 1.0030 |
| 1.002 | 1.0060 |

| $x$ | $x^3$ |
|---|---|
| 2.998 | 26.946 |
| 2.999 | 26.973 |
| 3.000 | 27.000 |
| 3.001 | 27.027 |
| 3.002 | 27.054 |

| $x$ | $x^3$ |
|---|---|
| 4.998 | 124.850 |
| 4.999 | 124.925 |
| 5.000 | 125.000 |
| 5.001 | 125.075 |
| 5.002 | 125.150 |

At $x = 1$, the values of $x^3$ are increasing by about 0.0030 for each increase of 0.001 in $x$, so the derivative appears to be $\frac{0.0030}{0.001} = 3$. At $x = 3$, the values increase by 0.027 over an $x$ increase of 0.001, so the value appears to be 27. At $x = 5$, the values increase by 0.075 for a change in $x$ of 0.001, so the derivative appears to be 75. The function $3x^2$ fits these data, so it is a good candidate for the derivative function (although it is not immediately obvious just from these calculations).

**12.** (a) $f'(1) \approx \frac{f(1.1) - f(1)}{0.1} = \frac{\ln(1.1) - \ln(1)}{0.1} \approx 0.95$.
$f'(2) \approx \frac{f(2.1) - f(2)}{0.1} = \frac{\ln(2.1) - \ln(2)}{0.1} \approx 0.49$.
$f'(3) \approx \frac{f(3.1) - f(3)}{0.1} = \frac{\ln(3.1) - \ln(3)}{0.1} \approx 0.33$.
$f'(4) \approx \frac{f(4.1) - f(4)}{0.1} = \frac{\ln(4.1) - \ln(4)}{0.1} \approx 0.25$.
$f'(5) \approx \frac{f(5.1) - f(5)}{0.1} = \frac{\ln(5.1) - \ln(5)}{0.1} \approx 0.20$.
(b) It looks like the derivative of $\ln(x)$ is $\frac{1}{x}$.

**13.**

**TABLE 2.5**

| $x$ | $f(x)$ | $x$ | $f(x)$ | $x$ | $f(x)$ |
|---|---|---|---|---|---|
| 1.998 | 2.6587 | 2.998 | 8.9820 | 3.998 | 21.3013 |
| 1.999 | 2.6627 | 2.999 | 8.9910 | 3.999 | 21.3173 |
| 2.000 | 2.6667 | 3.000 | 9.0000 | 4.000 | 21.3333 |
| 2.001 | 2.6707 | 3.001 | 9.0090 | 4.001 | 21.3493 |
| 2.002 | 2.6747 | 3.002 | 9.0180 | 4.002 | 21.3653 |

Near 2, the values of $f(x)$ seem to be increasing by 0.004 for each increase of 0.001 in $x$, so the derivative appears to be $\frac{0.004}{0.001} = 4$. Near 3, the values of $f(x)$ are increasing by 0.009 for each step of 0.001, so the derivative appears to be 9. Near 4, $f(x)$ increases by 0.016 for each step of 0.001, so the derivative appears to be 16. The pattern seems to be, then, that at a point $x$, the derivative of $f(x) = \frac{1}{3}x^3$ is $f'(x) = x^2$.

**14.**

**TABLE 2.6**

| $t$ | $g(t)$ | $t$ | $g(t)$ | $t$ | $g(t)$ |
|---|---|---|---|---|---|
| 0.998 | 1.994 | 1.998 | 5.990 | 2.998 | 11.986 |
| 0.999 | 1.997 | 1.999 | 5.995 | 2.999 | 11.993 |
| 1.000 | 2.000 | 2.000 | 6.000 | 3.000 | 12.000 |
| 1.001 | 2.003 | 2.001 | 6.005 | 3.001 | 12.007 |
| 1.002 | 2.006 | 2.002 | 6.010 | 3.002 | 12.014 |

Near 1, the values of $g(t)$ increase by 0.003 for each $t$ increase of 0.001, so the derivative appears to be 3. Near 2, the increase is 0.005 for each step of 0.001, so the derivative appears to be 5. Near 3, the increase is 0.007 for each step of 0.001, so the derivative appears to be 7. These values seem to be increasing in a linear manner as we go to higher and higher $t$ values, so we'll guess that the formula is $f'(t) = 2t + 1$.

**15.** Since $f'(x) > 0$ for $1 < x < 3$, $f(x)$ is increasing on this interval.
Since $f'(x) < 0$ for $x < 1$ or $x > 3$, $f(x)$ is decreasing on these intervals.

Since $f'(x) = 0$ for $x = 1$ and $x = 3$, the tangent to $f(x)$ will be horizontal at these $x$'s.
One of many possible shapes of $y = f(x)$ is shown in Figure 2.21.

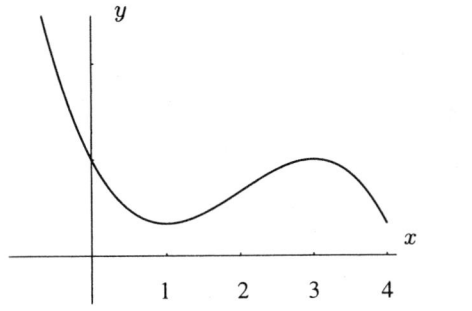

**Figure 2.21**

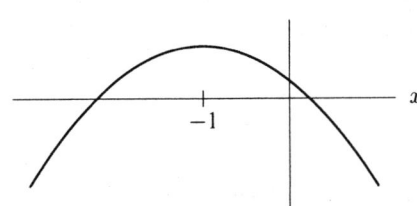

**Figure 2.22**

16. Since $f'(x) > 0$ for $x < -1$, $f(x)$ is increasing on this interval.
    Since $f'(x) < 0$ for $x > -1$, $f(x)$ is decreasing on this interval.
    Since $f'(x) = 0$ at $x = -1$, the tangent to $f(x)$ is horizontal at $x = -1$.
    One possible shape for $y = f(x)$ is shown in Figure 2.22.

17. (a) $x_3$  (b) $x_4$  (c) $x_5$  (d) $x_3$

18. (a) We estimate the derivative by the average rate of change over a 5-year interval.

$$f'(1960) \approx \frac{98.3 - 82.3}{1965 - 1960} = 3.2 \text{ million tons/year}$$

$$f'(1965) \approx \frac{118.3 - 98.3}{1970 - 1965} = 4 \text{ million tons/year}$$

$$f'(1970) \approx \frac{122.7 - 118.3}{1975 - 1970} = 0.88 \text{ million tons/year}$$

$$f'(1975) \approx \frac{139.1 - 122.7}{1980 - 1975} = 3.28 \text{ million tons/year}$$

$$f'(1980) \approx \frac{148.1 - 139.1}{1984 - 1980} = 2.25 \text{ million tons/year}$$

(b) The amount of solid waste generated per year increases each year.

19. (a) Using $f(x) = 5x + 2$, we have

$$\text{Average rate of change on } 2 \leq x \leq 5 = \frac{f(5) - f(2)}{5 - 2} = \frac{(5(5) + 2) - (5(2) + 2)}{3} = \frac{27 - 12}{3} = 5$$

$$\text{Average rate of change on } 2 \leq x \leq 3 = \frac{f(3) - f(2)}{3 - 2} = \frac{(5(3) + 2) - (5(2) + 2)}{1} = \frac{17 - 12}{1} = 5$$

$$\text{Average rate of change on } 2 \leq x \leq 2.1 = \frac{f(2.1) - f(2)}{2.1 - 2} = \frac{(5(2.1) + 2) - (5(2) + 2)}{0.1} = \frac{12.5 - 12}{0.1} = 5$$

(b) Since we are getting the same value 5 for the rate of change on any interval around $x = 2$, we can say that $f'(2) = 5$. Now, we also realize that $f'(2)$ coincides in value with the coefficient of $x$ in $f(x) = 5x + 2$.

If we graph the function $f(x) = 5x + 2$, we can see that the graph of $f(x) = 5x + 2$ is a line whose slope is 5. Since a line is its own tangent, the derivative at any point of a function whose graph is a line is the slope of that line.

**116** CHAPTER TWO /SOLUTIONS

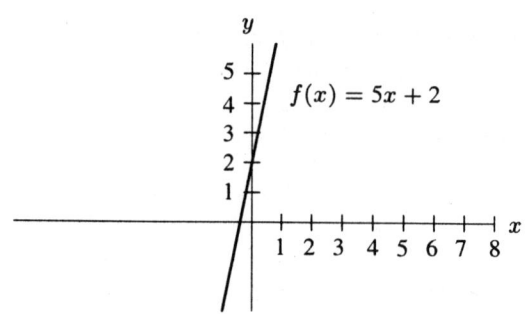

**Figure 2.23**

(c) Since the graph of $g(x) = mx + b$ is a line, then $g'(a) = m$ for any $a$.

20.
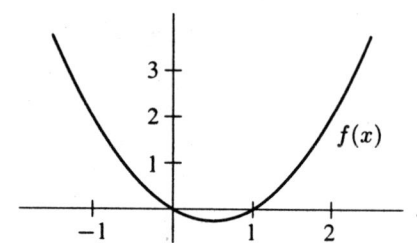 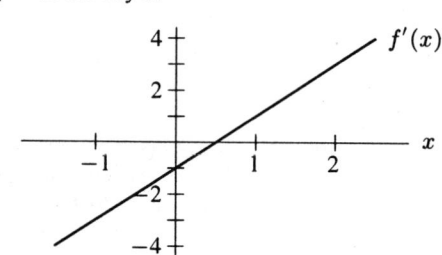

21.
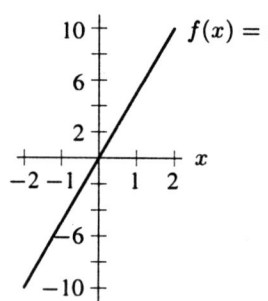 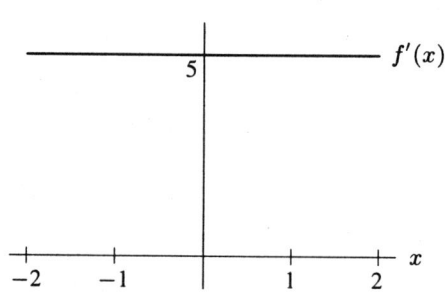

22.
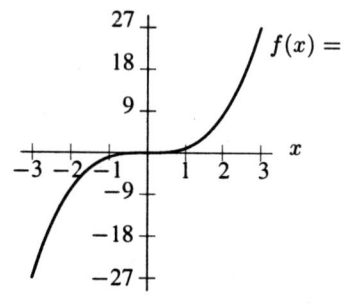 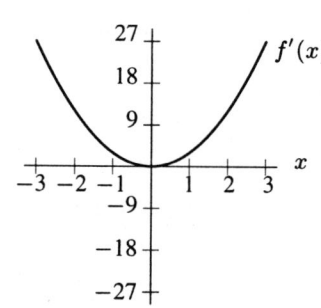

23.
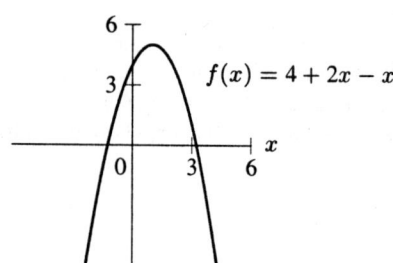 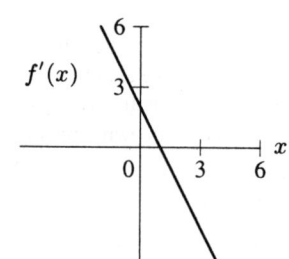

24.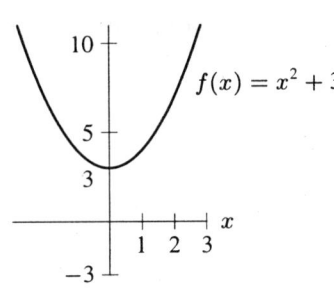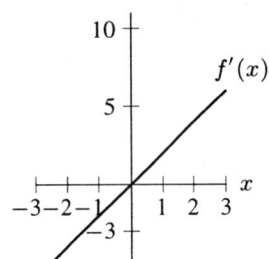

25.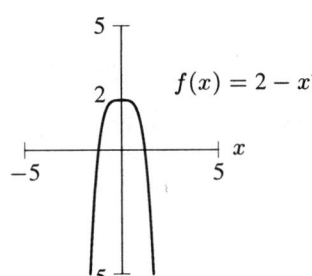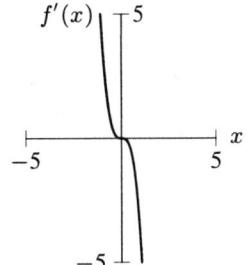

**Figure 2.24**

26.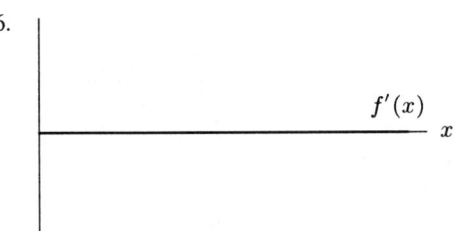

27.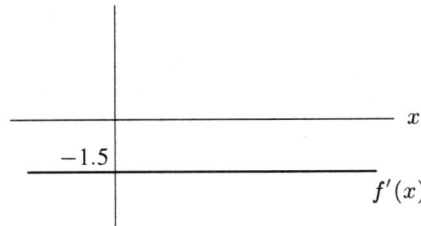

28.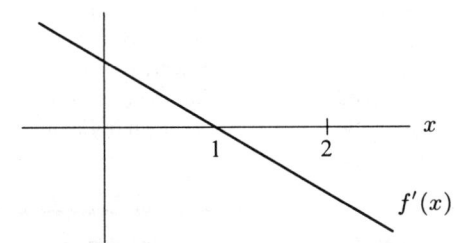

29.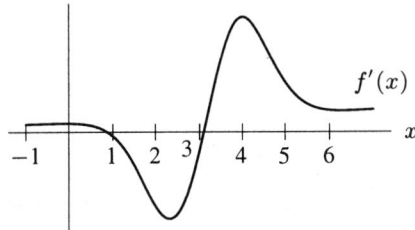

30.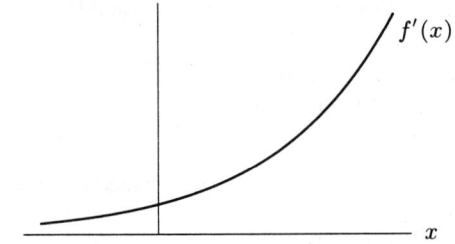

31.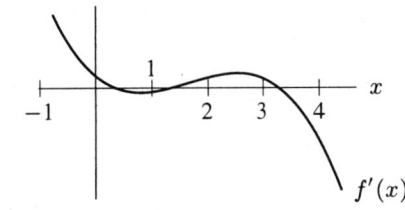

32. (a) We see from Figure 2.25 that the number of bacteria rises for the first five days and then decreases after that.

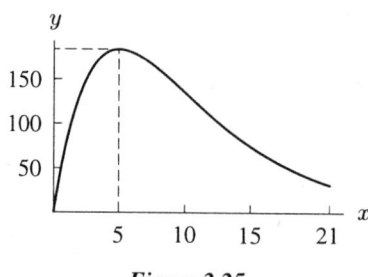

Figure 2.25

(b) The population of bacteria in the body is at a maximum 5 days into the illness. At this time there are approximately 163.8 bacteria per cubic centimeter.
(c) We are looking for the point at which the graph has the largest slope. From Figure 2.25 we see that this occurs at $t = 0$, at the very beginning of the illness. The population of bacteria appears to be decreasing fastest at $t = 10$.
(d) Using a small interval, we find that after 1 week, $(t = 7)$, the population is changing at a rate of approximately $-12$ bacteria per cm$^3$/day.

33. (a) A possible graph is shown in Figure 2.26

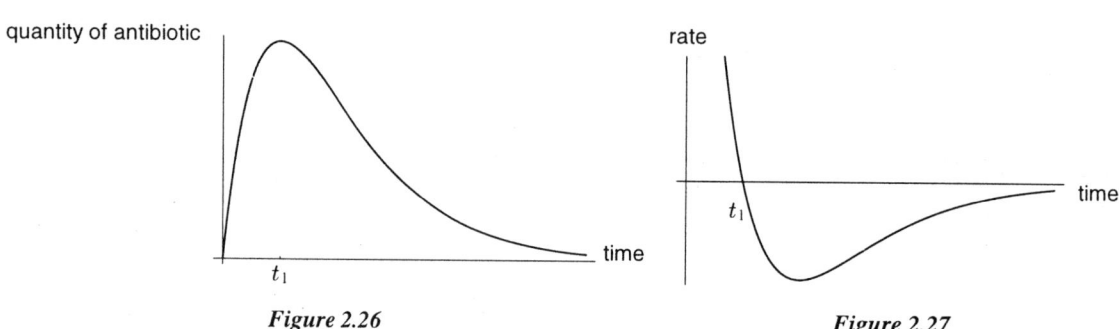

Figure 2.26          Figure 2.27

(b) The injection puts a reservoir of antibiotics in the muscle which begins to diffuse into the blood. As the antibiotic diffuses into the bloodstream, it begins to leave the blood either through normal metabolic action or absorption by some other organ. For a while the antibiotic diffuses into the blood faster than it is lost and its concentration rises, but as the reservoir in the muscle is drawn down, the diffusion rate into the blood decreases and eventually becomes less than the loss rate. After that, the concentration in the blood goes down. But the rate of decrease in concentration gets smaller as we approach the time when all the antibiotic is lost. This is shown in Figure 2.27.

## Solutions for Section 2.4

1. (Note that we are considering the average temperature of the yam, since its temperature is different at different points inside it.)

    (a) It is positive, because the temperature of the yam increases the longer it sits in the oven.
    (b) The units of $f'(20)$ are °F/min. $f'(20) = 2$ means that at time $t = 20$ minutes, the temperature $T$ increases by approximately 2°F for each additional minute in the oven.

2. (a) Since $f'(c)$ is negative, the function $P = f(c)$ is decreasing: pelican eggshells are getting thinner as the concentration, $c$, of PCBs in the environment is increasing.
    (b) The statement $f(200) = 0.28$ means that the thickness of pelican eggshells is 0.28 mm when the concentration of PCBs in the environment is 200 parts per million (ppm).
    
    The statement $f'(200) = -0.0005$ means that the thickness of pelican eggshells is decreasing (eggshells are becoming thinner) at a rate of 0.0005 mm per ppm of concentration of PCBs in the environment when the concentration of PCBs in the environment is 200 ppm.

3. The units of $f'(x)$ are feet/mile. The derivative, $f'(x)$, represents the rate of change of elevation with distance from the source, so if the river is flowing downhill everywhere, the elevation is always decreasing and $f'(x)$ is always negative. (In fact, there may be some stretches where the elevation is more or less constant, so $f'(x) = 0$.)

4. The units of $g'(t)$ are inches/year. The quantity $g'(10)$ represents how fast Amelia Earhart was growing at age 10, so $g'(10) > 0$. The quantity $g'(30)$ represents how fast she was growing at age 30, so $g'(30) = 0$ because she was probably not growing at that age.

5. Units of $C'(r)$ are dollars/percent. Approximately, $C'(r)$ means the additional amount needed to pay off the loan when the interest rate is increased by 1%. The sign of $C'(r)$ is positive, because increasing the interest rate will increase the amount it costs to pay off a loan.

6. Units of $P'(t)$ are dollars/year. The practical meaning of $P'(t)$ is the rate at which the monthly payments change as the duration of the mortgage increases. Approximately, $P'(t)$ represents the change in the monthly payment if the duration is increased by one year. $P'(t)$ is negative because increasing the duration of a mortgage decreases the monthly payments.

7. Since $B$ is measured in dollars and $t$ is measured in years, $dB/dt$ is measured in dollars per year. We can interpret $dB$ as the extra money added to your balance in $dt$ years. Therefore $dB/dt$ represents how fast your balance is growing, in units of dollars/year.

8. $f(10) = 240{,}000$ means that if the commodity costs \$10, then 240,000 units of it will be sold. $f'(10) = -29{,}000$ means that if the commodity costs \$10 now, each \$1 increase in price will cause a decline in sales of 29,000 units.

9. (a) This means that investing the \$1000 at 5% would yield \$1649 after 10 years.
   (b) Writing $g'(r)$ as $dB/dt$, we see that the units of $dB/dt$ are dollars per percent (interest). We can interpret $dB$ as the extra money earned if interest rate is increased by $dr$ percent. Therefore $g'(5) = \frac{dB}{dr}|_{r=5} \approx 165$ means that the balance, at 5% interest, would increase by about \$165 if the interest rate were increased by 1%. In other words, $g(6) \approx g(5) + 165 = 1649 + 165 = 1814$.

10. (a) Positive, since weight increases as the child gets older.
    (b) $f(8) = 45$ tells us that when the child is 8 years old, the child weighs 45 pounds.
    (c) The units of $f'(a)$ are lbs/year. $f'(a)$ tells the rate of growth in lbs/years at age $a$.
    (d) $f'(8) = 4$ tells us that the 8-year-old child is growing at about 4 lbs/year.
    (e) As $a$ increases, $f'(a)$ will decrease since the rate of growth slows down as the child grows up.

11. (a) The statement $f(20) = 0.36$ means that 20 minutes after smoking a cigarette, there will be 0.36 mg of nicotine in the body. The statement $f'(20) = -0.002$ means that 20 minutes after smoking a cigarette, nicotine is leaving the body at a rate 0.002 mg per minute. The units are 20 minutes, 0.36 mg, and $-0.002$ mg/minute.
    (b)
    $$f(21) \approx f(20) + \text{ change in } f \text{ in one minute}$$
    $$= 0.36 + (-0.002)$$
    $$= 0.358$$
    $$f(30) \approx f(20) + \text{ change in } f \text{ in 10 minutes}$$
    $$= 0.36 + (-0.002)(10)$$
    $$= 0.36 - 0.02$$
    $$= 0.34$$

12. Since we do not have information beyond $t = 25$, we will assume that the function will continue to change at the same rate. Therefore,
    $$f(26) \approx f(25) + f'(25)$$
    $$= 3.6 + (-0.2) = 3.4.$$
    Since $30 = 25 + 5$, then
    $$f(30) \approx f(25) + f'(25)(5)$$
    $$= 3.6 + (-0.2)(5) = 2.6.$$

**120** CHAPTER TWO /SOLUTIONS

13. (a) Since $t$ represents the number of days from now, we are told $f(0) = 80$ and $f'(0) = 0.50$.
    (b)
    $$f(10) \approx \text{value now} + \text{change in value in 10 days}$$
    $$= 80 + 0.50(10)$$
    $$= 80 + 5$$
    $$= 85.$$
    In 10 days, we expect that the mutual fund will be worth about $85 a share.

14. (a) From the table, we see that the debt was always increasing. Since we do not have information after 1993, we compute the derivative of the function in 1993 using the left estimate:
    $$f'(1993) \approx \frac{4351.2 - 4064.6}{1993 - 1992} = \frac{286.6}{1} = 286.6 \text{ billion/year}.$$
    The debt was increasing at a rate of about 286.6 billion/year in 1993.
    (b) To make estimates beyond 1993, we assume that the debt continues to climb at the same rate. Therefore:
    $$\text{Debt in 1994} = \text{Debt in 1993} + \text{Change in debt in one year}$$
    $$\approx 4351.2 + 286.6 = 4637.8 \text{ billion}.$$
    Since 2010 is 17 years beyond 1993,
    $$\text{Debt in 2010} \approx 4351.2 + 286.6(17) = 9223.4 \text{ billion}.$$
    The result for 1994 is probably a better estimate than that for the year 2010, since 1994 is closer to the years for which we have data.

15. Units of $g'(55)$ are mpg/mph. The statement $g'(55) = -0.54$ means that at 55 miles per hour the fuel efficiency (in miles per gallon, or mpg) of the car decreases at a rate of approximately one half mpg as the velocity increases by one mph.

16. Units of $dP/dt$ are barrels/year. $dP/dt$ is the change in quantity of petroleum per change in time (a year). This is negative. We could estimate it by finding the amount of petroleum used worldwide over a short period of time.

17. (a) The company hopes that increased advertising always brings in more customers instead of turning them away. Therefore, it hopes $f'(a)$ is always positive.
    (b) If $f'(100) = 2$, it means that if the advertising budget is $100,000, each extra dollar spent on advertising will bring in $2 worth of sales. If $f'(100) = 0.5$, each dollar above $100 thousand spent on advertising will bring in $0.50 worth of sales.
    (c) If $f'(100) = 2$, then as we saw in part (b), spending slightly more than $100,000 will increase revenue by an amount greater than the additional expense, and thus more should be spent on advertising. If $f'(100) = 0.5$, then the increase in revenue is less than the additional expense, hence too much is being spent on advertising. The optimum amount to spend, $a$, is an amount that makes $f'(a) = 1$. At this point, the increases in advertising expenditures just pay for themselves. If $f'(a) < 1$, too much is being spent; if $f'(a) > 1$, more should be spent.

18. (a)

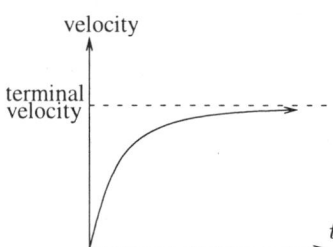

   (b) The graph should be concave down because wind resistance decreases your acceleration as you speed up, and so the slope of the graph of velocity is decreasing.
   (c) The slope represents the acceleration due to gravity.

19. Since $\frac{P(67) - P(66)}{67 - 66}$ is an estimate of $P'(66)$, we may think of $P'(66)$ as an estimate of $P(67) - P(66)$, and the latter is the number of people between 66 and 67 inches tall. Alternatively, since $\frac{P(66.5) - P(65.5)}{66.5 - 65.5}$ is a better estimate of $P'(66)$, we may regard $P'(66)$ as an estimate of the number of people of height between 65.5 and 66.5 inches. The units for $P'(x)$ are people per inch.

Since there were 250 million people at the 1990 census, we might guess that there are about 200 million full-grown persons in the US whose heights are distributed between $60''(5')$ and $75''(6'3'')$. There are probably quite a few people of height $66''$–perhaps $1\frac{1}{2}$ what you'd expect from an even, or uniform, distribution–because it's nearly average. An even distribution would yield $P'(66) = \frac{200 \text{ million}}{15''} \approx 13$ million per inch–so we can expect $P'(66)$ to be perhaps $13(1.5) \approx 20$.

$P'(x)$ is never negative because $P(x)$ is never decreasing. To see this, let's look at an example involving a particular value of $x$, say $x = 70$. The value $P(70)$ represents the number of people whose height is less than or equal to 70 inches, and $P(71)$ represents the number of people whose height is less than or equal to 71 inches. Since everyone shorter than 70 inches is also shorter than 71 inches, $P(70) \leq P(71)$. In general, $P(x)$ is 0 for small $x$, and increases as $x$ increases, and is eventually constant (for large enough $x$).

20. (a) Estimating derivatives using difference quotients (but other answers are possible):

$$P'(1900) \approx \frac{P(1910) - P(1900)}{10} = \frac{92.0 - 76.0}{10} = 1.6 \text{ million people per year}$$

$$P'(1945) \approx \frac{P(1950) - P(1940)}{10} = \frac{150.7 - 131.7}{10} = 1.9 \text{ million people per year}$$

$$P'(1990) \approx \frac{P(1990) - P(1980)}{10} = \frac{248.7 - 226.5}{10} = 2.22 \text{ million people per year}$$

(b) The population growth was maximal somewhere between 1950 and 1960.

(c) $P'(1950) \approx \frac{P(1960) - P(1950)}{10} = \frac{179.0 - 150.7}{10} = 2.83$ million people per year, so $P(1956) \approx P(1950) + P'(1950)(1956 - 1950) = 150.7 + 2.83(6) \approx 167.7$ million people.

(d) If the growth rate between 1990 and 2000 was the same as the growth rate from 1980 to 1990, then the total population should be about 271 million people in 2000.

21. (a) Clearly the population of the US at any instant is an integer that varies up and down every few seconds as a child is born, a person dies, or a new immigrant arrives. Since these events cannot usually be assigned to an exact instant, the population of the US at any given moment might actually be indeterminate. If we count in units of a thousand, however, the population appears to be a smooth function that has been rounded to the nearest thousand.

Major land acquisitions such as the Louisiana Purchase caused larger jumps in the population, but since the census is taken only every ten years and the territories acquired were rather sparsely populated, we cannot see these jumps in the census data.

(b) We can regard rate of change of the population for a particular time $t$ as representing an estimate of how much the population will increase during the year after time $t$.

(c) Many economic indicators are treated as smooth, such as the Gross National Product, the Dow Jones Industrial Average, volumes of trading, and the price of commodities like gold. But these figures only change in increments, and not continuously.

## Solutions for Section 2.5

1. (a) increasing, concave up
   (b) decreasing, concave down

2.

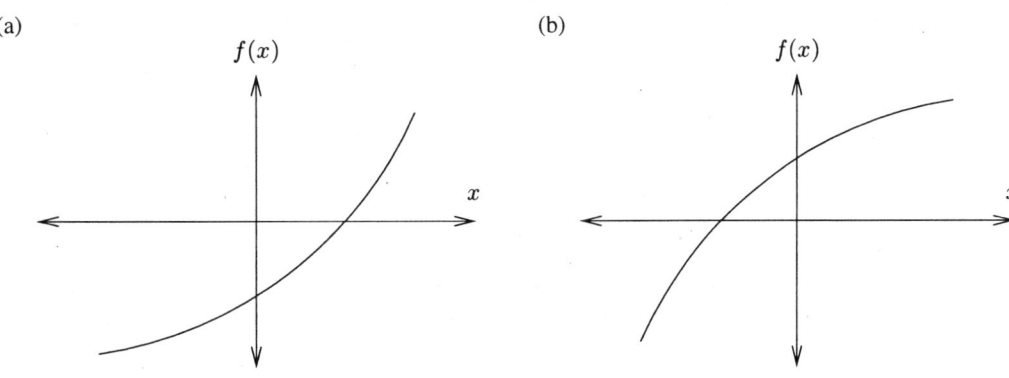

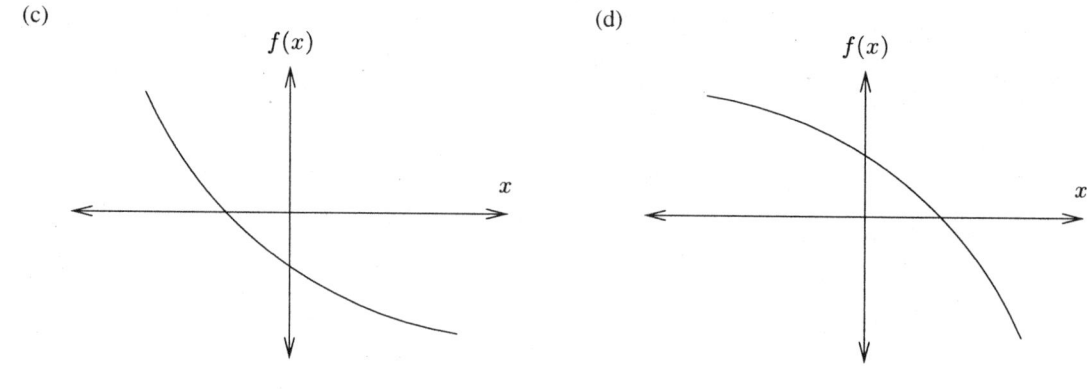

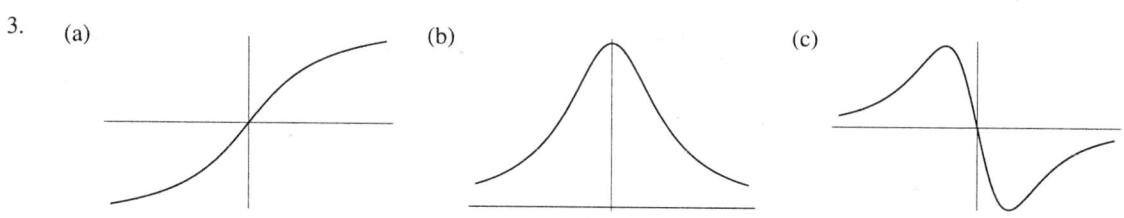

4. The derivative is positive on those intervals where the function is increasing and negative on those intervals where the function is decreasing. Therefore, the derivative is positive on the interval $-2.3 < t < -0.5$ and negative on the interval $-0.5 < t < 4$.

   The second derivative is positive on those intervals where the graph of the function is concave up and negative on those intervals where the graph of the function is concave down. Therefore, the second derivative is positive on the interval $0.5 < t < 4$ and negative on the interval $-2.3 < t < 0.5$.

5. The derivative is positive on those intervals where the function is increasing and negative on those intervals where the function is decreasing. Therefore, the derivative is positive on the intervals $0 < t < 0.4$ and $1.7 < t < 3.4$, and negative on the intervals $0.4 < t < 1.7$ and $3.4 < t < 4$.

   The second derivative is positive on those intervals where the graph of the function is concave up and negative on those intervals where the graph of the function is concave down. Therefore, the second derivative is positive on the interval $1 < t < 2.6$ and negative on the intervals $0 < t < 1$ and $2.6 < t < 4$.

6. The derivative, $s'(t)$, appears to be positive since $s(t)$ is increasing over the interval given. The second derivative also appears to be positive or zero since the function is concave up or possibly linear between $t = 1$ and $t = 3$, i.e., it is increasing at a non-decreasing rate.

7. The derivative of $w(t)$ appears to be negative since the function is decreasing over the interval given. The second derivative, however, appears to be positive since the function is concave up, i.e., it is decreasing at a decreasing rate.

8. This graph is increasing for all $x$, and is concave down to the left of 2 and concave up to the right of 2. One possible answer is shown in Figure 2.28:

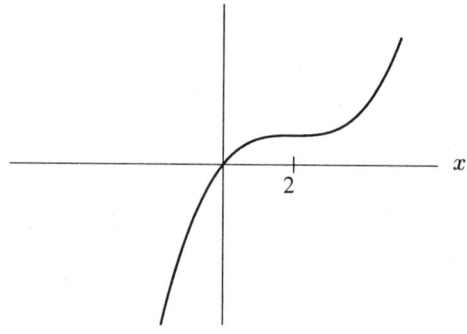

*Figure 2.28*

9. (a)

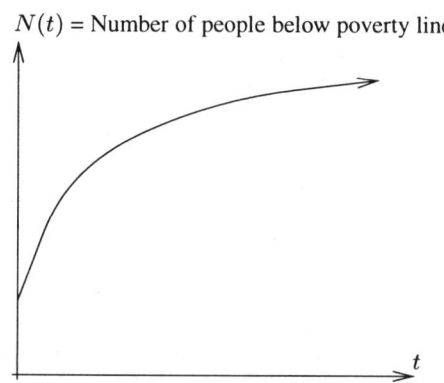

(b) $dN/dt$ is positive, since people are still slipping below the poverty line. $d^2N/dt^2$ is negative, since the rate at which people are slipping below the poverty line, $dN/dt$, is decreasing.

10. (a)

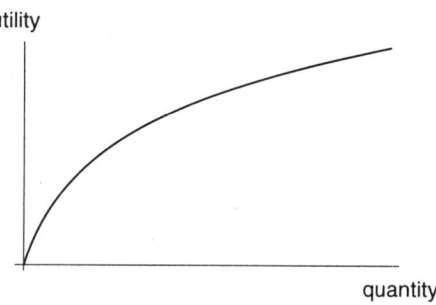

(b) As a function of quantity, utility is increasing but at a decreasing rate; the graph is increasing but concave down. So the derivative of utility is positive, but the second derivative of utility is negative.

11. (a) $dP/dt > 0$ and $d^2P/dt^2 > 0$.

(b) $dP/dt < 0$ and $d^2P/dt^2 > 0$ (but $dP/dt$ is close to zero).

12. Since all advertising campaigns are assumed to produce an increase in sales, a graph of sales against time would be expected to have a positive slope.

A positive second derivative means the rate at which sales are increasing is increasing. If a positive second derivative is observed during a new campaign, it is reasonable to conclude that this increase in the rate sales are increasing is caused by the new campaign–which is therefore judged a success. A negative second derivative means a decrease in the rate at which sales are increasing, and therefore suggests the new campaign is a failure.

13. (a) $f'(0.6) \approx \dfrac{f(0.8) - f(0.6)}{0.8 - 0.6} = \dfrac{4.0 - 3.9}{0.2} = 0.5.$  $f'(0.5) \approx \dfrac{f(0.6) - f(0.4)}{0.6 - 0.4} = \dfrac{0.4}{0.2} = 2.$

(b) Using the values of $f'$ from part (a), we get $f''(0.6) \approx \dfrac{f'(0.6) - f'(0.5)}{0.6 - 0.5} = \dfrac{0.5 - 2}{0.2} = \dfrac{-1.5}{0.2} = -7.5.$

(c) The maximum value of $f$ is probably near $x = 0.8$. The minimum value of $f$ is probably near $x = 0.3$.

14. (a) The EPA will say that the rate of discharge is still rising. The industry will say that the rate of discharge is increasing less quickly, and may soon level off or even start to fall.

(b) The EPA will say that the rate at which pollutants are being discharged is levelling off, but not to zero — so pollutants will continue to be dumped in the lake. The industry will say that the rate of discharge has decreased significantly.

15. (a) For 1986 we have
$$\frac{dP}{dt} \approx \frac{54.1 - 62.4}{1989 - 1986} = \frac{-8.3}{3} \approx -2.77 \text{ \%/year.}$$

For 1989 we have
$$\frac{dP}{dt} \approx \frac{48.0 - 54.1}{1992 - 1989} = \frac{-6.1}{3} \approx -2.03 \text{ \%/year.}$$

For 1992 we have
$$\frac{dP}{dt} \approx \frac{43.5 - 48.0}{1995 - 1992} = \frac{-4.5}{3} \approx -1.50 \text{ \%/year.}$$

For 1995 we have
$$\frac{dP}{dt} \approx \frac{41.8 - 43.5}{1998 - 1995} = \frac{-1.7}{3} \approx -0.57 \text{ \%/year.}$$

(b) Since $\frac{dP}{dt}$ is increasing from 1986 to 1998 we get that $\frac{d^2P}{dt^2}$ is positive.

(c) The values of $P$ and $\frac{dP}{dt}$ are troublesome because they indicate that the percent of students graduating is low, and that the number is getting smaller each year.

(d) Since $\frac{d^2P}{dt^2}$ is positive, the percent of students graduating is not decreasing as fast as it once was. Also, in 1995 $dP/dt$ has a magnitude of less than 1% a year so the level of drop-outs does in fact seem to be hitting its minimum at around 40%.

16. Since $f'$ is everywhere positive, $f$ is everywhere increasing. Hence the greatest value of $f$ is at $x_6$ and the least value of $f$ is at $x_1$. Directly from the graph, we see that $f'$ is greatest at $x_3$ and least at $x_2$. Since $f''$ gives the slope of the graph of $f'$, $f''$ is greatest where $f'$ is rising most rapidly, namely at $x_6$, and $f''$ is least where $f'$ is falling most rapidly, namely at $x_1$.

17. (a) B (where $f', f'' > 0$) and E (where $f', f'' < 0$)
    (b) A (where $f = f' = 0$) and D (where $f' = f'' = 0$)

## Solutions for Section 2.6

1. We know $MC \approx C(1{,}001) - C(1{,}000)$. Therefore, $C(1{,}001) \approx C(1{,}000) + MC$ or $C(1{,}001) \approx 5000 + 25 = 5025$ dollars.

    Since we do not know $MC(999)$, we will assume that $MC(999) = MC(1{,}000)$. Therefore:
    $$MC(999) = C(1{,}000) - C(999).$$
    Then:
    $$C(999) \approx C(1{,}000) - MC(999) = 5{,}000 - 25 = 4{,}975 \text{ dollars.}$$
    Alternatively, we can reason that
    $$MC(1{,}000) \approx C(1{,}000) - C(999),$$
    so
    $$C(999) \approx C(1{,}000) - MC(1{,}000) = 4{,}975 \text{ dollars.}$$
    Now for $C(1{,}000)$, we have
    $$C(1{,}100) \approx C(1{,}000) + MC \cdot 100.$$
    Since $1{,}100 - 1{,}000 = 100$,
    $$C(1{,}100) \approx 5{,}000 + 25 \times 100 = 5{,}000 + 2{,}500 = 7{,}500 \text{ dollars.}$$

2. (a) We can approximate $C(16)$ by adding $C'(15)$ to $C(15)$, since $C'(15)$ is an estimate of the cost of the $16^{\text{th}}$ item.
    $$C(16) \approx C(15) + C'(15) = \$2300 + \$108 = \$2408.$$
    (b) We approximate $C(14)$ by subtracting $C'(15)$ from $C(15)$, where $C'(15)$ is an approximation of the cost of producing the $15^{\text{th}}$ item.
    $$C(14) \approx C(15) - C'(15) = \$2300 - \$108 = \$2192.$$

3. Drawing in the tangent line at the point $(10000, C(10000))$ we get Figure 2.29. We see that each vertical increase of 2500 in the tangent line gives a corresponding horizontal increase of roughly 6000. Thus the marginal cost at the production level of 10,000 units is

$$C'(10,000) = \begin{array}{c}\text{Slope of tangent line}\\ \text{to } C(q) \text{ at } q = 10,000\end{array} = \frac{2500}{6000} = 0.42.$$

This tells us that after producing 10,000 units, it will cost roughly $0.42 to produce one more unit.

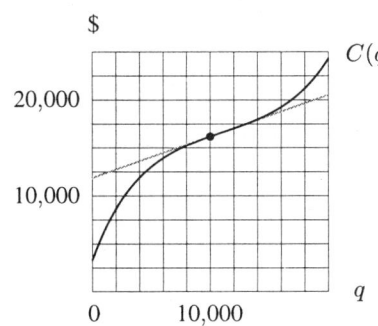

Figure 2.29

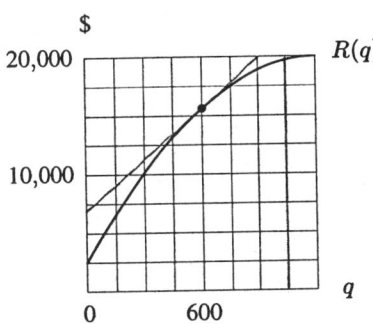

Figure 2.30

4. Drawing in the tangent line at the point $(600, R(600))$ we get Figure 2.30. We see that each vertical increase of 2500 in the tangent line gives a corresponding horizontal increase of roughly 150. The marginal revenue at the production level of 600 units is

$$R'(600) = \begin{array}{c}\text{Slope of tangent line}\\ \text{to } R(q) \text{ at } q = 600\end{array} = \frac{2500}{150} = 16.67.$$

This tells us that after producing 600 units, the revenue for producing the $601^{\text{st}}$ product will be roughly $16.67.

5. Marginal cost $= C'(q)$. Therefore, marginal cost at $q$ is the slope of the graph of $C(q)$ at $q$. We can see that the slope at $q = 5$ is greater than the slope at $q = 30$. Therefore, marginal cost is greater at $q = 5$. We see that, at $q = 20$, the slope is practically zero while at $q = 40$ the slope is positive. Therefore, marginal cost at $q = 40$ is greater than marginal cost at $q = 20$.

6. At $q = 50$, the slope of the revenue is larger than the slope of the cost. Thus, at $q = 50$, marginal revenue is greater than marginal cost and the $50^{\text{th}}$ bus should be added. At $q = 100$ the slope of revenue is less than the slope of cost. Thus, at $q = 100$ the marginal revenue is less than marginal cost and the $100^{\text{th}}$ bus should not be added.

7.

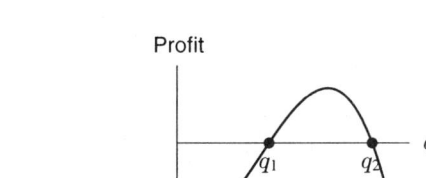 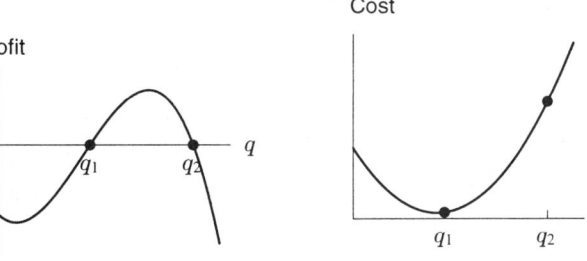 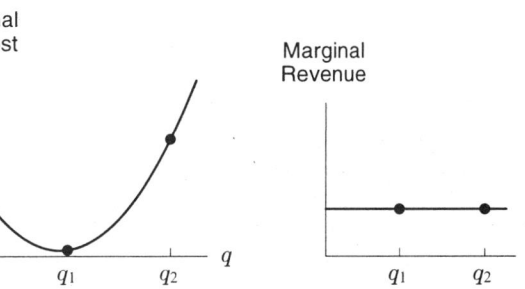

8. (a) $C(0)$ represents the fixed costs before production, that is, the cost of producing zero units, incurred for initial investments in equipment, etc.
   (b) The marginal cost decreases slowly, and then increases as quantity produced increases. See Problem 7, graph (b).
   (c) Concave down implies decreasing marginal cost, while concave up implies increasing marginal cost.
   (d) An inflection point of the cost function is (locally) the point of maximum or minimum marginal cost.
   (e) One would think that the more of an item you produce, the less it would cost to produce extra items. In economic terms, one would expect the marginal cost of production to decrease, so we would expect the cost curve to be concave down. In practice, though, it eventually becomes more expensive to produce more items, because workers and resources may become scarce as you increase production. Hence after a certain point, the marginal cost may rise again. This happens in oil production, for example.

9. (a) Since Profit = Revenue − Cost, we can calculate $\pi(q) = R(q) - C(q)$ for each of the $q$ values given:

| $q$ | 0 | 100 | 200 | 300 | 400 | 500 |
|---|---|---|---|---|---|---|
| $R(q)$ | 0 | 500 | 1000 | 1500 | 2000 | 2500 |
| $C(q)$ | 700 | 900 | 1000 | 1100 | 1300 | 1900 |
| $\pi(q)$ | −700 | −400 | 0 | 400 | 700 | 600 |

We see that maximum profit is $700 and it occurs when the production level $q$ is 400. See Figure 2.31.

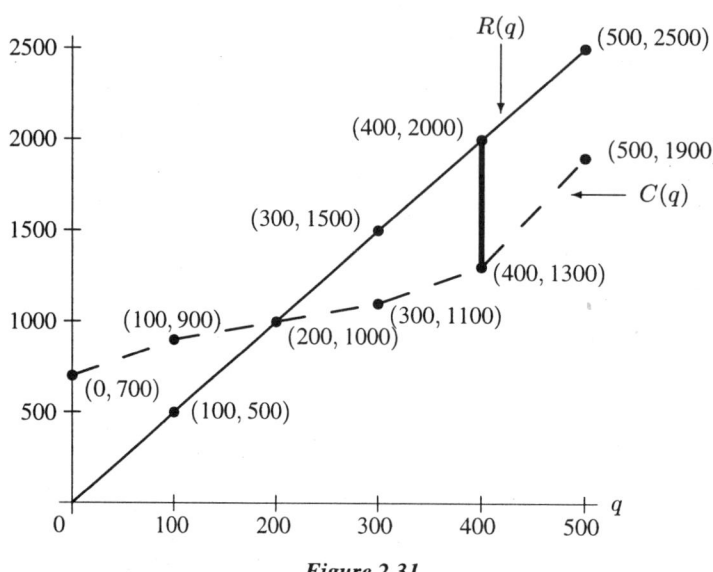

Figure 2.31

(b) Since revenue is $500 when $q = 100$, the selling price is $5 per unit.

(c) Since $C(0) = \$700$, the fixed costs are $700.

10. We have
$$C'(2000) \approx \frac{C(2500) - C(2000)}{2500 - 2000} = \frac{3825 - 3640}{500} = \$0.37/\text{ton}.$$
This means that recycling the 2001st ton of paper will cost around $0.37. The marginal cost is smallest at the point where the derivative of the function is smallest. Thus the marginal cost appears to be smallest on the interval $2500 \leq q \leq 3000$.

11. The profit is maximized at the point where the difference between revenue and cost is greatest. Thus the profit is maximized at approximately $q = 4000$.

12. (a) We know that marginal cost = $C'(q)$ or the slope of the graph of $C(q)$ at $q$. Similarly, marginal revenue = $R'(q)$ or the slope of the graph of $R(q)$ at $q$. From the graph, the slope of $R(q)$ at $q = 3000$ is greater than the slope of $C(q)$ at $q = 3000$. Therefore, marginal revenue is greater than marginal cost at $q = 3000$. The company should increase production.

(b) The graph of $C(q)$ is steeper than the graph of $R(q)$ at $q = 5000$. Therefore, marginal cost is greater than marginal revenue at $q = 5000$. The company should decrease production.

13. (a) The profit earned by the 51st is the revenue earned by the 51st item minus the cost of producing the 51st item. This can be approximated by
$$\pi'(50) = R'(50) - C'(50) = 84 - 75 = \$9.$$
Thus the profit earned from the 51st item will be approximately $9.

(b) The profit earned by the 91st item will be the revenue earned by the 91st item minus the cost of producing the 91st item. This can be approximated by
$$\pi'(90) = R'(90) - C'(90) = 68 - 71 = -\$3.$$
Thus, approximately three dollars are lost in the production of the 91st item.

(c) If $R'(78) > C'(78)$, production of a $79^{\text{th}}$ item would increase profit. If $R'(78) < C'(78)$, production of one less item would increase profit. Since profit is maximized at $q = 78$, we must have
$$C'(78) = R'(78).$$

## *Solutions for Chapter 2 Review*

**1.**

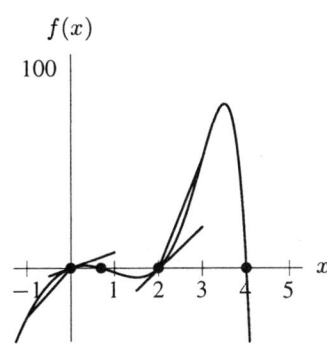

(a) The graph shows four different zeros in the interval, at $x = 0$, $x = 2$, $x = 4$ and $x \approx 0.7$.
(b) At $x = 0$ and $x = 2$, we see that the tangent has a positive slope so $f$ is increasing.
At $x = 4$, we notice that the tangent to the curve has negative slope, so $f$ is decreasing.
(c) Comparing the slopes of the secant lines at these values, we can see that the average rate of change of $f$ is greater on the interval $2 \le x \le 3$.
(d) Looking at the tangents of the function at $x = 0$ and $x = 2$, we see that the slope of the tangent at $x = 2$ is greater. Thus, the instantaneous rate of change of $f$ is greater at $x = 2$.

**2.** Using the interval $1 \le x \le 1.001$, we estimate
$$f'(1) \approx \frac{f(1.001) - f(1)}{0.001} = \frac{3.0033 - 3.0000}{0.001} = 3.3$$
The graph of $f(x) = 3^x$ is concave up so we expect our estimate to be greater than $f'(1)$.

**3.** 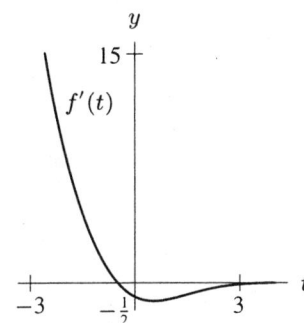   **4.**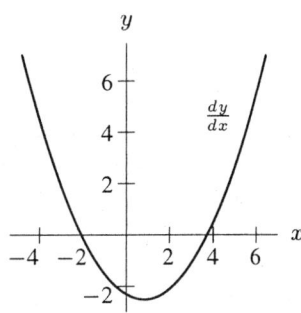

**5.** 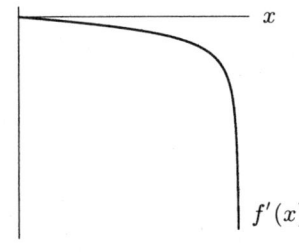   **6.**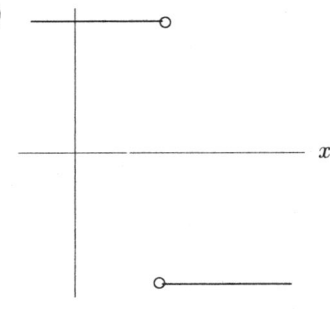

7.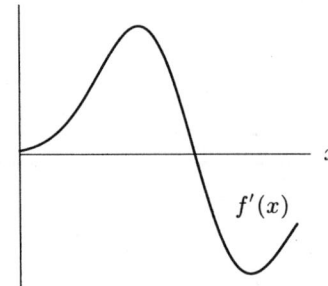

8.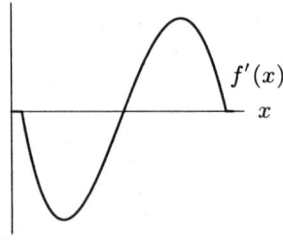

9.

10.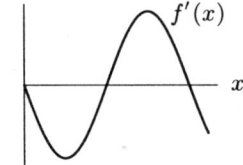

11. For $x < -2$, $f$ is increasing and concave up. For $-2 < x < 1$, $f$ is increasing and concave down. At $x = 1$, $f$ has a maximum. For $x > 1$, $f$ is decreasing and concave down. One such possible $f$ is shown to the right.

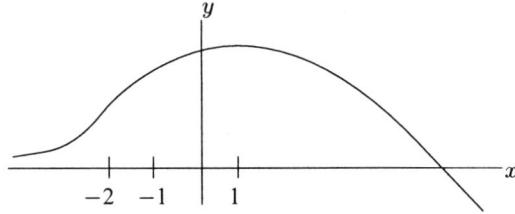

12. (a) $f'(x)$ is negative when the function is decreasing and positive when the function is increasing. Therefore, $f'(x)$ is positive at $C$ and $G$. $f'(x)$ is negative at $A$ and $E$. $f'(x)$ is zero at $B$, $D$, and $F$.
    (b) $f'(x)$ is the largest when the graph of the function is increasing the fastest (i.e. the point with the steepest positive slope). This occurs at point $G$. $f'(x)$ is the most negative when the graph of the function is decreasing the fastest (i.e. the point with the steepest negative slope). This occurs at point $A$.

13. Estimating the slope of the lines in Figure 2.32, we find that $f'(-2) \approx 1.0$, $f'(-1) \approx 0.3$, $f'(0) \approx -0.5$, and $f'(2) \approx -1$.

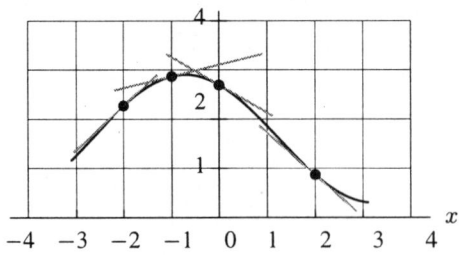

*Figure 2.32*

**14.** Using $\Delta x = 1$, we can say that

$$\begin{aligned} f(21) &= f(20) + \text{change in } f(x) \\ &\approx f(20) + f'(20)\Delta x \\ &= 68 + (-3)(1) \\ &= 65. \end{aligned}$$

Similarly, using $\Delta x = -1$,

$$\begin{aligned} f(19) &= f(20) + \text{change in } f(x) \\ &\approx f(20) + f'(20)\Delta x \\ &= 68 + (-3)(-1) \\ &= 71. \end{aligned}$$

Using $\Delta x = 5$, we can write

$$\begin{aligned} f(25) &= f(20) + \text{change in } f(x) \\ &\approx f(20) + (-3)(5) \\ &= 68 - 15 \\ &= 53. \end{aligned}$$

**15.** (a) The 12 represents the weight of the chemical; therefore, its units are pounds. The 5 represents the cost of the chemical; therefore, its units are dollars. The statement $f(12) = 5$ means that when the weight of the chemical is 12 pounds, the cost is 5 dollars.

(b) We expect the derivative to be positive since we expect the cost of the chemical to increase when the weight bought increases.

(c) Again, 12 is the weight of the chemical in pounds. The units of the 0.4 are dollars/pound since it is the rate of change of the cost as a function of the weight of the chemical bought. The statement $f'(12) = 0.4$ means that the cost is increasing at a rate of 0.4 dollars per pound when the weight is 12 pounds, or that an additional pound will cost an extra 40 cents.

**16.** The statement $f(12) = 37$ means that when $t = 12$, we have $P = 37$. This tells us that in 1994, 37% of households had a personal computer. The statement $f'(12) = 2$ tells us that in 1994, the percent of households with a personal computer is increasing at a rate of 2% a year.

**17.** (a) The function appears to be decreasing and concave down, and so we conjecture that $f'$ is negative and that $f''$ is negative.

(b) We use difference quotients to the right:
$f'(2) \approx \frac{137-145}{4-2} = -4$
$f'(8) \approx \frac{56-98}{10-8} = -21$.

**18.** (a) We have

$$f'(0.6) \approx \frac{f(0.8) - f(0.6)}{0.8 - 0.6} = \frac{0.1}{0.2} = 0.5$$

(b) Using the point $x = 0.6$, $y = 3.9$, with the slope 0.5, we have

$$\begin{aligned} y &= 0.5x + b \\ 3.9 &= 0.5(0.6) + b \\ b &= 3.6 \end{aligned}$$

so

$$y = 0.5x + 3.6$$

(c) We can use the tangent line from part (b), giving

$$\begin{aligned} f(0.7) &\approx 0.5(0.7) + 3.6 = 3.95 \\ f(1.2) &\approx 0.5(1.2) + 3.6 = 4.2 \\ f(1.4) &\approx 0.5(1.4) + 3.6 = 4.3 \end{aligned}$$

**130** CHAPTER TWO /SOLUTIONS

It's OK to be confident about $f(0.7)$ as 0.7 is close to 0.6 (and $f(0.8) = 4$, not too far off). However, $f(1.2)$ is likely to be less than $f(1.0) = 3.9$ as $f$ is decreasing from $0.8 < x < 1.0$. The estimate for $f(1.4)$ could be even further off.

19. (a) $f'(6.75) \approx \dfrac{f(7.0) - f(6.5)}{7.0 - 6.5} = \dfrac{8.2 - 10.3}{0.5} = -4.2.$

    $f'(7.0) \approx \dfrac{f(7.5) - f(7)}{7.5 - 7} = \dfrac{6.5 - 8.2}{0.5} = -3.4.$

    $f'(8.5) \approx \dfrac{f(9.0) - f(8.5)}{9.0 - 8.5} = \dfrac{3.2 - 4.1}{0.5} = -1.8.$

    (b) To estimate $f''$ at 7, we should have values for $f'$ at points near 7. We know from part (a) that $f'(6.75) \approx -4.2$ and estimate $f'(7.0) \approx -3.4$. Then
    $$f''(7) \approx \dfrac{f'(7) - f'(6.75)}{7 - 6.75} = \dfrac{-3.4 - (-4.2)}{0.25} = 3.2.$$

    (c) Since $f'(7) = -3.4$, we have
    $$y = -3.4x + b.$$
    Substituting $x = 7, y = 8.2$ gives
    $$8.2 = -3.4(7) + b \quad \text{so} \quad b = 32$$
    $$y = -3.4x + 32$$

    (d) We may use the tangent line from part (c) to approximate $f(6.8)$. In this case we get
    $$y \approx -3.4x + 32 = -3.4(6.8) + 32 = 8.88.$$

    ( We may also estimate $f(6.8)$ by assuming that the graph of $f$ is straight between the given points $(6.5, 10.3)$ and $(7.0, 8.2)$. This line has the equation $y = -4.2(x - 6.5) + 10.3$ and passes through $(6.8, 9.04)$, so we may estimate $f(6.8) \approx 9.04$. Here, we approximate using the secant line rather than the tangent line. )
    As we can see, the two estimates are fairly close.

20. Since 1986 is 6 years after 1980, the rate of growth in 1986 is the derivative of $P(t)$ at $t = 6$. To estimate $P'(t)$ at $t = 6$, we will take the interval between $t = 6$ and $t = 6.001$.
    $$P'(6) \approx \dfrac{P(6.001) - P(6)}{6.001 - 6} = \dfrac{68.4(1.026)^{6.001} - 68.4(1.026)^6}{0.001}$$
    $$P'(6) \approx \dfrac{79.79054 - 79.78849}{0.001} = \dfrac{0.00205}{0.001} = 2.05 \text{ millions of people/year}$$
    or
    $$P'(6) \approx 2,050,000 \text{ people per year}.$$

21. (a) The marginal cost at $q = 400$ is the slope of the tangent line to $C(q)$ at $q = 400$. Looking at the graph, we can estimate a slope of about 1. Thus, the marginal cost is \$1.
    (b) At $q = 500$, we can see that slope of the cost function is greater than the slope of the revenue function. Thus, the marginal cost is greater than the marginal revenue and thus the 500th item will incur a loss. So, the company should not produce the 500th item.
    (c) The quantity which maximizes profit is at the point where marginal costs are just about to exceed marginal revenue. This occurs when the slope of $R(q)$ equals $C(q)$, which occurs at $q = 400$. Thus, the company should produce 400 items.

22. (a) Since the traffic flow is the number of cars per hour, it is the slope of the graph of $C(t)$. It is greatest where the graph of the function $C(t)$ is the steepest and increasing. This happens at approximately $t = 3$ hours, or 7am.
    (b) By reading the values of $C(t)$ from the graph we see:
    $$\dfrac{\text{Average rate of change}}{\text{on } 1 \leq t \leq 2} = \dfrac{C(2) - C(1)}{2 - 1} = \dfrac{1000 - 400}{1} = \dfrac{600}{1} = 600 \text{ cars/hour,}$$
    $$\dfrac{\text{Average rate of change}}{\text{on } 2 \leq t \leq 3} = \dfrac{C(3) - C(2)}{3 - 2} = \dfrac{2000 - 1000}{1} = \dfrac{1000}{1} = 1000 \text{ cars/hour.}$$

A good estimate of $C'(2)$ is the average of the last two results. Therefore:

$$C'(2) \approx \frac{(600 + 1000)}{2} = \frac{1600}{2} = 800 \text{ cars/hour}.$$

(c) Since $t = 2$ is 6 am, the fact that $C'(2) \approx 800$ cars/hour means that the traffic flow at 6 am is about 800 cars/hour.

**23.** (a) The graph of $\pi(x)$ is shown in Figure 2.33:

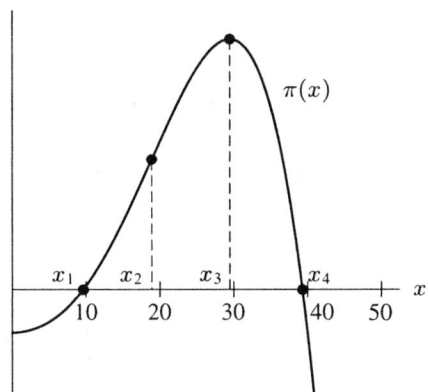
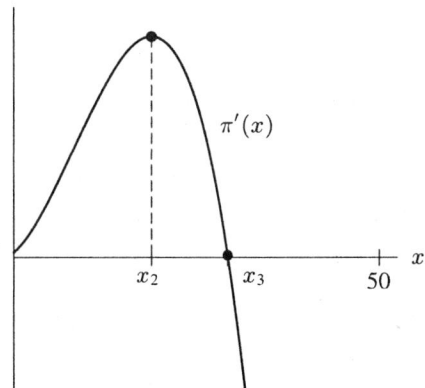

*Figure 2.33*  *Figure 2.34*

(b) The company will make a profit if $x_1 < x < x_4$. One can read the values $x_1$ and $x_4$ from the graph of $\pi(x)$:

$$x_1 \approx 9.6 \quad \text{and} \quad x_4 \approx 39.3.$$

(c) The function $\pi$ increases when $x$ is smaller than $x_3 \approx 29.4$ and decreases when $x > x_3$.

(d) The function $\pi'$ is increasing for values of $x$ smaller than $x_2 \approx 18.9$ and decreasing for $x > x_2$.

(e) The graph of $\pi'(x)$ in Figure 2.34 confirms the result of (d): $\pi'$ increases for $0 < x < x_2$ and decreases afterwards.

(f) The maximum rate of change of $\pi(x)$ corresponds to the maximum of the derivative $\pi'(x)$; therefore it is attained at $x_2 \approx 18.9$.

**24.** (a)

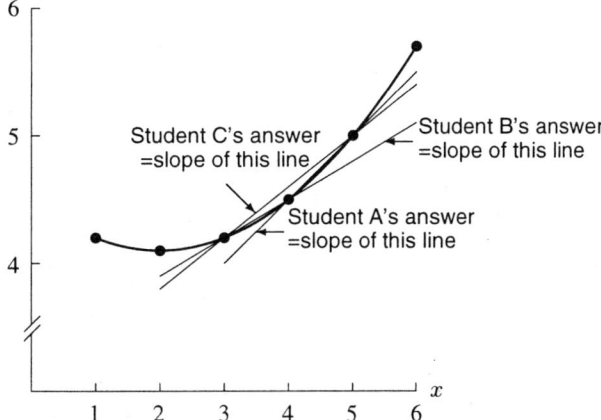

(b) The slope of $f$ appears to be somewhere between student A's answer and student B's, so student C's answer, halfway in between, is probably the most accurate.

**25.** (a) Slope of tangent line $\approx \frac{\sqrt{4.001} - \sqrt{4}}{0.001} = 0.249984$. Hence the slope of the tangent line is about 0.25.

(b) The tangent line has slope 0.25 and contains the point $(4, 2)$, so its equation is

$$y - 2 = 0.25(x - 4)$$
$$y - 2 = 0.25x - 1$$
$$y = 0.25x + 1$$

(c) We have $f(x) = kx^2$. If $(4, 2)$ is on the graph of $f$, then $f(4) = 2$, so $k \cdot 4^2 = 2$. Thus $k = \frac{1}{8}$, and $f(x) = \frac{1}{8}x^2$.

(d) To find where the graph of $f$ crosses the line $y = 0.25x + 1$, we solve:

$$\begin{aligned} \frac{1}{8}x^2 &= 0.25x + 1 \\ x^2 &= 2x + 8 \\ x^2 - 2x - 8 &= 0 \\ (x - 4)(x + 2) &= 0 \\ x &= 4 \text{ or } x = -2 \\ f(-2) &= \frac{1}{8}(4) = 0.5 \end{aligned}$$

Therefore, $(-2, 0.5)$ is the other point of intersection. (Of course, $(4, 2)$ is a point of intersection; we know that from the start.)

**26.** (a) The slope of the tangent line at $(0, 5)$ is zero: it is horizontal.
The slope of the tangent line at $(5, 0)$ is undefined: it is vertical.

(b) The slope appears to be about $\frac{3}{4}$.

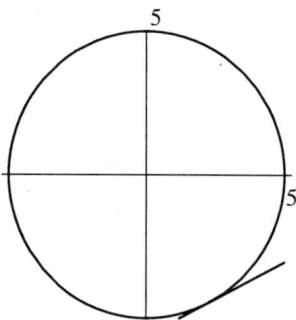

(c) Using symmetry we can determine: Slope at $(-3, 4)$: about $\frac{3}{4}$. Slope at $(-3, -4)$: about $-\frac{3}{4}$. Slope at $(3, 4)$: about $-\frac{3}{4}$.

**27.** (a) See (b).

(b)

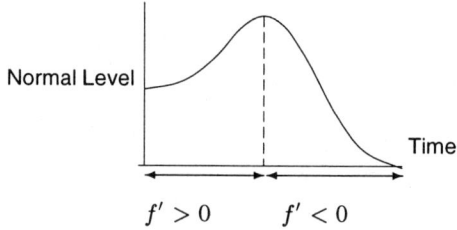

Concentration of Enzymes

(c) $f'$ is the rate at which the concentration is increasing or decreasing. $f'$ is positive at the start of the disease and negative toward the end. In practice, of course, one cannot measure $f'$ directly. Checking the value of $C$ in blood samples taken on consecutive days would tell us

$$f(t + 1) - f(t) = \frac{f(t + 1) - f(t)}{(t + 1) - t},$$

which is our estimate of $f'(t)$.

28. (a)

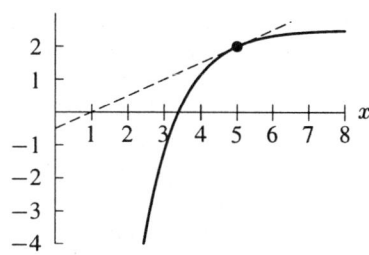

(b) Exactly one. There can't be more than one zero because $f$ is increasing everywhere. There does have to be one zero because $f$ stays below its tangent line (dotted line in above graph), and therefore $f$ must cross the $x$-axis.

(c) The equation of the (dotted) tangent line is $y = \frac{1}{2}x - \frac{1}{2}$, and so it crosses the $x$-axis at $x = 1$. Therefore the zero of $f$ must be between $x = 1$ and $x = 5$.

(d) As the values of $x$ become more and more negative the values of the function become more and more negative, because $f$ is increasing and concave down. Thus, as $x$ decreases, $f(x)$ decreases, at a faster and faster rate.

(e) Yes.

(f) No. The slope is decreasing since $f$ is concave down, so $f'(1) > f'(5)$, i.e. $f'(1) > \frac{1}{2}$.

29. (a) The statement $f(15) = 200$ tells us that when the price is \$15, we sell about 200 units of the product.

(b) The statement $f'(15) = -25$ tells us that if we increase the price by \$1 (from 15), we will sell about 25 fewer units of the product.

30. (a) The graph looks straight because the graph shows only a small part of the curve magnified greatly.

(b) The month is March: We see that about the 21$^{st}$ of the month there are twelve hours of daylight and hence twelve hours of night. This phenomenon (the length of the day equaling the length of the night) occurs at the equinox, midway between winter and summer. Since the length of the days is increasing, and Madrid is in the northern hemisphere, we are looking at March, not September.

(c) The slope of the curve is found from the graph to be about 0.04 (the rise is about 0.8 hours in 20 days or 0.04 hours/day). This means that the amount of daylight is increasing by about 0.04 hours (about $2\frac{1}{2}$ minutes) per calendar day, or that each day is $2\frac{1}{2}$ minutes longer than its predecessor.

## Solutions to the Projects

1. (a) A possible graph is shown below. At first, the yam heats up very quickly, since the difference in temperature between it and its surroundings is so large. As time goes by, the yam gets hotter and hotter, its rate of temperature increase slows down, and its temperature approaches the temperature of the oven as an asymptote. The graph is thus concave down. (We are considering the average temperature of the yam, since the temperature in its center and on its surface will vary in different ways.)

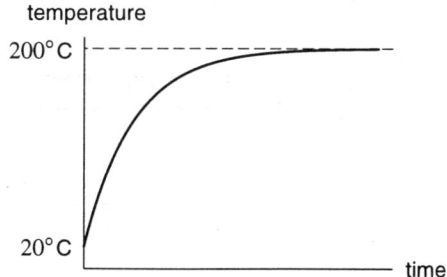

(b) If the rate of temperature increase were to remain 2°/min, in ten minutes the yam's temperature would increase 20°, from 120° to 140°. Since we know the graph is not linear, but concave down, the actual temperature is between 120° and 140°.

(c) In 30 minutes, we know the yam increases in temperature by 45° at an average rate of $45/30 = 1.5°$/min. Since the graph is concave down, the temperature at $t = 40$ is therefore between $120 + 1.5(10) = 135°$ and 140°.

(d) If the temperature increases at 2°/minute, it reaches 150° after 15 minutes, at $t = 45$. If the temperature increases at 1.5°/minute, it reaches 150° after 20 minutes, at $t = 50$. So $t$ is between 45 and 50 mins.

2. (a) Since illumination is concave up and temperature is concave down, the graph on the left side corresponds to the graph of illumination as a function of distance and the graph on the right side corresponds to the graph of temperature as a function of distance.

   (b) The illumination drops from 75% at $d = 2$ to 56% at $d = 5$. Since $T = 47$ when $d = 5$ and $T = 53.5$ when $d = 2$,

   $$\text{Average rate of change of temperature} = \frac{47 - 53.5}{5 - 2} = \frac{-6.5}{3} = -2.17° \text{ per foot.}$$

   (c) A good estimate of the illumination when the distance is 3.5 feet is the average of the values of the illumination at 3 feet and at 4 feet. Therefore:

   $$\text{Illumination at 3.5 feet} = \frac{67 + 60}{2} = 63.5\% < 65\%.$$

   Since illumination is concave up, the 63.5% is likely to be an overestimate, so you are not likely to be able to read the watch.

   (d) Let's represent the illumination as a function of the distance by $I(d)$ and the temperature as a function of the distance by $T(d)$. Therefore:

   $$I(7) = I(6) + \text{change in } I(d)$$
   $$\approx I(6) + \text{Average rate of change of } I(d)$$
   $$= 53\% + (-3\%) = 50\%.$$

   And

   $$T(7) = T(6) + \text{change in } T(d)$$
   $$\approx T(6) + \text{Average rate of change of } T(d)$$
   $$= 43.5°F + (-4.5°F) = 39°F.$$

   (e) Let's calculate the distance when $T(d) = 40$ (when we are cold):

   $$T(d) = T(6) + T'(6) \cdot (d - 6)$$
   $$40 = 43.5 + (-4.5)(d - 6)$$
   $$40 = 43.5 - 4.5d + 27$$
   $$40 = 70.5 - 4.5d$$
   $$-30.5 = -4.5d$$
   $$d = \frac{-30.5}{-4.5} = 6.78 \text{ feet.}$$

   From part (d) we know that the illumination is 50% (darkness) when the distance is 7 feet. Therefore, as we walk away from the candle, we first get cold and then we are in darkness.

## Solutions to Limits and the Definition of the Derivative

1. The answers are marked in Figure 2.35.

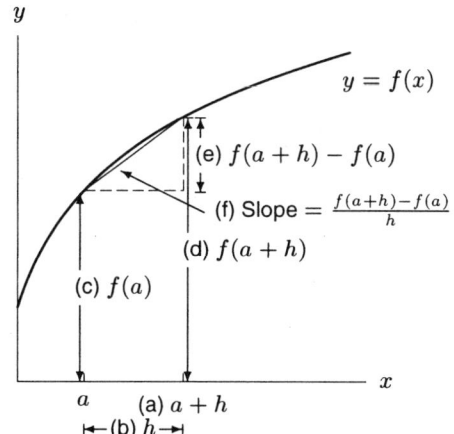

**Figure 2.35**

2. The answers are marked in Figure 2.36.

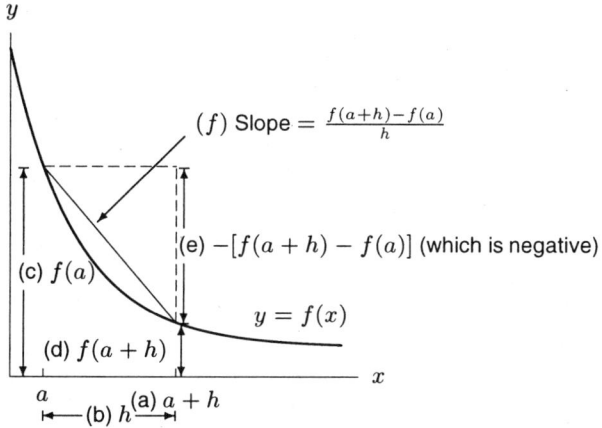

**Figure 2.36**

3. Figure 2.37 shows that as $x$ approaches 0 from either side, the value of $\dfrac{\sin x}{x}$ appears to approach 1, suggesting that $\lim\limits_{x \to 0} \dfrac{\sin x}{x} = 1$. Zooming in on the graph near $x = 0$ provides further support for this conclusion. Notice that $\dfrac{\sin x}{x}$ is undefined at $x = 0$.

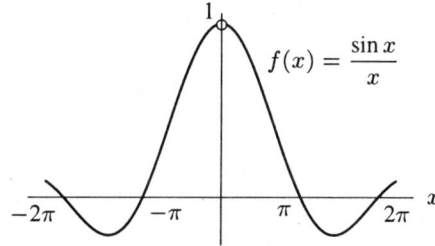

**Figure 2.37**

4. Figure 2.38 shows that as $x$ approaches 0 from either side, the values of $\dfrac{5^x - 1}{x}$ appear to approach 1.6, suggesting that

$$\lim_{x \to 0} \frac{5^x - 1}{x} \approx 1.6.$$

Zooming in on the graph near $x = 0$ provides further support for this conclusion. Notice that $\dfrac{5^x - 1}{x}$ is undefined at $x = 0$.

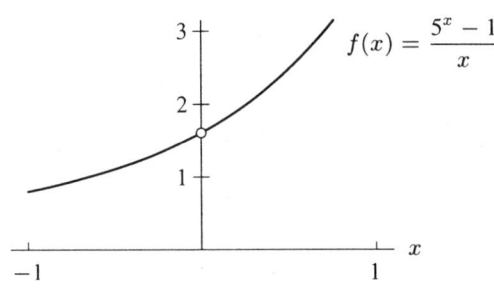

**Figure 2.38**

5. Using $h = 0.1, 0.01, 0.001$, we see

$$\frac{(3 + 0.1)^3 - 27}{0.1} = 27.91$$

$$\frac{(3 + 0.01)^3 - 27}{0.01} = 27.09$$

$$\frac{(3 + 0.001)^3 - 27}{0.001} = 27.009.$$

These calculations suggest that $\lim\limits_{h \to 0} \dfrac{(3 + h)^3 - 27}{h} = 27.$

6. Using $h = 0.1, 0.01, 0.001$, we see

$$\frac{7^{0.1} - 1}{0.1} = 2.148$$

$$\frac{7^{0.01} - 1}{0.01} = 1.965$$

$$\frac{7^{0.001} - 1}{0.001} = 1.948$$

$$\frac{7^{0.0001} - 1}{0.0001} = 1.946.$$

This suggests that $\lim\limits_{h \to 0} \dfrac{7^h - 1}{h} \approx 1.9\ldots$

7. Using $h = 0.1, 0.01, 0.001$, we see

| $h$ | $(e^{1+h} - e)/h$ |
|---|---|
| 0.01 | 2.7319 |
| 0.001 | 2.7196 |
| 0.0001 | 2.7184 |

These values suggest that $\lim\limits_{h \to 0} \dfrac{e^{1+h} - e}{h} = 2.7\ldots$. In fact, this limit is $e$.

## SOLUTIONS TO LIMITS AND THE DEFINITION OF THE DERIVATIVE

8. Using radians,

| $h$ | $(\cos h - 1)/h$ |
|---|---|
| 0.01 | $-0.005$ |
| 0.001 | $-0.0005$ |
| 0.0001 | $-0.00005$ |

These values suggest that $\lim_{h \to 0} \dfrac{\cos h - 1}{h} = 0$.

9. Yes, $f(x)$ is continuous on $0 \leq x \leq 2$.
10. No, $f(x)$ is not continuous on $0 \leq x \leq 2$, but it is continuous on the interval $0 \leq x \leq 0.5$.
11. No, $f(x)$ is not continuous on $0 \leq x \leq 2$, but it is continuous on $0 \leq x \leq 0.5$.
12. No, $f(x)$ is not continuous on $0 \leq x \leq 2$, but it is continuous on $0 \leq x \leq 0.5$.
13. Yes: $f(x) = x + 2$ is continuous for all values of $x$.
14. Yes: $f(x) = 2^x$ is continuous function for all values of $x$.
15. Yes: $f(x) = x^2 + 2$ is a continuous function for all values of $x$.
16. Yes: $f(x) = 1/(x-1)$ is a continuous function on any interval that does not contain $x = 1$.
17. No: $f(x) = 1/(x-1)$ is not continuous on any interval containing $x = 1$.
18. Yes: $f(x) = 1/(x^2 + 1)$ is continuous for all values of $x$ because the denominator is never 0.
19. This function is not continuous. Each time someone is born or dies, the number jumps by one.
20. Continuous
21. Since we can't make a fraction of a pair of pants, the number increases in jumps, so the function is not continuous.
22. Even though the car is stopping and starting, the distance traveled is a continuous function of time.
23. The time is not a continuous function of position as distance from your starting point, because every time you cross from one time zone into the next, the time jumps by 1 hour.
24. Using the definition of the derivative, we have

$$f'(x) = \lim_{h \to 0} \frac{f(x+h) - f(x)}{h}$$
$$= \lim_{h \to 0} \frac{5(x+h) - 5x}{h}$$
$$= \lim_{h \to 0} \frac{5x + 5h - 5x}{h}$$
$$= \lim_{h \to 0} \frac{5h}{h}.$$

As long as we let $h$ get close to zero without actually equaling zero, we can cancel the $h$ in the numerator and denominator, and we are left with $f'(x) = 5$.

25. Using the definition of the derivative, we have

$$f'(x) = \lim_{h \to 0} \frac{f(x+h) - f(x)}{h} = \lim_{h \to 0} \frac{3(x+h)^2 - 3x^2}{h}$$
$$= \lim_{h \to 0} \frac{3(x^2 + 2xh + h^2) - 3x^2}{h}$$
$$= \lim_{h \to 0} \frac{3x^2 + 6xh + 3h^2 - 3x^2}{h}$$
$$= \lim_{h \to 0} \frac{6xh + 3h^2}{h} = \lim_{h \to 0} \frac{h(6x + 3h)}{h}.$$

As $h$ gets very close to zero (but not equal to zero), we can cancel the $h$ in the numerator and denominator to leave the following:

$$f'(x) = \lim_{h \to 0} (6x + 3h).$$

As $h \to 0$, we have $f'(x) = 6x$.

# CHAPTER THREE

## Solutions for Section 3.1

1. (a) The velocity is 30 miles/hour for the first 2 hours, 40 miles/hour for the next 1/2 hour, and 20 miles/hour for the last 4 hours. The entire trip lasts $2 + 1/2 + 4 = 6.5$ hours, so we need a scale on our horizontal (time) axis running from 0 to 6.5. Between $t = 0$ and $t = 2$, the velocity is constant at 30 miles/hour, so the velocity graph is a horizontal line at 30. Likewise, between $t = 2$ and $t = 2.5$, the velocity graph is a horizontal line at 40, and between $t = 2.5$ and $t = 6.5$, the velocity graph is a horizontal line at 20. The graph is shown in Figure 3.1.

   (b) How can we visualize distance traveled on the velocity graph given in Figure 3.1? The velocity graph looks like the top edges of three rectangles. The distance traveled on the first leg of the journey is (30 miles/hour)(2 hours), which is the height times the width of the first rectangle in the velocity graph. The distance traveled on the first leg of the trip is equal to the area of the first rectangle. Likewise, the distances traveled during the second and third legs of the trip are equal to the areas of the second and third rectangles in the velocity graph. It appears that distance traveled is equal to the area under the velocity graph.

   In Figure 3.2, the area under the velocity graph in Figure 3.1 is shaded. Since this area is three rectangles and the area of each rectangle is given by Height × Width, we have

   $$\text{Total area} = (30)(2) + (40)(1/2) + (20)(4)$$
   $$= 60 + 20 + 80 = 160.$$

   The area under the velocity graph is equal to distance traveled.

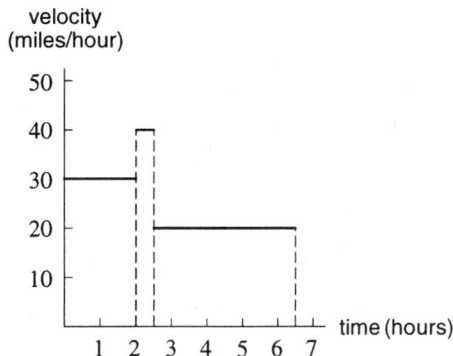

**Figure 3.1**: Velocity graph

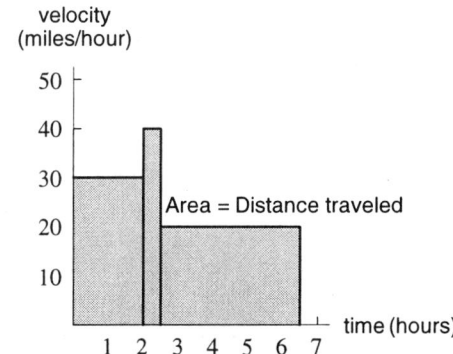

**Figure 3.2**: The area under the velocity graph gives distance traveled

2. Using the data in Table 3.3 of Example 3, we construct Figure 3.3.

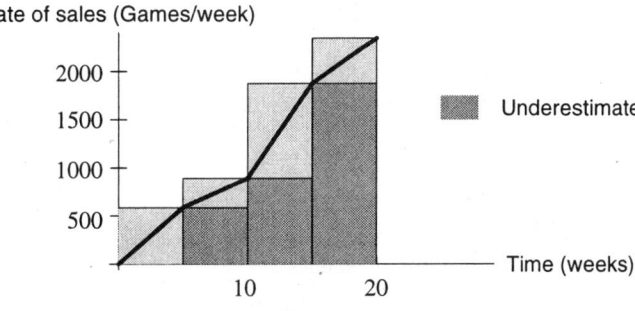

**Figure 3.3**

**140** CHAPTER THREE /SOLUTIONS

3. We use Distance = Rate × Time on each subinterval with $\Delta t = 3$.

$$\text{Underestimate} = 0 \cdot 3 + 10 \cdot 3 + 25 \cdot 3 + 45 \cdot 3 = 240,$$
$$\text{Overestimate} = 10 \cdot 3 + 25 \cdot 3 + 45 \cdot 3 + 75 \cdot 3 = 465.$$

We know that
$$240 \leq \text{Distance traveled} \leq 465.$$

A better estimate is the average. We have
$$\text{Distance traveled} \approx \frac{240 + 465}{2} = 352.5.$$

The car travels about 352.5 feet during these 12 seconds.

4. (a) Lower estimate $= (45)(2) + (16)(2) + (0)(2) = 122$ feet.
   Upper estimate $= (88)(2) + (45)(2) + (16)(2) = 298$ feet.
   (b)

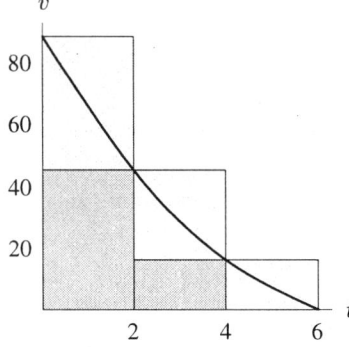

5. (a) Note that 15 minutes equals 0.25 hours. Lower estimate $= 11(0.25) + 10(0.25) = 5.25$ miles. Upper estimate $= 12(0.25) + 11(0.25) = 5.75$ miles.
   (b) Lower estimate $= 11(0.25) + 10(0.25) + 10(0.25) + 8(0.25) + 7(0.25) + 0(0.25) = 11.5$ miles. Upper estimate $= 12(0.25) + 11(0.25) + 10(0.25) + 10(0.25) + 8(0.25) + 7(0.25) = 14.5$ miles.

6. (a) An overestimate is 7 tons. An underestimate is 5 tons.
   (b) An overestimate is $7 + 8 + 10 + 13 + 16 + 20 = 74$ tons. An underestimate is $5 + 7 + 8 + 10 + 13 + 16 = 59$ tons.

7. (a) Based on the data, we will calculate the underestimate and the overestimate of the total change. A good estimate will be the average of both results.
   Underestimate of total change
   $$= 37 \cdot 10 + 41 \cdot 10 + 77 \cdot 10 + 77 \cdot 10 = 2320.$$
   77 was considered twice since we needed to calculate the area under the graph.
   Overestimate of total change
   $$= 41 \cdot 10 + 78 \cdot 10 + 78 \cdot 10 + 86 \cdot 10 = 2830.$$
   78 was considered twice since we needed to calculate the area over the graph.
   The average is: $(2320 + 2830)/2 = 2575$ million people.
   (b) The total change in the world's population between 1950 and 1990 is given by the difference between the populations in those two years. That is, the change in population equals

   5295 (population in 1990) − 2555 (population in 1950) = 2740 million people.

   Our estimate of 2575 million people and the actual difference of 2740 million people are on the same order of magnitude. Therefore, our estimate was a good one.

8. Sketch the graph of $v(t)$. See Figure 3.4. Adding up the areas using an overestimate with data every 1 second, we get $s \approx 2 + 5 + 10 + 17 + 26 = 60$ m. The actual distance traveled is less than 60 m.

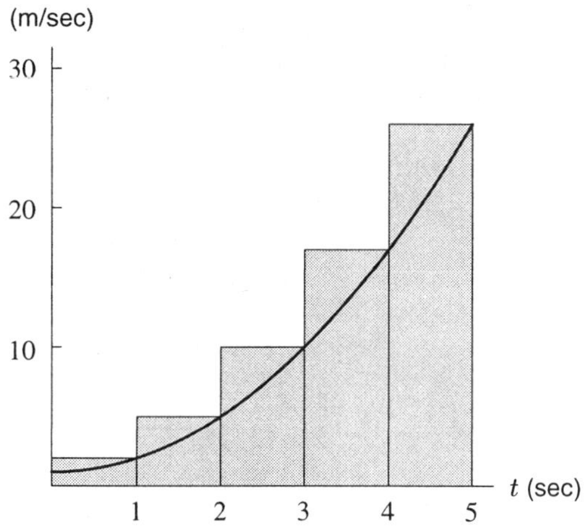

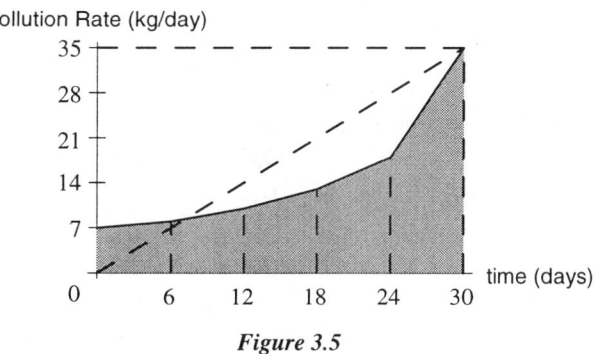

**Figure 3.4**

9. Just counting the squares (each of which has area 10), and allowing for the broken squares, we can see that the area under the curve from 0 to 6 is between 140 and 150. Hence the distance traveled is between 140 and 150 meters.

10. (a) Using rectangles under the curve, we get

$$\text{Acres defaced} \approx (1)(0.2 + 0.4 + 1 + 2) = 3.6 \text{ acres}.$$

   (b) Using rectangles above the curve, we get

$$\text{Acres defaced} \approx (1)(0.4 + 1 + 2 + 3.5) = 6.9 \text{ acres}.$$

   (c) It's between 3.6 and 6.9, so we'll guess the average: 5.25 acres.

11. (a) Let's begin by graphing the data given in the table; see Figure 3.5. The total amount of pollution entering the lake during the 30-day period is equal to the shaded area. The shaded area is roughly 40% of the rectangle measuring 30 units by 35 units. Therefore, the shaded area measures about $(0.40)(30)(35) = 420$ units. Since the units are kilograms, we estimate that 420 kg of pollution have entered the lake.

**Figure 3.5**

   (b) Using left and right sums, we have

$$\text{Underestimate} = (7)(6) + (8)(6) + (10)(6) + (13)(6) + (18)(6) = 336 \text{ kg}.$$
$$\text{Overestimate} = (8)(6) + (10)(6) + (13)(6) + (18)(6) + (35)(6) = 504 \text{ kg}.$$

## Solutions for Section 3.2

1. (a) If $\Delta t = 4$, then $n = 2$. We have:

   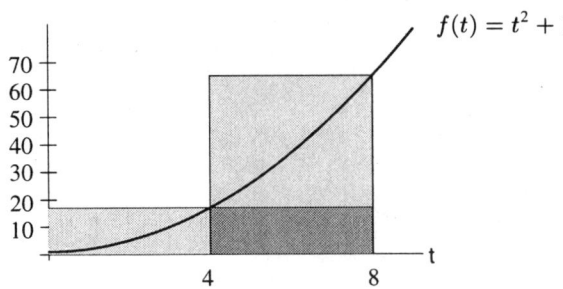

   Figure 3.6

   Underestimate of total change $= f(0)\Delta t + f(4)\Delta t = (1)(4) + (17)(4) = 4 + 68 = 72.$
   Overestimate of total change $= f(4)\Delta t + f(8)\Delta t = (17)(4) + (65)(4) = 68 + 260 = 328.$

   (b) If $\Delta t = 2$, then $n = 4$. We have:

   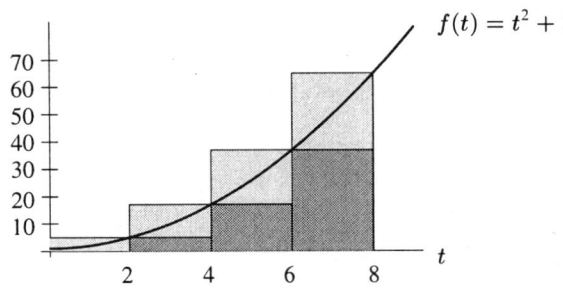

   Figure 3.7

   Underestimate of total change $= f(0)\Delta t + f(2)\Delta t + f(4)\Delta t + f(6)\Delta t$
   $= (1)(2) + (5)(2) + (17)(2) + (37)(2) = 120.$
   Overestimate of total change $= f(2)\Delta t + f(4)\Delta t + f(6)\Delta t + f(8)\Delta t$
   $= (5)(2) + (17)(2) + (37)(2) + (65)(2) = 248.$

   (c) If $\Delta t = 1$, then $n = 8$.

   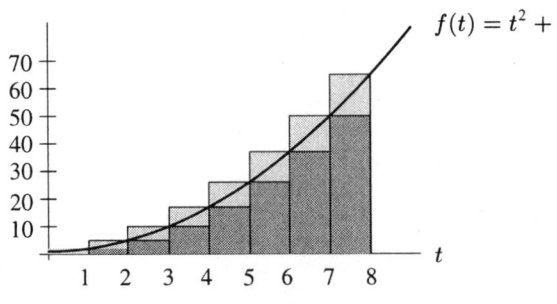

   Figure 3.8

Underestimate of total change

$$= f(0)\Delta t + f(1)\Delta t + f(2)\Delta t + f(3)\Delta t + f(4)\Delta t + f(5)\Delta t + f(6)\Delta t + f(7)\Delta t$$
$$= (1)(1) + (2)(1) + (5)(1) + (10)(1) + (17)(1) + (26)(1) + (37)(1) + (50)(1) = 148$$

Overestimate of total change

$$= f(1)\Delta t + f(2)\Delta t + f(3)\Delta t + f(4)\Delta t + f(5)\Delta t + f(6)\Delta t + f(7)\Delta t + f(8)\Delta t$$
$$= (2)(1) + (5)(1) + (10)(1) + (17)(1) + (26)(1) + (37)(1) + (50)(1) + (65)(1) = 212$$

2. When $n = 6$, the width of each rectangle is 1. We obtain the following estimates:

$$\text{Left-hand estimate} = (21)1 + (15.5)1 + (11)1 + (7.5)1 + (5)1 + (3.5)1 = 63.5,$$
and $\quad$ $\text{Right-hand estimate} = (15.5)1 + (11)1 + (7.5)1 + (5)1 + (3.5)1 + (3)1 = 45.5.$

The true value of the integral is between 45.5 and 63.5. A better estimate is the average:

$$\int_0^6 f(x)\,dx \approx \frac{45.5 + 63.5}{2} = 54.5.$$

3.

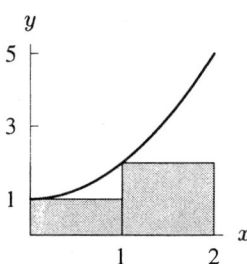

Figure 3.9: Left-hand sum with $n = 2$

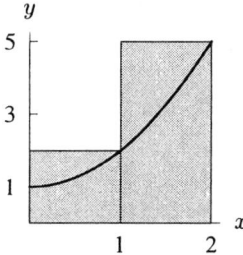

Figure 3.10: Right-hand sum with $n = 2$

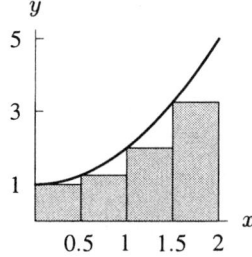

Figure 3.11: Left-hand sum with $n = 4$

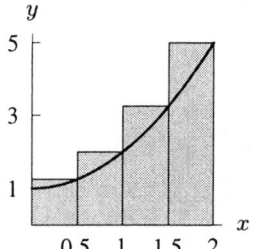

Figure 3.12: Right-hand sum with $n = 4$

For $n = 2$, the left-hand sum is equal to

$$f(0) \cdot 1 + f(1) \cdot 1 = 1 \cdot 1 + 2 \cdot 1$$
$$= 1 + 2 = 3$$

For $n = 2$, the right-hand sum is equal to

$$f(1) \cdot 1 + f(2) \cdot 1 = 2 \cdot 1 + 5 \cdot 1$$
$$= 2 + 5 = 7$$

For $n = 4$, the left-hand sum is equal to

$$f(0) \cdot \frac{1}{2} + f\left(\frac{1}{2}\right) \cdot \frac{1}{2} + f(1)\frac{1}{2} + f(\frac{3}{2})\frac{1}{2} = 1 \cdot \frac{1}{2} + \frac{5}{4} \cdot \frac{1}{2} + 2 \cdot \frac{1}{2} + \frac{13}{4} \cdot \frac{1}{2}$$
$$= \frac{1}{2} + \frac{5}{8} + 1 + \frac{13}{8}$$
$$= \frac{15}{4} = 3.75$$

For $n = 4$, the right-hand sum is equal to

$$f\left(\frac{1}{2}\right) \cdot \frac{1}{2} + f(1) \cdot \frac{1}{2} + f(\frac{3}{2})\frac{1}{2} + f(2)\frac{1}{2} = \frac{5}{4} \cdot \frac{1}{2} + 2 \cdot \frac{1}{2} + \frac{13}{4} \cdot \frac{1}{2} + 5 \cdot \frac{1}{2}$$
$$= \frac{5}{8} + 1 + \frac{13}{8} + \frac{5}{2}$$
$$= \frac{23}{4} = 5.75$$

In both the $n = 2$ and $n = 4$ cases, the left-hand sums are underestimates and the right-hand sums are overestimates.

4.

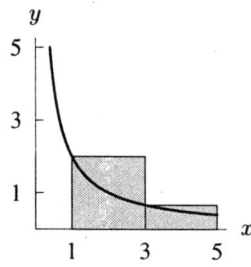

**Figure 3.13:** Left-hand sum with $n = 2$

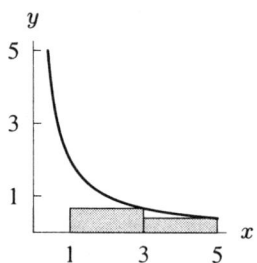

**Figure 3.14:** Right-hand sum with $n = 2$

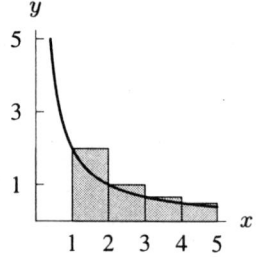

**Figure 3.15:** Left-hand sum with $n = 4$

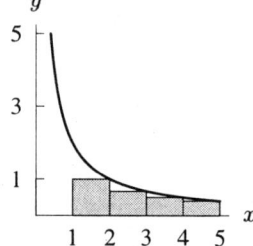

**Figure 3.16:** Right-hand sum with $n = 4$

For $n = 2$, the left-hand sum is equal to

$$f(1) \cdot 2 + f(3) \cdot 2 = 2 \cdot 2 + \frac{2}{3} \cdot 2$$
$$= 5.333$$

For $n = 2$, the right-hand sum is equal to

$$f(3) \cdot 2 + f(5) \cdot 2 = \frac{2}{3} \cdot 2 + 0.4 \cdot 2$$
$$= 2.133$$

For $n = 4$, the left-hand sum is equal to

$$f(1) \cdot 1 + f(2) \cdot 1 + f(3) \cdot 1 + f(4) \cdot 1 = 2 \cdot 1 + 1 \cdot 1 + 0.667 \cdot 1 + 0.5 \cdot 1$$
$$= 4.167$$

For $n = 4$, the right-hand sum is equal to

$$f(2) \cdot 1 + f(3) \cdot 1 + f(4) \cdot 1 + f(5) \cdot 1 = 1 \cdot 1 + 0.667 \cdot 1 + 0.5 \cdot 1 + 0.4 \cdot 1$$
$$= 2.567$$

In both the $n = 2$ and $n = 4$ cases, the left-hand sums are overestimates and the right-hand sums are underestimates.

5. (a) See Figure 3.17. $\int_0^1 x^3\, dx =$ area shaded, which is less than 0.5. Rough estimate is about 0.3.

  (b) $\int_0^1 x^3\, dx = 0.25$

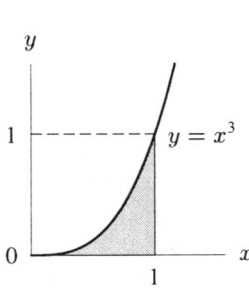

Figure 3.17

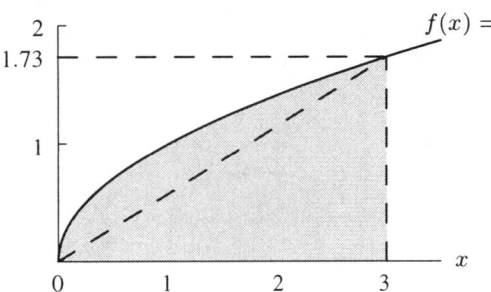

Figure 3.18

6. (a) See Figure 3.18. A rough estimate of the shaded area is 70% of the area of the rectangle 3 by 1.73. Thus,

$$\int_0^3 \sqrt{x}\, dx \approx 70\% \text{ of } (3)(1.73) = (0.70)(3)(1.73) = 3.63.$$

  (b) $\int_0^3 \sqrt{x}\, dx = 3.4641$

7. (a) See Figure 3.19.

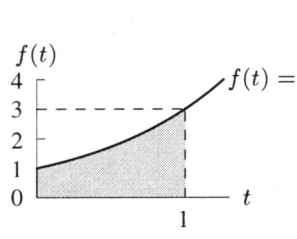

Figure 3.19

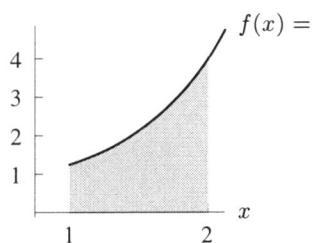

Figure 3.20

The shaded area is approximately 60% of the area of the rectangle 1 unit by 3 units. Therefore,

$$\int_0^1 3^t\, dt \approx (0.60)(1)(3) = 1.8.$$

  (b) $\int_0^1 3^t\, dt = 1.8205.$

8. (a) See Figure 3.20. The shaded area appears to be approximately 2 units, and so $\int_1^2 x^x\, dx \approx 2$.

  (b) $\int_1^2 x^x\, dx = 2.05045$

9. $\int_0^5 x^2 \, dx = 41.7$

10. $\int_1^4 \frac{1}{\sqrt{1+x^2}} \, dx = 1.2$

11. $\int_1^5 (3x+1)^2 \, dx = 448.0$

12. $\int_{1.1}^{1.7} 10(0.85)^t \, dt = 4.8$

13. $\int_1^2 2^x \, dx = 2.9$

14. $\int_1^2 (1.03)^t \, dt = 1.0$

15. The integral $\int_1^3 \ln x \, dx \approx 1.30$

16. $\int_{1.1}^{1.7} e^t \ln t \, dt = 0.865$

17. $\int_{-3}^3 e^{-t^2} \, dt \approx 2 \int_0^3 e^{-t^2} \, dt = 2(0.886) = 1.772$

18. In the left-hand sum the height of each rectangle is the value of the function at the left endpoint of the interval. See Figure 3.21. In the right-hand sum the height of each rectangle is the value of the function at the right endpoint of the interval. See Figure 3.22.

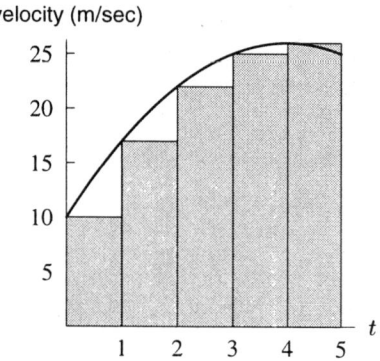

**Figure 3.21:** Left-hand sums for $v(t) = 10 + 8t - t^2$

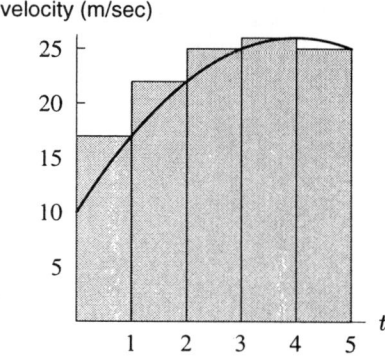

**Figure 3.22:** Right-hand sums for $v(t) = 10 + 8t - t^2$

The numerical values of the sums are found as follows. If we use $n = 5$, the width of each rectangle is 1 and, using $v(t) = 10 + 8t - t^2$, we compute the following estimates:

Left-hand estimate $= (v(0))(1) + (v(1))(1) + (v(2))(1) + (v(3))(1) + (v(4))(1)$
$= (10)(1) + (17)(1) + (22)(1) + (25)(1) + (26)(1)$
$= 100$ meters.

Right-hand estimate $= (v(1))(1) + (v(2))(1) + (v(3))(1) + (v(4))(1) + (v(5))(1)$
$= (17)(1) + (22)(1) + (25)(1) + (26)(1) + (25)(1)$
$= 115$ meters.

19. Since we are given a table of values, we must use Riemann sums to approximate the integral. Values are given every 0.2 units, so $\Delta t = 0.2$ and $n = 5$. Our best estimate is obtained by calculating the left-hand and right-hand sums, and then averaging the two.

Left-hand sum $= 25(0.2) + 23(0.2) + 20(0.2) + 15(0.2) + 9(0.2)$
$= 18.4$

Right-hand sum $= 23(0.2) + 20(0.2) + 15(0.2) + 9(0.2) + 2(0.2)$
$= 13.8$.

We average the two sums to obtain our best estimate of the integral:

$$\int_3^4 W(t) \, dt \approx \frac{18.4 + 13.8}{2} = 16.1.$$

20.
$$\text{Left-hand sum} = 50(3) + 48(3) + 44(3) + 36(3) + 24(3)$$
$$= 606$$
$$\text{Right-hand sum} = 48(3) + 44(3) + 36(3) + 24(3) + 8(3)$$
$$= 480$$
$$\text{Average} = \frac{606 + 480}{2} = 543$$

So we have $\int_0^{15} f(x)\,dx \approx 543$.

21. We estimate the integral by finding left- and right-hand sums and averaging them:
$$\text{Left-hand sum} = (100)(4) + (88)(4) + (72)(4) + (50)(4) = 1240,$$

and
$$\text{Right-hand sum} = (88)(4) + (72)(4) + (50)(4) + (28)(4) = 952.$$

We have
$$\int_{10}^{26} f(x)\,dx \approx \frac{1240 + 952}{2} = 1096.$$

22. (a) See Figure 3.23.
$$\text{Left Sum} = f(1)\Delta x + f(1.5)\Delta x$$
$$= (\ln 1)0.5 + \ln(1.5)0.5 = (\ln 1.5)(0.5)$$

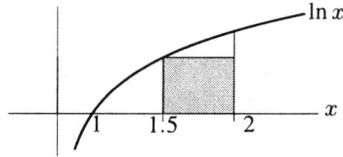
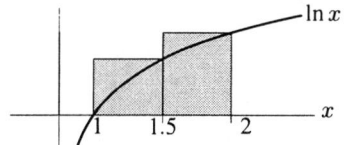

*Figure 3.23:* Left Sum

*Figure 3.24:* Right Sum

(b) See Figure 3.24.
$$\text{Right Sum} = f(1.5)\Delta x + f(2)\Delta x$$
$$= (\ln 1.5)0.5 + (\ln 2)0.5$$

(c) Right sum is an overestimate, left sum is an underestimate.

23.

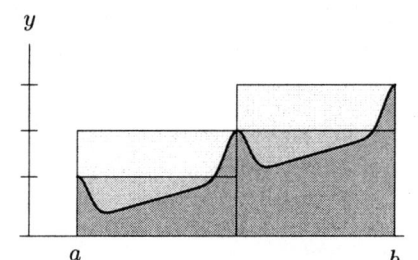

*Figure 3.25:* Integral vs. Left- and Right-Hand Sums

## Solutions for Section 3.3

1. There are 8 whole grid squares and 6 partial grid squares, each of which is about 1/2 a square. The area is about $8(1) + 6\left(\frac{1}{2}\right) = 11.0$

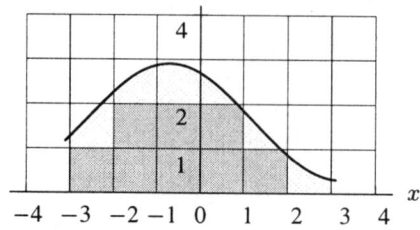

**Figure 3.26**

2. The integral represents the area below the graph of $f(x)$ but above the $x$-axis. Since each square has area 1, by counting squares and half-squares we find
$$\int_1^6 f(x)\,dx = 8.5.$$

3. $\int_0^{20} f(x)\,dx$ is equal to the area shaded. We can use Riemann sums to estimate the area, or we can count boxes in Figure 3.27. There are about 15 boxes and each box represents 4 square units, so the area shaded is about 60. We have $\int_0^{20} f(x)\,dx \approx 60$. See Figure 3.27.

4. To estimate the integral, we count the rectangles under the curve and above the $x$-axis for the interval $[0, 3]$. There are approximately 17 of these rectangles, and each has area 1, so
$$\int_0^3 f(x)\,dx \approx 17.$$

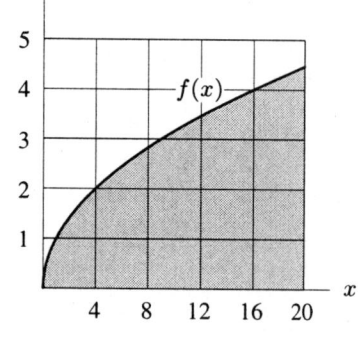

**Figure 3.27**

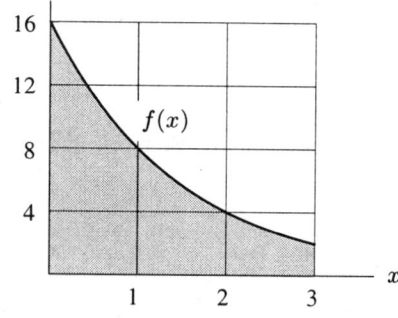

**Figure 3.28**

5. $\int_0^3 f(x)\,dx$ is equal to the area shaded. We can use Riemann sum to estimate this area, or we can count boxes. These are 3 whole boxes and about 4 half-boxes, for a total of 5 boxes. Since each box represent 4 square units, our estimated area is $5(4) = 20$. We have $\int_0^3 f(x)\,dx \approx 20$. See Figure 3.28.

6. The area $= \int_0^2 (x^3 + 2)dx$. Using technology to evaluate the integral, we see that Area $= \int_0^2 (x^3 + 2)dx = 8$. See Figure 3.29.

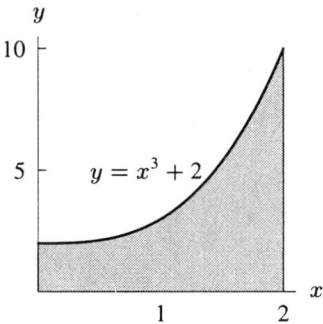

Figure 3.29

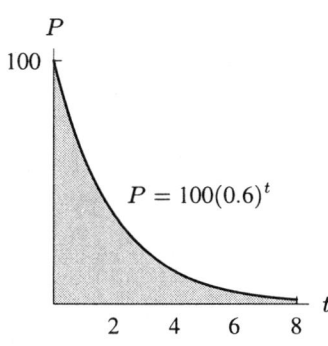

Figure 3.30

7. See Figure 3.30.
$$\text{Area} = \int_0^8 100(0.6)^t \, dt \approx 192.47$$

8. Since $f(x)$ is positive along the interval from 0 to 6 the area is simply $\int_0^6 (x^2 + 2)dx = 84$.

9. (a) Counting the squares yields an estimate of 16.5, each with area $= 1$, so the total shaded area is approximately 16.5.
   (b)
   $$\int_0^8 f(x)dx = \text{(shaded area above } x\text{-axis)} - \text{(shaded area below } x\text{-axis)}$$
   $$\approx 6.5 - 10 = -3.5$$
   (c) The answers in (a) and (b) are different because the shaded area below the $x$-axis is subtracted in order to find the value of the integral in (b).

10. (a) The area shaded between 0 and 1 is about the same as the area shaded between $-1$ and 0, so the area between 0 and 1 is about 0.25. Since this area lies below the $x$-axis, we estimate that
   $$\int_0^1 f(x)\, dx = -0.25.$$
   (b) Between $-1$ and 1, the area above the $x$-axis approximately equals the area below the $x$-axis, and so
   $$\int_{-1}^1 f(x)\, dx = (0.25) + (-0.25) = 0.$$
   (c) We estimate
   $$\text{Total area shaded} = 0.25 + 0.25 = 0.5.$$

11.

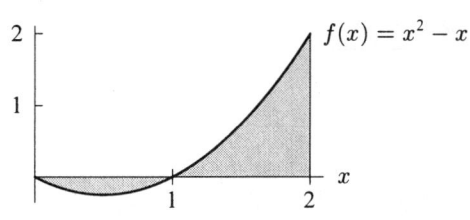

Figure 3.31

More area appears to be above the $x$-axis than below, so the integral is positive.

**150** CHAPTER THREE /SOLUTIONS

12.

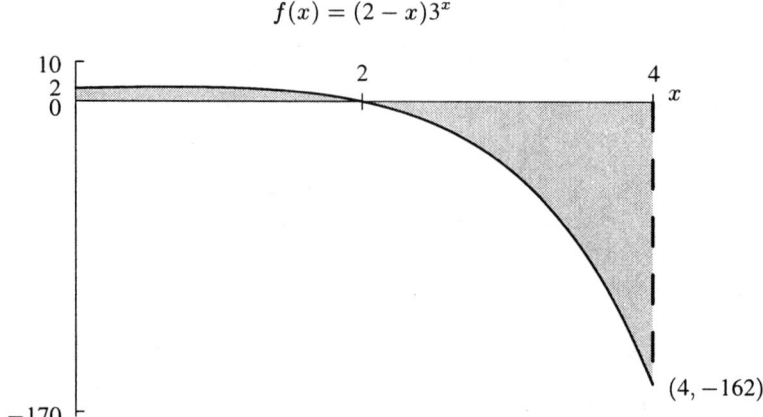

**Figure 3.32:** Figure not drawn to scale; the y-axis is distorted

Looking at the graph, more of the area seems to be below the $x$-axis than above it, so the integral is negative.

13. See Figure 3.33. More area appears to be above the $x$-axis than below it, so the integral is positive.

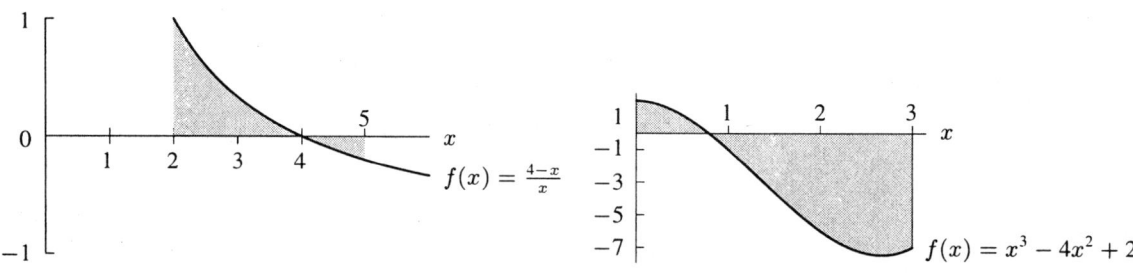

14. See Figure 3.34. More area appears to be below the $x$-axis than above, so the integral is negative.

15. The $x$ intercepts of $y = 4 - x^2$ are $x = -2$ and $x = 2$, and the graph is above the $x$-axis for $-2 \le x \le 2$. Thus, Area $= \int_{-2}^{2} (4 - x^2)\, dx \approx 10.67$. See Figure 3.35.

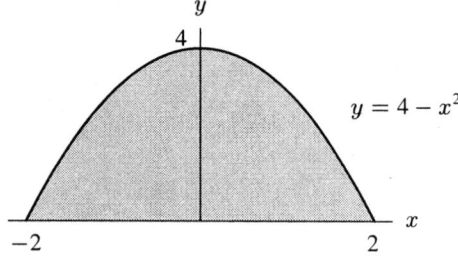

Figure 3.35

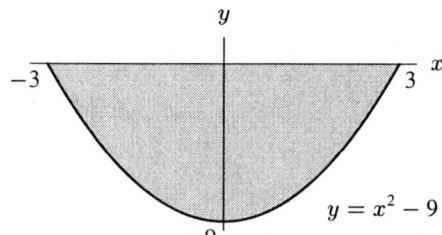

Figure 3.36

16. See Figure 3.36. The $x$ intercepts of $y = x^2 - 9$ are $x = -3$ and $x = 3$, and since the graph is below the $x$ axis for $-3 \le x \le 3$,

$$\text{Area} = -\int_{-3}^{3} (x^2 - 9)\, dx = 36.00.$$

17.

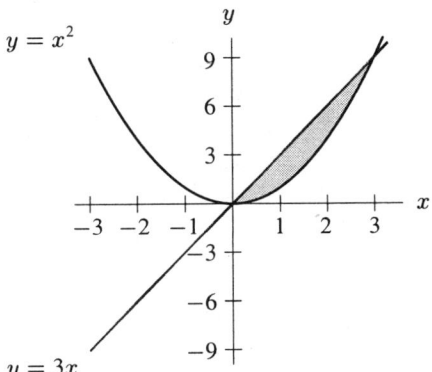

**Figure 3.37**

Inspection of the graph tells us that the curves intersect at $(0, 0)$ and $(3, 9)$, with $3x \geq x^2$ for $0 \leq x \leq 3$, so we can find the area by evaluating the integral

$$\int_0^3 (3x - x^2)dx.$$

Using technology to evaluate the integral, we see

$$\int_0^3 (3x - x^2)dx = 4.5.$$

So the area between the graphs is 4.5.

18. See Figure 3.39. Inspection of the graph tell us that the curves intersect at $(0, 0)$ and $(1, 1)$, with $\sqrt{x} \geq x$ for $0 \leq x \leq 1$, so we can find the area by evaluating the integral

$$\int_0^1 (\sqrt{x} - x)dx.$$

Using technology to evaluate the integral, we see

$$\int_0^1 (\sqrt{x} - x)dx \approx 0.1667.$$

So the area between the graphs is about 0.1667.

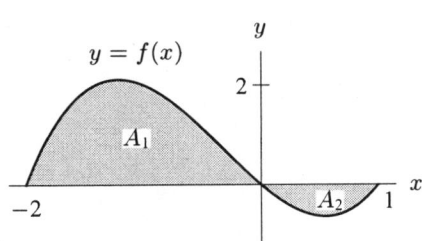

**Figure 3.38**

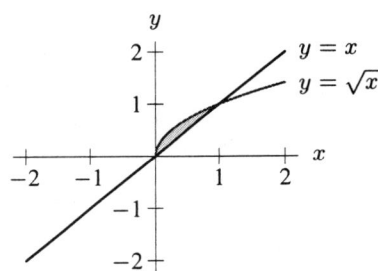

**Figure 3.39**

19. (a) See Figure 3.38.
   (b) $A_1 = \int_{-2}^0 f(x)\,dx = 2.667.$
   
   $A_2 = -\int_0^1 f(x)\,dx = 0.417.$
   
   So total area $= A_1 + A_2 \approx 3.084$. Note that while $A_1$ and $A_2$ are accurate to 3 decimal places, the quoted value for $A_1 + A_2$ is accurate only to 2 decimal places.
   
   (c) $\int_{-2}^1 f(x)\,dx = A_1 - A_2 = 2.250.$

20. The graph of $y = x^2 - 2$ is shown in Figure 3.40, and the relevant area is shaded. If you compute the integral $\int_0^3 (x^2-2)dx$, you find that
$$\int_0^3 (x^2 - 2)dx = 3.0.$$
However, since part of the area lies below the $x$-axis and part of it lies above the $x$-axis, this computation does not help us at all. (In fact, it is clear from the graph that the shaded area is more than 3.) We have to find the area above the $x$-axis and the area below the $x$-axis separately. We find that the graph crosses the $x$-axis at $x = 1.414$, and we compute the two areas separately:
$$\int_0^{1.414} (x^2 - 2)dx = -1.886 \quad \text{and} \quad \int_{1.414}^3 (x^2 - 2)dx = 4.886.$$
As we expect, we see that the integral between 0 and 1.414 is negative and the integral between 1.414 and 3 is positive. The total area shaded is the sum of the absolute values of the two integrals:
$$\text{Area shaded} = 1.886 + 4.886 = 6.772 \quad \text{square units}.$$

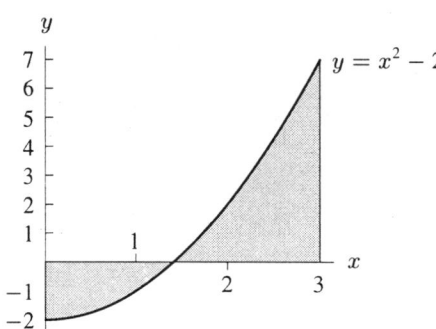

Figure 3.40

21.

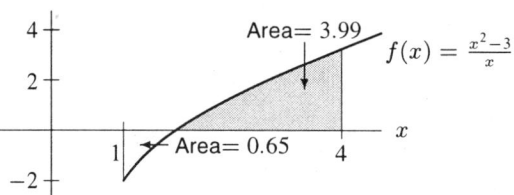

Figure 3.41

Using a calculator or computer, we see that
$$\int_1^4 \frac{x^2 - 3}{x} dx \approx 3.34.$$
The graph of $f(x) = \frac{x^2 - 3}{x}$ is shown in Figure 3.41. The function is negative to the left of $x = \sqrt{3} \approx 1.73$ and positive to the right of $x = 1.73$. We compute
$$\int_1^{1.73} \frac{x^2 - 3}{x} dx = -\text{Area below axis} \approx -0.65$$
and
$$\int_{1.73}^4 \frac{x^2 - 3}{x} dx = \text{Area above axis} \approx 3.99.$$
See Figure 3.41. Then
$$\int_1^4 \frac{x^2 - 3}{x} dx = \text{Area above axis} - \text{Area below axis} = 3.99 - 0.65 = 3.34.$$

22. Using a calculator or computer, we find

$$\int_1^4 (x - 3\ln x)\, dx \approx -0.136.$$

The function $f(x) = x - 3\ln x$ crosses the $x$-axis at $x \approx 1.86$. See Figure 3.42.

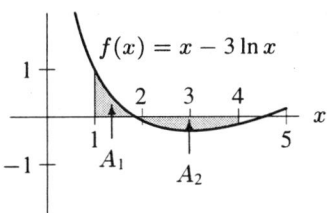

**Figure 3.42**

We find

$$\int_1^{1.86} (x - 3\ln x)\, dx = \text{Area above axis} \approx 0.347$$

$$\int_{1.86}^4 (x - 3\ln x)\, dx = -\text{Area below axis} \approx -0.483.$$

Thus, $A_1 \approx 0.347$ and $A_2 \approx 0.483$, so

$$\int_1^4 (x - 3\ln x)\, dx = \text{Area above axis} - \text{Area below axis} \approx 0.347 - 0.483 = -0.136.$$

23.

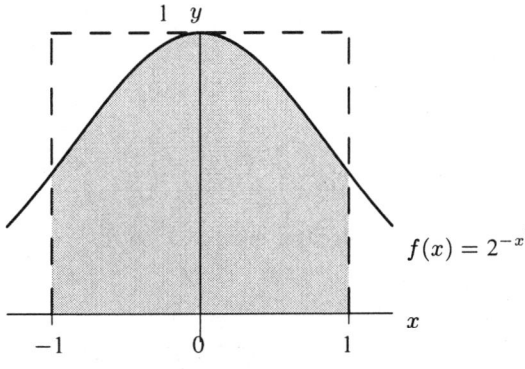

**Figure 3.43**

Since the shaded area is positive and less than the area of the rectangle measuring two units by one unit, we can say that

$$0 < \int_{-1}^1 2^{-x^2}\, dx < 2.$$

24. (a) $\int_{-1}^1 e^{x^2}\, dx > 0$, since $e^{x^2} > 0$, and $\int_{-1}^1 e^{x^2}\, dx$ represents the area below the curve $y = e^{x^2}$.

(b) Looking at the figure below, we see that $\int_0^1 e^{x^2}\, dx$ represents the area under the curve. This area is clearly greater than zero, but it is less than $e$ since it fits inside a rectangle of width 1 and height $e$ (with room to spare). Thus

$$0 < \int_0^1 e^{x^2}\, dx < e < 3.$$

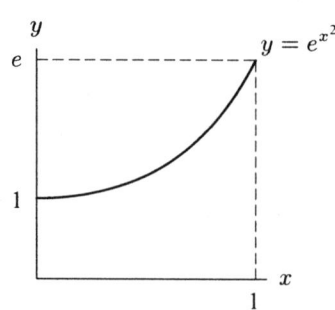

25. (a) 0, since the integrand is an odd function and the limits are symmetric around 0.
    (b) 0, since the integrand is an odd function and the limits are symmetric around 0.

26. (a)

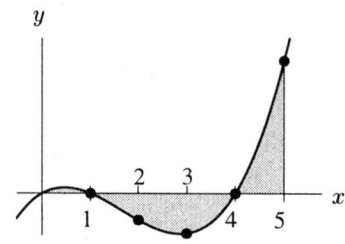

**Figure 3.44**: Graph of $y = x^3 - 5x^2 + 4x$

(b) $I_1 = \int_0^1 (x^3 - 5x^2 + 4x)dx$ is positive, since
$$x^3 - 5x^2 + 4x > 0 \quad \text{for} \quad 0 < x < 1.$$

$I_2 = \int_0^2 (x^3 - 5x^2 + 4x)dx$ seems to be negative, since the area under the $x$-axis is bigger than the area above the $x$-axis for $0 \leq x \leq 2$.

$I_3$ and $I_4$ will also be negative, since the area under the $x$-axis will also be larger than the area above the $x$-axis for $0 \leq x \leq 3$ and $0 \leq x \leq 4$. $I_5$ will also be negative, since the area above the $x$-axis will still be less than the area below the $x$-axis for $0 \leq x \leq 5$. From this we conclude that $I_1$ is the largest and $I_4$ is the smallest.

27. (a) $\int_{-3}^{0} f(x)\,dx = -2.$

    (b) $\int_{-3}^{4} f(x)\,dx = \int_{-3}^{0} f(x)\,dx + \int_{0}^{3} f(x)\,dx + \int_{3}^{4} f(x)\,dx = -2 + 2 - \dfrac{A}{2} = -\dfrac{A}{2}.$

28. (a) For $-2 \leq x \leq 2$, $f$ is symmetrical about the $y$-axis, so $\int_{-2}^{0} f(x)\,dx = \int_{0}^{2} f(x)\,dx$ and $\int_{-2}^{2} f(x)\,dx = 2\int_{0}^{2} f(x)\,dx.$

    (b) For any function $f$, $\int_{0}^{2} f(x)\,dx = \int_{0}^{5} f(x)\,dx - \int_{2}^{5} f(x)\,dx.$

    (c) Note that $\int_{-2}^{0} f(x)\,dx = \frac{1}{2}\int_{-2}^{2} f(x)\,dx$, so $\int_{0}^{5} f(x)\,dx = \int_{-2}^{5} f(x)\,dx - \int_{-2}^{0} f(x)\,dx = \int_{-2}^{5} f(x)\,dx - \frac{1}{2}\int_{-2}^{2} f(x)\,dx.$

29. (a) We know that $\int_{2}^{5} f(x)\,dx = \int_{0}^{5} f(x)\,dx - \int_{0}^{2} f(x)\,dx$. By symmetry, $\int_{0}^{2} f(x)\,dx = \frac{1}{2}\int_{-2}^{2} f(x)\,dx$, so $\int_{2}^{5} f(x)\,dx = \int_{0}^{5} f(x)\,dx - \frac{1}{2}\int_{-2}^{2} f(x)\,dx.$

    (b) $\int_{2}^{5} f(x)\,dx = \int_{-2}^{5} f(x)\,dx - \int_{-2}^{2} f(x)\,dx = \int_{-2}^{5} f(x)\,dx - 2\int_{-2}^{0} f(x)\,dx.$

    (c) Using symmetry again, $\int_{0}^{2} f(x)\,dx = \frac{1}{2}\left(\int_{-2}^{5} f(x)\,dx - \int_{2}^{5} f(x)\,dx\right).$

## Solutions for Section 3.4

1. $\int_a^b f(t)dt$ is measured in
$$\left(\dfrac{\text{miles}}{\text{hours}}\right) \cdot (\text{hours}) = \text{miles}.$$

2. The units of measurement are meters per second (which are units of velocity).

3. The units of measurement are dollars.

4. For any $t$, consider the interval $[t, t + \Delta t]$. During this interval, oil is leaking out at an approximately constant rate of $f(t)$ gallons/minute. Thus, the amount of oil which has leaked out during this interval can be expressed as

$$\text{Amount of oil leaked} = \text{Rate} \times \text{Time} = f(t) \Delta t$$

and the units of $f(t) \Delta t$ are gallons/minute × minutes = gallons. The total amount of oil leaked is obtained by adding all these amounts between $t = 0$ and $t = 60$. (An hour is 60 minutes.) The sum of all these infinitesimal amounts is the integral

$$\begin{array}{c}\text{Total amount of}\\ \text{oil leaked, in gallons}\end{array} = \int_0^{60} f(t)\, dt.$$

5. Change in income $= \int_0^{12} r(t)\, dt = \int_0^{12} 40(1.002)^t\, dt = \$485.80$

6. (a) Quantity used $= \int_0^5 f(t)\, dt$.
   (b) Using a left hand-sum, our approximation is

$$32(1.05)^0 + 32(1.05)^1 + 32(1.05)^2 + 32(1.05)^3 + 32(1.05)^4 \approx 176.82.$$

   Since $f$ is an increasing function, this represents an underestimate.
   (c) Each term is a lower estimate of one year's consumption of oil.

7. The area under A's curve is greater than the area under B's curve on the interval from 0 to 6, so $A$ had the most total sales in the first 6 months. On the interval from 0 to 12, the area under B's curve is greater than the area under $A$, so $B$ had the most total sales in the first year. At approximately nine months, A and B appear to have sold equal amounts. Counting the squares yields a total of about 250 sales in the first year for B and 170 sales in first year for A.

8. (a) The area under the curve is greater for species B for the first 5 years. Thus species B has a larger population after 5 years. After 10 years, the area under the graph for species B is still greater so species B has a greater population after 10 years as well.
   (b) Unless something happens that we cannot predict now, species A will have a larger population after 20 years. It looks like species A will continue to quickly increase, while species B will add only a few new plants each year.

9. (a) The distance traveled in the first 3 hours (from $t = 0$ to $t = 3$) is given by

$$\int_0^3 (40t - 10t^2)\, dt.$$

   (b) The shaded area in Figure 3.45 represents the distance traveled.

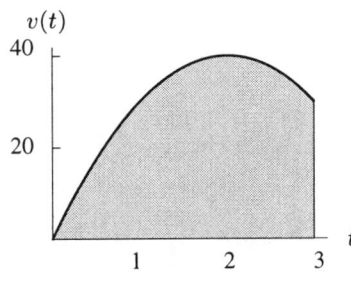

**Figure 3.45**

   (c) Using a calculator, we get

$$\int_0^3 (40t - 10t^2)\, dt = 90.$$

So the total distance traveled is 90 miles.

**156** CHAPTER THREE /SOLUTIONS

10. The total amount of antibodies produced is

$$\text{Total antibodies} = \int_0^4 r(t)\,dt \approx 1.417 \text{ thousand antibodies}$$

11. (a)

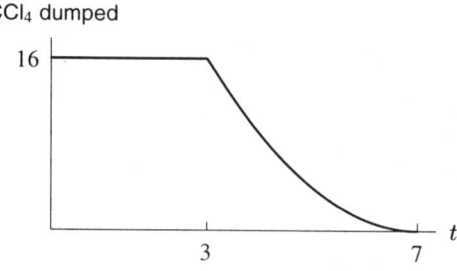

**Figure 3.46**

(b) 7 years, because $t^2 - 14t + 49 = (t-7)^2$ indicates that the rate of flow was zero after 7 years.

(c) 
$$\text{Area under the curve} = 3(16) + \int_3^7 (t^2 - 14t + 49)\,dt$$
$$= 69.3.$$

So $69\frac{1}{3}$ cubic yards of $CCl_4$ entered the waters from the time the EPA first learned of the situation until the flow of CCl4 was stopped.

12. You start traveling toward home. After 1 hour, you have traveled about 5 miles. (The distance traveled is the area under the first bump of the curve. The area is about 1/2 box; each box represents 10 miles.) You are now about $10 - 5 = 5$ miles from home. You then decide to turn around. You ride for an hour and a half away from home, in which time you travel around 13 miles. (The area under the next part of the curve is about 1.3 boxes below the axis.) You are now $5 + 13 = 18$ miles from home and you decide to turn around again at $t = 2.5$. The area above the curve from $t = 2.5$ to $t = 3.5$ has area 18 miles. This means that you ride for another hour, at which time you reach your house ($t = 3.5$.) You still want to ride, so you keep going past your house for a half hour. You make about 8 miles in this time and you decide to stop. So, at the end of the ride, you end up 8 miles away from home.

13. Notice that the area of a square on the graph represents $\frac{10}{6}$ miles. At $t = 1/3$ hours, $v = 0$. The area between the curve $v$ and the $t$-axis over the interval $0 \le t \le 1/3$ is $-\int_0^{1/3} v\,dt \approx \frac{5}{3}$. Since $v$ is negative here, she is moving toward the lake. At $t = \frac{1}{3}$, she is about $5 - \frac{5}{3} = \frac{10}{3}$ miles from the lake. Then, as she moves away from the lake, $v$ is positive for $\frac{1}{3} \le t \le 1$. At $t = 1$,

$$\int_0^1 v\,dt = \int_0^{1/3} v\,dt + \int_{1/3}^1 v\,dt \approx -\frac{5}{3} + 8 \cdot \frac{10}{6} = \frac{35}{3},$$

and the cyclist is about $5 + \frac{35}{3} = \frac{50}{3} = 16\frac{2}{3}$ miles from the lake. Since, starting from the moment $t = \frac{1}{3}$, she moves away from the lake, the cyclist will be farthest from the lake at $t = 1$. The maximal distance equals $16\frac{2}{3}$ miles.

14. Looking at the figure in the problem, we note that Product A has a greater peak concentration than Product B; Product A peaks much faster that Product B; Product B has greater overall bioavailability than Product A. Since Product B has greater bioavailability, it provides a greater amount of drug reaching the bloodstream. Therefore Product B provides relief for a longer time than Product A. On the other hand, Product A provides faster relief.

15. Looking at the figure in the problem, we note that Product B has a greater peak concentration than Product A; Product A peaks sooner than Product B; Product B has a greater overall bioavailability than Product A. Since we are looking for the product providing the faster response, Product A should be used as it peaks sooner.

16. (a) Distance $\approx 2(45) + 2(20) + 2(10) + 2(5) = 160$ ft.

(b)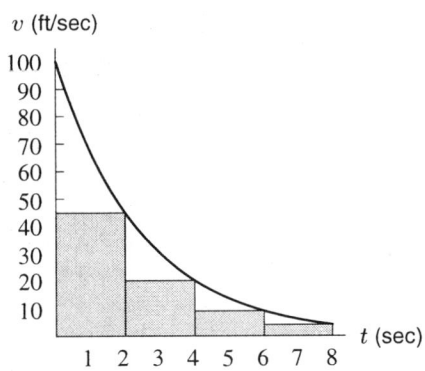

(c) This is an underestimate of total distance traveled, because the rectangles cover less area than is actually under the graph.

17. (a) The amount leaked between $t = 0$ and $t = 2$ is $\int_0^2 R(t)\,dt$.

(b) See Figure 3.47.

(c) The rectangular boxes on the diagram each have area $\frac{1}{16}$. Of these 45 are wholly beneath the curve, hence the area under the curve is certainly more than $\frac{45}{16} > 2.81$. There are 9 more partially beneath the curve, and so the desired area is completely covered by 54 boxes. Therefore the area is less than $\frac{54}{16} < 3.38$.

These are very safe estimates but far apart. We can do much better by estimating what fractions of the broken boxes are beneath the curve. Using this method, we can estimate the area to be about 3.2, which corresponds to 3.2 gallons leaking over two hours.

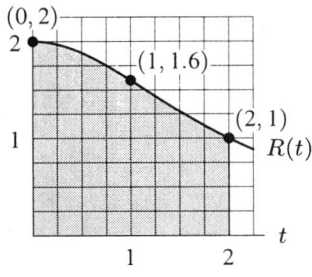

Figure 3.47

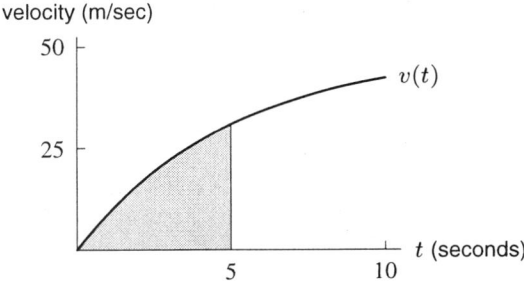

Figure 3.48: Velocity of a free-falling object

18. (a) Distance is the definite integral of velocity, so between time $t = 0$ and time $t = 5$,

$$\text{Distance traveled} = \int_0^5 49(1 - (0.8187)^t)\,dt.$$

(b) A graph of this velocity is given in Figure 3.48, and the distance traveled is represented by the shaded area under this velocity curve.

(c) We use a computer or calculator to see that

$$\int_0^5 49(1 - (0.8187)^t)\,dt \approx 90.14.$$

Thus, if you fall out of an airplane without a parachute, you fall about 90 meters during the first five seconds.

19. The distance you fall during the first 10 seconds is

$$\int_0^{10} 49\left(1 - (0.8187)^t\right)dt.$$

**158** CHAPTER THREE /SOLUTIONS

Using a calculator, we get

$$\int_0^{10} 49(1 - (0.1817)^t\,dt) \approx 278.$$

So the distance you fall in the first 10 seconds is approximately 278 meters. In Problem 18, we saw that you fell 90 meters between $t = 0$ and $t = 5$. Therefore, you fall approximately $278 - 90 = 188$ meters between $t = 5$ and $t = 10$.

## Solutions for Section 3.5

1. The units for the integral $\int_{800}^{900} C'(q)\,dq$ are $\left(\dfrac{\text{dollars}}{\text{tons}}\right) \cdot (\text{tons}) = \text{dollars}$.

   $\int_{800}^{900} C'(q)\,dq$ represents the cost of increasing production from 800 tons to 900 tons.

2. The total change in the net worth of the company from 1990 ($t = 0$) to 2000 ($t = 10$) is found using the Fundamental Theorem:

$$\text{Change in net worth} = f(10) - f(0) = \int_0^{10} f'(t)\,dt = \int_0^{10} (2000 - 12t^2)\,dt = 16{,}000 \text{ dollars}.$$

   The worth of the company in 2000 is the worth of the company in 1990 plus the change in worth between 1990 and 2000. Thus, in 2000,

$$\text{Net worth} = f(10) = f(0) + \text{Change in worth}$$
$$= \text{Worth in 1990} + \text{Change in worth between 1990 and 2000}$$
$$= 40{,}000 + 16{,}000$$
$$= \$56{,}000$$

3. By the Fundamental Theorem, we have

$$\text{Change in temperature} = f(10) - f(0) = \int_0^{10} f'(t)\,dt = \int_0^{10} -7(0.9)^t\,dt = -43.3°\text{C}.$$

   Since $f(0) = 90°$C, the temperature of the coffee, to one decimal place, when $t = 10$ is

$$f(10) = f(0) + (-43.3°\text{C}) = 90°\text{C} - 43.3°\text{C} = 46.7°\text{C}.$$

4. (a) The area under the curve of $P'(t)$ from 0 to $t$ gives the change in the value of the stock. Examination of the graph suggests that this area is greatest at $t = 5$, so we conclude that the stock is at its highest value at the end of the 5th week.
   (Some may also conclude that the area is greatest at $t = 1.5$, making the stock most valuable in the middle of the second week. Both are valid answers.)
   Since $P'(t) < 0$ from $t = 1.5$ to about $t = 3.8$, we know that the value of the stock decreases in this interval. This is the only interval in which the stock's value is decreasing, so the stock will reach its lowest value at the end of this interval, which is near the end of the fourth week.
   (b) We know that $P(t) - P(0)$ is the area under the curve of $P'(t)$ from 0 to $t$, so examination of the graph leads us to conclude that

$$P(4) < P(3) \approx P(0) < P(1) \approx P(2) < P(5).$$

5. (a) There are approximately 5.5 squares under the curve of $C'(q)$ from 0 to 30. Each square represents \$100, so the total variable cost to produce 30 units is around \$550. To find the total cost, we had the fixed cost

$$\text{Total cost} = \text{fixed cost} + \text{total variable cost}$$
$$= 10{,}000 + 550 = \$10{,}550.$$

   (b) There are approximately 1.5 squares under the curve of $C'(q)$ from 30 to 40. Each square represents \$100, so the additional cost of producing items 31 through 40 is around \$150.
   (c) Examination of the graph tells us that $C'(25) = 10$. This means that the cost of producing the 26th item is approximately \$10.

6. Since $C(0) = 500$, the fixed cost must be $500. The total *variable* cost to produce 20 units is

$$\int_0^{20} C'(q)\,dq = \int_0^{20} (q^2 - 16q + 70)\,dq = \$866.67 \text{ (using a calculator)}.$$

The *total* cost to produce 20 units is the fixed cost plus the variable cost of producing 20 units. Thus,

$$\text{Total cost} = \$500 + \$866.67 = \$1,366.67.$$

7. (a) Total variable cost in producing 400 units is

$$\int_0^{400} C'(q)\,dq.$$

We estimate this integral:

$$\text{Left–hand sum} = 25(100) + 20(100) + 18(100) + 22(100) = 8500;$$
$$\text{Right–hand sum} = 20(100) + 18(100) + 22(100) + 28(100) = 8800;$$

and so $\int_0^{400} C'(q)\,dq \approx \dfrac{8500 + 8800}{2} = \$8650.$

$$\text{Total cost} = \text{Fixed cost} + \text{Variable cost}$$
$$= \$10,000 + \$8650 = \$18,650.$$

(b) $C'(400) = 28$, so we would expect that the 401st unit would cost an extra $28.

8. The fixed cost is $C(0) = 1,000,000$.

$$\text{Total variable cost} = \int_0^{500} C'(x)\,dx = \int_0^{500} (4000 + 10x)\,dx = 3,250,000.$$

Therefore,

$$\text{Total cost} = \text{Fixed Cost} + \text{Total variable cost}$$
$$= 4,250,000 \text{ riyals}.$$

9. (a)

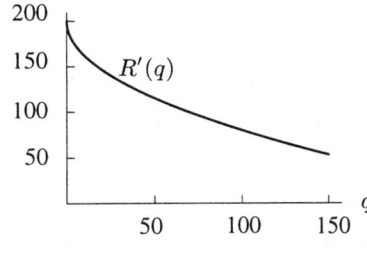

**Figure 3.49**

(b) By the Fundamental Theorem,

$$\int_0^{100} R'(q)\,dq = R(100) - R(0).$$

$R(0) = 0$ because no revenue is produced if no units are sold. Thus we get

$$R(100) = \int_0^{100} R'(q)\,dq \approx \$12,000.$$

(c) The marginal revenue in selling the 101st unit is given by $R'(100) = \$80/\text{unit}$. The total revenue in selling 101 units is:

$$R(100) + R'(100) = \$12,080.$$

10. (a) Using the Fundamental Theorem we get that the cost of producing 30 bicycles is

$$C(30) = \int_0^{30} C'(q)\,dq + C(0)$$

or

$$\int_0^{30} \frac{600}{0.3q+5}\,dq + \$2000 \approx \$4059.24$$

(b) If the bikes are sold for \$200 each the total revenue from 30 bicycles is

$$30 \cdot 200 = \$6000,$$

and so the total profit is

$$\$6000 - \$4059.24 = \$1940.76.$$

(c) The marginal profit on the 31$^{st}$ bicycle is the difference between the marginal cost of producing the 31$^{st}$ bicycle and the marginal revenue, which is the price. Thus the marginal profit is

$$200 - C'(30) = 200 - \frac{600}{14} = \$157.14$$

11. Using the Fundamental Theorem of Calculus with $f(x) = 4 - x^2$ and $a = 0$, we see that

$$F(b) - F(0) = \int_0^b F'(x)\,dx = \int_0^b (4 - x^2)\,dx.$$

We know that $F(0) = 0$, so

$$F(b) = \int_0^b (4 - x^2)\,dx.$$

Using Riemann sums to estimate the integral for values of $b$, we get

**TABLE 3.1**

| $b$ | 0.0 | 0.5 | 1.0 | 1.5 | 2.0 | 2.5 |
|---|---|---|---|---|---|---|
| $F(b)$ | 0 | 1.958 | 3.667 | 4.875 | 5.333 | 4.792 |

12. From the graph we see that $F'(x) < 0$ for $x < -2$ and $x > 2$, and that $F'(x) > 0$ for $-2 < x < 2$, so we conclude that $F(x)$ is decreasing for $x < -2$ and $x > 2$ and $F(x)$ is increasing for $-2 < x < 2$.

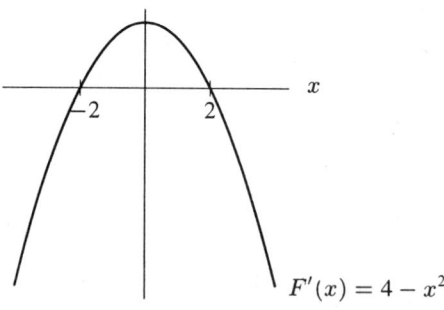

**Figure 3.50**

13. Since $F'(x)$ is positive for $0 < x < 2$ and $F'(x)$ is negative for $2 < x < 2.5$, $F(x)$ increases on $0 < x < 2$ and decreases on $2 < x < 2.5$. From this we conclude that $F(x)$ has a maximum at $x = 2$. From the process used in Problem 11 we see that the chart agrees with this assumption and that $F(2) = 5.333$.

14. (a) Using a calculator, $\int_1^4 (2t+1)\,dt = 18$.
    (b) By the Fundamental Theorem of Calculus $\int_a^b F'(t)\,dt = F(b) - F(a)$. Thus,
    $$\int_1^4 (2t+1)\,dt = F(4) - F(1) = (4^2 + 4) - (1^2 + 1) = 20 - 2 = 18.$$

15. (a) Using a calculator, $\int_1^4 (2t+3)\,dt = 24$.
    (b) Using the Fundamental Theorem of Calculus with $a = 1$ and $b = 4$ we get
    $$\int_1^4 (2t+3)\,dt = F(4) - F(1) = (4^2 + 12) - (1^2 + 3) = 28 - 4 = 24.$$
    This is the exact value of the integral.

16. (a) Using a calculator, $\int_1^4 \frac{1}{2\sqrt{x}}\,dx = 1$.
    (b) Using the Fundamental Theorem of Calculus with $a = 1$ and $b = 4$ we get
    $$\int_1^4 \frac{1}{2\sqrt{x}}\,dx = F(4) - F(1) = \sqrt{4} - \sqrt{1} = 2 - 1 = 1.$$

17. (a) Using a calculator, $\int_1^4 3x^2\,dx = 63$.
    (b) Using the Fundamental Theorem of Calculus with $a = 1$ and $b = 4$,
    $$\int_1^4 F'(x)\,dx = F(4) - F(1).$$
    Substituting $F(x)$ and $F'(x)$ yields
    $$\int_1^4 3x^2\,dx = F(4) - F(1) = 4^3 - 1^3 = 64 - 1 = 63$$
    which is the exact value of the integral.

## Solutions for Chapter 3 Review

1. The table gives the rate of emissions of nitrogen oxide in millions of metric tons per year. To find the total emissions, we use left-hand and right-hand Riemann sums. We have

    Left-hand sum $= (6.9)(10) + (9.4)(10) + (13.0)(10) + (18.5)(10) + (20.9)(10) = 687$.
    Right-hand sum $= (9.4)(10) + (13.0)(10) + (18.5)(10) + (20.9)(10) + (19.6)(10) = 814$.
    Average of left- and right-hand sums $= \dfrac{687 + 814}{2} = 750.5$.

    The total emissions of nitrogen oxide between 1940 and 1990 is about 750 million metric tons.

2. Taking left-hand sums we get
    $$5 \cdot (10.82 + 13.06 + 14.61 + 14.99 + 18.60 + 19.33) = 5 \cdot 91.41 = 457.05.$$
    Taking right-hand sums we get
    $$5 \cdot (13.06 + 14.61 + 14.99 + 18.60 + 19.33 + 22.46) = 5 \cdot 103.05 = 515.25.$$
    Taking the average we get that 486.15 quadrillion BTU is produced between the years 1960 and 1990 in the United States.

3. $\int_0^{10} 2^{-x}\,dx = 1.44$

4. $\int_1^5 (x^2 + 1)\,dx = 45.33$

5. $\int_0^1 \sqrt{1 + t^2}\,dt = 1.15$.

**162** CHAPTER THREE /SOLUTIONS

6. $\int_{-1}^{1} \frac{x^2+1}{x^2-4} dx = -0.75$.

7. $\int_{2}^{3} \frac{-1}{(r+1)^2} dr = -0.083$

8. $\int_{1}^{3} \frac{z^2+1}{z} dz = 5.10$

9. Calculating both the LHS and RHS and averaging the two, we get

$$\frac{1}{2}(5(100 + 82 + 69 + 60 + 53) + 5(82 + 69 + 60 + 53 + 49)) = 1692.5$$

10. By counting squares and fractions of squares, we find that the area under the graph appears to be around 310 (miles/hour) sec, within about 10. So the distance traveled was about $310 \left(\frac{5280}{3600}\right) \approx 455$ feet, within about $10 \left(\frac{5280}{3600}\right) \approx 15$ feet. (Note that 455 feet is about 0.086 miles)

11. Looking at the area under the graph we see that after 5 years Tree B is taller while Tree A is taller after 10 years.

12. (a) Since velocity is the rate of change of distance, we have

$$\text{Distance traveled} = \int_{0}^{5} (10 + 8t - t^2) \, dt.$$

This distance is the shaded area in Figure 3.51.

(b) A graph of this velocity function is given in Figure 3.51. Finding the distance traveled is equivalent to finding the area under this curve between $t = 0$ and $t = 5$. We estimate that this area is about 100 since the average height appears to be about 20 and the width is 5.

(c) We use a calculator or computer to calculate the definite integral

$$\text{Distance traveled} = \int_{0}^{5} (10 + 8t - t^2) \, dt \approx 108.33 \quad \text{meters}.$$

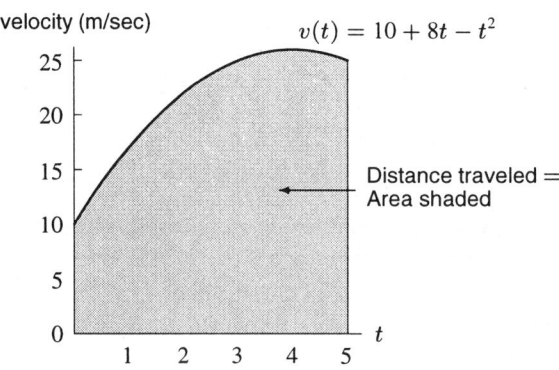

*Figure 3.51:* A velocity function

13. (a)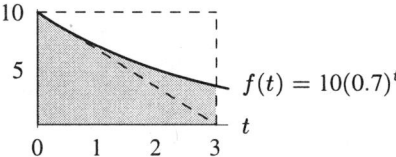

*Figure 3.52*

Since $f(t)$ is the rate of change of the oil leaking out of the tanker, then the total quantity of oil which has leaked out between $t = 0$ and $t = 3$ is given by $\int_{0}^{3} 10(0.7)^t \, dt$, which has the same value as the area under the graph of $f(t)$ between $t = 0$ and $t = 3$. Therefore, the total quantity of oil which leaked out between $t = 0$ and $t = 3$ equals

65% of the area of the rectangle 10 units by 3 units $= (0.65)(3)(10) = 19.5$ gallons.

(b) $\int_{0}^{3} 10(0.7)^t \, dt \approx 18.42 \quad$ gallons.

(c) Since the area under the graph of $f(t)$ between $t = 0$ and $t = 1$ is greater than the area under the graph of $f(t)$ between $t = 2$ and $t = 3$ we can say that more oil leaked during the first minute than during the third minute.

**14.** (a) The rate of ice formation is $\frac{dy}{dt} = \frac{\sqrt{t}}{2}$ inches/hour. So the amount of ice formed in 8 hours equals

$$y(8) - y(0) = \int_0^8 \frac{\sqrt{t}}{2} dt.$$

Now, $y(0) = 0$ because the ice starts forming at $t = 0$. Thus there are

$$\int_0^8 \frac{\sqrt{t}}{2} dt \approx 7.54 \text{ inches of ice in 8 hours.}$$

(b) At 8 hours, $\frac{dy}{dt} = \frac{\sqrt{8}}{2} = \sqrt{2}$ inches/hour. The ice is increasing at a rate of

$$\sqrt{2} = 1.41 \text{ inches/hour.}$$

**15.** Rate of income increase is $r(t) = 480(1.024)^t$ in dollars per year since 1987. In 1995, $t = 8$. Change in the average per capita income between 1987 and 1995 is given by

$$\text{Average 1995 income} - \text{Average 1987 income} = \int_0^8 r(t) \, dt$$

Since the average 1987 income is $26,000, we have

$$\text{Average 1995 income} = \int_0^8 r(t) \, dt + \$26,000$$
$$\approx \$4228 + \$26,000$$
$$= \$30,228$$

**16.** We are told that $C'(q) = q^2 - 50q + 700$ for $0 \leq q \leq 50$. We also know that

$$\int_0^{50} C'(q) dq = C(50) - C(0).$$

Since we are told that fixed costs are $500 we know that

$$C(0) = \$500.$$

Thus

$$C(50) = \int_0^{50} C'(q) dq + \$500 \approx \$14,667.$$

**17.** (a)

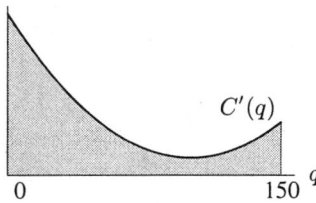

**Figure 3.53**

The total variable cost of producing 150 units is represented by the area under the graph of $C'(q)$ between 0 and 150, or

$$\int_0^{150} (0.005q^2 - q + 56) dq.$$

**164** CHAPTER THREE /SOLUTIONS

(b) An estimate of the total cost of producing 150 units is given by

$$20{,}000 + \int_0^{150} (0.005q^2 - q + 56)\,dq.$$

This represents the fixed cost ($20,000) plus the variable cost of producing 150 units, which is represented by the integral. Using a calculator, we see

$$\int_0^{150} (0.005q^2 - q + 56)\,dq \approx 2{,}775.$$

So the total cost is approximately

$$\$20{,}000 + \$2{,}775 = \$22{,}775.$$

(c) $C'(150) = 0.005(150)^2 - 150 + 56 = 18.5$. This means that the marginal cost of the 150th item is 18.5. In other words, the 151st item will cost approximately $18.50.

(d) $C(151)$ is the total cost of producing 151 items. This can be found by adding the total cost of producing 150 items (found in part (b)) and the additional cost of producing the 151st item ($C'(150)$, found in (c)). So we have

$$C(151) \approx 22{,}775 + 18.50 = \$22{,}793.50.$$

18.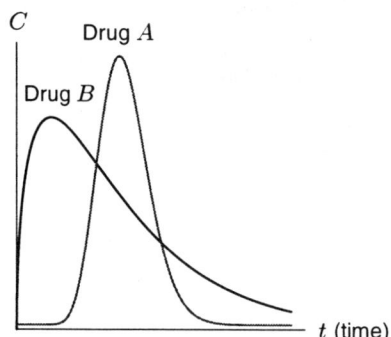

**Figure 3.54**

19. (a)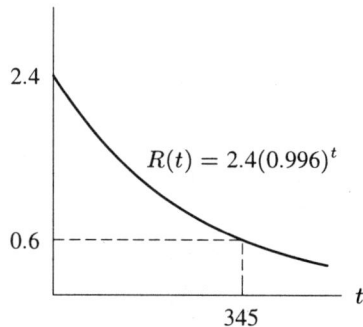

**Figure 3.55**

The amount when $t = 0$ is four times the maximum acceptable limit, or $4(0.6) = 2.4$. Therefore the level of radiation is given by

$$R(t) = 2.4(0.996)^t.$$

A graph of the function is shown in Figure 3.55 and we see that $R(t) = 0.6$ at about $t = 345$, so the level of radiation reaches an acceptable level after about 345 hours (or about 2 weeks).

(b) Total radiation emitted between $t = 0$ and $t = 345$ is

$$\int_0^{345} 2.4(0.996)^t\,dt \approx 449 \text{ millirems}.$$

20. (a) The equation $v = 6 - 2t$ implies that $v > 0$ (the car is moving forward) if $0 \le t \le 3$ and that $v < 0$ (the car is moving backwards) if $t > 3$. When $t = 3$, $v = 0$, so the car is not moving at the instant $t = 3$. The car arrives back at its starting point when $t = 6$.

(b) The car moves forward on the interval $0 \le t \le 3$, so it is farthest forward when $t = 3$. For all $t > 3$, the car is moving backward. There is no upper bound on the car's distance behind its starting point since it is moving backward for all $t > 3$.

21. (a) The acceleration is positive for $0 \le t < 40$ and for a tiny period before $t = 60$, since the slope is positive over these intervals. Just to the left of $t = 40$, it looks like the acceleration is approaching 0. Between $t = 40$ and a moment just before $t = 60$, the acceleration is negative.

(b) The maximum altitude was about 500 feet, when $t$ was a little greater than 40 (here we are estimating the area under the graph for $0 \le t \le 42$).

(c) The total change in altitude for the Montgolfiers and their balloon is the definite integral of their velocity, or the total area under the given graph (counting the part after $t = 42$ as negative, of course). As mentioned before, the total area of the graph for $0 \le t \le 42$ is about 500. The area for $t > 42$ is about 220. So subtracting, we see that the balloon finished 280 feet or so higher than where it began.

22. (a) The mouse changes direction (when its velocity is zero) at about times 17, 23, and 27.

(b) The mouse is moving most rapidly to the right at time 10 and most rapidly to the left at time 40.

(c) The mouse is farthest to the right when the integral of the velocity, $\int_0^t v(t)\,dt$, is most positive. Since the integral is the sum of the areas above the axis minus the areas below the axis, the integral is largest when the velocity is zero at about 17 seconds. The mouse is farthest to the left of center when the integral is most negative at 40 seconds.

(d) The mouse's speed decreases during seconds 10 to 17, from 20 to 23 seconds, and from 24 seconds to 27 seconds.

(e) The mouse is at the center of the tunnel at any time $t$ for which the integral from 0 to $t$ of the velocity is zero. This is true at time 0 and again somewhere around 35 seconds.

23. (a) The distance traveled is the integral of the velocity, so in $T$ seconds you fall

$$\int_0^T 49(1 - 0.8187^t)\,dt.$$

(b) We want the number $T$ for which

$$\int_0^T 49(1 - 0.8187^t)\,dt = 5000.$$

We can use technology to experiment with different values for $T$, and we find $T \approx 107$ seconds.

24. (a) Over the interval $[-1, 3]$, we estimate that the total change of the population is about 1.5, by counting boxes between the curve and the $x$-axis; we count about 1.5 boxes below the $x$-axis from $x = -1$ to $x = 1$ and about 3 above from $x = 1$ to $x = 3$. So the average rate of change is just the total change divided by the length of the interval, that is $1.5/4 = 0.375$ thousand/hour.

(b) We can estimate the total change of the algae population by counting boxes between the curve and the $x$-axis. Here, there is about 1 box above the $x$-axis from $x = -3$ to $x = -2$, about 0.75 of a box below the $x$-axis from $x = -2$ to $x = -1$, and a total change of about 1.5 boxes thereafter (as discussed in part (a)). So the total change is about $1 - 0.75 + 1.5 = 1.75$ thousands of algae.

25. (a) In the beginning, both birth and death rates are small; this is consistent with a very small population. Both rates begin climbing, the birth rate faster than the death rate, which is consistent with a growing population. The birth rate is then high, but it begins to decrease as the population increases.

(b)

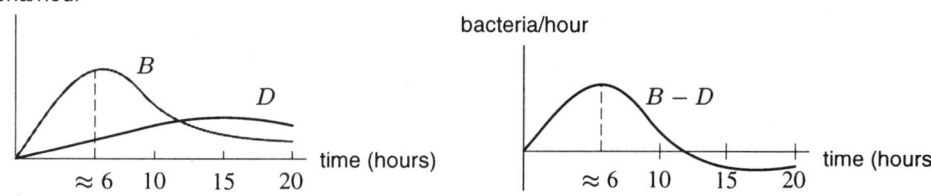

**Figure 3.56:** Difference between $B$ and $D$ is greatest at $t \approx 6$

The bacteria population is growing most quickly when $B - D$, the rate of change of population, is maximal; that happens when $B$ is farthest above $D$, which is at a point where the slopes of both graphs are equal. That point is $t \approx 6$ hours.

(c) Total number born by time $t$ is the area under the $B$ graph from $t = 0$ up to time $t$. See Figure 3.57.

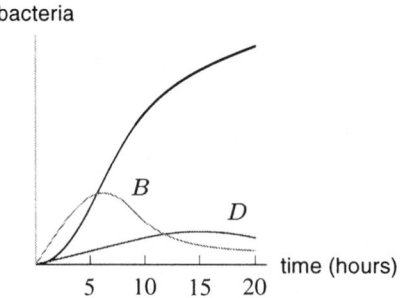

**Figure 3.57**: Number born by time $t$ is $\int_0^t B(x)\,dx$

Total number alive at time $t$ is the number born minus the number that have died, which is the area under the $B$ graph minus the area under the $D$ graph, up to time $t$. See Figure 3.58.

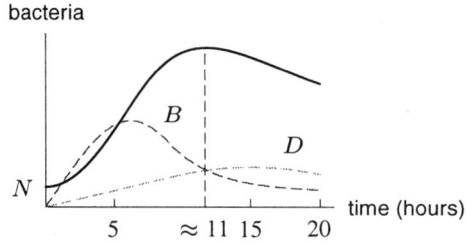

**Figure 3.58**: Number alive at time $t$ is $\int_0^t (B(x) - D(x))\,dx$

From Figure 3.58, we see that the population is at a maximum when $B = D$, that is, after about 11 hours. This stands to reason, because $B - D$ is the rate of change of population, so population is maximized when $B - D = 0$, that is, when $B = D$.

26. (a) Suppose $Q(t)$ is the amount of water in the reservoir at time $t$. Then

$$Q'(t) = \genfrac{}{}{0pt}{}{\text{Rate at which water}}{\text{in reservoir is changing}} = \genfrac{}{}{0pt}{}{\text{Inflow}}{\text{rate}} - \genfrac{}{}{0pt}{}{\text{Outflow}}{\text{rate}}$$

Thus the amount of water in the reservoir is increasing when the inflow curve is above the outflow, and decreasing when it is below. This means that $Q(t)$ is a maximum where the curves cross in July 1993 (as shown in Figure 3.59), and $Q(t)$ is decreasing fastest when the outflow is farthest above the inflow curve, which occurs about October 1993 (see Figure 3.59).

To estimate values of $Q(t)$, we use the Fundamental Theorem which says that the change in the total quantity of water in the reservoir is given by

$$Q(t) - Q(\text{Jan'}93) = \int_{\text{Jan}93}^{t} (\text{inflow rate} - \text{outflow rate})\,dt$$

or $\quad Q(t) = Q(\text{Jan'}93) + \int_{\text{Jan}93}^{t} (\text{inflow rate} - \text{outflow rate})\,dt.$

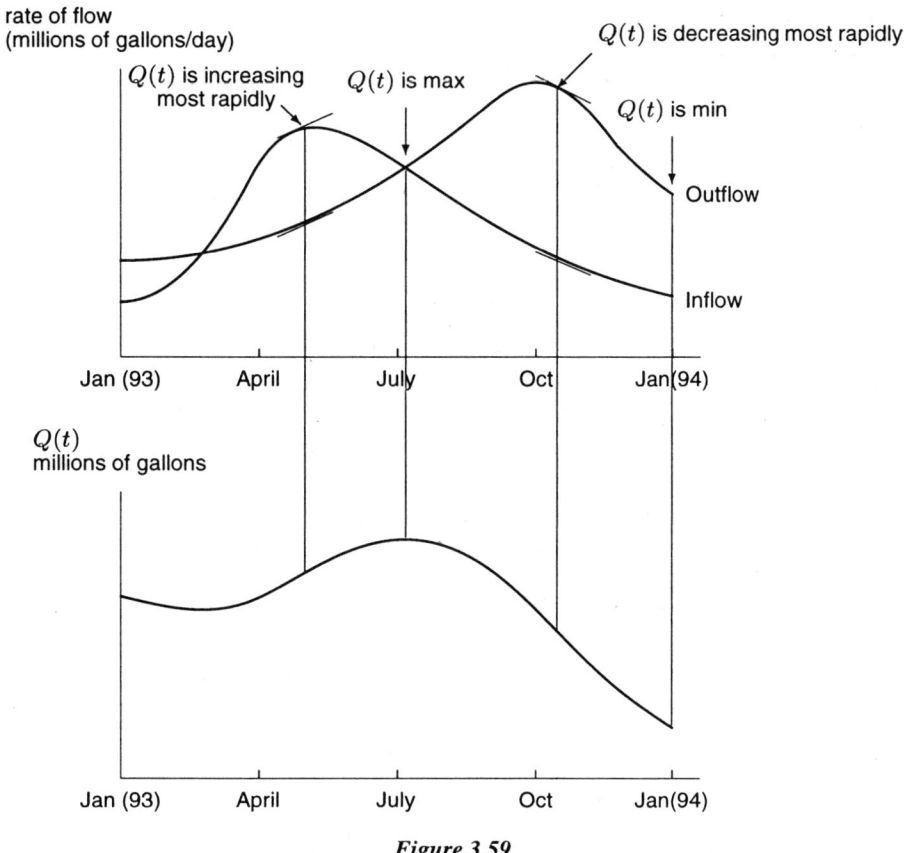

**Figure 3.59**

(b) See Figure 3.59. Maximum in July 1993. Minimum in Jan 1994.
(c) See Figure 3.59. Increasing fastest in May 1993. Decreasing fastest in Oct 1993.
(d) In order for the water to be the same as Jan '93 the total amount of water which has flowed into the reservoir must be 0. Referring to Figure 3.60, we have

$$\int_{Jan93}^{July94} (\text{inflow} - \text{outflow})dt = -A_1 + A_2 - A_3 + A_4 = 0$$

giving $A_1 + A_3 = A_2 + A_4$.

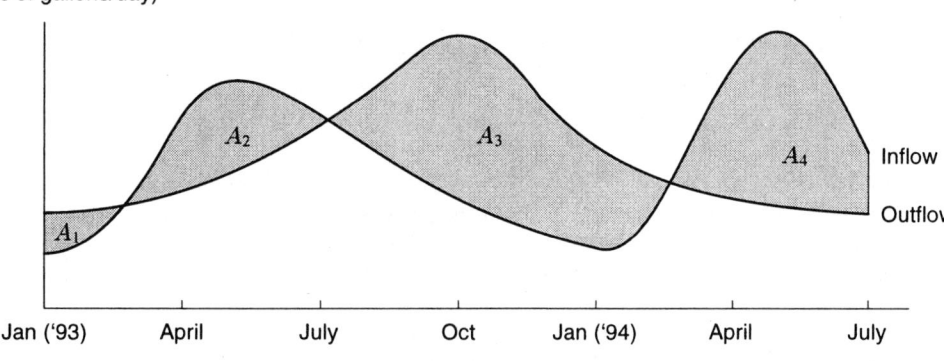

**Figure 3.60**

## Solutions to the Projects

1. (a) $CO_2$ is being taken out of the water during the day and returned at night. The pond must therefore contain some plants. (The data is in fact from pond water containing both plants and animals.)

   (b) Suppose $t$ is the number of hours past dawn. The graph in Figure 3.55 of the text shows that the $CO_2$ content changes at a greater rate for the first 6 hours of daylight, $0 < t < 6$, than it does for the final 6 hours of daylight, $6 < t < 12$. It turns out that plants photosynthesize more vigorously in the morning than in the afternoon. Similarly, $CO_2$ content changes more rapidly in the first half of the night, $12 < t < 18$, than in the 6 hours just before dawn, $18 < t < 24$. The reason seems to be that at night plants quickly use up most of the sugar that they synthesized during the day, and then their respiration rate is inhibited. So the constant rate hypothesis is false, if we assume plants are the main cause of $CO_2$ changes in the pond.

   (c) The question asks about the total quantity of $CO_2$ in the pond, rather than the rate at which it is changing. We will let $f(t)$ denote the $CO_2$ content of the pond water (in mmol/l) at $t$ hours past dawn. Then Figure 3.55 of the text is a graph of the derivative $f'(t)$. There are 2.600 mmol/l of $CO_2$ in the water at dawn, so $f(0) = 2.600$.

   The $CO_2$ content $f(t)$ decreases during the 12 hours of daylight, $0 < t < 12$, when $f'(t) < 0$, and then $f(t)$ increases for the next 12 hours. Thus, $f(t)$ is at a minimum when $t = 12$, at dusk. By the Fundamental Theorem,

   $$f(12) = f(0) + \int_0^{12} f'(t)\,dt = 2.600 + \int_0^{12} f'(t)\,dt.$$

   We must approximate the definite integral by a Riemann sum. From the graph in Figure 3.55 of the text, we estimate the values of the function $f'(t)$ in Table 3.2.

**TABLE 3.2** Rate, $f'(t)$, at which $CO_2$ is entering or leaving water

| $t$ | $f'(t)$ | $t$ | $f'(t)$ | $t$ | $f'(t)$ | $t$ | $f'(t)$ | $t$ | $f'(t)$ | $t$ | $f'(t)$ |
|---|---|---|---|---|---|---|---|---|---|---|---|
| 0 | 0.000 | 4 | −0.039 | 8 | −0.026 | 12 | 0.000 | 16 | 0.035 | 20 | 0.020 |
| 1 | −0.042 | 5 | −0.038 | 9 | −0.023 | 13 | 0.054 | 17 | 0.030 | 21 | 0.015 |
| 2 | −0.044 | 6 | −0.035 | 10 | −0.020 | 14 | 0.045 | 18 | 0.027 | 22 | 0.012 |
| 3 | −0.041 | 7 | −0.030 | 11 | −0.008 | 15 | 0.040 | 19 | 0.023 | 23 | 0.005 |

The left Riemann sum with $n = 12$ terms, corresponding to $\Delta t = 1$, gives

$$\int_0^{12} f'(t)\,dt \approx (0.000)(1) + (-0.042)(1) + (-0.044)(1) + \cdots + (-0.008)(1) = -0.346.$$

At 12 hours past dawn, the $CO_2$ content of the pond water reaches its lowest level, which is approximately

$$2.600 - 0.346 = 2.254 \text{ mmol/l}.$$

   (d) The increase in $CO_2$ during the 12 hours of darkness equals

   $$f(24) - f(12) = \int_{12}^{24} f'(t)\,dt.$$

   Using Riemann sums to estimate this integral, we find that about 0.306 mmol/l of $CO_2$ was released into the pond during the night. In part (c) we calculated that about 0.346 mmol/l of $CO_2$ was absorbed from the pond during the day. If the pond is in equilibrium, we would expect the daytime absorption to equal the nighttime release. These quantities are sufficiently close (0.346 and 0.306) that the difference could be due to measurement error, or to errors from the Riemann sum approximation.

   If the pond is in equilibrium, the area between the rate curve in Figure 3.55 of the text and the $t$-axis for $0 \leq t \leq 12$ will equal the area between the rate curve and the $t$-axis for $12 \leq t \leq 24$. In this experiment the areas do look approximately equal.

(e) We must evaluate

$$f(b) = f(0) + \int_0^b f'(t)\, dt = 2.600 + \int_0^b f'(t)\, dt$$

for the values $b = 0, 3, 6, 9, 12, 15, 18, 21, 24$. Left Riemann sums with $\Delta t = 1$ give the values for the $CO_2$ content in Table 3.3. The graph is shown in Figure 3.61.

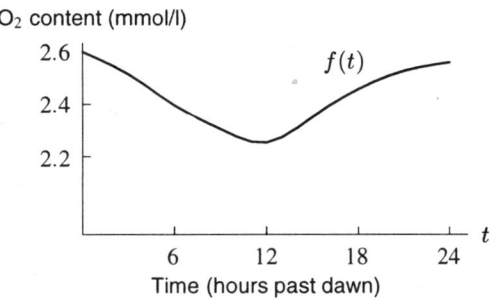

**Figure 3.61**: $CO_2$ content in pond water throughout the day

**TABLE 3.3** $CO_2$ content throughout the day

| $b$ (hours after dawn) | 0 | 3 | 6 | 9 | 12 | 15 | 18 | 21 | 24 |
|---|---|---|---|---|---|---|---|---|---|
| $f(b)$ ($CO_2$ content) | 2.600 | 2.514 | 2.396 | 2.305 | 2.254 | 2.353 | 2.458 | 2.528 | 2.560 |

2. (a) About 300 meter$^3$/sec.
   (b) About 250 meter$^3$/sec.
   (c) Looking at the graph, we can see that the 1996 flood reached its maximum just between March and April, for a high of about 1250 meter$^3$/sec. Similarly, the 1957 flood reached its maximum in mid-June, for a maximum flow rate of 3500 meter$^3$/sec.
   (d) The 1996 flood lasted about 1/3 of a month, or about 10 days. The 1957 flood lasted about 4 months.
   (e) The area under the controlled flood graph is about 2/3 box. Each box represents 500 meter$^3$/sec for one month. Since

   $$1 \text{ month} = 30 \frac{\text{days}}{\text{month}} \cdot 24 \frac{\text{hours}}{\text{day}} \cdot 60 \frac{\text{minutes}}{\text{hour}} \cdot 60 \frac{\text{seconds}}{\text{minute}}$$
   $$= 2.592 \cdot 10^6 \approx 3 \cdot 10^6 \text{ seconds},$$

   each box represents

   $$\text{Flow} \approx (500 \text{ meter}^3/\text{sec}) \cdot (2.6 \cdot 10^6 \text{ sec}) = 13 \cdot 10^8 \text{ meter}^3 \text{ of water.}$$

   So, for the artificial flood,

   $$\text{Additional flow} \approx \frac{2}{3} \cdot 13 \cdot 10^8 = 9 \cdot 10^8 \text{ meter}^3 \approx 10^9 \text{ meter}^3.$$

   (f) The 1957 flood released a volume of water represented by about 12 boxes above the 250 meter/sec baseline. Thus, for the natural flood,

   $$\text{Additional flow} \approx 12 \cdot 15 \cdot 10^8 = 1.8 \cdot 10^{10} \approx 2 \cdot 10^{10} \text{ meter}^3.$$

   So, the natural flood was nearly 20 times larger than the controlled flood and lasted much longer.

## Solutions to Problems on the Second Fundamental Theorem of Calculus

1. By the Second Fundamental Theorem, $G'(x) = x^3$.

2. By the Second Fundamental Theorem, $G'(x) = 3^x$.
3. By the Second Fundamental Theorem, $G'(x) = xe^x$.
4. The variable of integration does not affect the value of the integral, so by the Fundamental Theorem of Calculus, $G'(x) = \ln x$.
5. We find the changes in $f(x)$ between any two values of $x$ by counting the area between the curve of $f'(x)$ and the $x$-axis. Since $f'(x)$ is linear throughout, this is quite easy to do. From $x = 0$ to $x = 1$, we see that $f'(x)$ outlines a triangle of area $1/2$ below the $x$-axis (the base is 1 and the height is 1). By the Fundamental Theorem,

$$\int_0^1 f'(x)\,dx = f(1) - f(0),$$

so

$$f(0) + \int_0^1 f'(x)\,dx = f(1)$$

$$f(1) = 2 - \frac{1}{2} = \frac{3}{2}.$$

Similarly, between $x = 1$ and $x = 3$ we can see that $f'(x)$ outlines a rectangle below the $x$-axis with area $-1$, so $f(2) = 3/2 - 1 = 1/2$. Continuing with this procedure (note that at $x = 4$, $f'(x)$ becomes positive), we get the table below.

| $x$    | 0 | 1   | 2   | 3    | 4  | 5    | 6   |
|--------|---|-----|-----|------|----|------|-----|
| $f(x)$ | 2 | 3/2 | 1/2 | -1/2 | -1 | -1/2 | 1/2 |

6. (a) If $F(b) = \int_0^b 2^x\,dx$ then $F(0) = \int_0^0 2^x\,dx = 0$ since we are calculating the area under the graph of $f(x) = 2^x$ on the interval $0 \leq x \leq 0$, or on no interval at all.
   (b) Since $f(x) = 2^x$ is always positive, the value of $F$ will increase as $b$ increases. That is, as $b$ grows larger and larger, the area under $f(x)$ on the interval from 0 to $b$ will also grow larger.
   (c) Using a calculator or a computer, we get

$$F(1) = \int_0^1 2^x\,dx \approx 1.4,$$

$$F(2) = \int_0^2 2^x\,dx \approx 4.3,$$

$$F(3) = \int_0^3 2^x\,dx \approx 10.1.$$

7.

**TABLE 3.4**

| $x$    | 0 | 0.5  | 1    | 1.5  | 2    |
|--------|---|------|------|------|------|
| $I(x)$ | 0 | 0.50 | 1.09 | 2.03 | 3.65 |

8. Using the Fundamental Theorem, we know that the change in $F$ between $x = 0$ and $x = 0.5$ is given by

$$F(0.5) - F(0) = \int_0^{0.5} \sin t \cos t\,dt \approx 0.115.$$

Since $F(0) = 1.0$, we have $F(0.5) \approx 1.115$. The other values are found similarly, and are given in Table 3.5.

**TABLE 3.5**

| $b$    | 0 | 0.5     | 1       | 1.5    | 2       | 2.5     | 3       |
|--------|---|---------|---------|--------|---------|---------|---------|
| $F(b)$ | 1 | 1.11492 | 1.35404 | 1.4975 | 1.41341 | 1.17908 | 1.00996 |

9. Note that $\int_a^b g(x)\,dx = \int_a^b g(t)\,dt$. Thus, we have

$$\int_a^b \bigl(f(x) + g(x)\bigr)\,dx = \int_a^b f(x)\,dx + \int_a^b g(x)\,dx = 8 + 2 = 10.$$

10. Note that $\int_a^b (g(x))^2\,dx = \int_a^b (g(t))^2\,dt$. Thus, we have

$$\int_a^b \bigl((f(x))^2 - (g(x))^2\bigr)\,dx = \int_a^b (f(x))^2\,dx - \int_a^b (g(x))^2\,dx = 12 - 3 = 9.$$

11. We have

$$\int_a^b (f(x))^2\,dx - \left(\int_a^b f(x)\,dx\right)^2 = 12 - 8^2 = -52.$$

12. Note that $\int_a^b f(z)\,dz = \int_a^b f(x)\,dx$. Thus, we have

$$\int_a^b cf(z)\,dz = c\int_a^b f(z)\,dz = 8c.$$

# CHAPTER FOUR

## Solutions for Section 4.1

1. $\frac{dy}{dx} = 0$
2. $\frac{dy}{dx} = 3$
3. $\frac{dy}{dx} = 5$
4. $y' = 12x^{11}$.
5. $y' = -12x^{-13}$.
6. $y' = \frac{4}{3}x^{1/3}$.
7. $y' = 24t^2$
8. $y' = 12t^3 - 4t$
9. $f'(x) = -4x^{-5}$.
10. $f'(q) = 3q^2$
11. $f'(x) = C \cdot \frac{d}{dx}(x^2) = 2Cx$.
12. $y' = 2x + 5$.
13. $y' = 18x^2 + 8x - 2$.
14. $y' = -12x^3 - 12x^2 - 6$.
15. $\frac{dy}{dx} = 6x + 7$.
16. $\frac{dy}{dt} = 24t^2 - 8t + 12$.
17. $\frac{dy}{dq} = 8.4q - 0.5$.
18. $y' = 2ax + b$.
19. $y' = 2z - \frac{1}{2z^2}$.
20. $y' = 6t - \frac{6}{t^{3/2}} + \frac{2}{t^3}$.
21. $y' = 15t^4 - \frac{5}{2}t^{-1/2} - \frac{7}{t^2}$.

22. Figure 4.1 shows the graph of $f(x) = x^2 + 1$.

    We have $f'(x) = \frac{d}{dx}(x^2 + 1) = 2x$, thus, $f'(0) = 2(0) = 0$. We check this by seeing in that Figure 4.1 the tangent line at $x = 0$ has slope 0.

    We have $f'(1) = 2(1) = 2$, $f'(2) = 2(2) = 4$. and $f'(-1) = 2(-1) = -2$. Thus, the slope is positive at $x = 2$ and $x = 1$, and negative at $x = -1$.

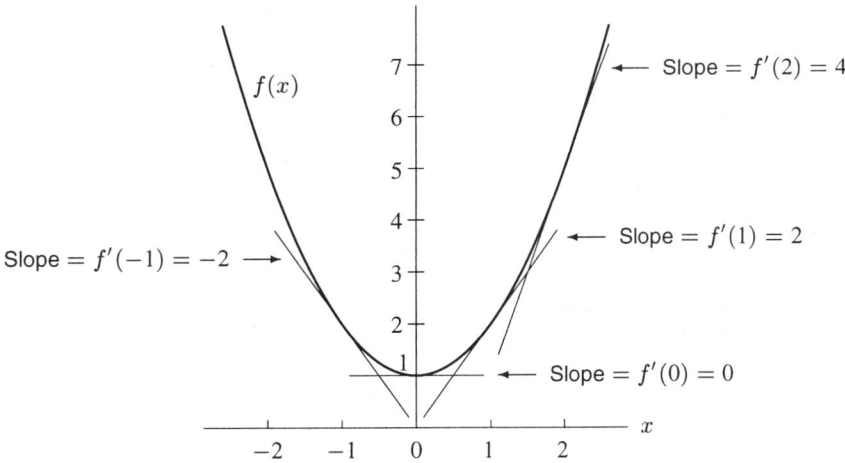

**Figure 4.1**: Using slopes to check values for derivatives

Moreover, it is greater at $x = 2$ than at $x = 1$. This agrees with the graph in Figure 4.1.

23. (a) $f'(t) = 2t - 4$.
    (b) $f'(1) = 2(1) - 4 = -2$
        $f'(2) = 2(2) - 4 = 0$

(c)

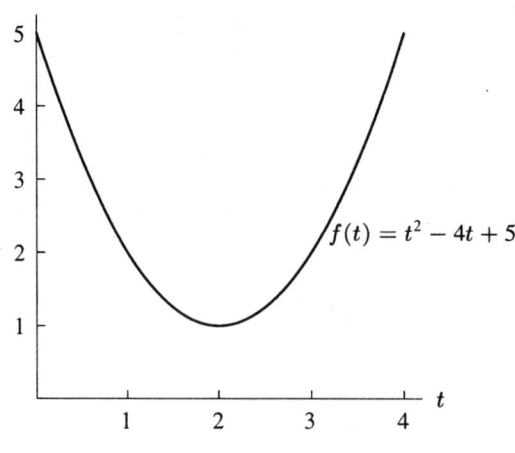

**Figure 4.2**

We see from part (b) that $f'(2) = 0$. This means that the slope of the line tangent to the curve at $x = 2$ is zero. From Figure 4.2, we see that indeed the tangent line is horizontal at the point $(2, 1)$. The fact that $f'(1) = -2$ means that the slope of the line tangent to the curve at $x = 1$ is $-2$. If we draw a line tangent to the graph at $x = 1$ (the point $(1, 2)$) we see that it does indeed have a slope of $-2$.

24. $f'(x) = 3x^2 - 8x + 7$, so $f'(0) = 7$, $f'(2) = 3$, and $f'(-1) = 18$.

25. $f'(x) = 2x + 3$, so $f'(0) = 3$, $f'(3) = 9$, and $f'(-2) = -1$.

26. (a)

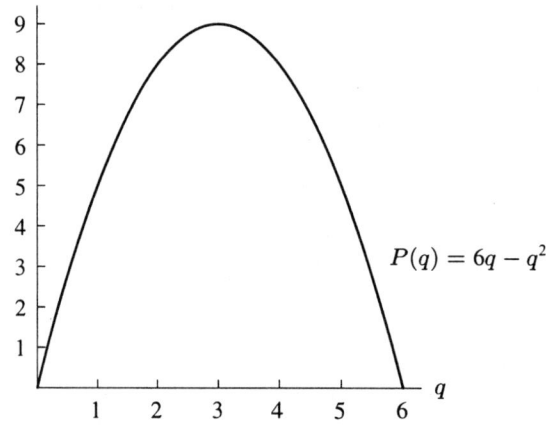

**Figure 4.3**

Since $P$ is increasing when $q = 1$, $P'(1)$ is positive. At $(3, 9)$ we observe that the function has a horizontal tangent line, and horizontal lines have slope zero. Thus $P'(3) = 0$. Finally, at $q = 4$ the function is decreasing; therefore the derivative $P'(4)$ is negative.

(b) The derivative of $P(q) = 6q - q^2$ is $P'(q) = 6 - 2q$. Therefore,

$$P'(1) = 6 - 2 = 4$$
$$P'(3) = 6 - 2(3) = 6 - 6 = 0$$
$$P'(4) = 6 - 2(4) = 6 - 8 = -2$$

27. The rate of change of the population is given by the derivative. For $P(t) = t^3 + 4t + 1$ the derivative is $P'(t) = 3t^2 + 4$. At $t = 2$, the rate of change of the population is $3(2)^2 + 4 = 12 + 4 = 16$, meaning the population is growing by 16 units per unit of time.

28. (a) $f'(x) = 3x^2 - 2(2x) + 3 + 0 = 3x^2 - 4x + 3$.
    (b) $f'(-1) = 3(-1)^2 - 4(-1) + 3 = 10$, $f'(0) = 3(0)^2 - 4(0) + 3 = 3$, $f'(1) = 3(1)^2 - 4(1) + 3 = 2$, $f'(2) = 3(2)^2 - 4(2) + 3 = 7$.

(c)

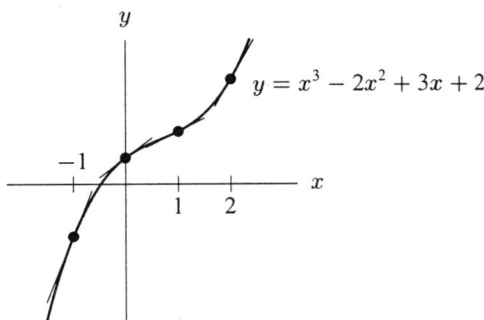

**Figure 4.4**

The slopes do match the answers we obtained in part (b), because the derivative of a function at a value $x$ is the same as the slope of the tangent line to the graph of the function at the point $(x, f(x))$.

29. $f'(t) = 6t^2 - 8t + 3$ and $f''(t) = 12t - 8$.

30. The derivative of $f(t)$ is $f'(t) = 4t^3 - 6t + 5$. The second derivative is the derivative of the derivative, and thus $f''(t) = 12t^2 - 6$.

31. (a) Since the power of $x$ will go down by one every time you take a derivative (until the exponent is zero after which the derivative will be zero), we can see immediately that $f^{(8)}(x) = 0$.
    (b) $f^{(7)}(x) = 7 \cdot 6 \cdot 5 \cdot 4 \cdot 3 \cdot 2 \cdot 1 \cdot x^0 = 5040$.

32.
$$f'(x) = 6x^2 - 4x \quad \text{so} \quad f'(1) = 6 - 4 = 2.$$
Thus the equation of the tangent line is $(y - 1) = 2(x - 1)$ or $y = 2x - 1$.

33.

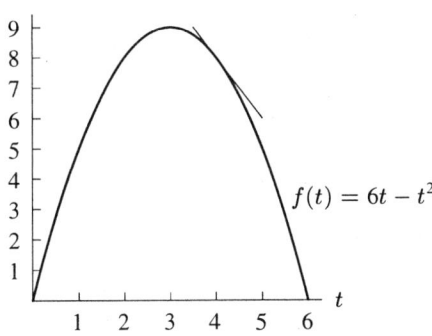

**Figure 4.5**

To find the equation of a line we need to have a point on the line and its slope. We know that this line is tangent to the curve $f(t) = 6t - t^2$ at $t = 4$. From this we know that both the curve and the line tangent to it will share the same point and the same slope. At $t = 4$, $f(4) = 6(4) - (4)^2 = 24 - 16 = 8$. Thus we have the point $(4, 8)$. To find the slope, we need to find the derivative. The derivative of $f(t)$ is $f'(t) = 6 - 2t$. The slope of the tangent line at $t = 4$ is $f'(4) = 6 - 2(4) = 6 - 8 = -2$. Now that we have a point and the slope, we can find an equation for the tangent line:
$$y = b + mt$$
$$8 = b + (-2)(4)$$
$$b = 16.$$
Thus, $y = -2t + 16$ is the equation for the line tangent to the curve at $t = 4$.

34. The marginal cost of producing the 25th item is $C'(25)$, where $C'(q) = 4q$, so the marginal cost is $100. This means that the cost of production increases by about $100 when we add one unit to a production level of 25 units.

35. (a) Let $R(p)$ represent revenue as a function of price. $R(p) = p(300 - 3p) = 300p - 3p^2$
    (b) The marginal revenue when the price is $10 is $R'(10)$, and $R'(p) = 300 - 6p$, so $R'(10) = 300 - 6(10) = 240$. The marginal revenue is 240. This means that revenues are going up at a rate of $240 per dollar of price increase when the price is $10.
    (c) $R'(p) = 300 - 6p$ is positive for $p < 50$ and negative for $p > 50$.

**176** CHAPTER FOUR /SOLUTIONS

36. After four months, there are $300(4)^2 = 4800$ mussels in the bay. The population is growing at the rate $Z'(4)$ mussels per month, where $Z'(t) = 600t$, so the rate of increase is 2400 mussels per month.

37. (a) The yield is $f(5) = 320 + 140(5) - 10(5)^2 = 770$ bushels per acre.
    (b) $f'(x) = 140 - 20x$, so $f'(5) = 40$ bushels per acre per pound of fertilizer. For each acre, yield will go up by about 40 bushels if an additional pound of fertilizer is used.
    (c) More should be used, because at this level of use, more fertilizer will result in a higher yield. Fertilizer's use should be increased until an additional unit results in a decrease in yield. i.e. until the derivative at that point becomes negative.

38. (a) The $p$-intercept is the value of $p$ when $q = 0$.
$$p = f(0) = 50 - 0.03(0)^2 = 50.$$

The $p$-intercept occurs at $p = 50$.
    The $q$-intercept is the value of $q$ such that $p = f(q) = 0$.
$$p = f(q) = 50 - 0.03q^2 = 0$$
$$-0.03q^2 = -50$$
$$q^2 = \frac{50}{0.03} \quad \text{and} \quad q \geq 0$$
$$q = \sqrt{\frac{50}{0.03}} \approx 40.825$$

The $q$-intercept occurs at $q \approx 40.825$.
    The $p$-intercept represents the price at which demand is zero. That is, when the price reaches 50 dollars, demand for the product will be zero. The $q$-intercept represents the demand for the product if the product were being given away free of charge. In this case, 40,825 units of the product would be consumed if the product were free ($p = 0$).

(b)
$$f(20) = 50 - 0.03(20)^2 = 50 - 0.03(400) = 50 - 12 = 38 \text{ dollars}.$$

This tells us that if the price per unit is $38, then a total of 20 units are demanded.

(c) To find $f'(20)$ we first find $f'(q) = 2(-0.03)q = -0.06q$. Therefore, $f'(20) = -0.06(20) = -1.2$ dollars per unit demanded. This tells us given a demand of 20 units, which, according to our answer to part (b), occurs when the unit price is $38, an increase of $1.20 in the price will result in the reduction of consumption by approximately 1 unit while a decrease in price by the same amount will lead to an increase of approximately 1 unit in sales.

39. (a) The marginal cost function describes the change in cost on the margin, i.e. the change in cost associated with each additional item produced. Thus it is the change in cost with respect to change in production ($dC/dq$). We recognize this as an expression for the derivative of $C(q)$ with respect to $q$. Thus, the marginal cost function equals $C'(q) = 0.08(3q^2) + 75 = 0.24q^2 + 75$.

(b)
$$C(50) = 0.08(50)^3 + 75(50) + 1000 = \$14,750.$$

$C(50)$ tells us how much it costs to produce 50 items. From above we can see that the company spends $14,750 to produce 50 items, and thus the units for $C(q)$ are dollars.
$$C'(50) = 0.24(50)^2 + 75 = \$675 \text{ per item}.$$

$C'(q)$ tells us the change in cost to produce one additional item of product. Thus at $q = 50$ costs will increase by $675 for each additional item of product produced, and thus the units are dollars/item.

40.
$$y' = 3x^2 - 18x - 16$$
$$5 = 3x^2 - 18x - 16$$
$$0 = 3x^2 - 18x - 21$$
$$0 = x^2 - 6x - 7$$
$$0 = (x+1)(x-7)$$
$$x = -1 \text{ or } x = 7.$$

When $x = -1, y = 7$; when $x = 7, y = -209$.
Thus, the two points are $(-1, 7)$ and $(7, -209)$.

41. (a) $f'(x) = 3x^2 - 12 = 3(x-2)(x+2)$, which is negative when $-2 < x < 2$. Thus $f(x)$ is decreasing for $-2 < x < 2$.
   (b) $f''(x) = 6x$, which is negative for $x < 0$, so the graph of $f(x)$ is concave down for $x < 0$.
   (c) $f(x)$ is both decreasing and concave down for $-2 < x < 0$.

42. (a) $R(q) = q(b + mq) = bq + mq^2$.
   (b) $R'(q) = b + 2mq$.

43. (a) Velocity $v(t) = \frac{dy}{dt} = \frac{d}{dt}(1250 - 16t^2) = -32t$.
   Since $t \geq 0$, the ball's velocity is negative. This is reasonable, since its height $y$ is decreasing.
   (b) The ball hits the ground when its height $y = 0$. This gives
   $$1250 - 16t^2 = 0$$
   $$t \approx \pm 8.84 \text{ seconds}$$
   We discard $t = -8.84$ because time $t$ is nonnegative. So the ball hits the ground about 8.84 seconds after its release, at which time its velocity is
   $$v(8.84) = -32(8.84) = -282.88 \text{ feet/sec} = -192.87 \text{ mph}.$$

44. (a) $A = \pi r^2$
   $\frac{dA}{dr} = 2\pi r$.
   (b) This is the formula for the circumference of a circle.
   (c) $A'(r) \approx \frac{A(r+h) - A(r)}{h}$ for small $h$. When $h > 0$, the numerator of the difference quotient denotes the area of the region contained between the inner circle (radius $r$) and the outer circle (radius $r + h$). See figure below. As $h$ approaches 0, this area can be approximated by the product of the circumference of the inner circle and the "width" of the region, i.e., $h$. Dividing this by the denominator, $h$, we get $A' = $ the circumference of the circle with radius $r$.

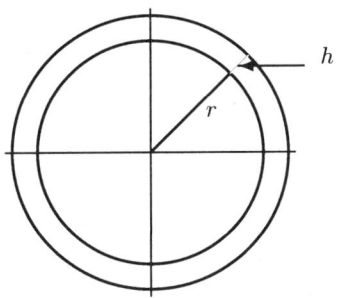

We can also think about the derivative of $A$ as the rate of change of area for a small change in radius. If the radius increases by a tiny amount, the area will increase by a thin ring whose area is simply the circumference at that radius times the small amount. To get the rate of change, we divide by the small amount and obtain the circumference.

45. $V = \frac{4}{3}\pi r^3$
   $\frac{dV}{dr} = 4\pi r^2 = $ surface area of a sphere.
   
   Our reasoning is similar to that of Problem 44. The difference quotient $\frac{V(r+h) - V(r)}{h}$ is the volume between two spheres divided by the change in radius. Furthermore, when $h$ is very small (and consequently $V(r+h) \approx V(r)$) this volume is like a coating of paint of depth $h$ applied to the surface of the sphere. The volume of the paint is about $h \cdot $ (Surface Area) for small $h$: dividing by $h$ gives back the surface area. Also, thinking about the derivative as the rate of change of the function for a small change in the variable, the answer seems clear. If you increase the radius of a sphere the tiniest amount, the volume will increase by a very thin layer whose volume will be the surface area at that radius multiplied by that tiniest amount.

## Solutions for Section 4.2

1. $y' = 10t + 4e^t$.
2. $f'(x) = 2e^x + 2x$.
3. $f'(x) = (\ln 2)2^x + 2(\ln 3)3^x$.
4. $\dfrac{dy}{dx} = 4(\ln 10)10^x - 3x^2$.
5. $\dfrac{dy}{dx} = 3 - 2(\ln 4)4^x$.
6. $\dfrac{dy}{dx} = \dfrac{1}{3}(\ln 3)3^x - \dfrac{33}{2}(x^{-\frac{3}{2}})$.
7. $f'(x) = 3x^2 + 3^x \ln 3$.
8. $\dfrac{dy}{dx} = 5 \cdot 5^t \ln 5 + 6 \cdot 6^t \ln 6$.
9. $P'(t) = Ce^t$.
10. $D' = -1/p$.
11. $R' = \dfrac{3}{q}$.
12. $y' = 2t + 5/t$
13. $y' = Ae^t$
14. $f'(x) = Ae^x - 2Bx$.
15. $\dfrac{dP}{dt} = 9t^2 + 2e^t$.
16. $P'(t) = 3000(\ln 1.02)(1.02)^t$.
17. $P'(t) = 12.41(\ln 0.94)(0.94)^t$.
18. $\dfrac{dy}{dx} = 5(\ln 2)(2^x) - 5$.
19. $R'(q) = 2q - 2/q$.
20. $\dfrac{dy}{dx} = 2x + 4 - 3/x$.
21. $Ae^t + \dfrac{B}{t}$.

22. $f(t) = 4 - 2e^t$, so $f'(t) = -2e^t$ and $f'(-1) = -2e^{-1} \approx -0.736$. $f'(0) = -2e^0 = -2$. $f'(1) = -2e^1 \approx -5.437$.
    As expected, the slopes of the line segments do match the derivatives found. See Figure 4.6.

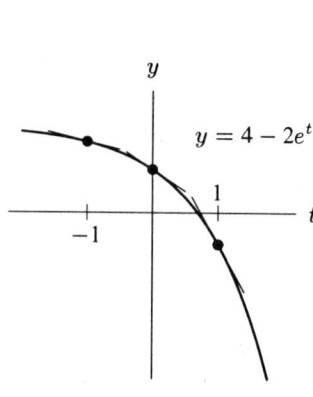

Figure 4.6

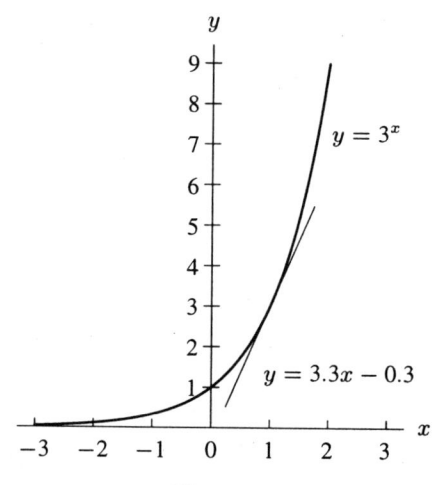

Figure 4.7

23. See Figure 4.7. The slope of the line tangent to the graph of the function at $x = 1$ will be the derivative of the function evaluated at $x = 1$. Since $y = 3^x$, $\dfrac{dy}{dx} = (\ln 3)3^x$. At $x = 1$, the derivative is $3 \ln 3 \approx 3.3$; this is the slope of the line tangent to the curve at $x = 1$. The function evaluated at $x = 1$ yields the point $(1, 3)$. Thus the equation of the line is
$$y - 3 = 3.3(x - 1)$$
$$y = 3.3x - 0.3$$

Thus the equation of the line is $y = 3.3x - 0.3$.

24. $C(500) = 1000 + 300 \ln(500) \approx 2864.38$; it costs about \$2864 to produce 500 units. $C'(q) = \dfrac{300}{q}$, $C'(500) = \dfrac{300}{500} = 0.6$.
    When the production level is 500, each additional unit costs about \$0.60 to produce.

25. (a) $f(x) = 1 - e^x$ crosses the $x$-axis where $0 = 1 - e^x$, which happens when $e^x = 1$, so $x = 0$. Since $f'(x) = -e^x$, $f'(0) = -e^0 = -1$.
(b) $y = -x$

26. Since $P = 1 \cdot (1.05)^t$, $\frac{dP}{dt} = \ln(1.05)1.05^t$. When $t = 10$,
$$\frac{dP}{dt} = (\ln 1.05)(1.05)^{10} \approx \$0.07947/\text{year} \approx 7.95¢/\text{year}.$$

27.
$$\frac{dP}{dt} = 35{,}000 \cdot (\ln 0.98)(0.98^t).$$

At $t = 23$, this is $35{,}000(\ln 0.98)(0.98^{23}) \approx -444.3 \frac{\text{people}}{\text{year}}$. (Note: the negative sign indicates that the population is decreasing.)

28. $\frac{dV}{dt} = 75(1.35)^t \ln 1.35 \approx 22.5(1.35)^t$.

29. (a) $P = 4.1(1 + 0.02)^t = 4.1(1.02)^t$ billion.
(b)
$$\frac{dP}{dt} = 4.1 \frac{d}{dt}(1.02)^t = 4.1(1.02)^t(\ln 1.02).$$
$$\left.\frac{dP}{dt}\right|_{t=0} = 4.1(1.02)^0 \ln 1.02 \approx 0.0812 \text{ billion people per year.}$$
$$\left.\frac{dP}{dt}\right|_{t=15} = 4.1(1.02)^{15} \ln 1.02 \approx 0.1093 \text{ billion people per year.}$$

$\frac{dP}{dt}$ is the rate of growth of the world's population; $\left.\frac{dP}{dt}\right|_{t=0}$ and $\left.\frac{dP}{dt}\right|_{t=15}$ are the rates of growth in the years 1975 and 1990, respectively.

30. (a) $P = 10.8(0.998)^{10} \approx 10.59$ million.
(b)
$$\frac{dP}{dt} = 10.8(\ln 0.998)(0.998)^t$$
so $\left.\frac{dP}{dt}\right|_{t=10} = 10.8(\ln 0.998)(0.998)^{10} \approx -0.02$ million/year.

Thus in 2000, Hungary's population will be decreasing by 20,000 people per year.

31. The derivative of $e^x$ is $\frac{d}{dx}(e^x) = e^x$. Thus the tangent line at $x = 0$, has slope $e^0 = 1$, and the tangent line is $y = x + 1$. A function which is always concave up will always stay above any of its tangent lines. Thus $e^x \geq x + 1$ for all $x$, as shown in Figure 4.8.

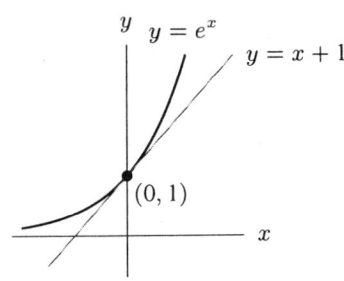

**Figure 4.8**

32. (a) $y = \ln x, y' = \frac{1}{x}; f'(1) = \frac{1}{1} = 1.$
$y - y_1 = m(x - x_1), y - 0 = 1(x - 1); y = g(x) = x - 1.$
(b) $g(1.1) = 1.1 - 1 = 0.1; g(2) = 2 - 1 = 1.$
(c) $f(1.1)$ and $f(2)$ are below $g(x) = x - 1$. $f(0.9)$ and $f(0.5)$ are also below $g(x)$. This would be true for any approximation of this function by a tangent line since $f$ is concave down ($f''(x) = -\frac{1}{x^2} < 0$ for all $x \neq 0$). See figure below. Thus, for a given $x$-value, the $y$-value given by the function is always below the value given by the tangent line.

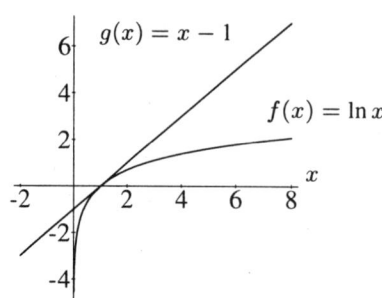

33. We are interested in when the derivative $\dfrac{d(a^x)}{dx}$ is positive and when it is negative. The quantity $a^x$ is always positive. However $\ln a > 0$ for $a > 1$ and $\ln a < 0$ for $0 < a < 1$. Thus the function $a^x$ is increasing for $a > 1$ and decreasing for $a < 1$.

34. The equation $2^x = 2x$ has solutions $x = 1$ and $x = 2$. (Check this by substituting these values into the equation). The graph below suggests that these are the only solutions, but how can we be sure?

Let's look at the slope of the curve $f(x) = 2^x$, which is $f'(x) = (\ln 2)2^x \approx (0.693)2^x$, and the slope of the line $g(x) = 2x$ which is 2. At $x = 1$, the slope of $f(x)$ is less than 2; at $x = 2$, the slope of $f(x)$ is more than 2. Since the slope of $f(x)$ is always increasing, there can be no other point of intersection. (If there were another point of intersection, the graph $f$ would have to "turn around".)

Here's another way of seeing this. Suppose $g(x)$ represents the position of a car going a steady 2 mph, while $f(x)$ represents a car which starts ahead of $g$ (because the graph of $f$ is above $g$) and is initially going slower than $g$. The car $f$ is first overtaken by $g$. All the while, however, $f$ is speeding up until eventually it overtakes $g$ again. Notice that the two cars will only meet twice (corresponding to the two intersections of the curve): once when $g$ overtakes $f$ and once when $f$ overtakes $g$.

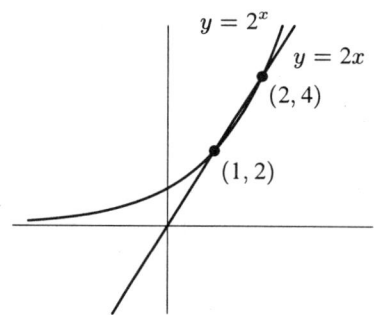

35. Since $y = 2^x$, $y' = (\ln 2)2^x$. At $(0, 1)$, the tangent line has slope $\ln 2$ so its equation is $y = (\ln 2)x + 1$. At $c$, $y = 0$, so $0 = (\ln 2)c + 1$, thus $c = -\frac{1}{\ln 2}$.

36.

$$g(x) = ax^2 + bx + c \qquad\qquad f(x) = e^x$$
$$g'(x) = 2ax + b \qquad\qquad f'(x) = e^x$$
$$g''(x) = 2a \qquad\qquad f''(x) = e^x$$

So, using $g''(0) = f''(0)$, etc., we have $2a = 1$, $b = 1$, and $c = 1$, and thus $g(x) = \frac{1}{2}x^2 + x + 1$, as shown in the figure below.

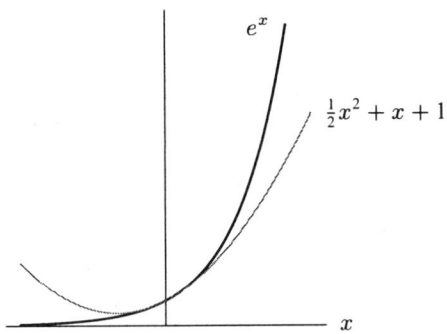

The two functions do look very much alike near $x = 0$. They both increase for large values of $x$, but $e^x$ increases much more quickly. For very negative values of $x$, the quadratic goes to $\infty$ whereas the exponential goes to 0. By choosing a function whose first few derivatives agreed with the exponential when $x = 0$, we got a function which looks like the exponential for $x$-values near 0.

37. (a) $V(4) = 25(0.85)^4 = 25(0.522) = 13{,}050$. Thus the value of the car after 4 years is \$13,050.
(b) We have a function of the form $f(t) = Ca^t$. We know that such functions have a derivative of the form $(C \ln a) \cdot a^t$. Thus, $V'(t) = 25(0.85)^t \cdot \ln 0.85 = -4.063(0.85)^t$. The units would be the change in value (in thousands of dollars) with respect to time (in years), or thousands of dollars/year.
(c) $V'(4) = -4.063(0.85)^4 = -4.063(0.522) = -2.121$. This means that at the end of the fourth year, the value of the car is decreasing by \$2121 per year.
(d) $V(t)$ is a positive decreasing function, so that the value of the automobile is positive and decreasing. $V'(t)$ is a negative function whose magnitude is decreasing, meaning the value of the automobile is always dropping, but the yearly loss of value is less as time goes on. The graphs of $V(t)$ and $V'(t)$ confirm that the value of the car decreases with time. What they do not take into account are the *costs* associated with owning the vehicle. At some time, $t$, it is likely that the costs of owning the vehicle will outweigh its value. At that time, it may no longer be worthwhile to keep the car.

## Solutions for Section 4.3

1. $f'(x) = 99(x + 1)^{98} \cdot 1 = 99(x + 1)^{98}$.

2. Let us begin by setting $u = q^2 + 1$. We now have $R = u^4$. Thus $\dfrac{dR}{dq} = \dfrac{dR}{du} \cdot \dfrac{du}{dq}$. Now $\dfrac{dR}{du} = 4u^3$, while $\dfrac{du}{dq} = 2q$. Substituting $q^2 + 1$ back in for $u$, we have

$$\frac{dR}{dq} = \frac{dR}{du} \cdot \frac{du}{dq} = 4(q^2 + 1)^3 \cdot (2q) = 8q(q^2 + 1)^3.$$

3. $w' = 100(t^2 + 1)^{99}(2t) = 200t(t^2 + 1)^{99}$.

4. $w' = 100(t^3+1)^{99}(3t^2) = 300t^2(t^3+1)^{99}$.

5. We need to find $\dfrac{dw}{dr}$. Let's begin by setting $u = 5r - 6$. We now have $w = u^3$. We know that $\dfrac{dw}{dr} = \dfrac{dw}{du} \cdot \dfrac{du}{dr}$. Now $\dfrac{dw}{du} = 3u^2$, while $\dfrac{du}{dr} = 5$. Thus, substituting $5r - 6$ back in for $u$, we get
$$\frac{dw}{dr} = \frac{dw}{du} \cdot \frac{du}{dr} = 3(5r-6)^2 \cdot 5 = 15(5r-6)^2.$$

6. $f'(t) = (e^{3t})(3) = 3e^{3t}$.

7. Letting $u = 0.7t$, we have $y = e^u$ and $\dfrac{du}{dt} = 0.7$. We know that the derivative of $e^u$ is $e^u$, and thus $\dfrac{dy}{du} = e^u$. Substituting $0.7t$ for $u$ back into our original equation yields
$$\frac{dy}{dt} = \frac{dy}{du} \cdot \frac{du}{dt} = e^{0.7t} \cdot (0.7) = 0.7e^{0.7t}.$$

8. $y' = -4e^{-4t}$.

9. $y' = \dfrac{3s^2}{2\sqrt{s^3+1}}$.

10. $w' = \dfrac{1}{2\sqrt{s}} e^{\sqrt{s}}$.

11. $P' = -0.2e^{-0.2t}$.

12. $\dfrac{dw}{dt} = -6te^{-3t^2}$.

13. $\dfrac{dy}{dt} = \dfrac{5}{5t+1}$.

14. $P' = 50(-0.6)e^{-0.6t} = -30e^{-0.6t}$.

15. $\dfrac{dP}{dt} = 200(0.12)e^{0.12t} = 24e^{0.12t}$.

16. $\dfrac{dy}{dx} = -3(2x) + 2(3e^{3x}) = -6x + 6e^{3x}$.

17. By the chain rule,
$$\frac{dC}{dq} = (12)((3)(3q^2-5)^2)(6q) = 216q(3q^2-5)^2.$$

18. $f'(x) = 6(e^{5x})(5) + (e^{-x^2})(-2x) = 30e^{5x} - 2xe^{-x^2}$.

19. $\dfrac{dy}{dt} = 5(5e^{5t+1}) = 25e^{5t+1}$.

20. $f'(x) = \dfrac{-1}{1-x} = \dfrac{1}{x-1}$.

21. $f'(t) = \dfrac{2t}{t^2+1}$.

22. $f'(x) = \dfrac{1}{1-e^{-x}} \cdot -e^{-x}(-1) = \dfrac{e^{-x}}{1-e^{-x}}$.

23. $f'(x) = \dfrac{1}{e^x+1} \cdot e^x$.

24. $f'(t) = 5 \cdot \dfrac{1}{5t+1} \cdot 5 = \dfrac{25}{5t+1}$.

25. $g'(t) = \dfrac{1}{4t+9}(4) = \dfrac{4}{4t+9}$.

26. $\dfrac{dy}{dx} = \dfrac{1}{3t+2}(3) = \dfrac{3}{3t+2}$.

27. $\dfrac{dQ}{dt} = 100(0.5)(t^2+5)^{-0.5}(2t) = 100t(t^2+5)^{-0.5}$.

28. $\dfrac{dy}{dx} = 5 + \dfrac{1}{x+2}(1) = 5 + \dfrac{1}{x+2}$.

29. $\dfrac{dy}{dx} = 2(5+e^x)e^x$.

30. $\dfrac{dP}{dx} = (0.5)(1+\ln x)^{-0.5}\left(\dfrac{1}{x}\right) = \dfrac{0.5}{x(1+\ln x)^{0.5}}$.

31. $y = e^{-2t}$, $y' = -2e^{-2t}$. At $t = 0$, $y = 1$ and $y' = -2$. Thus the tangent line at $(0, 1)$ is $y = -2t + 1$.

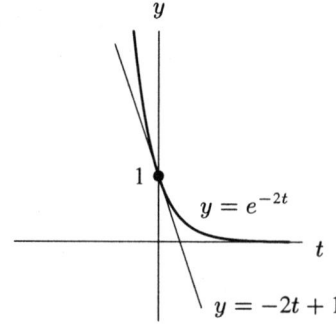

**Figure 4.9**

32.
$$f(x) = 6e^{5x} + e^{-x^2} \qquad f'(x) = 30e^{5x} - 2xe^{-x^2}$$
$$f(1) = 6e^5 + e^{-1} \qquad f'(1) = 30e^5 - 2(1)e^{-1}$$

$$y - y_1 = m(x - x_1)$$
$$y - (6e^5 + e^{-1}) = (30e^5 - 2e^{-1})(x - 1)$$
$$y - (6e^5 + e^{-1}) = (30e^5 - 2e^{-1})x - (30e^5 - 2e^{-1})$$
$$y = (30e^5 - 2e^{-1})x - 30e^5 + 2e^{-1} + 6e^5 + e^{-1}$$
$$\approx 4451.66x - 3560.81.$$

33. $f(p) = 10,000e^{-0.25p}$, $f(2) = 10,000e^{-0.5} \approx 6065$. If the product sells for $2, then 6065 units can be sold.

$$f'(p) = 10,000e^{-0.25p}(-0.25) = -2500e^{-0.25p}$$
$$f'(2) = -2,500e^{-0.5} \approx -1516$$

$f'(2) = -1516$ means that at a price of $2, a $1 increase in price will result in a decrease in quantity sold of 1516 units.

34.
$$C(q) = 1000 + 30e^{0.05q}$$
$$C(50) = 1000 + 30e^{2.5} \approx 1365$$

so it costs about $1365 to produce 50 units.

$$C'(q) = 30(0.05)e^{0.05q} = 1.5e^{0.05q}$$
$$C'(50) = 1.5e^{2.5} \approx 18.27$$

It costs about $18.27 to produce an additional unit when the production level is 50 units.

35. We have $P = f(t) = 5.3e^{0.018t}$, so $\dfrac{dp}{dt} = f'(t) = 5.3(e^{0.018t} \cdot 0.018) = 0.0954e^{0.018t}$. Therefore $f(0) = 5.3$ billion people, and $f'(0) = 0.095$ billion people per year. In 1990, the population of the world was 5.3 billion people and was increasing at a rate of 0.095 billion people per year. We also see that $f(10) = 5.3e^{0.018(10)} = 6.345$ billion people, and $f'(10) = 0.0954e^{0.018(10)} = 0.114$ billion people per year. In the year 2000, the model predicts that the population of the world will be 6.345 billion people and will be incresing at a rate of 0.114 billion people per year. (This is an increase of about 217 people every minute.)

36. (a)
$$\frac{dQ}{dt} = \frac{d}{dt}e^{-0.000121t}$$
$$= -0.000121e^{-0.000121t}$$

(b)

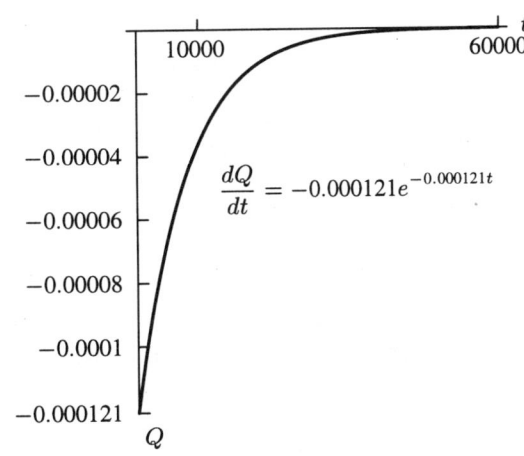

37. (a)
$$\frac{dH}{dt} = \frac{d}{dt}(40 + 30e^{-2t}) = 30(-2)e^{-2t} = -60e^{-2t}.$$

(b) Since $e^{-2t}$ is always positive, $\frac{dH}{dt} < 0$; this makes sense because the temperature of the soda is decreasing.

(c) The magnitude of $\frac{dH}{dt}$ is
$$\left|\frac{dH}{dt}\right| = \left|-60e^{-2t}\right| = 60e^{-2t} \leq 60 = \left|\frac{dH}{dt}\right|_{t=0},$$
since $e^{-2t} \leq 1$ for all $t \geq 0$ and $e^0 = 1$. This is just saying that at the moment that the can of soda is put in the refrigerator (at $t = 0$), the temperature difference between the soda and the inside of the refrigerator is the greatest, so the temperature of the soda is dropping the quickest.

38. (a) $\frac{dB}{dt} = P\left(1 + \frac{r}{100}\right)^t \ln\left(1 + \frac{r}{100}\right)$. The expression $\frac{dB}{dt}$ tells us how fast the amount of money in the bank is changing with respect to time for fixed initial investment $P$ and interest rate $r$.

(b) $\frac{dB}{dr} = Pt\left(1 + \frac{r}{100}\right)^{t-1}\frac{1}{100}$. The expression $\frac{dB}{dr}$ indicates how fast the amount of money changes with respect to the interest rate $r$, assuming fixed initial investment $P$ and time $t$.

## Solutions for Section 4.4

1. By the product rule, $f'(x) = 2x(x^3 + 5) + x^2(3x^2) = 2x^4 + 3x^4 + 10x = 5x^4 + 10x$. Alternatively, $f'(x) = (x^5 + 5x^2)' = 5x^4 + 10x$. The two answers should, and do, match.

2. By the product rule,
$$f'(x) = 2(3x - 2) + (2x + 1) \cdot 3 = 12x - 1.$$
Alternatively,
$$f'(x) = (6x^2 - x - 2)' = 12x - 1.$$
The two answers match.

3. $f'(x) = x \cdot e^x + e^x \cdot 1 = e^x(x + 1)$.
4. $f'(t) = (1)e^{-2t} + t(-2e^{-2t}) = e^{-2t} - 2te^{-2t}$.
5. $y' = 2^x + x(\ln 2)2^x = 2^x(1 + x \ln 2)$.
6. Differentiating with respect to $x$, we have
$$\frac{dy}{dx} = \frac{d}{dx}(5xe^{x^2}) = \left(\frac{d}{dx}(5x)\right)e^{x^2} + 5x\frac{d}{dx}(e^{x^2})$$
$$= (5)e^{x^2} + 5x(e^{x^2} \cdot 2x)$$
$$= 5e^{x^2} + 10x^2 e^{x^2}.$$

7. Differentiating with respect to $t$, we have
$$\frac{dy}{dt} = \frac{d}{dt}(t^2(3t+1)^3) = \frac{d}{dt}(t^2) \cdot (3t+1)^3 + t^2\frac{d}{dt}((3t+1)^3)$$
$$= (2t)(3t+1)^3 + t^2(3(3t+1)^2 \cdot 3)$$
$$= 2t(3t+1)^3 + 9t^2(3t+1)^2$$

8. Differentiating with respect to $x$, we have
$$\frac{dy}{dx} = \frac{d}{dx}(x \ln x) = \left(\frac{d}{dx}(x)\right)\ln x + x\frac{d}{dx}(\ln x)$$
$$= 1 \cdot \ln x + x \cdot \frac{1}{x}$$
$$= \ln x + 1.$$

9. $w' = (3t^2 + 5)(t^2 - 7t + 2) + (t^3 + 5t)(2t - 7)$.

10. $\dfrac{dy}{dt} = 2te^t + (t^2 + 3)e^t = e^t(t^2 + 2t + 3)$.

11. $\dfrac{dz}{dt} = (3t + 1)(5) + (3)(5t + 2) = 15t + 5 + 15t + 6 = 30t + 11$. We could have started by multiplying the factors to obtain $15t^2 + 11t + 2$, and then taken the derivative of the result.

12. $y' = (3t^2 - 14t)e^t + (t^3 - 7t^2 + 1)e^t = (t^3 - 4t^2 - 14t + 1)e^t$.

13. $\dfrac{dP}{dt} = (t^2)(\dfrac{1}{t}) + (2t)(\ln t) = t + 2t \ln t$.

14. Divide and then differentiate
$$f(x) = x + \dfrac{3}{x}$$
$$f'(x) = 1 - \dfrac{3}{x^2}.$$

15. $\dfrac{dR}{dq} = (3q)(-e^{-q}) + (e^{-q})(3) = -3qe^{-q} + 3e^{-q}$.

16. $y' = 1 \cdot e^{-t^2} + te^{-t^2}(-2t)$

17. $f'(z) = \dfrac{1}{2\sqrt{z}}e^{-z} - \sqrt{z}e^{-z}$.

18. $g'(p) = p\left(\dfrac{2}{2p+1}\right) + \ln(2p+1)(1) = \dfrac{2p}{2p+1} + \ln(2p+1)$.

19. $f'(t) = 1 \cdot e^{5-2t} + te^{5-2t}(-2) = e^{5-2t}(1 - 2t)$.

20.
$$f'(w) = (e^{w^2})(10w) + (5w^2 + 3)(e^{w^2})(2w)$$
$$= 2we^{w^2}(5 + 5w^2 + 3)$$
$$= 2we^{w^2}(5w^2 + 8).$$

21. $f'(x) = \dfrac{e^x \cdot 1 - x \cdot e^x}{(e^x)^2} = \dfrac{e^x(1-x)}{(e^x)^2} = \dfrac{1-x}{e^x}$.

22. Using the quotient rule, we have
$$\dfrac{dw}{dz} = \dfrac{d}{dz}\left(\dfrac{3z}{1+2z}\right) = \dfrac{3(1+2z) - 3z(2)}{(1+2z)^2} = \dfrac{3}{(1+2z)^2}.$$

23. Using the quotient rule gives
$$\dfrac{dz}{dt} = \dfrac{d}{dt}\left(\dfrac{1-t}{1+t}\right) = \dfrac{-1\cdot(1+t) - (1-t)\cdot 1}{(1+t)^2} = \dfrac{-1-t-1+t}{(1+t)^2} = \dfrac{-2}{(1+t)^2}.$$

24. Using the quotient rule,
$$\dfrac{dy}{dx} = \dfrac{d}{dx}\left(\dfrac{e^x}{1+e^x}\right) = \dfrac{e^x(1+e^x) - e^x(e^x)}{(1+e^x)^2} = \dfrac{e^x + e^{2x} - e^{2x}}{(1+e^x)^2} = \dfrac{e^x}{(1+e^x)^2}.$$

25. Using the quotient rule,
$$\dfrac{dw}{dy} = \dfrac{d}{dy}\left(\dfrac{3y + y^2}{5+y}\right) = \dfrac{(3+2y)(5+y) - (3y+y^2)\cdot 1}{(5+y)^2}$$
$$= \dfrac{15 + 13y + 2y^2 - 3y - y^2}{(5+y)^2} = \dfrac{15 + 10y + y^2}{(5+y)^2}.$$

26. Using the quotient rule, we have

$$\frac{dy}{dz} = \frac{d}{dz}\left(\frac{1+z}{\ln z}\right) = \frac{1 \cdot \ln z - (1+z)(1/z)}{(\ln z)^2} = \frac{z \ln z - 1 - z}{z(\ln z)^2}.$$

27. $f(x) = x^2 e^{-x}$, $f(0) = 0$

$f'(x) = 2xe^{-x} + x^2 e^{-x} \cdot (-1) = e^{-x}(2x - x^2)$, so $f'(0) = 0$. Thus the tangent line is $y = 0$ (the x-axis). See Figure 4.10.

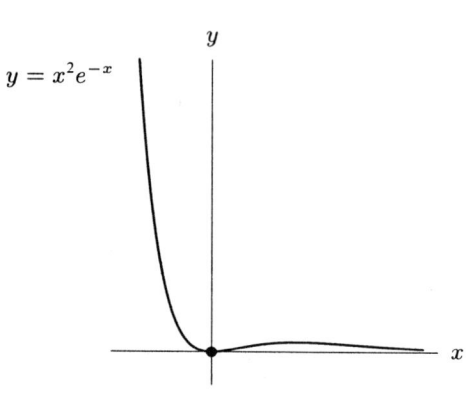

Figure 4.10

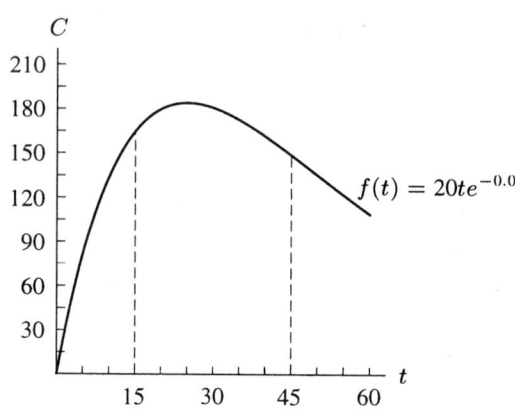

Figure 4.11

28. (a) $R(p) = p \cdot 1000e^{-0.02p} = 1000pe^{-0.02p}$.
    (b) $R'(p) = 1000e^{-0.02p} + 1000pe^{-0.02p}(-0.02) = e^{-0.02p}(1000 - 20p)$
    (c) $R(10) = 10,000e^{-0.2} \approx 8187$; you will have about 8187 dollars in revenue if you sell the product for $10.
        $R'(10) = e^{-0.2}(1000 - 200) \approx 655$; a one dollar increase in price over $10 will generate about $655 in additional revenue.

29. (a) See Figure 4.11. Looking at the graph of $C$, we can see that the see that at $t = 15$, $C$ is increasing. Thus, the slope of the curve at that point is positive, and so $f'(15)$ is also positive. At $t = 45$, the function is decreasing, i.e. the slope of the curve is negative, and thus $f'(45)$ is negative.
    (b) We begin by differentiating the function:

$$f'(t) = (20t)(-0.04e^{-0.04t}) + (e^{-0.04t})(20)$$
$$f'(t) = e^{-0.04t}(20 - 0.8t).$$

At $t = 30$,

$$f(30) = 20(30)e^{-0.04 \cdot (30)} = 600e^{-1.2} \approx 181 \text{ mg/ml}$$
$$f'(30) = e^{-1.2}(20 - (0.8)(30)) = e^{-1.2}(-4) \approx -1.2 \text{ mg/ml/min}.$$

These results mean the following: At $t = 30$, or after 30 minutes, the concentration of the drug in the body ($f(30)$) is about 181 mg/ml. The rate of change of the concentration ($f'(30)$) is about $-1.2$ mg/ml/min, meaning that the concentration of the drug in the body is dropping by 1.2 mg/ml each minute at $t = 30$ minutes.

30. To find the equation of the line tangent to the graph of $P(t) = t \ln t$ at $t = 2$ we must find the point $(2, P(2))$ as well as the slope of the tangent line at $t = 2$. $P(2) = 2(\ln 2) \approx 1.386$. Thus we have the point $(2, 1.386)$. To find the slope, we must first find $P'(t)$:

$$P'(t) = t\frac{1}{t} + \ln t(1) = 1 + \ln t.$$

At $t = 2$ we have

$$P'(2) = 1 + \ln 2 \approx 1.693$$

Since we now have the slope of the line and a point, we can solve for the equation of the line:

$$Q(t) - 1.386 = 1.693(t - 2)$$
$$Q(t) - 1.386 = 1.693t - 3.386$$
$$Q(t) = 1.693t - 2.$$

The equation of the tangent line is $Q(t) = 1.693t - 2$. We see our results displayed graphically in Figure 4.12.

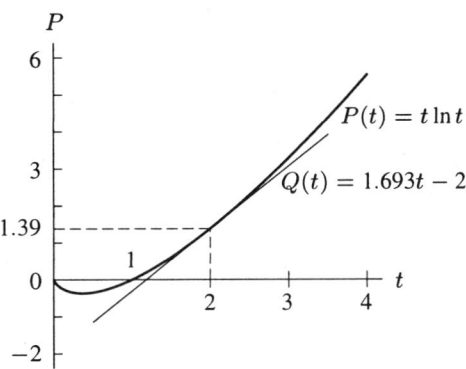

**Figure 4.12**

31. (a) $f(140) = 15{,}000$ says that 15,000 skateboards are sold when the cost is $140 per board.
    $f'(140) = -100$ means that if the price is increased from $140, roughly speaking, every dollar of increase will decrease the total sales by 100 boards.
    (b) $\dfrac{dR}{dp} = \dfrac{d}{dp}(p \cdot q) = \dfrac{d}{dp}\left(p \cdot f(p)\right) = f(p) + pf'(p).$
    So,
    $$\left.\dfrac{dR}{dp}\right|_{p=140} = f(140) + 140f'(140)$$
    $$= 15{,}000 + 140(-100) = 1000.$$
    (c) From (b) we see that $\left.\dfrac{dR}{dp}\right|_{p=140} = 1000 > 0$. This means that the revenue will increase by about $1000 if the price is raised by $1.

32. The fact that $f(80) = 0.05$ means that when the car is moving at 80 km/hr is it using up 0.05 liters of gasoline per kilometer traveled.

    $f'(v)$ is the rate of change of gasoline consumption with respect to speed. That is, $f'(v)$ tells us how the consumption of gasoline is changing as speeds vary. We are told that $f'(80) = 0.0005$. This means that a 1-kilometer increase in the speed of the vehicle will result in an increase in consumption of 0.0005 liters per km. Put another way, an increase in speed means that the vehicle will be burning more gasoline per km traveled than at lower speeds.

## Solutions for Section 4.5

1. $\dfrac{dy}{dx} = 5\cos x.$

2. $\dfrac{dP}{dt} = -\sin t.$

3. $\dfrac{dy}{dt} = 2t - 5\sin t.$

4. $\dfrac{dy}{dt} = A\cos t.$

5. $\dfrac{dy}{dx} = 5\cos x - 5.$

6. $R'(q) = 2q + 2\sin q.$

**188** CHAPTER FOUR /SOLUTIONS

7. $R' = 5\cos(5t)$.

8. $\dfrac{dW}{dt} = 4(-\sin(t^2)) \cdot 2t = -8t\sin(t^2)$.

9. $\dfrac{dy}{dt} = A(\cos(Bt)) \cdot B = AB\cos(Bt)$.

10. $\dfrac{dy}{dx} = 2x\cos(x^2)$.

11. $\dfrac{dy}{dt} = 2(-\sin(5t))(5) = -10\sin(5t)$.

12. $\dfrac{dy}{dt} = 6 \cdot 2\cos(2t) + 3(-4\sin(4t)) = 12(\cos(2t) - \sin(4t))$.

13. $f'(x) = \cos(3x) \cdot 3 = 3\cos(3x)$.

14. $z' = -4\sin(4\theta)$.

15. $f'(x) = (2x)(\cos x) + x^2(-\sin x) = 2x\cos x - x^2\sin x$.

16. $f'(x) = 2 \cdot [\sin(3x)] + 2x[\cos(3x)] \cdot 3 = 2\sin(3x) + 6x\cos(3x)$

17. Using the quotient rule, we get

$$\dfrac{d}{dt}\left(\dfrac{t^2}{\cos t}\right) = \dfrac{2t\cos t - t^2(-\sin t)}{(\cos t)^2}$$

$$= \dfrac{2t\cos t + t^2\sin t}{(\cos t)^2}.$$

18. Using the quotient rule, we have

$$\dfrac{d}{d\theta}\left(\dfrac{\sin\theta}{\theta}\right) = \dfrac{(\cos\theta)(\theta) - (\sin\theta)(1)}{\theta^2} = \dfrac{\theta\cos\theta - \sin\theta}{\theta^2}.$$

This problem can also be done by writing $(\sin\theta)/\theta = (\sin\theta)\theta^{-1}$ and using the product rule.

19. At $x = \pi$, $y = \sin\pi = 0$, and the slope $\dfrac{dy}{dx}\Big|_{x=\pi} = \cos x\Big|_{x=\pi} = -1$. Therefore the equation of the tangent line is $y = -(x - \pi) = -x + \pi$. See Figure 4.13.

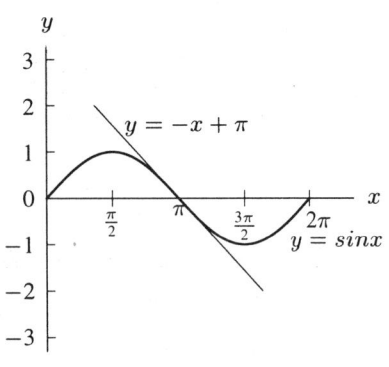

*Figure 4.13*

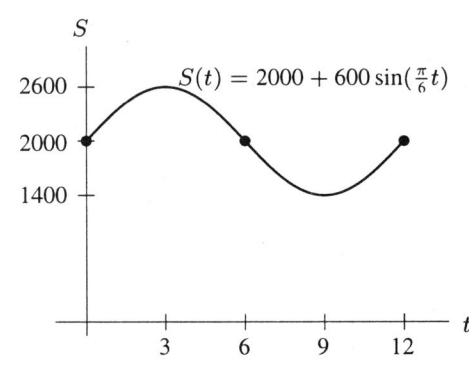

*Figure 4.14*

20. (a) Looking at the graph in Figure 4.14, we see that the maximum is $2600 per month and the minimum is $1400 per month. If $t = 0$ is January 1, then the sales are highest on April 1.

(b) $S(2)$ is the monthly sales on March 1,

$$S(2) = 2000 + 600\sin(\dfrac{\pi}{3})$$

$$= 2000 + 600\sqrt{3}/2 \approx 2519.62 \quad \text{dollars/month}$$

$S'(2)$ is the rate of change of monthly sales on March 1, and since

$$S'(t) = 600[\cos(\dfrac{\pi}{6}t)](\dfrac{\pi}{6})$$

$$= 100\pi\cos(\dfrac{\pi}{6}t),$$

We have,

$$S'(2) = 100\pi\cos(\dfrac{\pi}{3}) = 50\pi \approx 157.08$$

21. (a) $v(t) = \dfrac{dy}{dt} = \dfrac{d}{dt}(15 + \sin(2\pi t)) = 2\pi \cos(2\pi t)$.
    (b)

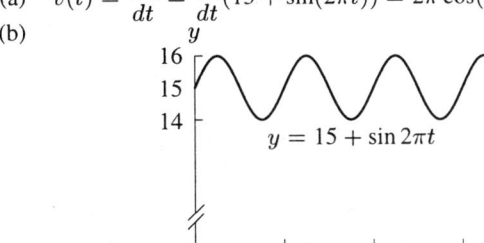
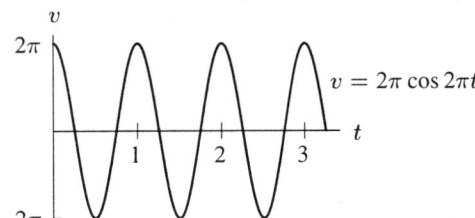

22. (a) $\dfrac{dy}{dt} = -\dfrac{4.9\pi}{6}\sin\left(\dfrac{\pi}{6}t\right)$. It represents the rate of change of the depth of the water.
    (b) $\dfrac{dy}{dt}$ is zero where the tangent line to the curve $y$ is horizontal. $\dfrac{dy}{dt} = 0$ occurs when $\sin(\dfrac{\pi}{6}t) = 0$, or at $t = 6, 12, 18$ and 24 (6 am, noon, 6 pm, and midnight). When $\dfrac{dy}{dt} = 0$, the depth of the water is no longer changing. Therefore, it has either just finished rising or just finished falling, and we know that the harbor's level is at a maximum or a minimum.

## Solutions for Chapter 4 Review

1. $f'(t) = 24t^3$.
2. $P'(t) = 2e^{2t}$.
3. $\dfrac{dW}{dr} = 3r^2 + 5$.
4. $\dfrac{dC}{dq} = 0.08e^{0.08q}$.
5. $f'(x) = 3x^2 - 6x + 5$.
6. $\dfrac{dy}{dt} = 5(-0.2)e^{-0.2t} = -e^{-0.2t}$.
7. $s'(t) = 2t(5t - 1) + (t^2 + 4) \cdot 5 = 15t^2 - 2t + 20$.
8. $\dfrac{d}{dt}e^{(1+3t)^2} = e^{(1+3t)^2}\dfrac{d}{dt}(1+3t)^2 = e^{(1+3t)^2} \cdot 2(1+3t) \cdot 3 = 6(1+3t)e^{(1+3t)^2}$.
9. $f'(x) = 2x + \dfrac{3}{x}$.
10. $Q'(t) = 5 + 3.6e^{1.2t}$.
11. $g'(z) = 3(z^2 + 5)^2 \cdot \dfrac{d}{dz}(z^2 + 5) = 3(z^2 + 5)^2(2z) = 6z(z^2 + 5)^2$.
12. $f'(x) = 6(3(5x-1)^2) \cdot \dfrac{d}{dx}(5x - 1) = 18(5x - 1)^2(5) = 90(5x - 1)^2$.
13. $f'(z) = \dfrac{1}{z^2+1}(2z) = \dfrac{2z}{z^2+1}$.
14. $\dfrac{dy}{dx} = e^{3x} + x \cdot 3e^{3x} = e^{3x}(1 + 3x)$.
15. $\dfrac{dq}{qp} = 100(-0.05)e^{-0.05p} = -5e^{-0.05p}$.
16. $\dfrac{dy}{dx} = 2x\ln x + x^2 \cdot \dfrac{1}{x} = x(2\ln x + 1)$.
17. $s'(t) = 2t + \dfrac{2}{t}$.
18. $\dfrac{dP}{dt} = 8t + 7\cos t$.
19. $R'(t) = 5(\sin t)^4 \cdot \dfrac{d}{dt}(\sin t) = 5(\sin t)^4(\cos t)$.

**20.** $h'(t) = \dfrac{1}{e^{-t} - t}\left(-e^{-t} - 1\right).$

**21.** $f'(x) = 2\cos(2x).$

**22.** $\dfrac{dy}{dx} = 2x\cos x + x^2(-\sin x) = 2x\cos x - x^2\sin x.$

**23.** $g'(x) = \dfrac{50xe^x - 25x^2 e^x}{e^{2x}} = \dfrac{50x - 25x^2}{e^x}.$

**24.** $h'(t) = \dfrac{(1)(t-4) - (1)(t+4)}{(t-4)^2} = \dfrac{t-4-t-4}{(t-4)^2} = \dfrac{-8}{(t-4)^2}.$

**25.** $\dfrac{dz}{dt} = \dfrac{3(5t+2) - (3t+1)5}{(5t+2)^2} = \dfrac{15t + 6 - 15t - 5}{(5t+2)^2} = \dfrac{1}{(5t+2)^2}.$

**26.** $z' = \dfrac{(2t+5)(t+3) - (t^2+5t+2)}{(t+3)^2} = \dfrac{t^2 + 6t + 13}{(t+3)^2}.$

**27.** $h'(p) = \dfrac{2p(3+2p^2) - 4p(1+p^2)}{(3+2p^2)^2} = \dfrac{6p + 4p^3 - 4p - 4p^3}{(3+2p^2)^2} = \dfrac{2p}{(3+2p^2)^2}.$

**28.** Since $f(1) = 2(1^3) - 5(1^2) + 3(1) - 5 = -5$, the point $(1, -5)$ is on the line. We use the derivative to find the slope. Differentiating gives
$$f'(x) = 6x^2 - 10x + 3,$$
and so the slope at $x = 1$ is
$$f'(1) = 6(1^2) - 10(1) + 3 = -1.$$

The equation of the tangent line is
$$y - (-5) = -1(x - 1)$$
$$y + 5 = -x + 1$$
$$y = -4 - x.$$

**29.** (a) $H'(2) = r'(2) + s'(2) = -1 + 3 = 2.$
(b) $H'(2) = 5s'(2) = 5(3) = 15.$
(c) $H'(2) = r'(2)s(2) + r(2)s'(2) = -1 \cdot 1 + 4 \cdot 3 = 11.$
(d) $H'(2) = \dfrac{r'(2)}{2\sqrt{r(2)}} = \dfrac{-1}{2\sqrt{4}} = -\dfrac{1}{4}.$

**30.** If the distance $s(t) = 20e^{\frac{t}{2}}$, then the velocity, $v(t)$, is given by
$$v(t) = s'(t) = \left(20e^{\frac{t}{2}}\right)' = \left(\dfrac{1}{2}\right)\left(20e^{\frac{t}{2}}\right) = 10e^{\frac{t}{2}}.$$

**31.** (a) $q(10) = 5000e^{-0.8} \approx 2247$ units.
(b) $q' = 5000(-0.08)e^{-0.08p} = -400e^{-0.08p}$, $q'(10) = -400e^{-0.8} \approx -180$. This means that at a price of \$10, a \$1 increase in price will result in a decrease in quantity demanded by 180 units.

**32.** $R(p) = p \cdot q = 5000pe^{-0.08p}$

$R(10) = 50{,}000e^{-0.8} \approx 22{,}466$; revenues of about \$22,466 can be expected when the selling price is \$10.
$R'(p) = 5000e^{-0.08p} + 5000p(-0.08)e^{-0.08p} = (5000 - 400p)e^{-0.08p}$

$R'(10) = 1000e^{-0.8} \approx 449$; if price is increased by one dollar over \$10, revenue will increase by about \$449.

**33.** If $a = e$, the only solution is $(0,1)$.

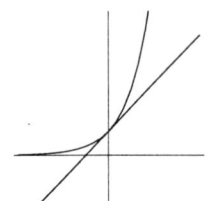

If $1 < a < e$, there are two solutions as illustrated:

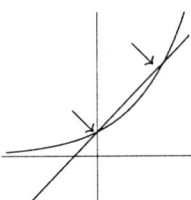

and if $a > e$, there are also two solutions.

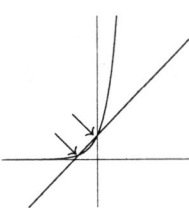

One way to prove the above is to compare the slopes of the lines. For example, $e^x$ will have slope greater than 1 for all $x > 0$ and less than 1 for all $x < 0$, so it cannot meet the line $1 + x$ at any other points. Similar arguments can be made for the other cases.

**34.**
$$\frac{dy}{dt} = -7.5(0.507)\sin(0.507t) = -3.80\sin(0.507t)$$

(a) When $t = 6$, $\frac{dy}{dt} = -3.80\sin(0.507 \cdot 6) = -0.38$ meters/hour. So it's falling at 0.38 meters/hour.

(b) When $t = 9$, $\frac{dy}{dt} = -3.80\sin(0.507 \cdot 9) = 3.76$ meters/hour. So it's rising at 3.76 meters/hour.

(c) When $t = 12$, $\frac{dy}{dt} = -3.80\sin(0.507 \cdot 12) = 0.75$ meters/hour. So it's rising at 0.75 meters/hour.

(d) When $t = 18$, $\frac{dy}{dt} = -3.80\sin(0.507 \cdot 18) = -1.12$ meters/hour. So it's falling at 1.12 meters/hour.

**35.** (a) To find the temperature of the yam when it was placed in the oven, we need to evaluate the function at $t = 0$. In this case, the temperature of the yam to begin with equals $350(1 - 0.7e^0) = 350(0.3) = 105°$.

(b) By looking at the function we see that the temperature which the yam is approaching is $350°$. That is, if the yam were left in the oven for a long period of time (i.e. as $t \to \infty$) the temperature would move closer and closer to $350°$ (because $e^{-0.008t}$ would approach zero, and thus $1 - 0.7e^{-0.008t}$ would approach 1). Thus, the temperature of the oven is $350°$.

(c) The yam's temperature will reach $175°$ when $Y(t) = 175$. Thus, we must solve for $t$:

$$Y(t) = 175$$
$$175 = 350(1 - 0.7e^{-0.008t})$$
$$\frac{175}{350} = 1 - 0.7e^{-0.008t}$$
$$0.7e^{-0.008t} = 0.5$$
$$e^{-0.008t} = 5/7$$
$$\ln e^{-0.008t} = \ln 5/7$$
$$-0.008t = \ln 5/7$$
$$t = \frac{\ln 5/7}{-0.008} \approx 42 \text{ minutes}.$$

Thus the yam's temperature will be $175°$ approximately 42 minutes after it is put into the oven.

(d) The rate at which the temperature is increasing is given by the derivative of the function.

$$Y(t) = 350(1 - 0.7e^{-0.008t}) = 350 - 245e^{-0.008t}.$$

Therefore,
$$Y'(t) = 0 - 245(-0.008e^{-0.008t}) = 1.96e^{-0.008t}.$$

At $t = 20$, the rate of change of the temperature of the yam is given by $Y'(20)$:

$$Y'(20) = 1.96e^{-0.008(20)} = 1.96e^{-.16} = 1.96(0.8521) \approx 1.67 \text{ degrees/minute}.$$

Thus, at $t = 20$ the yam's temperature is increasing by about 1.67 degrees each minute.

**36.** Decreasing means $f'(x) < 0$:
$$f'(x) = 4x^3 - 12x^2 = 4x^2(x-3),$$
so $f'(x) < 0$ when $x < 3$ and $x \neq 0$. Concave up means $f''(x) > 0$:
$$f''(x) = 12x^2 - 24x = 12x(x-2)$$
so $f''(x) > 0$ when
$$12x(x-2) > 0$$
$$x < 0 \quad \text{or} \quad x > 2.$$
So, both conditions hold for $x < 0$ or $2 < x < 3$.

**37.** (a) We have $p(x) = x^2 - x$. We see that $p'(x) = 2x - 1 < 0$ when $x < \frac{1}{2}$. So $p$ is decreasing when $x < \frac{1}{2}$.

(b) We have $p(x) = x^{1/2} - x$, so
$$p'(x) = \frac{1}{2}x^{-1/2} - 1 < 0$$
$$\frac{1}{2}x^{-1/2} < 1$$
$$x^{-1/2} < 2$$
$$x^{1/2} > \frac{1}{2}$$
$$x > \frac{1}{4}.$$

Thus $p(x)$ is decreasing when $x > \frac{1}{4}$.

(c) We have $p(x) = x^{-1} - x$, so
$$p'(x) = -1x^{-2} - 1 < 0$$
$$-x^{-2} < 1$$
$$x^{-2} > -1,$$

which is always true where $x^{-2}$ is defined since $x^{-2} = 1/x^2$ is always positive. Thus $p(x)$ is decreasing for $x < 0$ and for $x > 0$.

**38.** The slopes of the tangent lines to $y = x^2 - 2x + 4$ are given by $y' = 2x - 2$. A line through the origin has equation $y = mx$. So, at the tangent point, $x^2 - 2x + 4 = mx$ where $m = y' = 2x - 2$.
$$x^2 - 2x + 4 = (2x - 2)x$$
$$x^2 - 2x + 4 = 2x^2 - 2x$$
$$-x^2 + 4 = 0$$
$$-(x+2)(x-2) = 0$$
$$x = 2, -2.$$

Thus, the points of tangency are $(2, 4)$ and $(-2, 12)$. The lines through these points and the origin are $y = 2x$ and $y = -6x$, respectively. Graphically:

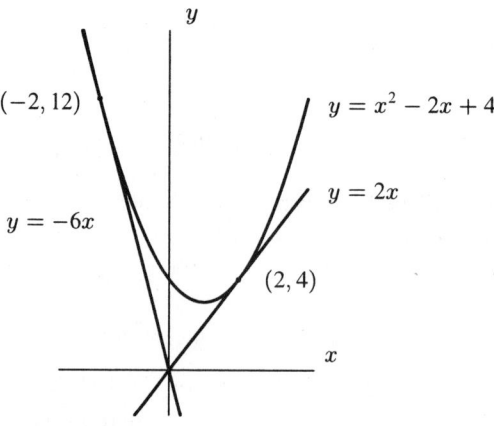

*Figure 4.15*

39. The tangent lines to $f(x) = \sin x$ have slope $\frac{d}{dx}(\sin x) = \cos x$. The tangent line at $x = 0$ has slope $f'(0) = \cos 0 = 1$ and goes through the point $(0, 0)$. Consequently, its equation is $y = g(x) = x$. The approximate value of $\sin \frac{\pi}{6}$ given by this equation is then $g(\frac{\pi}{6}) = \frac{\pi}{6} \approx 0.524$.

Similarly, the tangent line at $x = \frac{\pi}{3}$ has slope $f'(\frac{\pi}{3}) = \cos \frac{\pi}{3} = \frac{1}{2}$ and goes through the point $(\frac{\pi}{3}, \frac{\sqrt{3}}{2})$. Consequently, its equation is $y = h(x) = \frac{1}{2}x + \frac{3\sqrt{3}-\pi}{6}$. The approximate value of $\sin \frac{\pi}{6}$ given by this equation is then $h(\frac{\pi}{6}) = \frac{6\sqrt{3}-\pi}{12} \approx 0.604$.

The actual value of $\sin \frac{\pi}{6}$ is $\frac{1}{2}$, so the approximation from 0 is better than that from $\frac{\pi}{3}$. This is because the slope of the function changes less between $x = 0$ and $x = \frac{\pi}{6}$ than it does between $x = \frac{\pi}{6}$ and $x = \frac{\pi}{3}$. This is illustrated below.

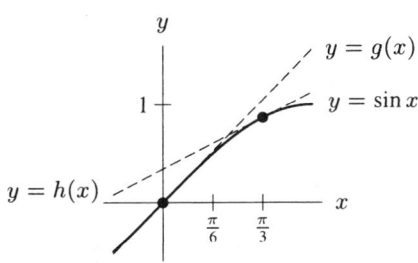

40. (a) If the museum sells the painting and invests the proceeds $P(t)$ at time $t$, then $t$ years have elapsed since 2000, and the time span up to 2020 is $20 - t$. This is how long the proceeds $P(t)$ are earning interest in the bank. Each year the money is in the bank it earns 5% interest, which means the amount in the bank is multiplied by a factor of 1.05. So, at the end of $(20 - t)$ years, the balance is given by

$$B(t) = P(t)(1 + 0.05)^{20-t} = P(t)(1.05)^{20-t}.$$

(b)
$$B(t) = P(t)(1.05)^{20}(1.05)^{-t} = (1.05)^{20} \frac{P(t)}{(1.05)^t}.$$

(c) By the quotient rule,
$$B'(t) = (1.05)^{20} \left[ \frac{P'(t)(1.05)^t - P(t)(1.05)^t \ln 1.05}{(1.05)^{2t}} \right].$$

So,
$$B'(10) = (1.05)^{20} \left[ \frac{5000(1.05)^{10} - 150{,}000(1.05)^{10} \ln 1.05}{(1.05)^{20}} \right]$$
$$= (1.05)^{10}(5000 - 150{,}000 \ln 1.05)$$
$$\approx -3776.63.$$

41. All of the functions go through the origin. They will look the same if they have the same tangent line, or equivalently, the same slope at $x = 0$. Therefore for each function we find the derivative and evaluate it at $x = 0$:

For $y = x$, $\quad y' = 1$, $\quad$ so $y'(0) = 1$.
For $y = \sqrt{x}$, $\quad y' = \frac{1}{2\sqrt{x}}$, $\quad$ so $y'(0)$ is undefined.
For $y = x^2$, $\quad y' = 2x$, $\quad$ so $y'(0) = 0$.
For $y = x^3 + \frac{1}{2}x^2$, $\quad y' = 3x^2 + x$, $\quad$ so $y'(0) = 0$.
For $y = x^3$, $\quad y' = 3x^2$, $\quad$ so $y'(0) = 0$.
For $y = \ln(x + 1)$, $\quad y' = \frac{1}{x+1}$, $\quad$ so $y'(0) = 1$.
For $y = \frac{1}{2}\ln(x^2 + 1)$, $\quad y' = \frac{x}{x^2+1}$, $\quad$ so $y'(0) = 0$.
For $y = \sqrt{2x - x^2}$, $\quad y' = \frac{1-x}{\sqrt{2x-x^2}}$, $\quad$ so $y'(0)$ is undefined.

So near the origin, functions with $y'(0) = 1$ will all be indistinguishable resembling the line $y = x$. These functions are:

$$y = x \quad \text{and} \quad y = \ln(x + 1).$$

Functions with $y'(0) = 0$ will be indistinguishable near the origin and resemble the line $y = 0$ (a horizontal line). These functions are:

$$y = x^2, \qquad y = x^3 + \frac{1}{2}x^2, \qquad y = x^3, \qquad \text{and} \qquad y = \frac{1}{2}\ln(x^2 + 1).$$

Functions that have undefined derivatives at $x = 0$ look like vertical lines at the origin. These functions are

$$y = \sqrt{x} \qquad \text{and} \qquad y = \sqrt{2x - x^2}.$$

42. If $f(x) = x^n$, then $f'(x) = nx^{n-1}$. This means $f'(1) = n \cdot 1^{n-1} = n \cdot 1 = n$, because any power of 1 equals 1.

43. Since $f(x) = ax^n$, $f'(x) = anx^{n-1}$. We know that $f'(2) = (an)2^{n-1} = 3$, and $f'(4) = (an)4^{n-1} = 24$. Therefore,

$$\frac{f'(4)}{f'(2)} = \frac{24}{3}$$

$$\frac{(an)4^{n-1}}{(an)2^{n-1}} = \left(\frac{4}{2}\right)^{n-1} = 8$$

$$2^{n-1} = 8, \text{ and thus } n = 4.$$

Substituting $n = 4$ into the expression for $f'(2)$, we get $3 = a(4)(8)$, or $a = 3/32$.

44. (a) $H(x) = F(G(x))$
$H(4) = F(G(4)) = F(2) = 1$
(b) $H(x) = F(G(x))$
$H'(x) = F'(G(x)) \cdot G'(x)$
$H'(4) = F'(G(4)) \cdot G'(4) = F'(2) \cdot 6 = 5 \cdot 6 = 30$
(c) $H(x) = G(F(x))$
$H(4) = G(F(4)) = G(3) = 4$
(d) $H(x) = G(F(x))$
$H'(x) = G'(F(x)) \cdot F'(x)$
$H'(4) = G'(F(4)) \cdot F'(4) = G'(3) \cdot 7 = 8 \cdot 7 = 56$
(e) $H(x) = \frac{F(x)}{G(x)}$
$H'(x) = \frac{G(x) \cdot F'(x) - F(x) \cdot G'(x)}{[G(x)]^2}$
$H'(4) = \frac{G(4) \cdot F'(4) - F(4) \cdot G'(4)}{[G(4)]^2} = \frac{2 \cdot 7 - 3 \cdot 6}{2^2} = \frac{14 - 18}{4} = \frac{-4}{4} = -1$

45. (a) $P(12) = 10e^{0.6(12)} = 10e^{7.2} \approx 13{,}394$ zebra mussels. There are 13,394 zebra mussels in the area after 12 months.
(b) We differentiate to find $P'(t)$, and then substitute in to find $P'(12)$:

$$P'(t) = 10(e^{0.6t})(0.6) = 6e^{0.6t}$$

$$P'(12) = 6e^{0.6(12)} \approx 8{,}037 \text{ mussels/month}.$$

The population is growing at a rate of approximately 8037 zebra mussels per month.

46. The population of Mexico is given by the formula

$$M = 84(1 + 0.026)^t = 84(1.026)^t \text{ million}$$

and that of the US by

$$U = 250(1 + 0.007)^t = 250(1.007)^t \text{ million},$$

where $t$ is measured in years ($t = 0$ corresponds to the year 1990). So,

$$\left.\frac{dM}{dt}\right|_{t=0} = 84\frac{d}{dt}(1.026)^t\bigg|_{t=0} = 84(1.026)^t \ln(1.026)\bigg|_{t=0} \approx 2.156$$

and

$$\left.\frac{dU}{dt}\right|_{t=0} = 250\frac{d}{dt}(1.007)^t\bigg|_{t=0} = 250(1.007)^t \ln(1.007)\bigg|_{t=0} \approx 1.744$$

Since $\left.\frac{dM}{dt}\right|_{t=0} > \left.\frac{dU}{dt}\right|_{t=0}$, the population of Mexico was growing faster in 1990.

**47.** Since we're given that the instantaneous rate of change of $T$ at $t = 30$ is 2, we want to choose $a$ and $b$ so that the derivative of $T$ agrees with this value. Differentiating, $T'(t) = ab \cdot e^{-bt}$. Then we have

$$2 = T'(30) = abe^{-30b} \text{ or } e^{-30b} = \frac{2}{ab}$$

We also know that at $t = 30$, $T = 120$, so

$$120 = T(30) = 200 - ae^{-30b} \text{ or } e^{-30b} = \frac{80}{a}$$

Thus $\frac{80}{a} = e^{-30b} = \frac{2}{ab}$, so $b = \frac{1}{40} = 0.025$ and $a = 169.36$.

## Solutions to the Projects

**1.** (a) Assuming that $T(1) = 98.6 - 2 = 96.6$, we get

$$96.6 = 68 + 30.6e^{-k \cdot 1}$$
$$28.6 = 30.6e^{-k}$$
$$0.935 = e^{-k}.$$

So
$$k = -\ln(0.935) \approx 0.067.$$

(b) We're looking for a value of $t$ which gives $T'(t) = -1$. First we find $T'(t)$:

$$T(t) = 68 + 30.6e^{-0.067t}$$
$$T'(t) = (30.6)(-0.067)e^{-0.067t} \approx -2e^{-0.067t}.$$

Setting this equal to $-1°$F per hour gives

$$1 = 2e^{-0.067t}$$
$$\ln(0.5) = -0.067t.$$

Thus, when $t \approx 10.3$ hours, we have $T'(t) \approx -1°$F per hour.

(c) The coroner's rule of thumb predicts that in 24 hours the body temperature will decrease $25°$F, to about $73.6°$F. The formula predicts a temperature of

$$T(24) = 68 + 30.6e^{-0.067 \cdot 24} \approx 74.1°\text{F}.$$

**2.** (a)

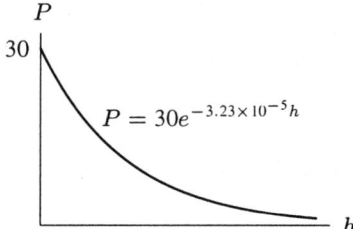

(b)
$$\frac{dP}{dh} = 30e^{-3.23 \times 10^{-5}h}(-3.23 \times 10^{-5})$$

so
$$\left.\frac{dP}{dh}\right|_{h=0} = -30(3.23 \times 10^{-5}) = -9.69 \times 10^{-4}$$

Hence, at $h = 0$, the slope of the tangent line is $-9.69 \times 10^{-4}$, so the equation of the tangent line is

$$y - 30 = (-9.69 \times 10^{-4})(h - 0)$$
$$y = (-9.69 \times 10^{-4})h + 30.$$

(c) The rule of thumb says

$$\left(\begin{array}{c}\text{Drop in pressure from}\\ \text{sea level to height } h\end{array}\right) = \frac{h}{1000}$$

But since the pressure at sea level is 30 inches of mercury, this drop in pressure is also $(30 - P)$, so

$$30 - P = \frac{h}{1000}$$

giving

$$P = 30 - 0.001h.$$

(d) The equations in (b) and (c) are almost the same: both have $P$ intercepts of 30, and the slopes are almost the same ($9.69 \times 10^{-4} \approx 0.001$). The rule of thumb calculates values of $P$ which are very close to the tangent lines, and therefore yields values very close to the curve.

(e) The tangent line is slightly below the curve, and the rule of thumb line, having a slightly more negative slope, is slightly below the tangent line (for $h > 0$). Thus, the rule of thumb values are slightly smaller.

## Solutions to Problems on Establishing the Derivative Formulas

1. Using the definition of the derivative, we have

$$f'(x) = \lim_{h \to 0} \frac{f(x+h) - f(x)}{h}$$
$$= \lim_{h \to 0} \frac{2(x+h) + 1 - (2x + 1)}{h}$$
$$= \lim_{h \to 0} \frac{2x + 2h + 1 - 2x - 1}{h}$$
$$= \lim_{h \to 0} \frac{2h}{h}.$$

As long as $h$ is very close to, but not actually equal to, zero we can say that $\lim_{h \to 0} \frac{2h}{h} = 2$, and thus conclude that $f'(x) = 2$.

2. Using the definition of the derivative, we have

$$f'(x) = \lim_{h \to 0} \frac{f(x+h) - f(x)}{h} = \lim_{h \to 0} \frac{5(x+h)^2 - 5x^2}{h}$$
$$= \lim_{h \to 0} \frac{5(x^2 + 2xh + h^2) - 5x^2}{h} = \lim_{h \to 0} \frac{5x^2 + 10xh + 5h^2 - 5x^2}{h}$$
$$= \lim_{h \to 0} \frac{10xh + 5h^2}{h} = \lim_{h \to 0} \frac{h(10x + 5h)}{h}.$$

As $h$ gets close to zero, but not equal to zero, we can cancel the $h$'s in the numerator and denominator to obtain the following limit which is equal to $f'(x)$: $\lim_{h \to 0}(10x + 5h) = 10x$. Thus, $f'(x) = 10x$.

3. Using the definition of the derivative, we have

$$f'(x) = \lim_{h \to 0} \frac{f(x+h) - f(x)}{h}$$
$$= \lim_{h \to 0} \frac{2(x+h)^2 + 3 - (2x^2 + 3)}{h}$$
$$= \lim_{h \to 0} \frac{2x^2 + 4xh + 2h^2 + 3 - 2x^2 - 3}{h}$$
$$= \lim_{h \to 0} \frac{4xh + 2h^2}{h}$$
$$= \lim_{h \to 0} \frac{h(4x + 2h)}{h}$$

As $h$ gets close to zero (but not equal to zero), we can cancel the $h$ in the numerator and denominator to obtain the following:
$$f'(x) = \lim_{h \to 0}(4x + 2h) = 4x$$
Thus, we get $f'(x) = 4x$.

4. Using the definition of the derivative, we have
$$f'(x) = \lim_{h \to 0} \frac{f(x+h) - f(x)}{h} = \lim_{h \to 0} \frac{(x+h)^2 + (x+h) - (x^2 + x)}{h}$$
$$= \lim_{h \to 0} \frac{x^2 + 2xh + h^2 + x + h - x^2 - x}{h}$$
$$= \lim_{h \to 0} \frac{2xh + h^2 + h}{h} = \lim_{h \to 0} \frac{h(2x + h + 1)}{h}.$$

As $h$ approaches, but does not equal, zero we can cancel $h$'s in the numerator and denominator to obtain the following limit equal to $f'(x)$:
$$\lim_{h \to 0}(2x + h + 1) = 2x + 1.$$
Thus, $f'(x) = 2x + 1$.

5. The definition of the derivative states that
$$f'(x) = \lim_{h \to 0} \frac{f(x+h) - f(x)}{h}.$$

Using this definition, we have
$$f'(x) = \lim_{h \to 0} \frac{4(x+h)^2 + 1 - (4x^2 + 1)}{h}$$
$$= \lim_{h \to 0} \frac{4x^2 + 8xh + 4h^2 + 1 - 4x^2 - 1}{h}$$
$$= \lim_{h \to 0} \frac{8xh + 4h^2}{h}$$
$$= \lim_{h \to 0} \frac{h(8x + 4h)}{h}.$$

As long as $h$ approaches, but does not equal, zero we can cancel it out of the numerator and denominator. The derivative now becomes
$$\lim_{h \to 0}(8x + 4h) = 8x.$$
Thus, $f'(x) = 6x$ as we stated above.

6. Using the definition of the derivative, we have
$$f'(x) = \lim_{h \to 0} \frac{f(x+h) - f(x)}{h} = \lim_{h \to 0} \frac{(x+h)^4 - x^4}{h}$$
$$= \lim_{h \to 0} \frac{x^4 + 4x^3h + 6x^2h^2 + 4xh^3 + h^4 - x^4}{h}$$
$$= \lim_{h \to 0} \frac{4x^3h + 6x^2h^2 + 4xh^3 + h^4}{h}$$
$$= \lim_{h \to 0} \frac{h(4x^3 + 6x^2h + 4xh^2 + h^3)}{h}.$$

We can divide top and bottom by $h$ as long as $h \neq 0$; thus we get $\lim_{h \to 0}(4x^3 + 6x^2h + 4xh^2 + h^3)$ which goes to $4x^3$ as $h \to 0$. Thus, $f'(x) = 4x^3$.

7. Using the definition of the derivative, we have
$$f'(x) = \lim_{h \to 0} \frac{f(x+h) - f(x)}{h} = \lim_{h \to 0} \frac{(x+h)^5 - x^5}{h}$$

$$= \lim_{h \to 0} \frac{x^5 + 5x^4h + 10x^3h^2 + 10x^2h^3 + 5xh^4 + h^5 - x^5}{h}$$

$$= \lim_{h \to 0} \frac{5x^4h + 10x^3h^2 + 10x^2h^3 + 5xh^4 + h^5}{h}$$

$$= \lim_{h \to 0} \frac{h(5x^4 + 10x^3h + 10x^2h^2 + 5xh^3 + h^4)}{h}.$$

As $h \to 0$ but does not equal it, we can safely factor $h$ out of the numerator and denominator and cancel, leaving us with the following limit which equals $f'(x)$:

$$\lim_{h \to 0}(5x^4 + 10x^3h + 10x^2h^2 + 5xh^3 + h^4) = 5x^4.$$

Thus, $f'(x) = 5x^4$.

8. (a)

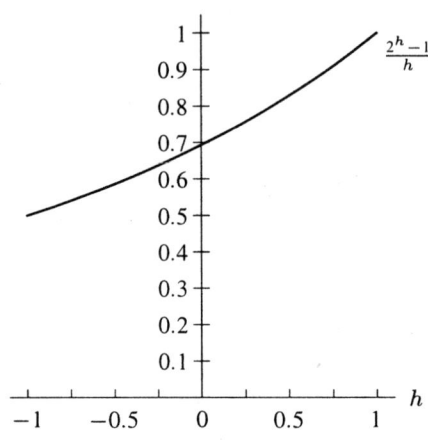

Figure 4.16

The graph of the function $\dfrac{2^h - 1}{h}$ is given in Figure 4.16. We can see that as $h$ gets closer to zero, the value of the function approaches 0.6931. Thus, we can conclude that

$$\lim_{h \to 0} \frac{2^h - 1}{h} \approx 0.6931.$$

You may also want to consider plugging some values of $h$ close to zero into your calculator so that you can observe that the function is indeed approaching 0.6931 as the values of $h$ get closer and closer to zero.

(b) Using the definition of the derivative and the results from part (a), we have

$$f'(x) = \lim_{h \to 0} \frac{f(x+h) - f(x)}{h} = \lim_{h \to 0} \frac{2^{x+h} - 2^x}{h}$$

$$= \lim_{h \to 0} \frac{2^x \cdot 2^h - 2^x}{h}$$

$$= \lim_{h \to 0} \frac{2^x(2^h - 1)}{h}$$

$$= 2^x \cdot \left(\lim_{h \to 0} \frac{2^h - 1}{h}\right).$$

From part (a) we know that $\lim_{h \to 0} \dfrac{2^h - 1}{h} \approx 0.6931$, and thus

$$f'(x) = 2^x \cdot \left(\lim_{h \to 0} \frac{2^h - 1}{h}\right) \approx (0.6931)2^x.$$

## Solutions to Practice Problems on Differentiation

1. $f'(t) = 2t + 4t^3$
2. $g'(x) = 20x^3$
3. $y' = 15x^2 + 14x - 3$
4. $s'(t) = -12t^{-3} + 9t^2 - 2t^{-1/2}$
5. $f'(x) = -2x^{-1} + 5\left(\frac{1}{2}x^{-1/2}\right) = \frac{-2}{x} + \frac{5}{2\sqrt{x}}$
6. $P'(t) = 100e^{0.05t}(0.05) = 5e^{0.05t}$
7. $f'(x) = 10e^{2x} - 2 \cdot 3^x (\ln 3)$
8. $P'(t) = 1{,}000(1.07)^t (\ln 1.07) \approx 68(1.07)^t$
9. $D'(p) = 2pe^{p^2} + 10p$
10. $y' = t^2 \left(5e^{5t}\right) + 2t \left(e^{5t}\right) = 5t^2 e^{5t} + 2te^{5t}$
11. $y' = 2x\sqrt{x^2 + 1} + x^2 \left(\frac{1}{2}(x^2+1)^{-1/2} \cdot 2x\right) = 2x\sqrt{x^2+1} + \frac{x^3}{\sqrt{x^2+1}}$
12. $f'(x) = \frac{2x}{x^2 + 1}$
13. $s'(t) = \frac{16}{2t + 1}$
14. $g'(w) = 2w \ln w + w^2 \left(\frac{1}{w}\right) = 2w \ln w + w$
15. $f'(x) = 2^x (\ln 2) + 2x$
16. $P'(t) = \frac{1}{2}(t^2 + 4)^{-1/2}(2t) = \frac{t}{\sqrt{t^2 + 4}}$
17. $C'(q) = 3(2q + 1)^2 \cdot 2 = 6(2q + 1)^2$
18. $g'(x) = 5(x + 3)^2 + 5x(2(x + 3)) = 5(x + 3)^2 + 10x(x + 3)$
19. $P'(t) = bke^{kt}$
20. $f'(x) = 2ax + b$
21. $y' = 2x \ln(2x + 1) + \frac{2x^2}{2x + 1}$
22. $f'(t) = 3(e^t + 4)^2 (e^t) = 3e^t (e^t + 4)^2$
23. $f'(x) = 10 \cos(2x)$
24. $W'(r) = 2r \cos r - r^2 \sin r$
25. $g'(t) = 15 \cos(5t)$
26. $y' = 3e^{3t} \sin(2t) + 2e^{3t} \cos(2t)$
27. $y' = 2e^x + 3 \cos x$
28. $f'(t) = 6t - 4.$
29. $y' = 17 + 12x^{-1/2}.$
30. $g'(x) = -\frac{1}{2}(5x^4 + 2).$
31. The power rule gives $f'(x) = 20x^3 - \frac{2}{x^3}.$

32. $\dfrac{dy}{dx} = \dfrac{2e^{2x}(x^2+1) - e^{2x}(2x)}{(x^2+1)^2} = \dfrac{2e^{2x}(x^2+1-x)}{(x^2+1)^2}$

33. Either notice that $f(x) = \dfrac{x^2 + 3x + 2}{x+1}$ can be written as $f(x) = \dfrac{(x+2)(x+1)}{x+1}$ which reduces to $f(x) = x + 2$, giving $f'(x) = 1$, or use the quotient rule which gives

$$\begin{aligned} f'(x) &= \dfrac{(x+1)(2x+3) - (x^2+3x+2)}{(x+1)^2} \\ &= \dfrac{2x^2 + 5x + 3 - x^2 - 3x - 2}{(x+1)^2} \\ &= \dfrac{x^2 + 2x + 1}{(x+1)^2} \\ &= \dfrac{(x+1)^2}{(x+1)^2} \\ &= 1. \end{aligned}$$

34. $y' = 2\left(\dfrac{x^2+2}{3}\right)\left(\dfrac{2x}{3}\right) = \dfrac{4}{9}x\left(x^2+2\right)$

35. $\dfrac{d}{dx}\sin(2-3x) = \cos(2-3x)\dfrac{d}{dx}(2-3x) = -3\cos(2-3x)$.

36. $f(z) = \dfrac{z}{3} + \dfrac{1}{3}z^{-1} = \dfrac{1}{3}\left(z + z^{-1}\right)$, so $f'(z) = \dfrac{1}{3}\left(1 - z^{-2}\right) = \dfrac{1}{3}\left(\dfrac{z^2-1}{z^2}\right)$.

37. $q'(r) = \dfrac{3(5r+2) - 3r(5)}{(5r+2)^2} = \dfrac{15r + 6 - 15r}{(5r+2)^2} = \dfrac{6}{(5r+2)^2}$

38. $\dfrac{dy}{dx} = \ln x + x\left(\dfrac{1}{x}\right) - 1 = \ln x$

39. $j'(x) = \dfrac{ae^{ax}}{(e^{ax} + b)}$

40. $g'(t) = \dfrac{(t+4) - (t-4)}{(t+4)^2} = \dfrac{8}{(t+4)^2}$.

41. $h'(w) = 5(w^4 - 2w)^4(4w^3 - 2)$

42. Using the product and chain rules gives $h'(w) = 3w^2 \ln(10w) + w^3 \dfrac{10}{10w} = 3w^2 \ln(10w) + w^2$.

43. Using the chain rule gives $f'(x) = \dfrac{\cos x - \sin x}{\sin x + \cos x}$.

44. We can write $w(r) = (r^4 + 1)^{1/2}$, so
$w'(r) = \dfrac{1}{2}(r^4 + 1)^{-1/2}(4r^3) = \dfrac{2r^3}{\sqrt{r^4 + 1}}$.

45. $h'(w) = 6w^{-4} + \dfrac{3}{2}w^{-1/2}$

46. We can write $h(x) = \left(\dfrac{x^2+9}{x+3}\right)^{1/2}$, so

$h'(x) = \dfrac{1}{2}\left(\dfrac{x^2+9}{x+3}\right)^{-1/2}\left[\dfrac{2x(x+3) - (x^2+9)}{(x+3)^2}\right] = \dfrac{1}{2}\sqrt{\dfrac{x+3}{x^2+9}}\left[\dfrac{x^2+6x-9}{(x+3)^2}\right]$.

47. Using the product rule gives $v'(t) = 2te^{-ct} - ce^{-ct}t^2 = (2t - ct^2)e^{-ct}$.

48. Using the quotient rule gives

$$\begin{aligned} f'(x) &= \dfrac{1 + \ln x - x(\frac{1}{x})}{(1 + \ln x)^2} \\ &= \dfrac{\ln x}{(1 + \ln x)^2}. \end{aligned}$$

49. Using the chain rule, $g'(\theta) = (\cos\theta)e^{\sin\theta}$.

50. $p'(t) = 4e^{4t+2}$.

51. $j'(x) = \dfrac{3x^2}{a} + \dfrac{2ax}{b} - c$

52. $\dfrac{d}{dz}\left(\dfrac{z^2+1}{\sqrt{z}}\right) = \dfrac{d}{dz}(z^{\frac{3}{2}} + z^{-\frac{1}{2}}) = \dfrac{3}{2}z^{\frac{1}{2}} - \dfrac{1}{2}z^{-\frac{3}{2}} = \dfrac{\sqrt{z}}{2}(3 - z^{-2})$.

53. $h'(r) = \dfrac{d}{dr}\left(\dfrac{r^2}{2r+1}\right) = \dfrac{(2r)(2r+1) - 2r^2}{(2r+1)^2} = \dfrac{2r(r+1)}{(2r+1)^2}$.

54. $g'(x) = \dfrac{d}{dx}(2x - x^{-1/3} + 3^x - e) = 2 + \dfrac{1}{3x^{\frac{4}{3}}} + 3^x \ln 3$.

55. $f'(t) = \dfrac{d}{dt}\left(2te^t - \dfrac{1}{\sqrt{t}}\right) = 2e^t + 2te^t + \dfrac{1}{2t^{3/2}}$.

56.
$$\dfrac{dw}{dz} = \dfrac{(-3)(5+3z) - (5-3z)(3)}{(5+3z)^2}$$
$$= \dfrac{-15 - 9z - 15 + 9z}{(5+3z)^2} = \dfrac{-30}{(5+3z)^2}$$

57. $f'(x) = \dfrac{3x^2}{9}(3\ln x - 1) + \dfrac{x^3}{9}\left(\dfrac{3}{x}\right) = x^2 \ln x - \dfrac{x^2}{3} + \dfrac{x^2}{3} = x^2 \ln x$

58. $g'(x) = \dfrac{d}{dx}\left(x^{\frac{1}{2}} + x^{-1} + x^{-\frac{3}{2}}\right) = \dfrac{1}{2}x^{-\frac{1}{2}} - x^{-2} - \dfrac{3}{2}x^{-\frac{5}{2}}$.

59.
$$\dfrac{dy}{dz} = 3(x^2+5)^2(2x)(3x^3-2)^2 + (x^2+5)^3[2(3x^3-2)(9x^2)]$$
$$= 3(2x)(x^2+5)^2(3x^3-2)[(3x^3-2) + (x^2+5)(3x)]$$
$$= 6x(x^2+5)^2(3x^3-2)[6x^3 + 15x - 2]$$

60. Using the quotient rule gives
$$f'(x) = \dfrac{(-2x)(a^2+x^2) - (2x)(a^2-x^2)}{(a^2+x^2)^2}$$
$$= \dfrac{-4a^2 x}{(a^2+x^2)^2}.$$

61. Using the quotient rule gives
$$w'(r) = \dfrac{2ar(b+r^3) - 3r^2(ar^2)}{(b+r^3)^2}$$
$$= \dfrac{2abr - ar^4}{(b+r^3)^2}.$$

62. Using the product rule gives
$$H'(t) = 2ate^{-ct} - c(at^2+b)e^{-ct}$$
$$= (-cat^2 + 2at - bc)e^{-ct}.$$

63. Since $g(w) = 5(a^2 - w^2)^{-2}$, $g'(w) = -10(a^2 - w^2)^{-3}(-2w) = \dfrac{20w}{(a^2-w^2)^3}$

# CHAPTER FIVE

## Solutions for Section 5.1

1. We find a critical point by noting where $f'(x) = 0$ or $f'$ is undefined. Since the curve is smooth throughout, $f'$ is always defined, so we look for where $f'(x) = 0$, or equivalently where the tangent line to the graph is horizontal. These points are shown in Figure 5.1:

   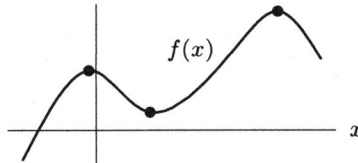

   **Figure 5.1**

   As we can see, there are three critical points. The leftmost one is a local maximum, because points near it are all lower; similarly, the middle critical point is surrounded by higher points, and is a local minimum. The critical point to the right is a local maximum.

2. We find a critical point by noting where $f'(t) = 0$ or $f'$ is undefined. Since the curve is smooth throughout, $f'$ is always defined, so we look for where $f'(t) = 0$, or equivalently where the tangent line to the graph is horizontal. These points are shown in Figure 5.2. As we can see, there are four labeled critical points. Critical point A is a local maximum because points near it are all lower; similarly, point B is a local minimum, point C is a local maximum, and point D is a local minimum.

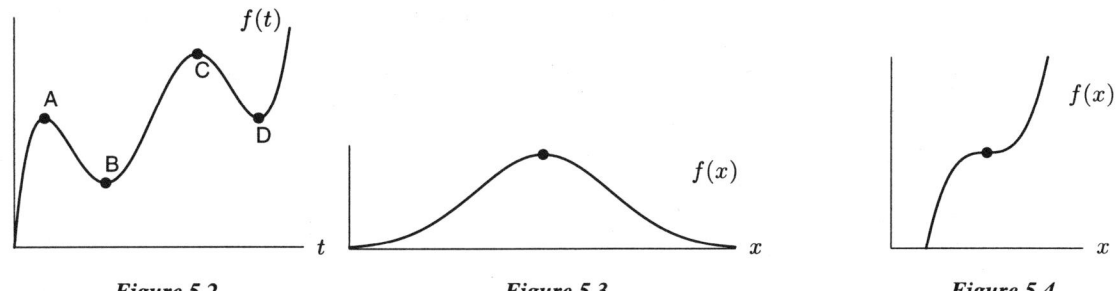

**Figure 5.2**     **Figure 5.3**     **Figure 5.4**

3. We find a critical point by noting where $f'(x) = 0$ or $f'$ is undefined. Since the curve is smooth throughout, $f'$ is always defined, so we look for where $f'(x) = 0$, or equivalently where the tangent line to the graph is horizontal. These points are shown in Figure 5.3. As we can see, there is one critical point. Since it is higher than nearby points, it is a local maximum.

4. We find a critical point by noting where $f'(x) = 0$ or $f'$ is undefined. Since the curve is smooth throughout, $f'$ is always defined, so we look for where $f'(x) = 0$, or equivalently where the tangent line to the graph is horizontal. These points are shown in Figure 5.4. As we can see, there is one critical point. Since some nearby points (those to the left) are lower, this point is not a local minimum; since nearby points to the right are higher, it is not a local maximum. So the one critical point is neither a local minimum nor a local maximum.

5. (a)          (b)

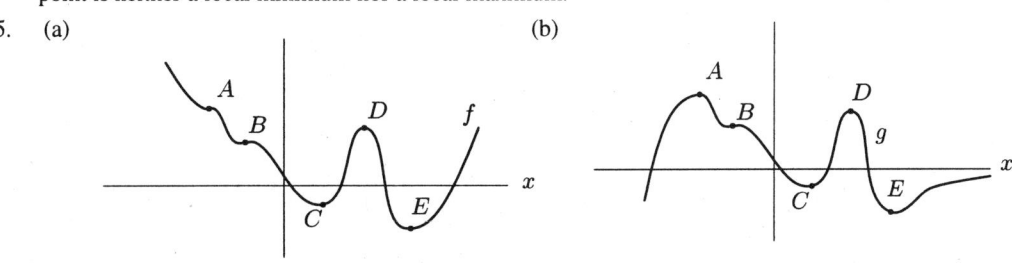

6. There was a critical point after the first eighteen hours when temperature was at its highest point, a local maximum for the temperature function.

7. There are several possibilities. The price could have been increasing during the last few days of June, reaching a high point on July 1, then going back down during the first few days of July. In this case there was a local maximum in the price on July 1.

   The price could have been decreasing during the last few days of June, reaching a low point on July 1, then going back up during the first few days of July. In this case there was a local minimum in the price on July 1.

   It is also possible that there was neither a local maximum nor a local minimum in the price on July 1. This could have happened two ways. On the one hand, the price could have been rising in late June, then held steady with no change around July 1, after which the price increased some more. On the other hand, the price could have been falling in late June, then held steady with no change around July 1, after which the price fell some more. The key feature in these critical point scenarios is that there was no appreciable change in the price of the stock around July 1.

8. (a) The demand for the product is increasing when $f'(t)$ is positive, and decreasing when $f'(t)$ is negative. Inspection of the table suggests that demand is increasing during weeks 0 to 2 and weeks 6 to 10, and decreasing during weeks 3 to 5.

   (b) Since $f'(t) = 4 > 0$ during week 2 and $f'(t) = -2 < 0$ during week 3, the demand for the product changes from increasing to decreasing near the end of week 2 or the beginning of week 3. Thus the demand has a local maximum during this time period. Since $f'(t) = -1 < 0$ during week 5 and $f'(t) = 3 > 0$ during week 6, the demand for the product changes from decreasing to increasing near the end of week 5 or the beginning of week 6. Thus the demand has a local minimum during this time period.

9. A critical point of $f$ requires $f'(x) = 0$ or $f'$ undefined. Since $f'$ is clearly defined over the relevant range, we find where $f'(x) = 0$, that is, where the graph of $f'$ crosses the $x$-axis. These points are shown and labeled in Figure 5.5:

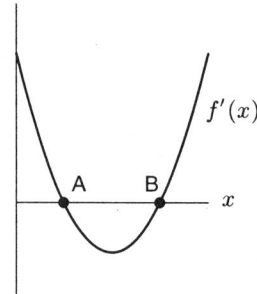

*Figure 5.5*

To the left of critical point $A$, we see that $f' > 0$ and $f$ is increasing; to the right of the critical point, we see that $f' < 0$ and $f$ is decreasing. So there is a local maximum at $A$.

To the left of critical point $B$, we see that $f' < 0$ and $f$ is decreasing; to the right of the critical point, we see that $f' > 0$ and $f$ is increasing. So there is a local minimum at $B$.

The sketch of $f(x)$ in Figure 5.6 shows $A$ is a local maximum and $B$ is a local minimum:

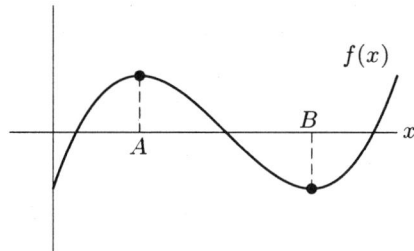

**Figure 5.6**

10. A critical point of $f$ would require $f'(x) = 0$ or $f'$ undefined. Since $f'$ is clearly defined over the relevant range, we wish to find where $f'(x) = 0$, or where the graph shown crosses the x-axis. These points are shown and labeled in Figure 5.7.

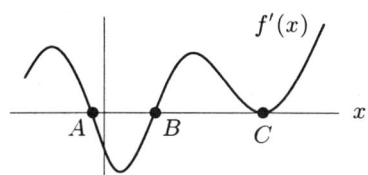

**Figure 5.7**: The critical points of $f(x)$

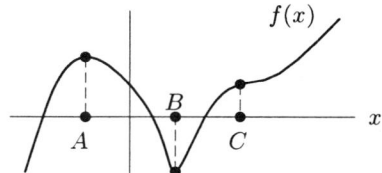

**Figure 5.8**: A possible graph of $f(x)$

To the left of critical point A, $f' > 0$ and $f$ is increasing; to the right, $f' < 0$ and $f$ is decreasing. So there is a local maximum at $A$.

To the left of critical point B, $f' < 0$ and $f$ is decreasing; to the right, $f' > 0$ and $f$ is increasing. So there is a local minimum at $B$.

To both the left and right of critical point $C$, $f' > 0$ and so $f$ increases on both sides of point $C$. So, point $C$ is neither a local maximum nor a local minimum.

A sketch of $f(x)$ is shown in Figure 5.8.

11. (a) One possible answer is shown in Figure 5.9.
    (b) One possible answer is shown in Figure 5.10

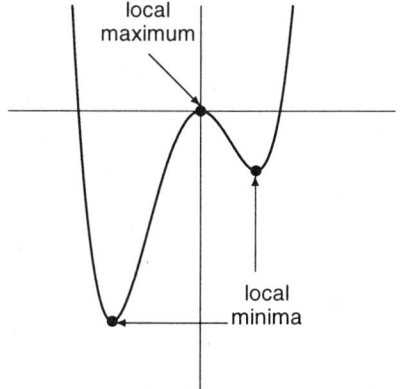

**Figure 5.9**

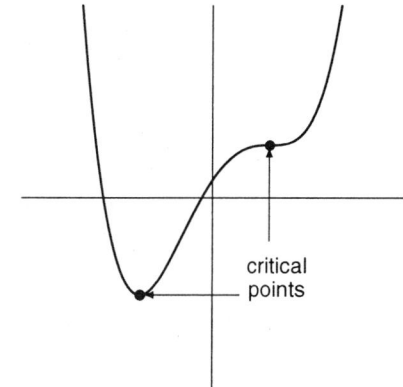

**Figure 5.10**

12. Since $f(x)$ has its minimum at $x = 3$, then $x = 3$ must be a critical point. So $f'(3) = 0$.

$$f'(x) = 2x + a \quad \text{so} \quad f'(3) = 6 + a.$$

Since $f'(3) = 0$, then $6 + a = 0$ or $a = -6$. Since $(3, 5)$ is a point on the graph of $f(x)$ we must have $f(3) = 5$:

$$f(3) = 3^2 + a(3) + b = 9 + 3a + b = 5.$$

Since we know that $a = -6$, we have:

$$f(3) = 9 + 3(-6) + b = b - 9 = 5 \quad \text{so} \quad b = 14.$$

Thus, we have found that $a = -6, b = 14$, giving us $f(x) = x^2 - 6x + 14$.

13. If the minimum of $f(x)$ is at $(-2, -3)$, then the derivative of $f$ must be equal to 0 there. In other words, $f'(-2) = 0$. If

$$f(x) = x^2 + ax + b, \quad \text{then}$$
$$f'(x) = 2x + a$$
$$f'(-2) = 2(-2) + a = -4 + a = 0$$

so $a = 4$. Since $(-2, -3)$ is on the graph of $f(x)$ we know that $f(-2) = -3$. So

$$f(-2) = (-2)^2 + a(-2) + b = -3$$
$$a = 4, \text{ so} \quad (-2)^2 + 4(-2) + b = -3$$
$$4 - 8 + b = -3$$
$$-4 + b = -3$$
$$b = 1$$

so $a = 4$ and $b = 1$, and $f(x) = x^2 + 4x + 1$.

14. We wish to have $f'(3) = 0$. Differentiating to find $f'(x)$ and then solving $f'(3) = 0$ for $a$ gives:

$$f'(x) = x(ae^{ax}) + 1(e^{ax}) = e^{ax}(ax + 1)$$
$$f'(3) = e^{3a}(3a + 1) = 0$$
$$3a + 1 = 0$$
$$a = -\frac{1}{3}.$$

Thus, $f(x) = xe^{-x/3}$.

15. Using the product rule on the function $f(x) = axe^{bx}$, we have $f'(x) = ae^{bx} + abxe^{bx} = ae^{bx}(1 + bx)$. We want $f(\frac{1}{3}) = 1$, and since this is to be a maximum, we require $f'(\frac{1}{3}) = 0$. These conditions give

$$f(1/3) = a(1/3)e^{b/3} = 1,$$
$$f'(1/3) = ae^{b/3}(1 + b/3) = 0.$$

Since $ae^{(1/3)b}$ is non-zero, we can divide both sides of the second equation by $ae^{(1/3)b}$ to obtain $0 = 1 + \frac{b}{3}$. This implies $b = -3$. Plugging $b = -3$ into the first equation gives us $a(\frac{1}{3})e^{-1} = 1$, or $a = 3e$. How do we know we have a maximum at $x = \frac{1}{3}$ and not a minimum? Since $f'(x) = ae^{bx}(1 + bx) = (3e)e^{-3x}(1 - 3x)$, and $(3e)e^{-3x}$ is always positive, it follows that $f'(x) > 0$ when $x < \frac{1}{3}$ and $f'(x) < 0$ when $x > \frac{1}{3}$. Since $f'$ is positive to the left of $x = \frac{1}{3}$ and negative to the right of $x = \frac{1}{3}$, $f(\frac{1}{3})$ is a local maximum.

16. (a) A critical point occurs when $f'(x) = 0$. Since $f'(x)$ changes sign between $x = 2$ and $x = 3$, between $x = 6$ and $x = 7$, and between $x = 9$ and $x = 10$, we expect critical points at around $x = 2.5$, $x = 6.5$, and $x = 9.5$.
    (b) Since $f'(x)$ goes from positive to negative at $x \approx 2.5$, a local maximum should occur there. Similarly, $x \approx 6.5$ is a local minimum and $x \approx 9.5$ a local maximum.

17. Local maximum for some $\theta$, with $1.1 < \theta < 1.2$
    Local minimum for some $\theta$, with $1.5 < \theta < 1.6$
    Local maximum for some $\theta$, with $2.0 < \theta < 2.1$

18. (a) In Figure 5.11, we see that $f(\theta) = \theta - \sin\theta$ has a zero at $\theta = 0$. To see if it has any other zeros near the origin, we use our calculator to zoom in. (See Figure 5.12.) No extra root seems to appear no matter how close to the origin we zoom. However, zooming can never tell you for sure that there is not a root that you have not found yet.

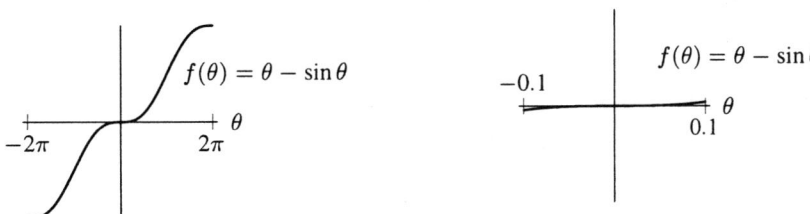

**Figure 5.11:** Graph of $f(\theta)$

**Figure 5.12:** Graph of $f(\theta)$ Zoomed In

(b) Using the derivative, $f'(\theta) = 1 - \cos\theta$, we can argue that there is no other zero. Since $\cos\theta < 1$ for $0 < \theta \leq 1$, we know $f'(\theta) > 0$ for $0 < \theta \leq 1$. Thus, $f$ increases for $0 < \theta \leq 1$. Consequently, we conclude that the only zero of $f$ is the one at the origin. If $f$ had another zero at $x_0$, with $x_0 > 0$, then $f$ would have to "turn around", and recross the $x$-axis at $x_0$. But if this were the case, $f'$ would be nonpositive somewhere, which we know is not the case.

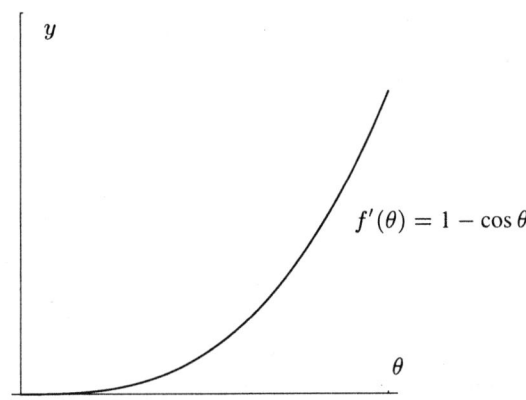

**Figure 5.13:** Graph of $f'(\theta)$

19. Since $f$ is differentiable everywhere, $f'$ must be zero (not undefined) at any critical points; thus, $f'(3) = 0$. Since $f$ has exactly one critical point, $f'$ may change sign only at $x = 3$. Thus $f$ is always increasing or always decreasing for $x < 3$ and for $x > 3$. Using the information in parts (a) through (d), we determine whether $x = 3$ is a local minimum, local maximum, or neither.

(a) $x = 3$ is a local maximum because $f(x)$ is increasing when $x < 3$ and decreasing when $x > 3$.

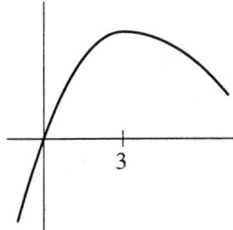

(b) $x = 3$ is a local minimum because $f(x)$ heads to infinity to either side of $x = 3$.

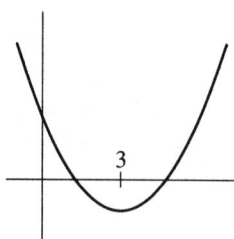

(c) $x = 3$ is neither a local minimum nor maximum, as $f(1) < f(2) < f(4) < f(5)$.

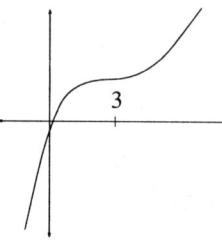

(d) $x = 3$ is a local minimum because $f(x)$ is decreasing to the left of $x = 3$ and must increase to the right of $x = 3$, as $f(3) = 1$ and eventually $f(x)$ must become close to 3.

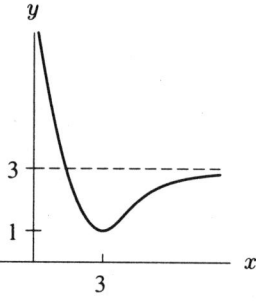

20. The derivative of $f(x) = x^5 + x + 7$ is $f'(x) = 5x^4 + 1$. Since $f'(x)$ is defined for all $x$ and $f'(x) \neq 0$ for any value of $x$, there are no critical points for the function. Furthermore, $f'(x)$ is positive for all $x$, so the function is increasing over its entire domain, and hence can only cross the $x$-axis once. Since $f(x) \to +\infty$ as $x \to +\infty$ and $f(x) \to -\infty$ as $x \to -\infty$, the graph of $f$ must cross the $x$-axis at least once, so we conclude that $f(x)$ has one real root.

## Solutions for Section 5.2

1. We find an inflection point by noting where the concavity changes, or equivalently where the tangent line passes from above the graph to below or vice versa. Such points are shown below in Figure 5.14:

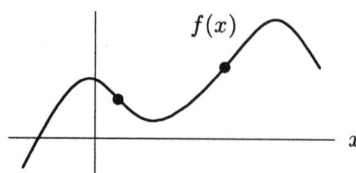

**Figure 5.14**

There are two inflection points.

2. We find an inflection point by noting where the concavity changes, or equivalently where the tangent line passes from above the graph to below or vice versa. Such points are shown below in Figure 5.15. There are three inflection points.

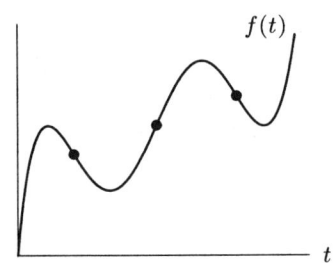

Figure 5.15

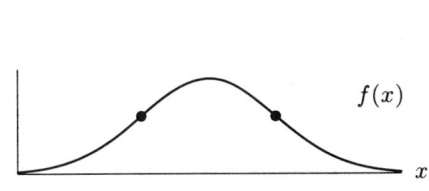

Figure 5.16

3. We find an inflection point by noting where the concavity changes, or equivalently where the tangent line passes from above the graph to below or vice versa. Such points are shown in Figure 5.16. There are two inflection points.

4. We find an inflection point by noting where the concavity changes, or equivalently where the tangent line passes from above the graph to below or vice versa.

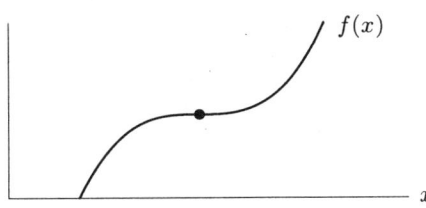

Figure 5.17

Looking at the above Figure 5.17, we see that in fact the concavity changes only at the critical point. So there is one inflection point, which is at the same point as the critical point.

5. From the graph of $f(x)$ in the figure below, we see that the function must have two inflection points. We calculate $f'(x) = 4x^3 + 3x^2 - 6x$, and $f''(x) = 12x^2 + 6x - 6$. Solving $f''(x) = 0$ we find that:

$$x_1 = -1 \quad \text{and} \quad x_2 = \frac{1}{2}.$$

Since $f''(x) > 0$ for $x < x_1$, $f''(x) < 0$ for $x_1 < x < x_2$, and $f''(x) > 0$ for $x_2 < x$, it follows that both points are inflection points.

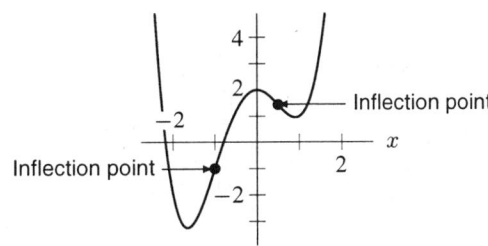

6. A critical point will occur whenever $f'(x) = 0$ or $f'$ is undefined. Since $f'(x)$ is always defined, we set

$$f'(x) = 2x - 5 = 0$$
$$2x = 5$$
$$x = \frac{5}{2}$$

To find the inflection points of $f(x)$, we find where $f''(x)$ goes from negative to positive or vice versa. For a point to satisfy this condition, it must have at least $f''(x) = 0$ or $f''$ undefined. Since $f''(x) = 2$, $f''(x)$ is always defined and never equal to zero, so $f(x)$ cannot have inflection points.

So $x = \frac{5}{2}$ is a critical point of $f(x)$, and there are no inflection points.

To identify the nature of the critical point $x = \frac{5}{2}$ that we have found, we can look at a graph of $f(x)$ for values of $x$ near the critical point. Such a graph is shown in Figure 5.18:

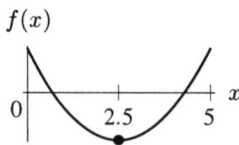

**Figure 5.18**

From the graph we can clearly see that $f(\frac{5}{2})$ is a local minimum of $f$.

7. $f'(x) = 6x^2 + 6x - 36$. To find critical points, we set $f'(x) = 0$. Then
$$6(x^2 + x - 6) = 6(x + 3)(x - 2) = 0.$$

Therefore, the critical points of $f$ are $x = -3$ and $x = 2$. To the left of $x = -3$, $f'(x) > 0$. Between $x = -3$ and $x = 2$, $f'(x) < 0$. To the right of $x = 2$, $f'(x) > 0$. Thus $f(-3)$ is a local maximum, $f(2)$ a local minimum. To find the inflection points of $f(x)$ we look for the points at which $f''(x)$ goes from negative to positive or vice-versa. Since $f''(x) = 12x + 6$, $x = -1/2$ is an inflection point.

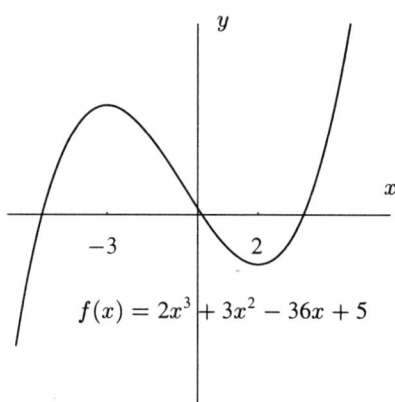

8. $f'(x) = 12x^3 - 12x^2$. To find critical points, we set $f'(x) = 0$. This implies $12x^2(x - 1) = 0$. So the critical points of $f$ are $x = 0$ and $x = 1$. To the left of $x = 0$, $f'(x) < 0$. Between $x = 0$ and $x = 1$, $f'(x) < 0$. To the right of $x = 1$, $f'(x) > 0$. Therefore, $f(1)$ is a local minimum, but $f(0)$ is not a local extremum. To find the inflection points of $f(x)$ we look for points at which $f''(x)$ goes from negative to positive or vice-versa. At any such point $f''(x)$ is either zero or undefined. Since $f''(x) = 36x^2 - 24x = 12x(3x - 2)$, our candidate points are $x = 0$ and $x = 2/3$. Both $x = 0$ and $x = 2/3$ are inflection points.

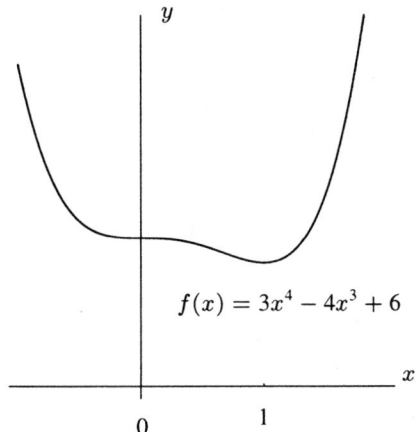

9. The derivative of $f(x)$ is $f'(x) = 4x^3 - 16x$. The critical points of $f(x)$ will be points for which $f'(x) = 0$. Factoring, we get

$$4x^3 - 16x = 0$$
$$4x(x^2 - 4) = 0$$
$$4x(x - 2)(x + 2) = 0$$

So the critical points of $f$ will be $x = 0, x = 2,$ and $x = -2$. We see that

$$f'(x) < 0 \quad \text{for} \quad x < -2$$
$$f'(x) > 0 \quad \text{for} \quad -2 < x < 0$$
$$f'(x) < 0 \quad \text{for} \quad 0 < x < 2$$
$$f'(x) > 0 \quad \text{for} \quad x > 2$$

So we conclude that $f(x)$ has local minima at $x = -2$ and $x = 2$, and has a local maximum at $x = 0$. From the graph, we see that this is correct.

To find the inflection points of $f$ we look for the points at which $f''(x)$ changes sign. At any such point $f''(x)$ is either zero or undefined. Since $f''(x) = 12x^2 - 16$ our candidate points are $x = \pm 2/\sqrt{3}$. At both of these points $f''(x)$ changes sign, so both of these points are inflection points.

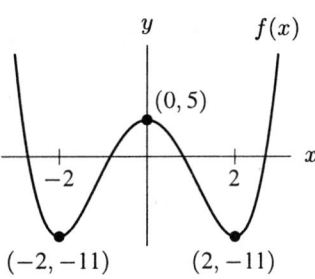

**Figure 5.19**

10. We first find the local extrema of $f(x) = x^4 - 4x^3 + 10$. Since $f'(x) = 4x^3 - 12x^2$, setting the derivative equal to 0 and factoring yields

$$4x^3 - 12x^2 = 0$$
$$4x^2(x - 3) = 0$$

So $x = 0$ and $x = 3$ are the critical points of $f(x)$. Furthermore,

$$f'(x) < 0 \text{ for } x < 0$$
$$f'(x) < 0 \text{ for } 0 < x < 3 \text{ and}$$
$$f'(x) > 0 \text{ for } x > 3.$$

From this we conclude that $f(x)$ does not have a local minimum or local maximum at $x = 0$, and that $f(x)$ has a local minimum at $x = 3$.

We now find the inflection points of $f(x)$.

$$f''(x) = 12x^2 - 24x.$$

Setting the second derivative equal to 0 and factoring yields

$$12x^2 - 24x = 0$$
$$12x(x - 2) = 0$$

So $x = 0$ and $x = 2$ may be points of inflection of $f(x)$. Furthermore,

$$f''(x) > 0 \text{ for } x < 0$$
$$f''(x) < 0 \text{ for } 0 < x < 2 \text{ and}$$
$$f''(x) > 0 \text{ for } x > 2.$$

Since $f''(x)$ changes sign at both $x = 0$ and $x = 2$, both are points of inflection for $f(x)$. Furthermore, $f(x)$ is concave down on the interval $0 < x < 2$ and concave up elsewhere. Using all of this information, we can now sketch the graph, as follows:

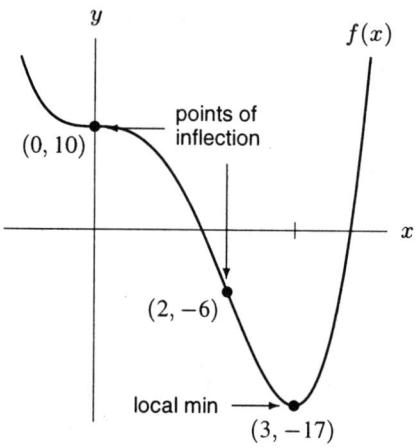

**Figure 5.20**

11.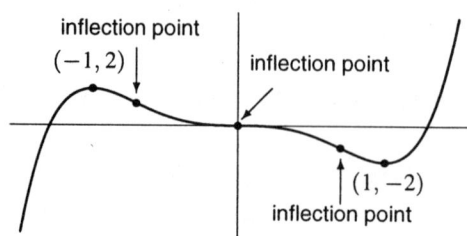

We see that $f'(x) = 15x^4 - 15x^2$ and $f''(x) = 60x^3 - 30x$. Since $f'(x) = 15x^2(x^2 - 1)$, the critical points are $x = 0, \pm 1$. Since $15x^2$ is always positive, $f'(x) < 0$ when $x^2 - 1 < 0$, or when $|x| < 1$. $f'(x) \geq 0$ otherwise. This tells us that $f(x)$ increases for $|x| > 1$, and decreases for $|x| < 1$. Thus $f$ has a local minimum at $x = 1$ and a local maximum at $x = -1$.

To find possible inflection points, we determine when $f''(x) = 0$. Since $f''(x) = 30x(2x^2 - 1)$, $f''(x) = 0$ when $x = 0$ or $x = \pm 1/\sqrt{2}$. Since $f''(x)$ changes sign at each of these points each of them are inflection points.

12. One possible answer is shown in Figure 5.21.

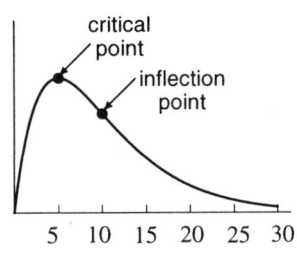

**Figure 5.21**

13. (a) One possible answer is shown in Figure 5.22.
    (b) This function is concave down at each local maximum and concave up at each local minimum, so it changes concavity at least three times. This function has at least 3 inflection points. See Figure 5.23

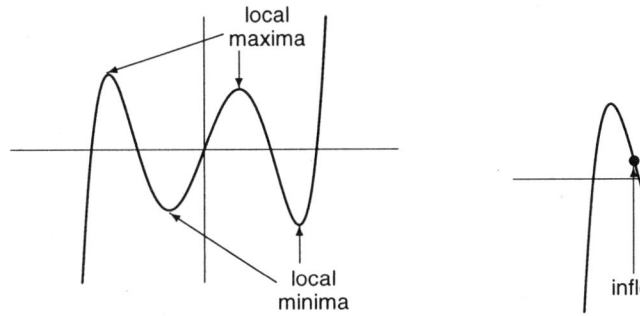 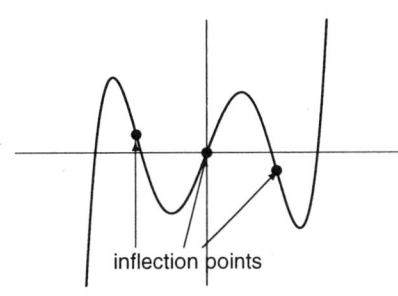

**Figure 5.22**     **Figure 5.23**

14. (a) 
$$P(t) = \frac{2000}{1 + e^{(5.3 - 0.4t)}}$$

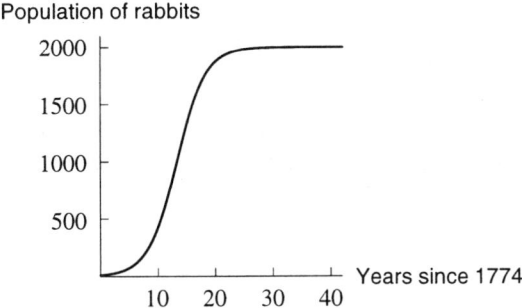

From the graph we see that the population levels off at about 2000 rabbits.

(b) The population appears to have been growing fastest when there were about 1000 rabbits, approximately 13 years after Captain Cook left the original rabbits on the island.

(c) The inflection point coincides with the point of most rapid increase of the rabbit population, that is, the inflection point occurs approximately 13 years after 1774 when the rabbit population is about 1000 rabbits.

(d) The rabbits reproduce quickly, so their population initially grew very rapidly. Limited food and space availability and perhaps predators on the island probably account for the population being unable to grow past 2000.

**214** CHAPTER FIVE /SOLUTIONS

15. (a) An inflection point occurs whenever the concavity of $f(x)$ changes. If the graph shown is that of $f(x)$, then an inflection point will occur whenever its concavity changes, or equivalently when the tangent line moves from above the curve to below or vice-versa. Such points are shown in Figure 5.24.

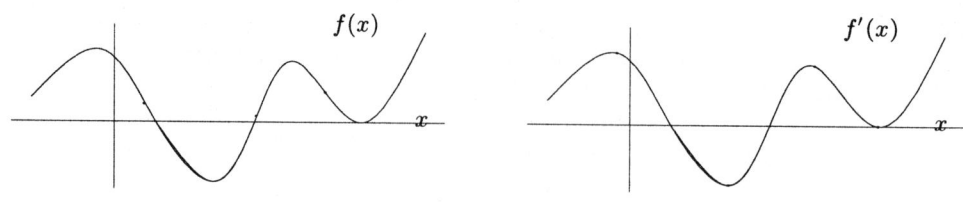

**Figure 5.24**        **Figure 5.25**

(b) To find inflection points of the function $f$ we must find points where $f''$ changes sign. However, because $f''$ is the derivative of $f'$, any point where $f''$ changes sign will be a local maximum or minimum on the graph of $f'$. Such points are shown in Figure 5.25.

(c) The inflection points of $f$ are the points where $f''$ changes sign. If the graph shown is that of $f''(x)$, then we are looking for where the given graph passes from above the x-axis to below, or vice versa. Such points are shown in Figure 5.26:

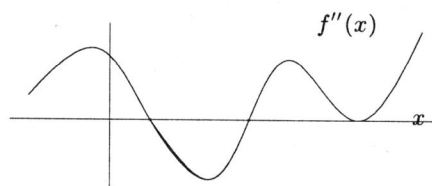

**Figure 5.26**

16.

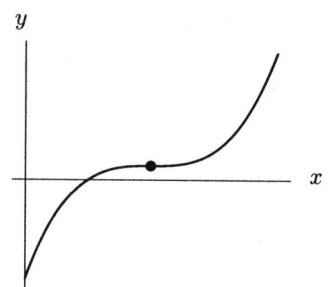

**Figure 5.27**

17. We have $f'(x) = 3x^2 - 36x - 10$ and $f''(x) = 6x - 36$. The inflection point occurs where $f''(x) = 0$, hence $6x - 36 = 0$. The inflection point is at $x = 6$. A graph is shown in Figure 5.28.

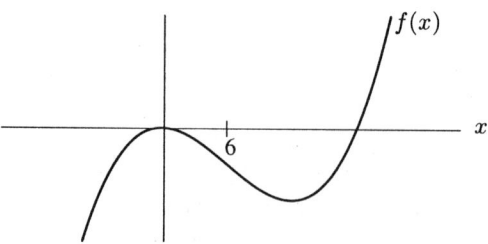

**Figure 5.28**

18. (a) Critical point.
    (b) Inflection point.
19. (a) There was a critical point at 6 pm when the temperature was at a local minimum.
    (b) The graph of temperature was decreasing but concave up in the morning. In the early afternoon the graph was decreasing but concave down. There was an inflection point at noon when the northerly wind started blowing. By 6 pm when the temperature was at a local minimum, the graph must have been concave up again so there must have been a second inflection point between noon and 6 pm. See Figure 5.29.

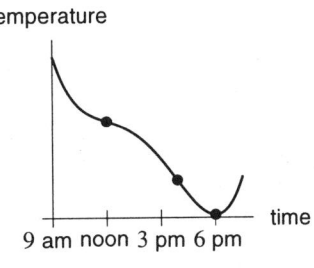

**Figure 5.29**

20. (a) Since the volume of water in the container is proportional to its depth, and the volume is increasing at a constant rate,

    $$d(t) = \text{Depth at time } t = Kt,$$

    where $K$ is some positive constant. So the graph is linear, as shown. Since initially no water is in the container, we have $d(0) = 0$, and the graph starts from the origin.

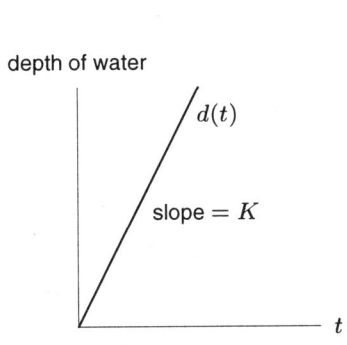
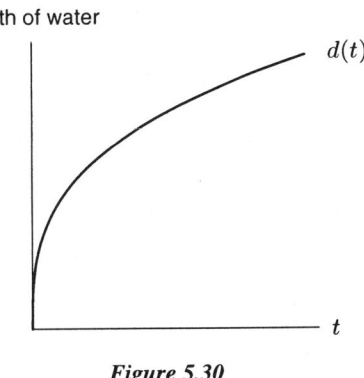

**Figure 5.30**

    (b) As time increases, the additional volume needed to raise the water level by a fixed amount increases. Thus, although the depth, $d(t)$, of water in the cone at time $t$, continues to increase, it does so more and more slowly. This means $d'(t)$ is positive but decreasing, i.e., $d(t)$ is concave down. See Figure 5.30.

21.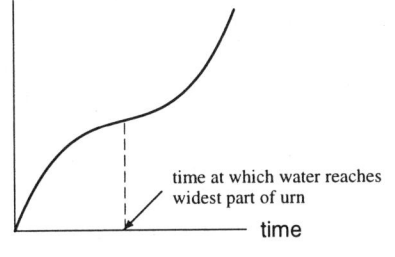

22.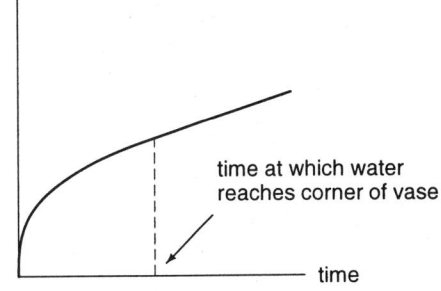

23. (a) This is one of many possible graphs.

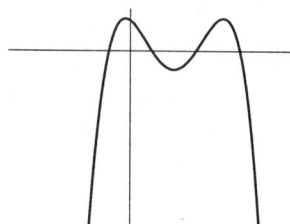

(b) Since $f$ must have a bump between each pair of zeros, $f$ could have at most four zeros.
(c) $f$ could well have no zeros at all. To see this, consider the graph of the above function shifted vertically downwards.
(d) $f$ must have at least two inflection points. Since $f$ has 3 maxima or minima, it has 3 critical points. Consequently $f'$ will have 3 corresponding zeros. Between each consecutive pair of these zeroes $f'$ must have a local maximum or minimum. Thus $f'$ will have one local maximum and one local minimum, which implies that $f''$ will have two zeros. These values, where the second derivative is zero, correspond to points of inflection on the graph of $f$.
(e) The 3 critical points are zeros of $f'$, so degree($f'$) $\geq$ 3. Thus degree($f$) $\geq$ 4.
(f) For example:
$$f(x) = \frac{-2}{15}(x+1)(x-1)(x-3)(x-5)$$
will look something like the graph in part (a). Many other answers are possible.

24. (a)

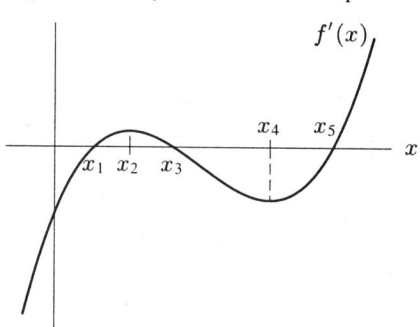

(b) $f'(x)$ changes sign at $x_1$, $x_3$, and $x_5$.
(c) $f'(x)$ has local extrema at $x_2$ and $x_4$.

25. The local maxima and minima of $f$ correspond to places where $f'$ is zero and changes sign or, possibly, to the endpoints of intervals in the domain of $f$. The points at which $f$ changes concavity correspond to local maxima and minima of $f'$. The change of sign of $f'$, from positive to negative corresponds to a maximum of $f$ and change of sign of $f'$ from negative to positive corresponds to a minimum of $f$.

26.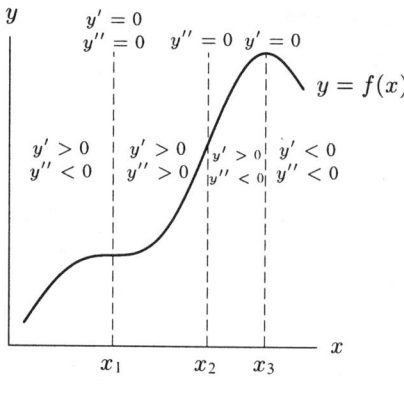

27.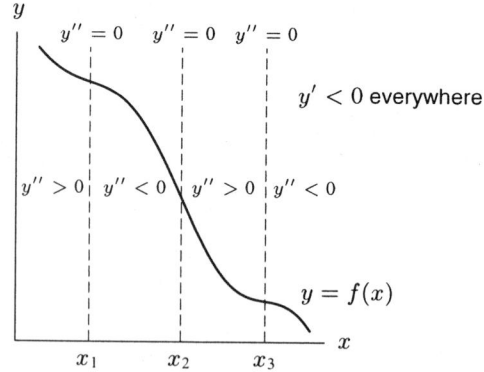

28.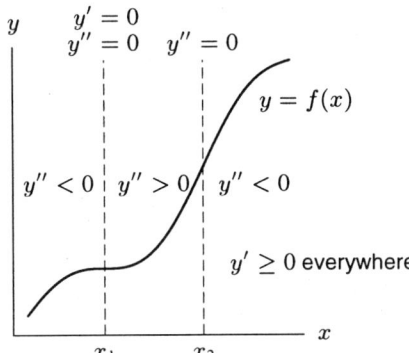

29.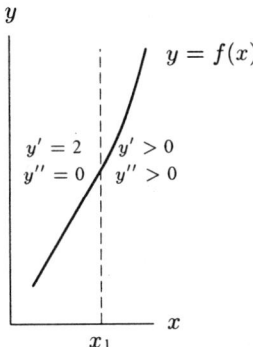

## Solutions for Section 5.3

1.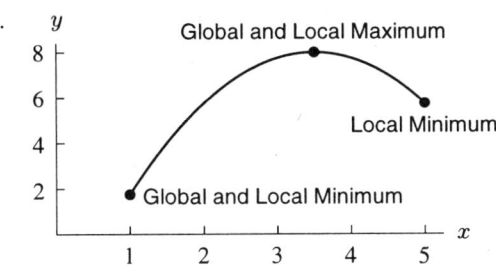

2.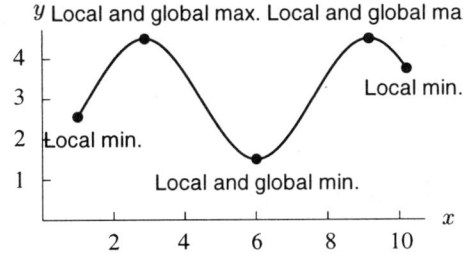

The global maximum is achieved at the two local maxima, which are at the same height.

3.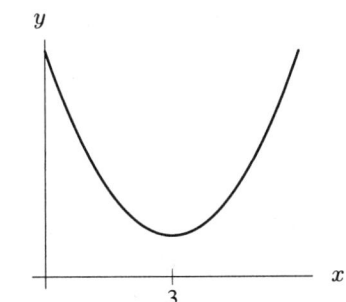

Figure 5.31

4.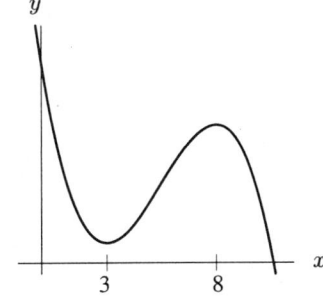

Figure 5.32

5.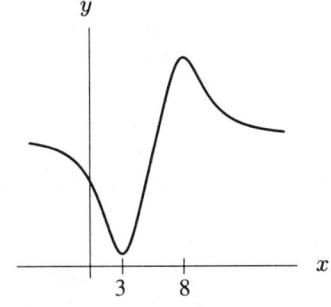

Figure 5.33

6.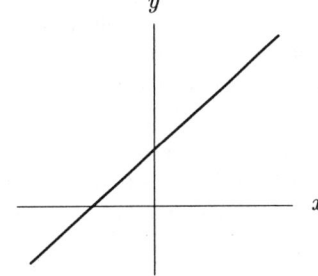

Figure 5.34

7.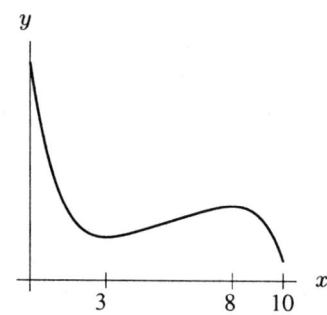

**Figure 5.35**

8.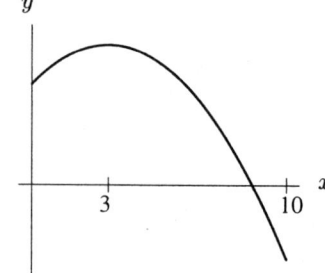

**Figure 5.36**

9.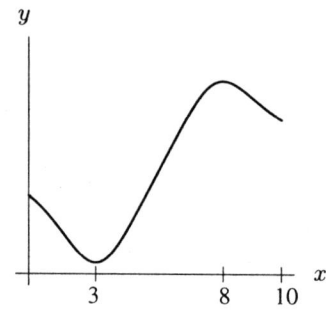

**Figure 5.37**

10.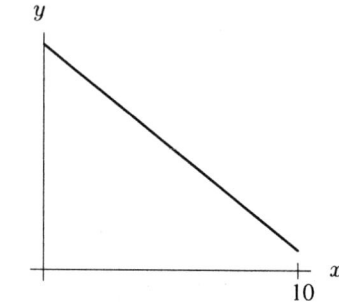

**Figure 5.38**

11. Using a computer to plot pictures of both the function, $f(x) = x^3 - e^x$, and its derivative, $f'(x) = 3x^2 - e^x$, we find that the derivative crosses the $x$-axis three times in the intervals we are interested in.

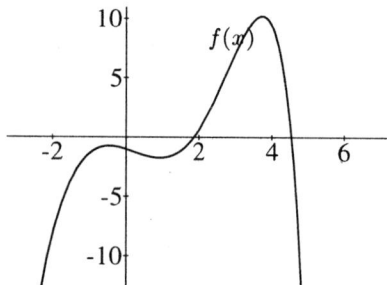 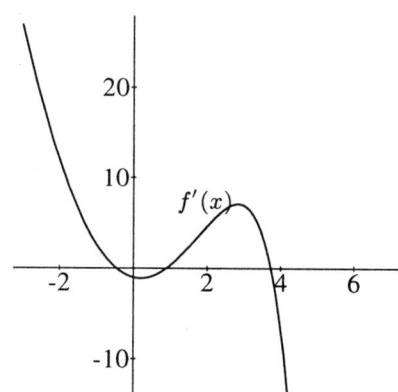

Through trial and error, we obtain approximations: local maximum at $x \approx 3.73$, local minimum at $x \approx 0.91$ and local maximum at $x \approx -0.46$. We can use the approximate values at these points, along with a picture as a guide, to find the global maximum and minimum on any interval.

(a) We find the global minimum and maximum on the interval $-1 \leq x \leq 4$ by examining the critical points above as well as the endpoints. Since $f(-1) = -1.3679$, $f(-0.46) = -0.7286$, $f(0.91) = -1.7308$, $f(3.73) = 10.2160$, $f(4) = 9.4018$, we see $x \approx 0.91$ gives a global minimum on the interval and $x \approx 3.73$ gives a global maximum.

(b) We find the global minimum and maximum on the interval $-3 \leq x \leq 2$ by examining the critical points above as well as the endpoints. Since $f(-3) = -27.0498$, $f(-0.46) = -0.7286$, $f(0.91) = -1.7308$, $f(2) = 0.6109$, we see $x = -3$ gives a global min and $x = 2$ a global max. (Even though $x \approx -0.46$ gives a local maximum, it does not give the greatest maximum on this interval; even though $x \approx 0.91$ gives a local minimum, it is not the smallest minimum on this interval.)

12. (a) We have $f'(x) = 10x^9 - 10 = 10(x^9 - 1)$. This is zero when $x = 1$, so $x = 1$ is a critical point of $f$. For values of $x$ less than 1, $x^9$ is less than 1, and thus $f'(x)$ is negative when $x < 1$. Similarly, $f'(x)$ is positive for $x > 1$. Thus $f(1) = -9$ is a local minimum.

    We also consider the endpoints $f(0) = 0$ and $f(2) = 1004$. Since $f'(0) < 0$ and $f'(2) > 0$, we see $x = 0$ and $x = 2$ are local maxima.
   (b) Comparing values of $f$ shows that the global minimum is at $x = 1$, and the global maximum is at $x = 2$.

13. (a) $f'(x) = 1 - 1/x$. This is zero only when $x = 1$. Now $f'(x)$ is positive when $1 < x \leq 2$, and negative when $0.1 < x < 1$. Thus $f(1) = 1$ is a local minimum. The endpoints $f(0.1) \approx 2.4026$ and $f(2) \approx 1.3069$ are local maxima.
   (b) Comparing values of $f$ shows that $x = 0.1$ gives the global maximum and $x = 1$ gives the global minimum.

14. (a) Profit is maximized when $R(q) - C(q)$ is as large as possible. This occurs at $q = 2500$, where profit = $7500 - 5500 = \$2000$.
   (b) We see that $R(q) = 3q$ and so the price is $p = 3$, or \$3 per unit.
   (c) Since $C(0) = 3000$, the fixed costs are \$3000.

15. (a) At $q = 5000$, $MR > MC$, so the marginal revenue to produce the next item is greater than the marginal cost. This means that the company will make money by producing additional units, and production should be increased.
   (b) Profit is maximized where $MR = MC$, and where the profit function is going from increasing ($MR > MC$) to decreasing ($MR < MC$). This occurs at $q = 8000$.

16. First find marginal revenue and marginal cost.

$$MR = R'(q) = 450$$

$$MC = C'(q) = 6q$$

Setting $MR = MC$ yields $6q = 450$, so marginal cost is equal to marginal revenue when

$$q = \frac{450}{6} = 75 \text{ units.}$$

Is profit maximized at $q = 75$? Profit $= R(q) - C(q)$;

$$R(75) - C(75) = 450(75) - (10{,}000 + 3(75)^2)$$
$$= 33{,}750 - 26{,}875 = \$6875.$$

Testing $q = 74$ and $q = 76$:

$$R(74) - C(74) = 450(74) - (10{,}000 + 3(74)^2)$$
$$= 33{,}300 - 26{,}428 = \$6872.$$

$$R(76) - C(76) = 450(76) - (10{,}000 + 3(76)^2)$$
$$= 34{,}200 - 27{,}328 = \$6872.$$

Since profit at $q = 75$ is more than profit at $q = 74$ and $q = 76$, we conclude that profit is maximized locally at $q = 75$. The only endpoint we need to check is $q = 0$.

$$R(0) - C(0) = 450(0) - (10{,}000 + 3(0)^2)$$
$$= -\$10{,}000.$$

This is clearly not a maximum, so we conclude that the profit is maximized globally at $q = 75$, and the total profit at this production level is \$6,875.

17. First find the marginal revenue and marginal cost. Note that each product sells for \$588, so revenue is given by $R(q) = 588q$.

$$MR = R'(q) = 588$$
$$MC = C'(q) = 3q^2 - 120q + 1200$$

Setting $MR = MC$ yields

$$3q^2 - 120q + 1200 = 588$$
$$\text{so} \quad 3q^2 - 120q + 612 = 0$$

This factors to
$$3(q-34)(q-6) = 0$$
so $MR = MC$ at $q = 34$ and $q = 6$. We now find the profit at these points:

$$R(6) - C(6) = 588(6) - \left[(6)^3 - 60(6)^2 + 1200(6) + 1000\right]$$
$$= 3{,}528 - 6{,}256 = -\$2{,}728$$

$$R(34) - C(34) = 588(34) - \left[(34)^3 - 60(34)^2 + 1200(34) + 1000\right]$$
$$= 19{,}992 - 11{,}744 = \$8{,}248$$

We must also try the endpoints

$$R(0) - C(0) = 588(0) - \left[(0)^3 - 60(0)^2 + 1200(0) + 1000\right]$$
$$= -\$1{,}000$$

$$R(50) - C(50) = 588(50) - \left[(50)^3 - 60(50)^2 + 1200(50) + 1000\right]$$
$$= 29{,}400 - 36{,}000 = -\$6{,}600$$

From this we see that profit is maximized at $q = 34$ units. The total cost at $q = 34$ is $C(34) = \$11{,}744$. The total revenue at $q = 34$ is $R(34) = \$19{,}992$ and the total profit is

$$19{,}992 - 11{,}744 = \$8{,}248.$$

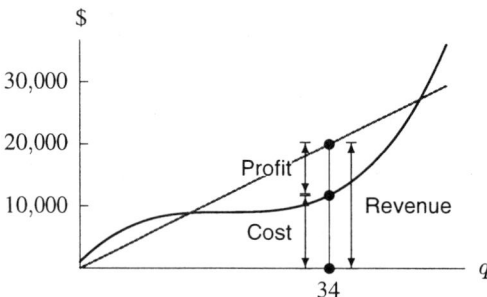

**Figure 5.39**

18. The profit function is positive when $R(q) > C(q)$, and negative when $C(q) > R(q)$. It's positive for $5 < q < 10$, and negative from $0 < q < 5$ and $10 < q$. Profit is maximized when $R(q) > C(q)$ and $R'(q) = C'(q)$ which occurs at about $q = 8$.

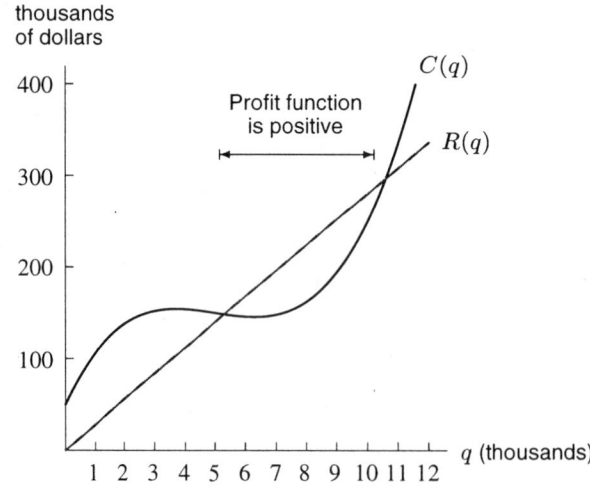

**Figure 5.40**

19.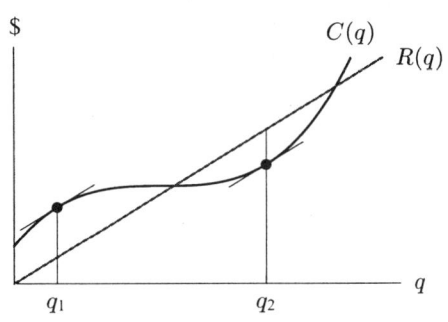

**Figure 5.41**

(a) We find $q_1$ and $q_2$ by checking to see where the slope of the tangent line to $C(q)$ is equal to the slope of $R(q)$. Because the slopes of $C(q)$ and $R(q)$ represent marginal cost and marginal revenue, respectively, at $q$'s where the slopes are equal, the cost of producing an additional unit of $q$ exactly equals the revenue gained from selling an additional unit. In other words, $q_1$ and $q_2$ are levels of production at which the additional or marginal profit from producing an additional unit of $q$ is zero.

(b) We can think of the vertical distance between the cost and revenue curves as representing the firm's total profits. The quantity $q_1$ is the production level at which total profits are at a minimum. We see this by noticing that for points slightly to the left of $q_1$, the slope of $C(q)$ is slightly greater than the slope of $R(q)$. This means that the cost of producing an additional unit of $q$ is greater than the revenue earned from selling it. So for points to the left of $q_1$, additional production decreases profits. For points slightly to the right of $q_1$, the slope of $C(q)$ is less than the slope of $R(q)$. Thus additional production results in an increase in profits. The production level $q_1$ is the level at which the firm ceases to take a loss on each additional item and begins to make a profit on each additional item. Note that the total profit is still negative, and remains so until the graphs cross, i.e, where total cost equals total revenue.

Similar reasoning applies for $q_2$, except it is the level of production at which profits are maximized. For points slightly to the left of $q_2$, the slope of $C(q)$ is less than the slope of $R(q)$. Thus the cost of producing an additional unit is less than the revenue gained from selling it. So by selling an additional unit, the firm can increase profits. For points to the right of $q_2$, the slope of $C(q)$ is greater than slope of $R(q)$. This means that the profit from producing and selling an additional unit of $q$ will be negative, decreasing total profits. The point $q_2$ is the level of production at which the firm stops making a profit on each additional item sold and begins to take a loss.

At both points $q_1$ and $q_2$, note that the vertical distance between $C(q)$ and $R(q)$ is at a local maximum. This represents the fact that $q_1$ and $q_2$ are local profit minimum and local profit maximum points.

20. Since marginal revenue is larger than marginal cost around $q = 2000$, as you produce more of the product your revenue increases faster than your costs, so profit goes up, and maximal profit will occur at a production level above 2000.

21. (a) The fixed cost is 0 because $C(0) = 0$.
(b) Profit, $\pi(q)$, is equal to money from sales, $7q$, minus total cost to produce those items, $C(q)$.

$$\pi = 7q - 0.01q^3 + 0.6q^2 - 13q$$
$$\pi' = -0.03q^2 + 1.2q - 6$$

$$\pi' = 0 \quad \text{if} \quad q = \frac{-1.2 \pm \sqrt{(1.2)^2 - 4(0.03)(6)}}{-0.06} \approx 5.9 \quad \text{or} \quad 34.1.$$

Now $\pi'' = -0.06q + 1.2$, so $\pi''(5.9) > 0$ and $\pi''(34.1) < 0$. This means $q = 5.9$ is a local min and $q = 34.1$ a local max. We now evaluate the endpoint, $\pi(0) = 0$, and the points nearest $q = 34.1$ with integer $q$-values:

$$\pi(35) = 7(35) - 0.01(35)^3 + 0.6(35)^2 - 13(35) = 245 - 148.75 = 96.25,$$

$$\pi(34) = 7(34) - 0.01(34)^3 + 0.6(34)^2 - 13(34) = 238 - 141.44 = 96.56.$$

So the (global) maximum profit is $\pi(34) = 96.56$. The money from sales is $238, the cost to produce the items is $141.44, resulting in a profit of $96.56.

(c) The money from sales is equal to price×quantity sold. If the price is raised from \$7 by \$x to \$(7+x), the result is a reduction in sales from 34 items to (34 − 2x) items. So the result of raising the price by \$x is to change the money from sales from (7)(34) to (7 + x)(34 − 2x) dollars. If the production level is fixed at 34, then the production costs are fixed at \$141.44, as found in part (b), and the profit is given by:

$$\pi(x) = (7+x)(34-2x) - 141.44$$

This expression gives the profit as a function of change in price $x$, rather than as a function of quantity as in part (b). We set the derivative of $\pi$ with respect to $x$ equal to zero to find the change in price that maximizes the profit:

$$\frac{d\pi}{dx} = (1)(34-2x) + (7+x)(-2) = 20 - 4x = 0$$

So $x = 5$, and this must give a maximum for $\pi(x)$ since the graph of $\pi$ is a parabola which opens downwards. The profit when the price is \$12 ($= 7 + x = 7 + 5$) is thus $\pi(5) = (7+5)(34 - 2(5)) - 141.44 = \$146.56$. This is indeed higher than the profit when the price is \$7, so the smart thing to do is to raise the price by \$5.

22. We first need to find an expression for $R(q)$, or revenue in terms of quantity sold. We know that $R(q) = pq$, where $p$ is the price of one item. Here $p = 45 - 0.01q$, so we make the substitution

$$R(q) = (45 - .01q)q = 45q - 0.01q^2.$$

This is the function we want to maximize. Finding the derivative and setting it equal to 0 yields

$$R'(q) = 0$$
$$45 - 0.02q = 0$$
$$0.02q = 45 \text{ so}$$
$$q = 2250.$$

Is this a maximum?

$$R'(q) > 0 \text{ for } q < 2250 \text{ and}$$
$$R'(q) < 0 \text{ for } q > 2250.$$

So we conclude that $R(q)$ has a local maximum at $q = 2250$. Testing $q = 0$, the only endpoint, $R(0) = 0$, which is less than $R(2250) = \$50{,}625$. So we conclude that revenue is maximized at $q = 2250$. The price of each item at this production level is

$$p = 45 - .01(2250) = \$22.50$$

and total revenue is

$$pq = \$22.50(2250) = \$50{,}625,$$

which agrees with the above answer.

23. Note that profit $= \pi(q) = R(q) - C(q)$, so to find profit, we must find an expression for $R(q)$. If $p$ is the price of a single item,

$$R(q) = p \cdot q.$$

Substituting $p = b_1 - a_1 q$ gives

$$R(q) = (b_1 - a_1 q) \cdot q$$
$$= b_1 q - a_1 q^2 \quad \text{and}$$
$$\pi(q) = R(q) - C(q)$$
$$= b_1 q - a_1 q^2 - b_2 - a_2 q.$$

Finding the derivative and setting it equal to 0 yields

$$\pi'(q) = b_1 - 2a_1 q - a_2 = 0$$
$$\text{so } q = \frac{-a_2 + b_1}{2a_1}$$

is a critical point for $\pi(q)$. Using the second derivative test, we see $\pi$ is concave down,

$$\pi''(q) = -2a_1$$

Since $a_1$ is positive, $-2a_1$ will always be negative, so, since $\pi''(q) < 0$ for all $q$, $q = (-a_2 + b_1)/(2a_1)$ is a local maximum. This $q$ is the only critical point, so it is the global maximum for $\pi(q)$.

24. We first need to find an expression for revenue in terms of price. At a price of $8, 1500 tickets are sold. For each $1 above $8, 75 fewer tickets are sold. This suggests the following formula for $q$, the quantity sold for any price $p$.

$$q = 1500 - 75(p - 8)$$
$$= 1500 - 75p + 600$$
$$= 2100 - 75p.$$

We know that $R = pq$, so substitution yields

$$R(p) = p(2100 - 75p) = 2100p - 75p^2$$

To maximize revenue, we find the derivative of $R(p)$ and set it equal to 0.

$$R'(p) = 2100 - 150p = 0$$
$$150p = 2100$$

so $p = \frac{2100}{150} = 14$. Does $R(p)$ have a maximum at $p = 14$? Using the first derivative test,

$$R'(p) > 0 \text{ if } p < 14 \text{ and}$$
$$R'(p) < 0 \text{ if } p > 14.$$

So $R(p)$ has a local maximum at $p = 14$. Since this is the only critical point for $p \geq 0$, it must be a global maximum. So we conclude that revenue is maximized when the price is $14.

25. Let $x$ equal the number of chairs ordered in excess of 300, so $0 \leq x \leq 100$.

$$\text{Revenue} = R = (90 - 0.25x)(300 + x)$$
$$= 27,000 - 75x + 90x - 0.25x^2 = 27,000 + 15x - 0.25x^2$$

At a critical point $dR/dx = 0$. Since $dR/dx = 15 - 0.5x$, we have $x = 30$, and the maximum revenue is $27,225 since the graph of $R$ is a parabola which opens downwards. The minimum is $0 (when no chairs are sold).

26. For each month,

$$\text{Profit} = \text{Revenue} - \text{Cost}$$
$$\pi = pq - wL = pcK^\alpha L^\beta - wL$$

The variable on the right is $L$, so at the maximum

$$\frac{d\pi}{dL} = \beta p c K^\alpha L^{\beta-1} - w = 0$$

Now $\beta - 1$ is negative, since $0 < \beta < 1$, so $1 - \beta$ is positive and we can write

$$\frac{\beta p c K^\alpha}{L^{1-\beta}} = w$$

giving

$$L = \left(\frac{\beta p c K^\alpha}{w}\right)^{\frac{1}{1-\beta}}$$

Since $\beta - 1$ is negative, when $L$ is just above 0, the quantity $L^{\beta-1}$ is huge and positive, so $d\pi/dL > 0$. When $L$ is large, $L^{\beta-1}$ is small, so $d\pi/dL < 0$. Thus the value of $L$ we have found gives a global maximum, since it is the only critical point.

27. Since the function is positive, the graph lies above the $x$-axis. If there is a global maximum at $x = 3$, $t'(x)$ must be positive, then negative. Since $t'(x)$ and $t''(x)$ have the same sign for $x < 3$, they must both be positive, and thus the graph must be increasing and concave up. Since $t'(x)$ and $t''(x)$ have opposite signs for $x > 3$ and $t'(x)$ is negative, $t''(x)$ must again be positive and the graph must be decreasing and concave up. A possible sketch of $y = t(x)$ is shown in the figure below.

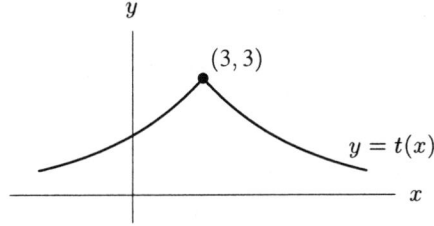

28. (a) $\pi(q)$ is maximized when $R(q) > C(q)$ and they are as far apart as possible:

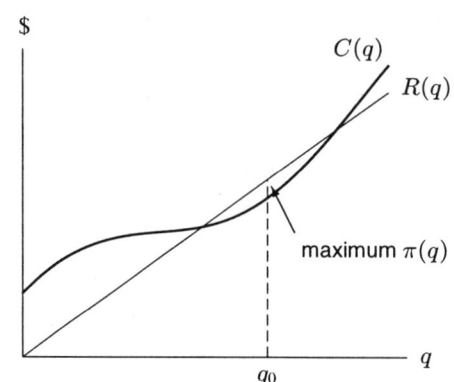

(b) $\pi'(q_0) = R'(q_0) - C'(q_0) = 0$ implies that $C'(q_0) = R'(q_0) = p$.

Graphically, the slopes of the two curves at $q_0$ are equal. This is plausible because if $C'(q_0)$ were greater than $p$ or less than $p$, the maximum of $\pi(q)$ would be to the left or right of $q_0$, respectively. In economic terms, if the cost were rising more quickly than revenues, the profit would be maximized at a lower quantity (and if the cost were rising more slowly, at a higher quantity).

(c)

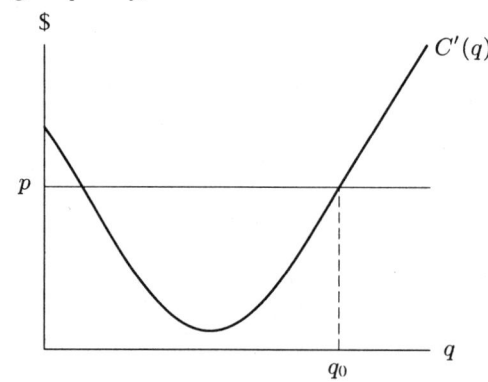

29. (a) We know that Profit = Revenue − Cost, so differentiating with respect to $q$ gives:

Marginal Profit = Marginal Revenue − Marginal Cost.

We see from the figure in the problem that just to the left of $q = a$, marginal revenue is less than marginal cost, so marginal profit is negative there. To the right of $q = a$ marginal revenue is greater than marginal cost, so marginal profit is positive there. At $q = a$ marginal profit changes from negative to positive. This means that profit is decreasing to the left of $a$ and increasing to the right. The point $q = a$ corresponds to a local minimum of profit, and does not maximize profit. It would be a terrible idea for the company to set its production level at $q = a$.

(b) We see from the figure in the problem that just to the left of $q = b$ marginal revenue is greater than marginal cost, so marginal profit is positive there. Just to the right of $q = b$ marginal revenue is less than marginal cost, so marginal profit is negative there. At $q = b$ marginal profit changes from positive to negative. This means that profit is increasing to the left of $b$ and decreasing to the right. The point $q = b$ corresponds to a local maximum of profit. In fact, since the area between the $MC$ and $MR$ curves in the figure in the text between $q = a$ and $q = b$ is bigger than the area between $q = 0$ and $q = a$, $q = b$ is in fact a global maximum.

## Solutions for Section 5.4

1. (a) (i) The average cost of quantity $q$ is given by the formula $a(q)/q$. So average cost at $q = 30$ is given by $C(30)/30$. From the graph, we see that $C(30) \approx 260$, so $a(q) \approx \frac{260}{30} \approx \$8.67$ per unit. To interpret this graphically, note that $a(q) = \frac{C(q)}{q} = \frac{C(q)-0}{q-0}$. This is exactly the formula for the slope of a line from the origin to a point $(q, C(q))$ on the curve. So $a(30)$ is the slope of a line connecting $(0,0)$ to $(30, C(30))$. Such a line is shown below in Figure 5.42:

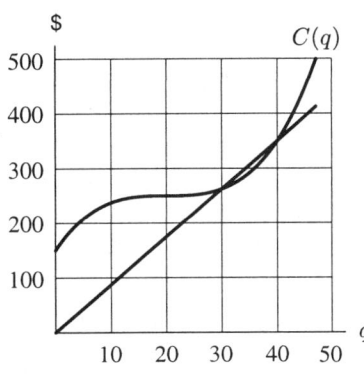

**Figure 5.42**

(ii) The marginal cost is given by $C'(q)$. This derivative is equal to the slope of a tangent line to $C(q)$ at $q = 30$. To estimate this slope, it will be easiest to draw the tangent line, as shown in Figure 5.43:

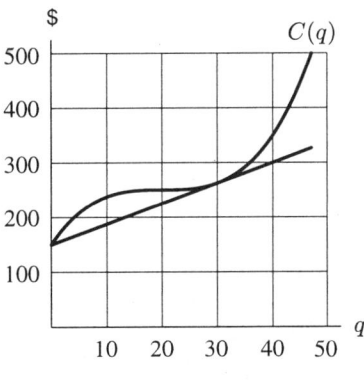

**Figure 5.43**

From this plot, we see approximately that the point $(40, 300)$ and the point $(0, 160)$ are on this line, so its slope is approximately $\frac{300-160}{40-0} = 3.5$. $C'(30) \approx \$3.50$ per unit.

(b) We know that $a(q)$ is minimized where $a(q) = C'(q)$. Using the graphical interpretations froms parts (b) and (c), this is equivalent to saying that the tangent line has the same slope as the line connecting the point on the curve to the origin. Since these two lines share a point, specifically the point $(q, C(q))$ on the curve, and have the same slope, they are in fact the same line. So $a(q)$ is minimized where the line passing from $(q, C(q))$ to the origin is also tangent to the curve. To find such points, a variety of lines passing through the origin and the curve are shown in Figure 5.44:

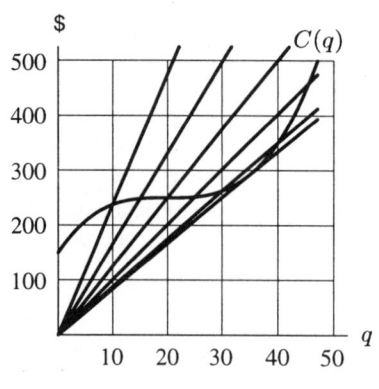

**Figure 5.44**

From this plot, we see that the line with the desired properties intersects the curve at $q \approx 35$. So $q \approx 35$ units minimizes $a(q)$.

2. (a) The average cost of quantity $q$ is given by the formula $C(q)/q$. So average cost at $q = 10{,}000$ is given by $C(10{,}000)/10{,}000$. From the graph, we see that $C(10{,}000) \approx 16{,}000$, so $a(q) \approx \frac{16{,}000}{10{,}000} \approx \$1.60$ per unit. The economic interpretation of this is that \$1.60 is each unit's share of the total cost of producing 10,000 units.

   (b) To interpret this graphically, note that $a(q) = \frac{C(q)}{q} = \frac{C(q)-0}{q-0}$. This is exactly the formula for the slope of a line from the origin to a point $(q, C(q))$ on the curve. So $a(10{,}000)$ is the slope of a line connecting $(0,0)$ to $(10{,}000, C(10{,}000))$. Such a line is shown below in Figure 5.45.

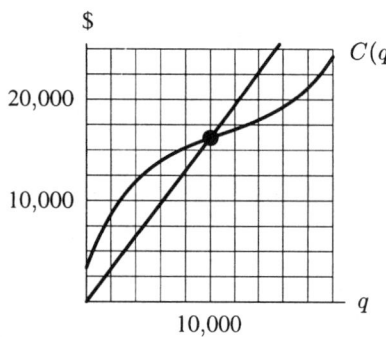

**Figure 5.45**

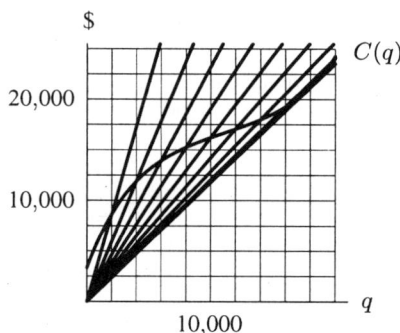

**Figure 5.46**

   (c) We know that $a(q)$ is minimized where $a(q) = C'(q)$. Using the graphical interpretations froms parts (b) and (c), this is equivalent to saying that the tangent line has the same slope as the line connecting the point on the curve to the origin. Since these two lines share a point, specifically the point $(q, C(q))$ on the curve, and have the same slope, they are in fact the same line. So $a(q)$ is minimized where the line passing from $(q, C(q))$ to the origin is also tangent to the curve. To find such points, a variety of lines passing through the origin and the curve are shown in Figure 5.46.

   From this plot, we see that the line with the desired properties intersects the curve at $q \approx 18{,}000$. So $q \approx 18{,}000$ units minimizes $a(q)$.

3. (a) The marginal cost of producing the $q_0^{\text{th}}$ item is simply $C'(q_0)$, or the slope of the graph at $q_0$. Since the slope of the cost function is always 12, the marginal cost of producing the $100^{\text{th}}$ item and the marginal cost of producing the $1000^{\text{th}}$ item is \$12.

   (b) The average cost at $q = 100$ is given by

   $$a(100) = \frac{C(100)}{100} = \frac{3700}{100} = \$37.$$

   The average cost at $q = 1000$ is given by

   $$a(1000) = \frac{C(1000)}{1000} = \frac{14{,}500}{1000} = \$14.50.$$

4. The cost function is $C(q) = 1000 + 20q$. The marginal cost function is the derivative $C'(q) = 20$, so the marginal cost to produce the 200$^{\text{th}}$ unit is $20 per unit. The average cost of producing 200 units is given by

$$a(200) = \frac{C(200)}{200} = \frac{5000}{200} = \$25/\text{unit}$$

5. (a) The marginal cost tells us that additional units produced would cost about $10 each, which is below the average cost, so producing them would reduce average cost.
   (b) It is impossible to determine the effect on profit from the information given. Profit depends on both cost and revenue, $\pi = R - C$, but we have no information on revenue.

6. If the minimum average cost occurs at a production level of 15000 units, the line from the origin to the curve is tangent to the curve at that point. The slope of this line is 25, so the cost at 15000 units is $25(15000) = 375000$. See Figure 5.47.

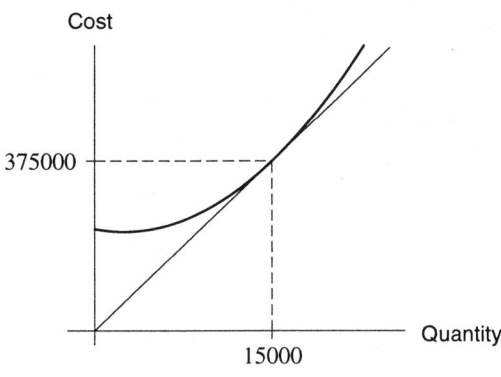

*Figure 5.47*

7. (a) Profit equals revenue minus cost. Your monthly revenue is

$$1200 \text{ slippers} \times \$20/\text{slipper} = \$24000,$$

your monthly cost equals 1200 slippers × $2/slipper = $2400. Since you are earning a monthly profit of $24000 − $2400 = $21600, you are making money.
   (b) Since additional units produced cost about $3 each, which is above the average cost, producing them increases average cost. Since additional slippers cost about $3 to produce and can be sold for $20, you can increase your profit by making and selling them. This is a case where marginal revenue, which is $20 per slipper, is greater than marginal cost, which is $3 per slipper.
   (c) You should recommend increase in production, since that will increase profit. The fact that average cost of production will increase is irrelevant to your decision.

8. (a) $N = 100 + 20x$, graphed in Figure 5.48.

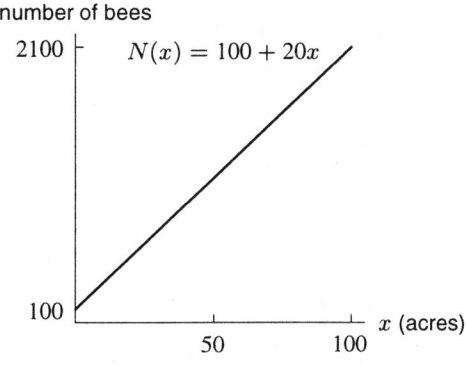

*Figure 5.48*

**228** CHAPTER FIVE /SOLUTIONS

(b) $N'(x) = 20$ and its graph is just a horizontal line. This means that rate of increase of the number of bees with acres of clover is constant — each acre of clover brings 20 more bees.

On the other hand, $N(x)/x = 100/x + 20$ means that the average number of bees per acre of clover approaches 20 as more acres are put under clover. See Figure 5.49. As $x$ increases, $100/x$ decreases to 0, so $N(x)/x$ approaches 20 (i.e. $N(x)/x \to 20$). Since the total number of bees is 20 per acre plus the original 100, the average number of bees per acre is 20 plus the 100 shared out over $x$ acres. As $x$ increases, the 100 are shared out over more acres, and so its contribution to the average becomes less. Thus the average number of bees per acre approaches 20 for large $x$.

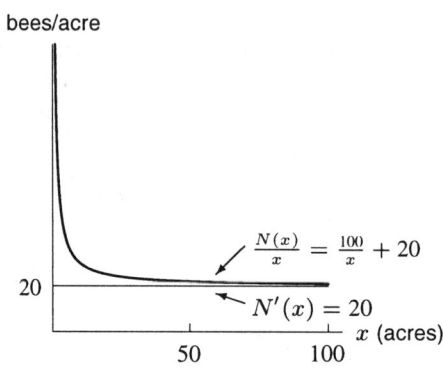

**Figure 5.49**

9. The graph of the average cost function is shown in Figure 5.50.

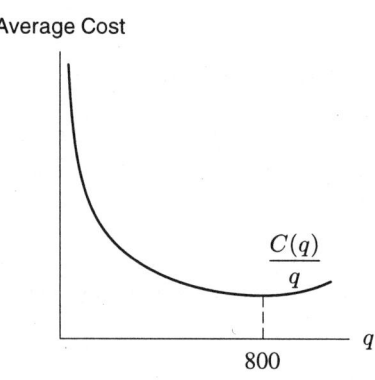

**Figure 5.50**

10. (a) We know that the average cost is given by
$$a(q) = \frac{C(q)}{q}.$$
Thus the average cost is
$$a(q) = 0.04q^2 - 3q + 75 + \frac{96}{q}.$$

(b) Average cost is graphed in Figure 5.51.
(c) Looking at the graph of the average value we see that it is decreasing for $q < 38$. Likewise we see that $a(q)$ is increasing for $q > 38$.
(d) Thus the average cost hits its minimum at about $q = 38$ with the value of about $21.

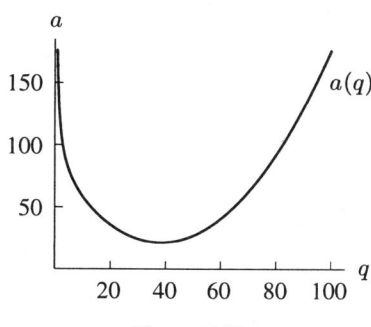

Figure 5.51

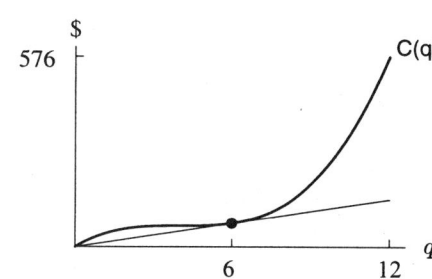

Figure 5.52

11. (a) The line connecting the origin and the graph of $C(q)$ in Figure 5.52 appears to have minimum slope at $q = 6$. Therefore we conclude that average cost is minimized at about $q = 6$.
(b) The average cost of the first $q$ items is given by

$$a(q) = \frac{C(q)}{q} = \frac{q^3 - 12q^2 + 48q}{q} = q^2 - 12q + 48$$

We want to minimize $a(q)$.

$$a'(q) = 2q - 12.$$

Setting this equal to 0 and solving yields $q = 6$. Is this our minimum?

$$a'(q) < 0 \text{ if } q < 6 \text{ and}$$
$$a'(q) > 0 \text{ if } q > 6$$

so $q = 6$ is a local minimum for $a(q)$. We also test the endpoints, $q = 0$ and $q = 12$.

$$a(0) = 48 \text{ and}$$
$$a(12) = 48.$$

Both are greater than $a(6) = 12$, so the average cost is least at $q = 6$.

12. (a) $a(q) = C(q)/q$, so $C(q) = 0.01q^3 - 0.6q^2 + 13q$.
(b) Taking the derivative of $C(q)$ gives an expression for the marginal cost:

$$C'(q) = MC(q) = 0.03q^2 - 1.2q + 13.$$

To find the smallest $MC$ we take its derivative and find the value of $q$ that makes it zero. So: $MC'(q) = 0.06q - 1.2 = 0$ when $q = 1.2/0.06 = 20$. This value of $q$ must give a minimum because the graph of $MC(q)$ is a parabola opening upwards. Therefore the minimum marginal cost is $MC(20) = 1$. So the marginal cost is at a minimum when the additional cost per item is $1.
(c) $a'(q) = 0.02q - 0.6$
Setting $a'(q) = 0$ and solving for $q$ gives $q = 30$ as the quantity at which the average is minimized, since the graph of $a$ is a parabola which opens upwards. The minimum average cost is $a(30) = 4$ dollars per item.
(d) The marginal cost at $q = 30$ is $MC(30) = 0.03(30)^2 - 1.2(30) + 13 = 4$. This is the same as the average cost at this quantity. Note that since $a(q) = C(q)/q$, we have $a'(q) = (qC'(q) - C(q))/q^2$. At a critical point, $q_0$, of $a(q)$, we have

$$0 = a'(q_0) = \frac{q_0 C'(q_0) - C(q_0)}{q_0^2},$$

so $C'(q_0) = C(q_0)/q_0 = a(q_0)$. Therefore $C'(30) = a(30) = 4$ dollars per item.
Another way to see why the marginal cost at $q = 30$ must equal the minimum average cost $a(30) = 4$ is to view $C'(30)$ as the approximate cost of producing the $30^{\text{th}}$ or $31^{\text{st}}$ good. If $C'(30) < a(30)$, then producing the $31^{\text{st}}$ good would lower the average cost, i.e. $a(31) < a(30)$. If $C'(30) > a(30)$, then producing the $30^{\text{th}}$ good would raise the average cost, i.e. $a(30) > a(29)$. Since $a(30)$ is the global minimum, we must have $C'(30) = a(30)$.

230    CHAPTER FIVE /SOLUTIONS

13. (a) Since the graph is concave down, the average cost gets smaller as $q$ increases. This is because the cost per item gets smaller as $q$ increases. There is no value of $q$ for which the average cost is minimized since for any $q_0$ larger than $q$ the average cost at $q_0$ is less than the average cost at $q$. Graphically, the average cost at $q$ is the slope of the line going through the origin and through the point $(q, C(q))$. Figure 5.53 shows how as $q$ gets larger, the average cost decreases.

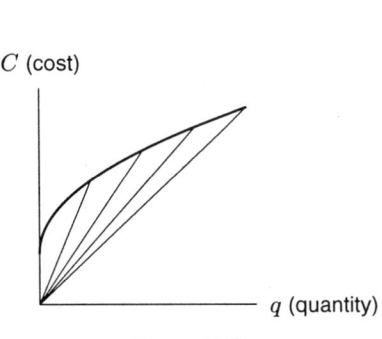

Figure 5.53

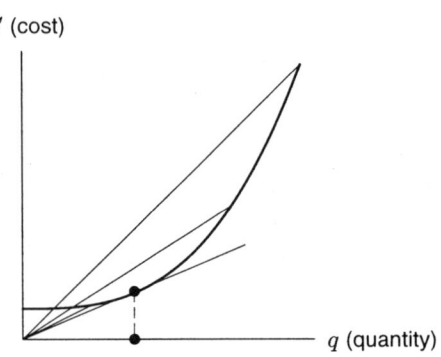

Figure 5.54

(b) The average cost will be minimized at some $q$ for which the line through $(0, 0)$ and $(q, c(q))$ is tangent to the cost curve. This point is shown in Figure 5.54.

14. (a) Differentiating $C(q)$ gives
$$C'(q) = \frac{K}{a} q^{(1/a)-1}, \quad C''(q) = \frac{K}{a}\left(\frac{1}{a} - 1\right) q^{(1/a)-2}.$$
If $a > 1$, then $C''(q) < 0$, so $C$ is concave down.

(b) We have
$$a(q) = \frac{C(q)}{q} = \frac{Kq^{1/a} + F}{q}$$
$$C'(q) = \frac{K}{a} q^{(1/a)-1}$$
so $a(q) = C'(q)$ means
$$\frac{Kq^{1/a} + F}{q} = \frac{K}{a} q^{(1/a)-1}.$$
Solving,
$$Kq^{1/a} + F = \frac{K}{a} q^{1/a}$$
$$K\left(\frac{1}{a} - 1\right) q^{1/a} = F$$
$$q = \left[\frac{Fa}{K(1-a)}\right]^a.$$

15. Since $a(q) = C(q)/q$, we use the product rule to find
$$da/dq = C'(q)(q^{-1}) + C(q)(-q^{-2}) = \frac{C'(q)}{q} - \frac{C(q)}{q^2} = \frac{C'(q) - a(q)}{q}$$
Since marginal cost is $C'$, the hypothesis $C'(q) < a(q)$ means that $C'(q) - a(a) < 0$, and thus $da/dq < 0$.

16. Since $a(q) = C(q)/q$, we use the product rule to find
$$da/dq = C'(q)(q^{-1}) + C(q)(-q^{-2}) = \frac{C'(q)}{q} - \frac{C(q)}{q^2} = \frac{C'(q) - a(q)}{q}$$
Since marginal cost is $C'$, the hypothesis $C'(q) > a(q)$ means that $C'(q) - a(a) > 0$, and thus $da/dq < 0$.

## Solutions for Section 5.5

1. The effect on demand is approximately $E$ times the change in price. A price increase cause a decrease in demand and a price decrease cause an increase in demand.

    (a) Demand decreases by about $2(3\%) = 6\%$.
    (b) Demand increases by about $2(3\%) = 6\%$.

2. The effect on demand is approximately $E$ times the change in price. A price increase causes a decrease in demand and a price decrease causes an increase in demand.

    (a) Demand will decrease by about $0.5(3\%) = 1.5\%$.
    (b) Demand will increase by about $0.5(3\%) = 1.5\%$.

3. (a) We have $E = |p/q \cdot dq/dp| = |\text{dollars/tons} \cdot \text{tons/dollars}|$. All the units cancel, and so elasticity has no units.
    (b) We have $E = |p/q \cdot dq/dp| = |\text{yen/liters} \cdot \text{liters/yen}|$. All the units cancel, and so elasticity has no units.
    (c) Elasticity has no units. This is why it makes sense to compare elasticities of different products valued in different ways and measured in different units. Changing units of measurement will not change the value of elasticity.

4. Table 5.5 of Section 5.5 gives the elasticity of jewelry as 2.60. Since $E > 1$, the demand is elastic, and a change in price will cause a larger percentage change in demand. So if the price of jewelry goes up, fewer people will buy it. This is reasonable since jewelry is a luxury item.

5. Table 5.5 of Section 5.5 gives the elasticity of milk as 0.31. Here $0 \leq E \leq 1$, so the demand is inelastic, meaning that changes in price will not change the demand so much. This is expected for milk; as a staple food item, people will continue to buy it even if the price goes up.

6. Demand for high-definition TV's will be elastic, since it is not a necessary item. If the prices are too high, people will not choose to buy them, so price changes will cause relatively large demand changes.

7. Elasticity will be high. As soon as the price of a brand is raised, many people will switch to another brand causing a drop in sales.

8. Elasticity will be low. Nearly everyone regards telephone service as a necessity. Since there is not another company they can turn to, they will keep their telephone service even if the company raises the price. The number of sales will change very little.

9. (a) There were good substitutes for slaves in city occupations - including free blacks. There were no good substitutes in the countryside.
    (b) They were from the countryside, where there was no satisfactory substitute for slaves.

10. (a) If the price of yams is $2/pound, the quantity sold will be

    $$q = 5000 - 10(2)^2 = 5000 - 40 = 4960$$

    so 4960 pounds will be sold.

    (b) Elasticity of demand is given by

    $$E = \left|\frac{p}{q} \cdot \frac{dq}{dp}\right| = \left|\frac{p}{q} \cdot \frac{d}{dp}(5000 - 10p^2)\right| = \left|\frac{p}{q} \cdot (-20p)\right| = \frac{20p^2}{q}$$

    Substituting $p = 2$ and $q = 4960$ yields

    $$E = \frac{20(2)^2}{4960} = \frac{80}{4960} = 0.016.$$

    Since $E < 1$ the demand is inelastic, so it would be more accurate to say "People want yams and will buy them no matter what the price."

11. (a) At a price of $2/pound, the quantity sold will be

    $$q = 5000 - 10(2)^2 = 5000 - 40 = 4960$$

    so the total revenue will be

    $$R = pq = 2 \cdot 4960 = \$9,920$$

**232** CHAPTER FIVE /SOLUTIONS

(b) We know that $R = pq$, and that $q = 5000 - 10p^2$, so we can substitute for $q$ to find $R(p)$

$$R(p) = p(5000 - 10p^2) = 5000p - 10p^3$$

To find the price that maximizes revenue we take the derivative and set it equal to 0.

$$R'(p) = 0$$
$$5000 - 30p^2 = 0$$
$$30p^2 = 5000$$
$$p^2 = 166.67$$
$$p = \pm 12.91$$

We disregard the negative answer, so $p = 12.91$ is the only critical point. Is it the maximum? We use the first derivative test.

$$R'(p) > 0 \text{ if } p < 12.91 \text{ and}$$
$$R'(p) < 0 \text{ if } p > 12.91$$

So $R(p)$ has a local maximum at $p = 12.91$. We also test the function at $p = 0$, which is the only endpoint.

$$R(0) = 5000(0) - 10(0)^3 = 0$$
$$R(12.91) = 5000(12.91) - 10(12.91)^3 = 64{,}550 - 21{,}516.85 = \$43{,}033.15$$

So we conclude that revenue is maximized at price of \$12.91/pound.

(c) At a price of \$12.91/pound the quantity sold is

$$q = 5000 - 10(12.91)^2 = 5000 - 1666.68 = 3333.32$$

so the total revenue is

$$R = pq = (3333.32)(12.91) = \$43{,}033.16$$

which agrees with part (b).

(d)
$$E = \left| \frac{p}{q} \cdot \frac{dq}{dp} \right| = \left| \frac{p}{q} \cdot \frac{d}{dp}(5000 - 10p^2) \right| = \left| \frac{p}{q} \cdot (-20p) \right| = \frac{20p^2}{q}$$

Substituting $p = 12.91$ and $q = 3333.32$ yields

$$E = \frac{20(12.91)^2}{3333.32} = \frac{3333.36}{3333.32} \approx 1$$

which agrees with the result that maximum revenue occurs when $E = 1$.

12. $E = \left| \frac{p}{q} \frac{dq}{dp} \right| = \left| \frac{p}{q} \cdot \frac{d}{dp}(2000 - 5p) \right| = \left| \frac{p}{q} \cdot (-5) \right|$, so $E = \frac{5p}{q}$. At a price of \$20.00, the number of items produced is $q = 2000 - 5(20) = 1900$, so at $p = 20$, we have

$$E = \frac{5(20)}{(1900)} \approx 0.05.$$

Since $0 \leq E < 1$, this product has inelastic demand–a 1% change in price will only decrease demand by 0.05%.

13. (a) We use

$$E = \left| \frac{p}{q} \cdot \frac{dq}{dp} \right|.$$

So we approximate $dq/dp$ at $p = 1.00$.

$$\frac{dq}{dp} \approx \frac{2440 - 2765}{1.25 - 1.00} = \frac{-325}{0.25} = -1300$$

so
$$E = \left| \frac{p}{q} \cdot \frac{dq}{dp} \right| \approx \left| \frac{1.00}{2765} \cdot (-1300) \right| = 0.470$$

Since $E = 0.470 < 1$, demand for the candy is inelastic at $p = 1.00$.

(b) At $p = 1.25$,
$$\frac{dq}{dp} \approx \frac{1980 - 2440}{1.50 - 1.25} = \frac{-460}{0.25} = -1840$$
so
$$E = \left|\frac{p}{q} \cdot \frac{dq}{dp}\right| \approx \left|\frac{1.25}{2440} \cdot (-1840)\right| = 0.943$$

At $p = 1.5$,
$$\frac{dq}{dp} \approx \frac{1660 - 1980}{1.75 - 1.50} = \frac{-320}{0.25} = -1280$$
so
$$E = \left|\frac{p}{q} \cdot \frac{dq}{dp}\right| \approx \left|\frac{1.50}{1980} \cdot (-1280)\right| = 0.970$$

At $p = 1.75$,
$$\frac{dq}{dp} \approx \frac{1175 - 1660}{2.00 - 1.75} = \frac{-485}{0.25} = -1940$$
so
$$E = \left|\frac{p}{q} \cdot \frac{dq}{dp}\right| \approx \left|\frac{1.75}{1660} \cdot (-1940)\right| = 2.05$$

At $p = 2.00$,
$$\frac{dq}{dp} \approx \frac{800 - 1175}{2.25 - 2.00} = \frac{-375}{0.25} = -1500$$
so
$$E = \left|\frac{p}{q} \cdot \frac{dq}{dp}\right| \approx \left|\frac{2.00}{1175} \cdot (-1500)\right| = 2.55$$

At $p = 2.25$,
$$\frac{dq}{dp} \approx \frac{430 - 800}{2.50 - 2.25} = \frac{-370}{0.25} = -1480$$
so
$$E = \left|\frac{p}{q} \cdot \frac{dq}{dp}\right| \approx \left|\frac{2.25}{800} \cdot (-1480)\right| = 4.16$$

Examination of the elasticities for each of the prices suggests that elasticity gets larger as price increases. In other words, at higher prices, an increase in price will cause a larger drop in demand than the same size price increase at a lower price level. This can be explained by the fact that people will not pay too much for candy, as it is somewhat of a "luxury" item.

(c) Elasticity is approximately equal to 1 at $p = \$1.25$ and $p = \$1.50$.

(d) 

**TABLE 5.1**

| $p(\$)$ | $q$ | Revenue $= p \cdot q$ (\$) |
|---|---|---|
| 1.00 | 2765 | 2765 |
| 1.25 | 2440 | 3050 |
| 1.50 | 1980 | 2970 |
| 1.75 | 1660 | 2905 |
| 2.00 | 1175 | 2350 |
| 2.25 | 800 | 1800 |
| 2.50 | 430 | 1075 |

We can see that revenue is maximized at $p = \$1.25$, with $p = \$1.50$ a close second, which agrees with part (c).

14. The revenue is maximized by finding the critical point of the revenue function:
$$R = pq = p(1000 - 2p^2) = 1000p - 2p^3.$$

Differentiate to find the critical points:
$$\frac{dR}{dp} = 1000 - 6p^2 = 0$$
$$p^2 = \frac{1000}{6}$$
$$p \approx 12.91$$

To maximize revenues, the price of the product should be $12.91.

15. (a) Since $q = k/p$, we have $dq/dp = -k/p^2$ and $p/q = p/(k/p) = p^2/k$. Therefore

$$E = \left|\frac{p}{q} \cdot \frac{dq}{dp}\right| = \left|\frac{p^2}{k} \cdot \frac{-k}{p^2}\right| = 1$$

Elasticity is a constant equal to 1, independent of the price.

(b) Elasticity equal to 1 corresponds to critical points of the revenue function $R$. Since $R = pq = k$ is constant, $dR/dp = 0$, and so all prices are critical points of the revenue function.

16.
$$E = |p/q \cdot dq/dp| = |p^{r+1}/k \cdot -kr/p^{r+1}| = r$$

17. Demand is elastic at all prices. No matter what the price is, you can increase revenue by lowering the price. In the end, you would lower your prices all the way to zero. This is not a realistic example, but it is mathematically possible. It would correspond, for instance, to the demand equation $q = 1/p^2$, which gives revenue $R = pq = 1/p$ which is decreasing for all prices $p > 0$.

18. Demand is inelastic at all prices. No matter what the price is, you can increase revenue by raising the price, so there is no actual price for which your revenue is maximized. This is not a realistic example, but it is mathematically possible. It would correspond, for instance, to the demand equation $q = 1/\sqrt{p}$, which gives revenue $R = pq = \sqrt{p}$ which is increasing for all prices $p > 0$.

19. Since $R = pq$, we have $dR/dp = p(dq/dp) + q$. We are assuming that

$$E = \left|\frac{p}{q} \cdot \frac{dq}{dp}\right| > 1$$

so, removing the absolute values

$$-E = \frac{p}{q}\frac{dq}{dp} < -1.$$

Multiplication by $q$ gives

$$p\frac{dq}{dp} < -q$$

and hence

$$\frac{dR}{dp} = p\frac{dq}{dp} + q < 0$$

20. Since $R = pq$, we have $dR/dp = p(dq/dp) + q$. We are assuming that

$$0 \le E = \left|\frac{p}{q} \cdot \frac{dq}{dp}\right| < 1$$

so, removing the absolute values

$$0 \ge -E = \frac{p}{q}\frac{dq}{dp} > -1$$

Multiplication by $q$ gives

$$p\frac{dq}{dp} > -q$$

and hence

$$\frac{dR}{dp} = p\frac{dq}{dp} + q > 0$$

21. The approximation $E_{\text{cross}} \approx \left|\frac{\Delta q/q}{\Delta p/p}\right|$ shows that the cross-price elasticity measures the ratio of the fractional change in quantity of chicken demanded to the fractional change in the price of beef. Thus, for example, a 1% increase in the price of beef will stimulate a $E_{\text{cross}}$% increase in the demand for chicken, presumably because consumers will react to the price rise in beef by switching to chicken. The cross-price elasticity measures the strength of this switch.

22. The approximation $E_{\text{income}} \approx \left|\frac{\Delta q/q}{\Delta I/I}\right|$ shows that the income elasticity measures the ratio of the fractional change in quantity of the product demanded to the fractional change in the income of the consumer. Thus, for example, a 1% increase in income will translate into an $E_{\text{income}}$% increase in the quantity purchased. After an increase in income, the consumer will tend to buy more. The income elasticity measures the strength of this tendency.

23. (a) We see that $d_1 = p_0$. In addition, we have

$$\frac{dp}{dq} = \text{Slope of demand curve} = \frac{\text{Rise}}{\text{Run}} = \frac{-d_2}{q_0}$$

so $d_2 = -q_0 \cdot dp/dq$. Therefore

$$\frac{d_1}{d_2} = \frac{p_0}{-q_0 \cdot \frac{dp}{dq}} = -\frac{p_0}{q_0} \cdot \frac{dq}{dp} = E$$

(b) For prices near the maximum possible, we have $d_1 > d_2$ and so elasticity $E = d_1/d_2$ is greater than 1 for small quantities. For prices $p$ near 0 (that is, near the $q$-axis), we have $d_1 < d_2$ and so $E = d_1/d_2$ is less than 1 for large quantities. Elasticity equals 1 where $d_1 = d_2$ at exactly half the maximum possible quantity. See Figure 5.55.

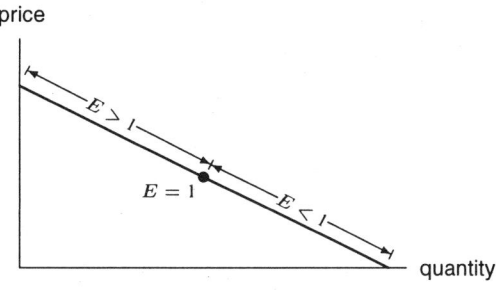

**Figure 5.55**

## Solutions for Section 5.6

1. Substituting $t = 0, 10, 20, \ldots, 70$ into the function $P = 3.9(1.03)^t$ gives the values in Table 5.2. Notice that the agreement is very close, reflecting the fact that an exponential function models the growth well over the period 1790–1860.

**TABLE 5.2** *Predicted versus actual US population 1790–1860, in millions. (exponential model)*

| Year | Actual | Predicted | Year | Actual | Predicted |
|---|---|---|---|---|---|
| 1790 | 3.9 | 3.9 | 1830 | 12.9 | 12.7 |
| 1800 | 5.3 | 5.2 | 1840 | 17.1 | 17.1 |
| 1810 | 7.2 | 7.0 | 1850 | 23.2 | 23.0 |
| 1820 | 9.6 | 9.5 | 1860 | 31.4 | 30.9 |

2. Substituting $t = 0, 10, 20, 30, \ldots$ into the function

$$P = \frac{185}{1 + 48e^{-0.032t}}$$

gives the values in Table 5.3. Notice that the agreement is close between the predicted and actual values.

**TABLE 5.3**

| Year | Actual | Predicted | Year | Actual | Predicted | Year | Actual | Predicted |
|---|---|---|---|---|---|---|---|---|
| 1790 | 3.9 | 3.8 | 1840 | 17.1 | 17.3 | 1890 | 62.9 | 62.6 |
| 1800 | 5.3 | 5.2 | 1850 | 23.2 | 23.0 | 1900 | 76.0 | 76.4 |
| 1810 | 7.2 | 7.0 | 1860 | 31.4 | 30.3 | 1910 | 92.0 | 91.1 |
| 1820 | 9.6 | 9.5 | 1870 | 38.6 | 39.3 | 1920 | 105.7 | 105.8 |
| 1830 | 12.9 | 12.9 | 1880 | 50.2 | 50.1 | 1930 | 122.8 | 119.8 |
|  |  |  |  |  |  | 1940 | 131.6 | 132.6 |

3. (a) Substituting the value $t = 0$ we get

$$N(0) = \frac{400}{1 + 399e^{-0.4(0)}}$$
$$= \frac{400}{1 + 399(1)}$$
$$= 1$$

The fact that $N(0) = 1$ tells us that at the moment the rumor begins spreading, there is only one person who knows the content of the rumor.

(b) Substituting $t = 2$ we get

$$N(2) = \frac{400}{1 + 399e^{-0.4(2)}}$$
$$\approx \frac{400}{1 + 399(0.449)}$$
$$\approx 2$$

Substituting in $t = 10$ we get

$$N(10) = \frac{400}{1 + 399e^{-0.4(10)}}$$
$$= \frac{400}{1 + 7.308}$$
$$\approx 48$$

(c) The graph of $N(t)$ is shown in Figure 5.56.

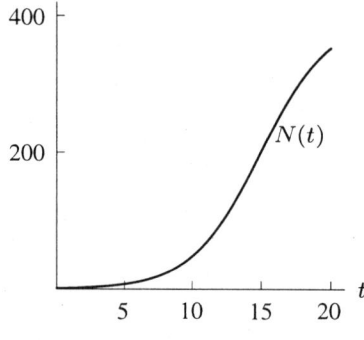

**Figure 5.56**

(d) We are asked to find the time $t$ at which 200 people will have heard the rumor. In other words, we are asked to solve

$$200 = \frac{400}{1 + 399e^{-0.4t}}.$$

Solving we get

$$1 + 399e^{-0.4t} = \frac{400}{200} = 2$$
$$399e^{-0.4t} = 1$$
$$e^{-0.4t} = \frac{1}{399}$$
$$\ln e^{-0.4t} = \ln \frac{1}{399}$$
$$-0.4t \approx -5.989$$
$$t = \frac{-5.989}{-0.4} \approx 15.$$

Thus after 15 hours half the people will have heard the rumor.

We are asked to solve for the time at which virtually everyone will have heard the rumor. Since the value of $N(t)$ only makes sense after rounding to the nearest integer (i.e., it does not make sense to say that 2.5 people heard the rumor,) asking for the time at which all 400 people will have heard the rumor is equivalent to asking for the time $t$ such that
$$N(t) = 399.5.$$

Solving this we get

$$399.5 = N(t) = \frac{400}{1 + 399e^{-0.4t}}$$
$$1 + 399e^{-0.4t} = \frac{400}{399.5} \approx 1.00125$$
$$399e^{-0.4t} \approx 0.00125$$
$$e^{-0.4t} \approx \frac{0.00125}{399}$$
$$\ln e^{-0.4t} \approx \ln \frac{0.00125}{399}$$
$$-0.4t \approx -12.7$$
$$t \approx \frac{-12.7}{-0.4} = 31.75$$

Thus after approximately 32 hours virtually everyone will have heard the rumor.

(e) The rumor is spreading fastest at $L/2 = 400/2 = 200$ or when 200 people have already heard the rumor, so after about 15 hours.

4. (a) As $t$ gets very very large, $e^{-0.08t} \to 0$ and the function becomes $P \approx 40/1$. Thus, this model implies that when $t$ is very large, the population is 40 billion.

(b) A graph of $P$ against $t$ is shown in Figure 5.57.

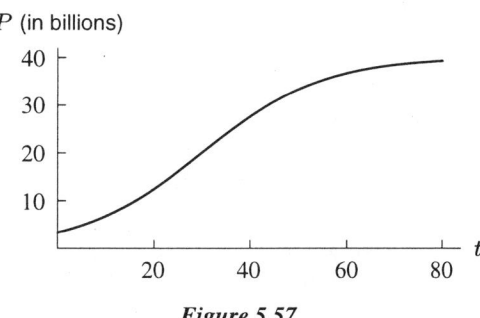

**Figure 5.57**

(c) We are asked to find the time $t$ such that $P(t) = 20$. Solving we get

$$20 = P(t) = \frac{40}{1 + 11e^{-0.08t}}$$
$$1 + 11e^{-0.08t} = \frac{40}{20} = 2$$
$$11e^{-0.08t} = 1$$
$$e^{-0.08t} = \frac{1}{11}$$
$$\ln e^{-0.08t} = \ln \frac{1}{11}$$
$$-0.08t \approx -2.4$$
$$t \approx \frac{-2.4}{-0.08} = 30$$

Thus 30 years from 1990 (the year 2020) the population of the world should be 20 billion.

We are asked to find the time $t$ such that $P(t) = 39.9$. Solving we get

$$39.9 = P(t) = \frac{40}{1 + 11e^{-0.08t}}$$

$$1 + 11e^{-0.08t} = \frac{40}{39.9} = 1.00251$$

$$11e^{-0.08t} = 0.00251$$

$$e^{-0.08t} = \frac{0.00251}{11}$$

$$\ln e^{-0.08t} = \ln \frac{0.00251}{11}$$

$$-0.08t \approx -8.39$$

$$t \approx \frac{-8.39}{-0.08} \approx 105$$

Thus 105 years from 1990 (the year 2095) the population of the world should be 39.9 billion.

5. (a) If we graph the data, we see that it looks like logistic growth. (See Figure 5.58.) But logistic growth also makes sense from a common-sense viewpoint. As VCRs "catch on," the percentage of households which have them will at first grow exponentially but then slow down after more and more people have them. Eventually, nearly everyone who will buy one already has and the percentage levels off.

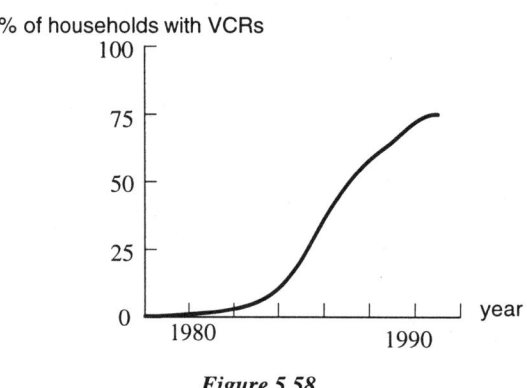

Figure 5.58

(b) The point of diminishing returns happens around 36%. This predicts a carrying capacity of 70% which is pretty close to the 71.9% we see in 1990 and 1991.
(c) 75%
(d) The limiting value predicts the percentage of households with a television set that will eventually have VCRs. This model predicts that there will never be a time when more than 75% of the households with TVs also have VCRs, which seems reasonable.

6. Sales of a new product could very well follow a logistic curve. At first, sales will grow exponentially as more and more people hear of the product and decide to buy it. Eventually, though, everyone will know about the product and while people may still buy it, not as many will (sales will slow down). Eventually, it's possible that everyone who would buy the product already has, at which point sales will stop. It behooves the seller to notice the point of diminishing returns so they don't make more of their product than people will want to buy.

7. (a) The data is plotted in Figure 5.59.

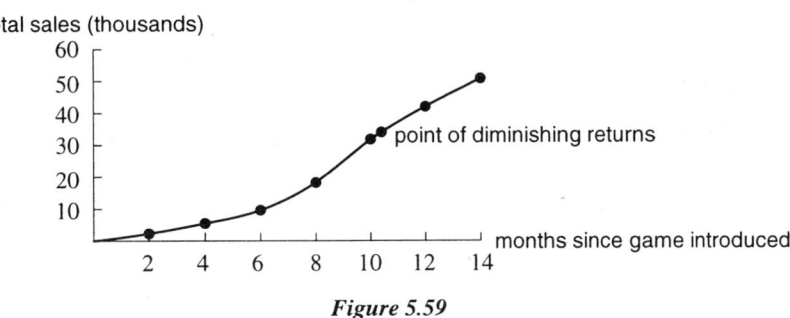

**Figure 5.59**

(b) The point of diminishing returns happens when total sales reach about 34,000. This predicts total sales of about $2(34) = 68,000$.

8. (a) At $t = 0$, which corresponds to 1935, we have

$$P = \frac{1}{1 + 3e^{-0.0275(0)}} = 0.25$$

showing that 25% of the land was in use in 1935.

(b) This model predicts that as $t$ gets very large, $P$ approaches 1. That is, the model predicts that in the long run, all the land will be used for farming.

(c) To solve this graphically, enter the function into a graphing calculator and trace the resulting curve until it reaches a height of 0.5, which occurs when $t = 39.9 \approx 40$. Since $t = 0$ corresponds to 1935, $t = 40$ corresponds to $1935 + 40 = 1975$. According to this model, the Tojolobal were using half their land in 1975. See Figure 5.60.

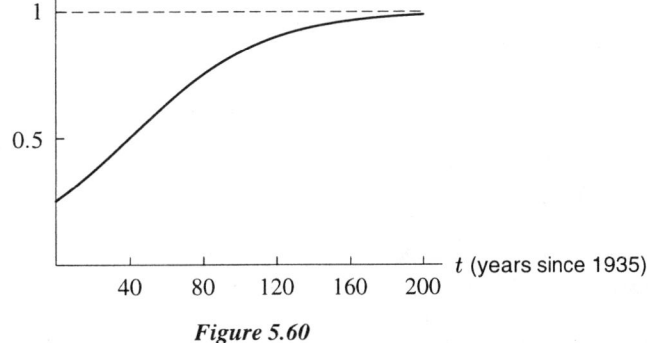

**Figure 5.60**

(d) The point of diminishing returns occurs when $P = L/2$ or at one-half the carrying capacity. In this case, $P = 1/2$ in 1975, as shown in part (c).

9. (a) We use $k = 1.78$ as a rough approximation. We let $L = 5000$ since the problem tells us that 5000 people eventually get the virus. This means the limiting value is 5000.

(b) We know that

$$P(t) = \frac{5000}{1 + Ce^{-1.78t}} \quad \text{and} \quad P(0) = 10$$

so

$$10 = \frac{5000}{1 + Ce^0} = \frac{5000}{1 + C}$$
$$10(1 + C) = 5000$$
$$1 + C = 500$$
$$C = 499.$$

(c) We have $P(t) = \dfrac{5000}{1 + 499e^{-1.78t}}$. This function is graphed in Figure 5.61.

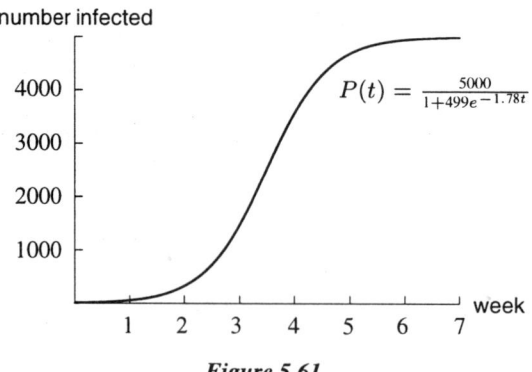

**Figure 5.61**

(d) The point of diminishing returns appears to be at the point $(3.5, 2500)$; that is, after 3 and a half weeks and when 2500 people are infected.

10. (a, b) The graph of $P(t)$ with carrying capacity $L$ and point of diminishing returns $t_0$ is in Figure 5.62. The derivative $P'(t)$ is also shown.

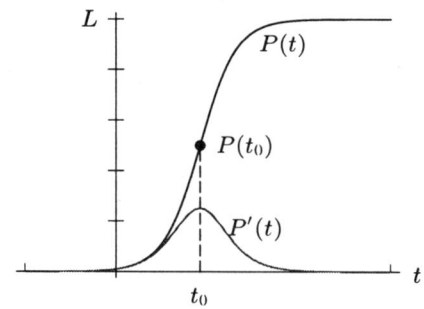

**Figure 5.62**

(c) Keeping track of rate of sales is the same as keeping track of the derivative $P'(t)$. The point of diminishing returns happens when the concavity of $P(t)$ changes, which is the when the derivative, $P'(t)$, switches from increasing to decreasing. This happens when $P'(t)$ reaches its maximum at $t = t_0$

11. (a) The fact that $f'(15) = 11$ means that the slope of the curve at the inflection point, $(15, 50)$ is 11. In terms of dose and response, this large slope tells us that the range of doses for which this drug is both safe and effective is small.

(b) As we can see from Figure 5.57 in the text, a dose-response curve starts out concave up (slope increasing) and switches to concave down (slope decreasing) at an inflection point. Since the slope at the inflection point $(15, 50)$ is 11, and the slope is increasing before the inflection point, $f'(10)$ is less than 11. Since the slope is decreasing after the inflection point, $f'(20)$ is also less than 11.

12. (a) From 1960 to 1975, the curve is concave up. This says the world electrical generating capacity of nuclear power plants was increasing at an ever increasing rate from 1960 to 1975.

(b) The concavity of the curve changes around 1982 which corresponds to the time approximately 170 gigawatts of nuclear power was being generated. The limiting value predicted from the point of diminishing returns being 170 is $2(170) = 340$ gigawatts. The graph implies the limiting value is 325 or 330 megawatts which isn't very far from 340.

(c) A logistic model for this data predicts that nuclear power will remain about the same after 1993.

13. (a) The dose-response curve for product $C$ crosses the minimum desired response line last, so it requires the largest dose to achieve the desired response. The dose-response curve for product $B$ crosses the minimum desired response line first, so it requires the smallest dose to achieve the desired response.

(b) The dose-response curve for product $A$ levels off at the highest point, so it has the largest maximum response. The dose-response curve for product $B$ levels off at the lowest point, so it has the smallest maximum response.

(c) Product $C$ is the safest to administer because its slope in the safe and effective region is the least, so there is a broad range of dosages for which the drug is both safe and effective.

14. (a) The dose-response curve for this drug is shown if Figure 5.63.

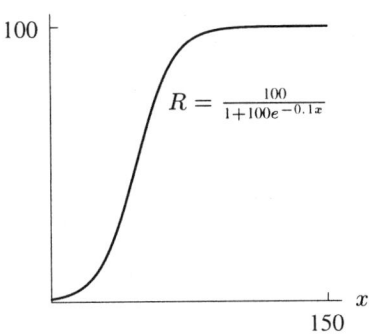

**Figure 5.63**

(b) First solve for $x$:

$$R = \frac{100}{1 + 100e^{-0.1x}}$$
$$R + 100Re^{-0.1x} = 100$$
$$100Re^{-0.1x} = 100 - R$$
$$e^{-0.1x} = \frac{100 - R}{100R}$$
$$-0.1x = \ln\left(\frac{100 - R}{100R}\right)$$
$$x = -10\ln\left(\frac{100 - R}{100R}\right).$$

Substituting the value of $R = 50$ we get

$$x = -10\ln\left(\frac{100 - 50}{100 \cdot 50}\right) \approx 46.0517019 \approx 46.05$$

(c) Evaluate the formula for $x$ obtained in part (b) for $R = 20$ and $R = 70$.
For $R = 20$,
$$x = -10\ln\left(\frac{100 - 20}{100 \cdot 20}\right) \approx 32.18875825 \approx 32.12$$

For $R = 70$,
$$x = -10\ln\left(\frac{100 - 70}{100 \cdot 70}\right) \approx 54.5246805 \approx 54.52$$

So the range of doses that is both safe and effective is between 32.12 mg and 54.52 mg.

15. If the derivative of the dose-response curve is smaller, the slope is not as steep. Since the slope is not as steep, the response increases less at the same dosage. Therefore, there is a wider range of dosages that are both safe and effective, and consequently the dosage given to the patient does not have to be as exact.

16. The range of safe and effective doses begins at 10 mg where the drug is effective for 100 percent of patients. It ends at 18 mg where the percent lethal curve begins to rise above zero.

17. When 50 mg of the drug is administered, it is effective for 85 percent of the patients and lethal for 6 percent.

18.

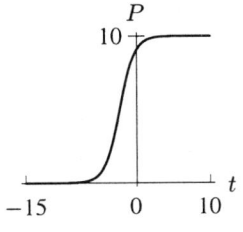

**Figure 5.64:** $C = 0.1$

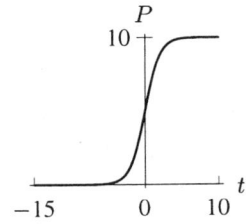

**Figure 5.65:** $C = 1$

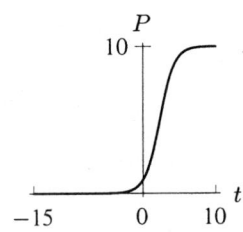

**Figure 5.66:** $C = 10$

See Figures 5.64–5.66. The value of $C$ appears to affect where the curve cross the vertical axis. When $C = 1$, the graph crosses the vertical axis at $t = 0$ or at the point of diminishing returns. If $C > 1$, the graph crosses the the vertical axis before the point of diminishing returns; the higher $C$ is, the sooner the graph crosses the vertical axis. If $C < 1$, the curve crosses the vertical axis after it changes concavity; the smaller $C$ is, the closer to the limiting value the graph is before crossing the vertical axis.

19. (a) Before differentiating, we multiply out, giving

$$\frac{dP}{dt} = kP\left(1 - \frac{P}{L}\right) = kP - \frac{kP^2}{L}.$$

Since $k$ and $L$ are constant, the chain rule gives

$$\frac{d^2P}{dt^2} = \frac{d}{dt}\left(\frac{dP}{dt}\right) = \frac{d}{dt}\left(kP - \frac{kP^2}{L}\right)$$
$$= k\frac{dP}{dt} - \frac{2kP}{L}\frac{dP}{dt}.$$

(b) Setting $d^2P/dt^2 = 0$ gives

$$0 = k\frac{dP}{dt} - \frac{2kP}{L}\frac{dP}{dt} = \frac{kdP}{dt}\left(1 - \frac{2P}{L}\right).$$

So

$$1 - \frac{2P}{L} = 0,$$

that is, when

$$P = \frac{L}{2},$$

we have $d^2P/dt^2 = 0$; that is, we are at the point of diminishing returns.

## Solutions for Section 5.7

1.

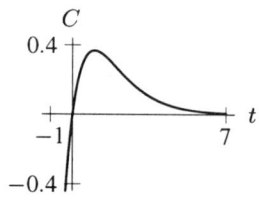

**Figure 5.67:** $a = 1$

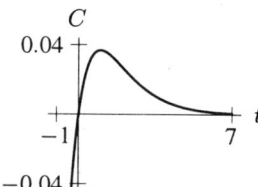

**Figure 5.68:** $a = -10$

**Figure 5.69:** $a = 0.1$

The parameter $a$ apparently affects the height and direction of $C = ate^{-bt}$. If $a$ is positive, the "hump" is above the $t$-axis. If $a$ is negative, it's below the $t$-axis. If $a > 0$, the larger the value of $a$, the larger the maximum value of $C$. If $a < 0$, the more negative the value of $a$, the smaller the minimum value of $C$.

2. (a) See Figure 5.70.
(b) The surge function $y = ate^{bt}$ changes from increasing to decreasing at $t = \frac{1}{b}$. For this function $b = 0.2$ so the peak is at $\frac{1}{0.2} = 5$ hours. We can now substitute this into the formula to compute the peak concentration:

$$C = 12.4(5)e^{-0.2(5)} = 22.8085254 \text{ ng/ml} \approx 22.8 \text{ ng/ml}.$$

(c) Tracing along the graph of $C = 12.4te^{-0.2t}$, we see it crosses the line $C = 10$ at $t \approx 1$ hour and at $t \approx 14.4$ hours. Thus, the drug is effective for $1 \leq t \leq 14.4$ hours.

(d) The drug drops below $C = 4$ for $t > 20.8$ hours. Thus, it is safe to take the other drug after 20.8 hours.

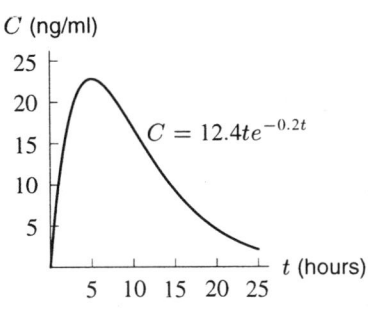

Figure 5.70

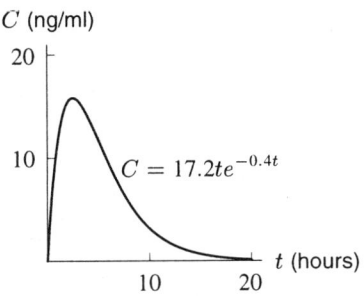

Figure 5.71

3. (a) See Figure 5.71. The first dose becomes ineffective when
$$10 = 17.2te^{-0.4t}.$$
To find this value of $t$, trace along the curve $C = 17.2te^{-0.4t}$, giving $t \approx 5.7$ hours. Thus, the second dose should be given after about 5.7 hours.

(b) By tracing, we find that the second dose becomes effective about 0.8 hours after it is given. We want it to become effective when $t \approx 5.7$, so the second dose should be given at $t \approx 4.9$ hours.

4. For cigarettes the peak concentration is approximately 17 ng/ml, the time until peak concentration is about 10 minutes, and the nicotine is at first eliminated quickly, but after about 45 minutes it is eliminated more slowly.

For chewing tobacco, the peak concentration is approximately 14 ng/ml and the time until peak concentration is about 30 minutes (when chewing stops). The nicotine is eliminated at a slow, somewhat erratic rate.

For nicotine gum the peak concentration is about 10 ng/ml and the time until peak concentration is approximately 45 minutes. The nicotine is eliminated at a very slow but steady rate.

5.
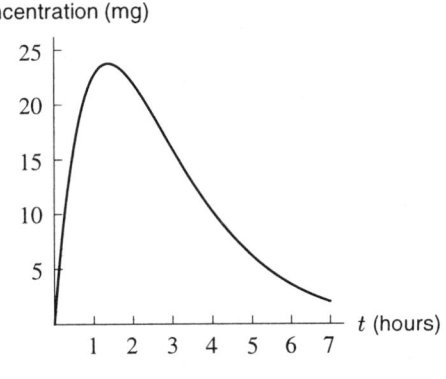

Figure 5.72

6. (a) Products $A$ and $B$ have much higher peak concentrations than products $C$ and $D$. Product $A$ reaches its peak concentration slightly before products $B, C$, and $D$, which all take about the same time to reach peak concentration.

(b) If the minimum effective concentration were low, perhaps 0.2, and the maximum safe concentration were also low, perhaps 1.0, then products $C$ and $D$ would be the preferred drugs since they do not enter the unsafe range while being well within the effective range.

(c) If the minimum effective concentration were high, perhaps 1.2, and the maximum safe concentration were also high, perhaps 2.0, then product $A$ would be the preferred drug since it does not enter the unsafe range and it is the only drug that is in the effective range for a substantial amount of time.

7. When bemetizide is taken with triamterene, rather than by itself, the peak concentration is lower, the time it takes to reach peak concentration is about the same, the time until onset of effectiveness is slightly less, and the duration of effectiveness is less. If bemetizide became unsafe in a concentration greater than 70 ng/ml, then it would be wise to take it with triamterene.

## 244 CHAPTER FIVE /SOLUTIONS

8.  (a) Varying the $a$ parameter changes the peak concentration proportionally, but does not change the time to reach peak concentration.
    (b) The value of the peak concentration decreases as $b$ increases and the time until peak concentration decreases as $b$ increases.
    (c) As $a$ got larger, the time until peak concentration would decrease because it is only affected by the constant in the exponent. The value of the peak concentration would not be affected as $a$ changed because the peak always occurs at $t = \frac{1}{b}$ and if $a = b$, the value of the peak concentration is $c = a \cdot \frac{1}{a} e^{-a \cdot \frac{1}{a}} = e^{-1}$, independent of $a$.

9.  Food dramatically increases the value of the peak concentration but does not affect the time it takes to reach the peak concentration. The effect of food is stronger during the first 8 hours.

10. Large quantities of water dramatically increase the value of the peak concentration, but do not change the amount of time it takes to reach the peak concentration. The effect of the volume of water taken with the drug wears off after approximately 6 hours.

11. In general, a faster dissolution rate corresponds to a larger peak concentration. Dissolution rate does not affect time to reach peak concentration, as the four products have different dissolution rates but virtually the same time to reach peak concentration.

## Solutions for Chapter 5 Review

1.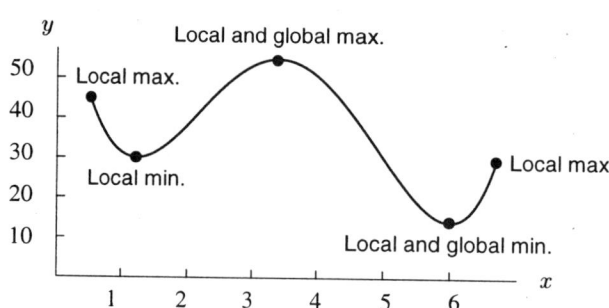

2.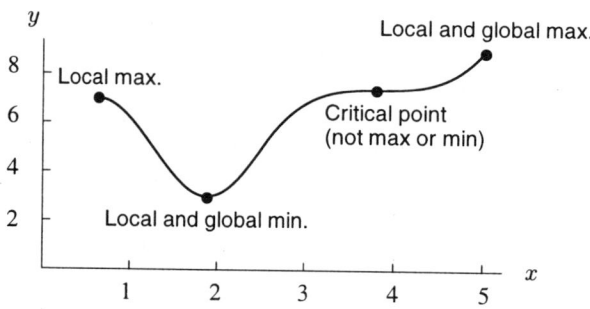

3. We want to maximize the height, $y$, of the grapefruit above the ground, as shown in the figure below. Using the derivative we can find exactly when the grapefruit is at the highest point. We can think of this in two ways. By common sense, at the peak of the grapefruit's flight, the velocity, $dy/dt$, must be zero. Alternately, we are looking for a global maximum of $y$, so we look for critical points where $dy/dt = 0$. We have

$$\frac{dy}{dt} = -32t + 50 = 0 \quad \text{and so} \quad t = \frac{-50}{-32} \approx 1.56 \text{ sec.}$$

Thus, we have the time at which the height is a maximum; the maximum value of $y$ is then

$$y \approx -16(1.56)^2 + 50(1.56) + 5 = 44.1 \text{ feet.}$$

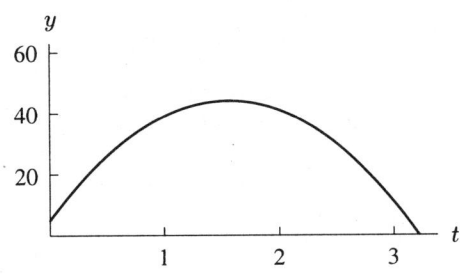
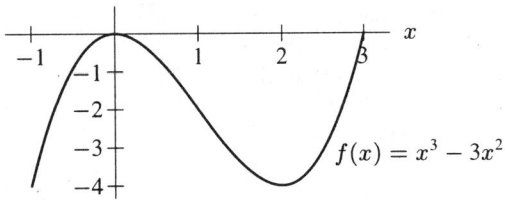

Figure 5.73

4. (a) Differentiating $f(x) = x^3 - 3x^2$ produces $f'(x) = 3x^2 - 6x$. A second differentiation produces $f''(x) = 6x - 6$.
   (b) $f'(x)$ is defined for all $x$ and $f'(x) = 0$ when $x = 0, 2$. Thus 0 and 2 are the critical points of $f$.
   (c) $f''(x)$ is defined for all $x$ and $f''(x) = 0$ when $x = 1$. When $x < 1$, $f''(x) < 0$ and when $x > 1$, $f''(x) > 0$. Thus the concavity of the graph of $f$ changes at $x = 1$. Hence $x = 1$ is an inflection point.
   (d) $f(-1) = -4$, $f(0) = 0$, $f(2) = -4$, $f(3) = 0$. So $f$ has a local maximum at $x = 0$, a local minimum at $x = 2$, global maxima at $x = 0$ and $x = 3$, and global minima at $x = -1$ and $x = 2$.
   (e) Plotting the function $f(x)$ for $-1 \leq x \leq 3$ gives the graph shown in Figure 5.73.

5. (a) Differentiating $f(x) = x + \sin x$ produces $f'(x) = 1 + \cos x$. A second differentiation produces $f''(x) = -\sin x$.
   (b) $f'(x)$ is defined for all $x$ and $f'(x) = 0$ when $x = \pi$. Thus $\pi$ is the critical point of $f$.
   (c) $f''(x)$ is defined for all $x$ and $f''(x) = 0$ when $x = 0$, $x = \pi$ and $x = 2\pi$. Since the concavity of $f$ changes at each of these points they are all inflection points.
   (d) $f(0) = 0$, $f(2\pi) = 2\pi$, $f(\pi) = \pi$. So $f$ has a global minimum at $x = 0$ and a global maximum at $x = 2\pi$.
   (e) Plotting the function $f(x)$ for $0 \leq x \leq 2\pi$ gives the graph shown in Figure 5.74:

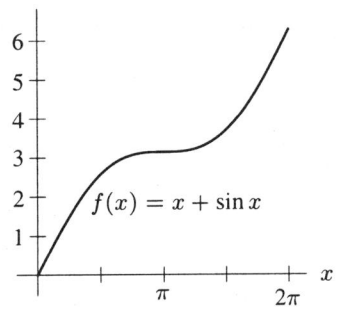

Figure 5.74

6. (a) Differentiating $f(x) = e^{-x} \sin x$ produces $f'(x) = -e^{-x} \sin x + e^{-x} \cos x$. A second differentiation produces $f''(x) = -2e^{-x} \cos x$.
   (b) $f'(x)$ is defined for all $x$ and $f'(x) = 0$ when $x = \pi/4$ and when $x = 5\pi/4$. Thus $\pi/4$ and $5\pi/4$ are the critical points of $f$.
   (c) $f''(x)$ is defined for all $x$ and $f''(x) = 0$ when $x = \pi/2, 3\pi/2$. Since the concavity of $f$ changes at both of these points they are both inflection points.
   (d) $f(0) = 0$, $f(2\pi) = 0$, $f(\pi/4) = e^{-\pi/4} \sin \pi/4$ and $f(5\pi/4) = e^{-5\pi/4} \sin(5\pi/4)$. So $f$ has a global maximum at $x = \pi/4$ and a global minimum at $x = 5\pi/4$.
   (e) Plotting the function $f(x)$ for $0 \leq x \leq 2\pi$ gives the graph shown in Figure 5.75.

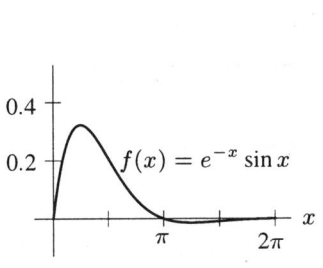
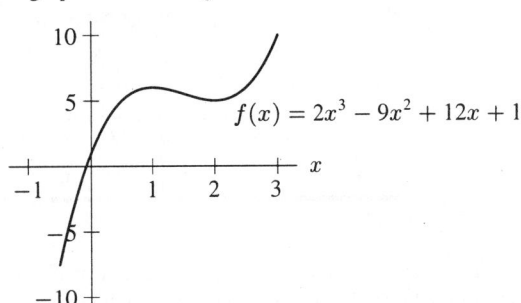

Figure 5.75

Figure 5.76

**246** CHAPTER FIVE /SOLUTIONS

7. (a) Differentiating $f(x) = 2x^3 - 9x^2 + 12x + 1$ produces $f'(x) = 6x^2 - 18x + 12$. A second differentiation produces $f''(x) = 12x - 18$.
   (b) $f'(x)$ is defined for all $x$ and $f'(x) = 0$ when $x = 1, 2$. Thus $x = 1, 2$ are critical points.
   (c) $f''(x)$ is defined for all $x$ and $f''(x) = 0$ when $x = \frac{3}{2}$. Since the concavity of $f$ changes at this point, it is an inflection point.
   (d) $f(-0.5) = -7.5$, $f(3) = 10$, $f(1) = 6$, $f(2) = 5$. So $f$ has a local maximum at $x = 1$ and a local minimum at $x = 2$, a global maximum at $x = 3$ and a global minimum at $x = -0.5$
   (e) Plotting the function $f(x)$ for $-0.5 \leq x \leq 3$ gives the graph shown in Figure 5.76.

8. (a) Differentiating $f(x) = x^3 - 3x^2 - 9x + 15$ produces $f'(x) = 3x^2 - 6x - 9$. A second differentiation produces $f''(x) = 6x - 6$.
   (b) $f'(x)$ is defined for all $x$ and $f'(x) = 0$ when $x = -1, 3$. Thus $x = -1, 3$ are critical points.
   (c) $f''(x)$ is defined for all $x$ and $f''(x) = 0$ when $x = 1$. Since the concavity of $f$ changes at this point, it is an inflection point.
   (d) $f(-5) = -140$, $f(4) = -5$, $f(-1) = 2$, $f(3) = -12$. So $f$ has a global maximum at $x = -1$ and a global minimum at $x = -5$, and a local minimum at $x = 3$
   (e) Plotting the function $f(x)$ for $-5 \leq x \leq 4$ gives the graph shown in Figure 5.77:

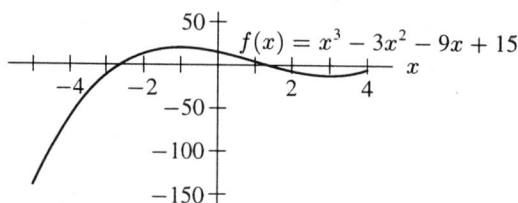

*Figure 5.77*

9. (a) Increasing for $x > 0$, decreasing for $x < 0$.
   (b) $f(0)$ is a local and global minimum, and $f$ has no global maximum.

10. (a) Increasing for all $x$.
    (b) No maxima or minima.

11. (a) Decreasing for $x < 0$, increasing for $0 < x < 4$, and decreasing for $x > 4$.
    (b) $f(0)$ is a local minimum, and $f(4)$ is a local maximum.

12. (a) Decreasing for $x < -1$, increasing for $-1 < x < 0$, decreasing for $0 < x < 1$, and increasing for $x > 1$.
    (b) $f(-1)$ and $f(1)$ are local minima, $f(0)$ is a local maximum.

13. We know that the maximum (or minimum) profit can occur when

    Marginal cost = Marginal revenue    or    $MC = MR$.

    From the table it appears that $MC = MR$ at $q \approx 2500$ and $q \approx 4500$. To decide which one corresponds to the maximum profit, look at the marginal profit at these points. Since

    Marginal profit = Marginal revenue − Marginal cost

    (or $M\pi = MR - MC$), we compute marginal profit at the different values of $q$ in Table 5.4:

    **TABLE 5.4**

    | $q$ | 1000 | 2000 | 3000 | 4000 | 5000 | 6000 |
    |---|---|---|---|---|---|---|
    | $M\pi = MR - MC$ | −22 | −4 | 4 | 7 | −5 | −22 |

    From the table, at $q \approx 2500$, we see that profit changes from decreasing to increasing, so $q \approx 2500$ gives a local minimum. At $q \approx 4500$, profit changes from increasing to decreasing, so $q \approx 4500$ is a local maximum. See Figure 5.78. Therefore, the global maximum occurs at $q = 4500$ or at the endpoint $q = 1000$.

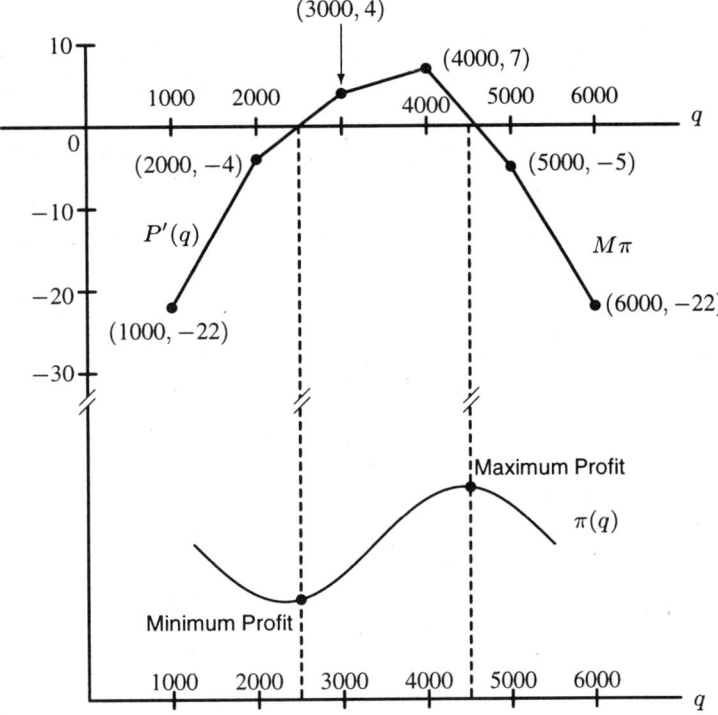

**Figure 5.78**

**14.**

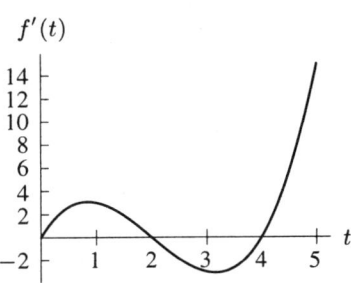

**Figure 5.79**

A graph of the derivative function is shown in Figure 5.79. Looking at the figure we see that

$f(t)$ is increasing when $0 < t < 2$ or when $t > 4$

$f(t)$ is decreasing when $2 < t < 4$

$f(t)$ has a local maximum at $t = 2$ but the global maximum occurs at $t = 5$

$f(t)$ has a minimum at $t = 4$.

**15.** To find the intercepts of $f(x)$, we first find the $y$-intercept, which occurs at $f(0) = \sin(0^2) = 0$. To find the $x$-intercepts on the given interval, we use a calculator to graph $f(x)$ as shown in Figure 5.80:

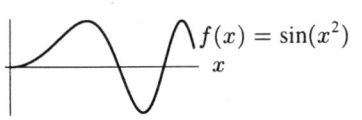

**Figure 5.80**

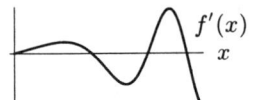

**Figure 5.81**

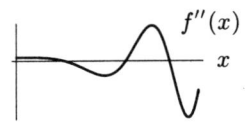

**Figure 5.82**

We can use a calculator's root-finding capability to find where $f(x) = 0$. We get:

$$x = 0, \quad \text{and} \quad x = 1.77, \quad \text{and} \quad x = 2.51,$$

so our intercepts are
$$(0, 0), \quad (1.77, 0), \quad (2.51, 0).$$

To find critical points, we look for where $f'(x) = 0$ or $f'$ is undefined. Using a calculator's differentiation and graphing features provides a graph of $f'(x)$ shown in Figure 5.81.

Since in Figure 5.81 all the critical points $(x, f(x))$ have $f'(x) = 0$, we use the calculator's root-finding capability to find the critical points where $f'(x) = 0$:
$$x = 0, \quad x = 1.25, \quad x = 2.17, \quad \text{and} \quad x = 2.80.$$

Writing these out as coordinates, the critical points are at
$$(0, 0), \quad (1.25, 1), \quad (2.17, -1), \quad (2.80, 1).$$

Similarly, inflection points occur where $f''(x)$ changes from negative to positive or vice versa. We can look for such points on a graph of $f''(x)$, shown in Figure 5.82.

We can use the calculator's root-finding capability on the $f''(x)$ to get these inflection points:
$$x = 0.81, \quad \text{and} \quad x = 1.81, \quad \text{and} \quad x = 2.52$$

The inflections points have coordinates $(0.81, f(0.81))$, etcetera and so they are
$$(0.81, 0.61), \quad (1.81, -0.13), \quad (2.52, 0.07).$$

16.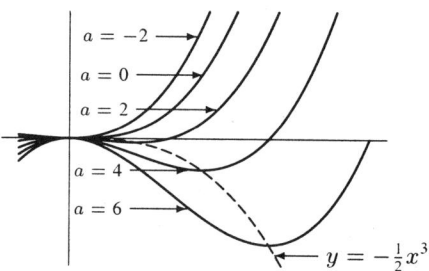

To solve for the critical points, we set $\frac{dy}{dx} = 0$. Since $\frac{d}{dx}\left(x^3 - ax^2\right) = 3x^2 - 2ax$, we want $3x^2 - 2ax = 0$, so $x = 0$ or $x = \frac{2}{3}a$. At $x = 0$, we have $y = 0$. This first critical point is independent of $a$ and lies on the curve $y = -\frac{1}{2}x^3$. At $x = \frac{2}{3}a$, we calculate $y = -\frac{4}{27}a^3 = -\frac{1}{2}\left(\frac{2}{3}a\right)^3$. Thus the second critical point also lies on the curve $y = -\frac{1}{2}x^3$.

17. (a) Profit $= \pi = R - C$; profit is maximized when the slopes of the two graphs are equal, at around $q = 350$. See Figure 5.83.

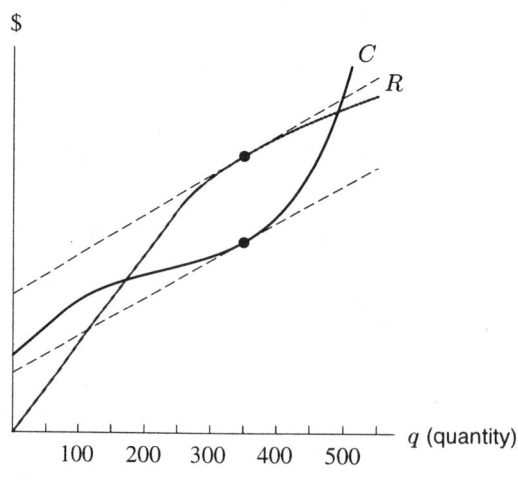

**Figure 5.83**

(b) The graphs of $MR$ and $MC$ are the derivatives of the graphs of $R$ and $C$. Both $R$ and $C$ are increasing everywhere, so $MR$ and $MC$ are everywhere positive. The cost function is concave down and then concave up, so $MC$ is decreasing and then increasing. The revenue function is linear and then concave down, so $MR$ is constant and then decreasing. See Figure 5.84.

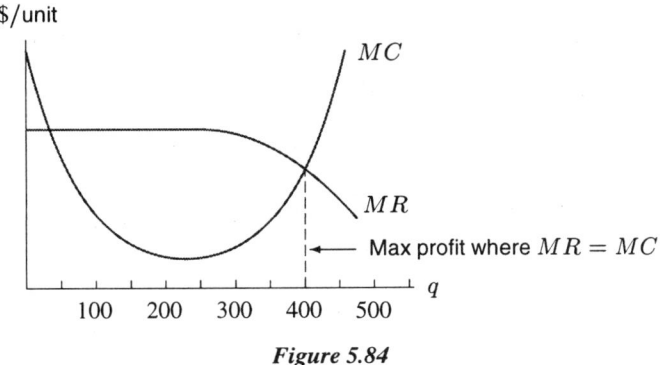

**Figure 5.84**

18. From the given graph, we can sketch the graph of marginal profit = $M\pi = MR - MC$: see Figure 5.85.

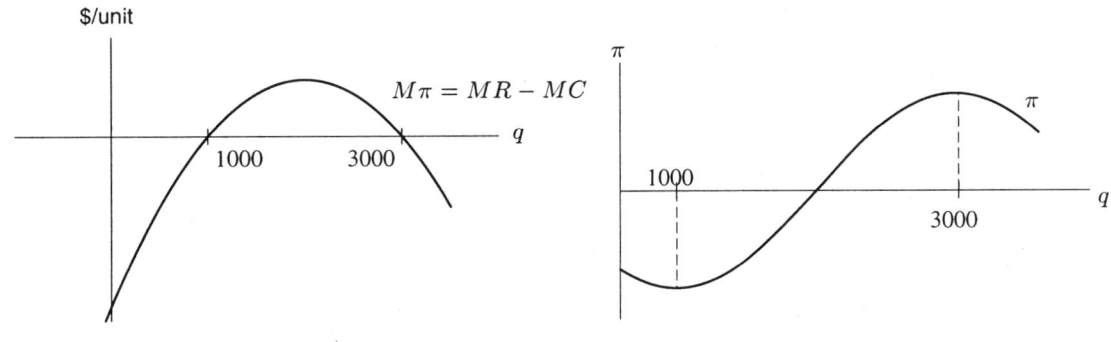

**Figure 5.85**          **Figure 5.86**

And, from the graph of marginal profit, we can make a sketch showing the shape of the shape of the graph of the profit function, since the marginal profit curve is the graph of the derivative of the profit function. See Figure 5.86. Because we don't know the value of the profit when $q = 0$, the graph in Figure 5.86 may be shifted vertically.

Since $M\pi < 0$ for $q < 1000$ and for $q > 3000$, while $M\pi > 0$ for $1000 < q < 3000$ we see that profits appear to be at a minimum when $q = 1000$ and at a maximum for $q = 3000$. We see from Figure 5.86 that there is a local maximum at $q = 3000$. Therefore, the profit has a global maximum either at $q = 3,000$ or at the endpoint $q = 0$.

Because the profit function could be shifted vertically downward, notice that the maximum profit could be zero or negative.

19. (a) $a(q)$ is represented by the slope of the line from the origin to the graph. For example, the slope of line (1) through $(0,0)$ and $(q, C(q))$ is $\frac{C(q)}{q} = a(q)$.
    (b) $a(q)$ is minimal where $a(q)$ is tangent to the graph (line (2)).

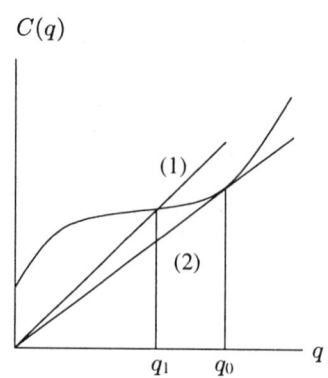

(c) We have $a(q) = \frac{C(q)}{q}$; by the quotient rule

$$a'(q) = \frac{qC'(q) - C(q)}{q^2}$$
$$= \frac{C'(q) - \frac{C(q)}{q}}{q}$$
$$= \frac{1}{q}(C'(q) - a(q)).$$

Thus if $q = q_0$, then $a'(q_0) = \frac{1}{q}(C'(q_0) - a(q_0)) = 0$, so that $C'(q_0) = a(q_0)$; or, the average cost is minimized when it equals the marginal cost.

(d)

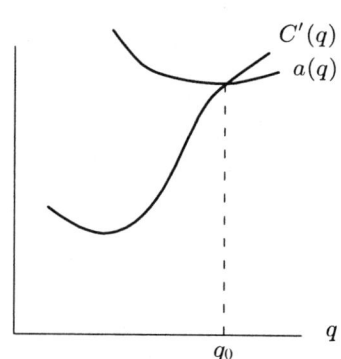

20. To maximize revenue, we first must find an expression for revenue in terms of price. We know that $R(p) = pq$, where $p$=price and $q$=quantity sold. We now need to find an expression for $q$ in terms of $p$. Using the information given, we find that

$$q = 4000 + \frac{(4.00 - p)}{0.25}(200)$$

Simplification of $q$ yields

$$q = 4000 + 800(4 - p)$$
$$= 4000 + 3200 - 800p$$
$$= 7200 - 800p$$

We can now get an expression for revenue in terms of price.

$$R(p) = qp = (7200 - 800p)p$$
$$= 7200p - 800p^2$$

We want to maximize this function in terms of $p$. First find the critical points by finding the derivative.

$$R'(p) = 7200 - 1600p$$

Setting $R'(p) = 0$ and solving for $p$ yields

$$7200 - 1600p = 0$$
$$1600p = 7200$$
$$p = 4.5$$

Since $R'(p) > 0$ for $p < 4.5$ and $R'(p) < 0$ for $p > 4.5$, we conclude that revenue has a local maximum at $p = 4.5$. Since this is the only critical point, we conclude that it is the global maximum. So revenue is maximized at a price of $4.50. The quantity sold at this amount is given by

$$q = 7200 - 800(4.50) = 3600$$

and the total revenue is

$$R(4.5) = 7200(4.5) - 800(4.5)^2 = \$16,200.$$

21. (a) The IV method reaches peak concentration the fastest, it in fact begins at its peak.
    The P-IM method reaches peak concentration the slowest.
    (b) The IV method has the largest peak concentration.
    The PO method has the smallest peak concentration.
    (c) The IV method wears off the fastest.
    The P-IM method wears off the slowest.
    (d) The P-IM method has the longest effective duration.
    The IV method has the shortest effective duration.
    (e) It is effective for approximately 5 hours.

22.

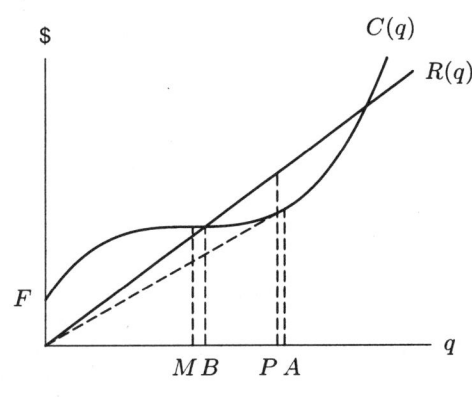

Figure 5.87

23. (a) To maximize benefit (surviving young), we pick 10, because that's the highest point of the benefit graph.
    (b) To optimize the vertical distance between the curves, we can either do it by inspection or note that the slopes of the two curves will be the same where the difference is maximized. Either way, one gets approximately 9.

24. (a) At higher speeds, more energy is used so the graph rises to the right. The initial drop is explained by the fact that the energy it takes a bird to fly at very low speeds is greater than that needed to fly at a slightly higher speed. When it flies slightly faster, the amount of energy consumed decreases. But when it flies at very high speeds, the bird consumes a lot more energy (this is analogous to our swimming in a pool).
    (b) $f(v)$ measures energy per second; $a(v)$ measures energy per meter. A bird traveling at rate $v$ will in 1 second travel $v$ meters, and thus will consume $v \cdot a(v)$ joules of energy in that 1 second period. Thus $v \cdot a(v)$ represents the energy consumption per second, and so $f(v) = v \cdot a(v)$.
    (c) Since $v \cdot a(v) = f(v)$, $a(v) = f(v)/v$. But this ratio has the same value as the slope of a line passing from the origin through the point $(v, f(v))$ on the curve (see figure). Thus $a(v)$ is minimal when the slope of this line is minimal. To find the value of $v$ minimizing $a(v)$, we solve $a'(v) = 0$. By the quotient rule,

$$a'(v) = \frac{vf'(v) - f(v)}{v^2}.$$

**252** CHAPTER FIVE /SOLUTIONS

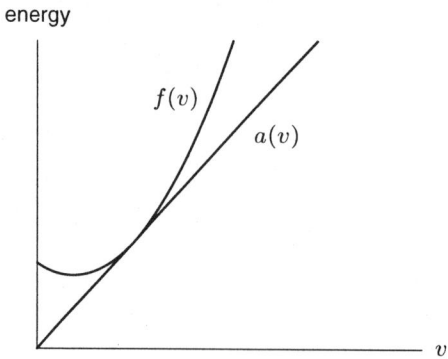

Thus $a'(v) = 0$ when $vf'(v) = f(v)$, or when $f'(v) = f(v)/v = a(v)$. Since $a(v)$ is represented by the slope of a line through the origin and a point on the curve, $a(v)$ is minimized when this line is tangent to $f(v)$, so that the slope $a(v)$ equals $f'(v)$.

(d) The bird should minimize $a(v)$ assuming it wants to go from one particular point to another, i.e. where the distance is set. Then minimizing $a(v)$ minimizes the total energy used for the flight.

25. Since marginal revenue equals $dR/dq$ and $R = pq$, we have, using the product rule,

$$\frac{dR}{dq} = \frac{d(pq)}{dq} = p \cdot 1 + \frac{dp}{dq} \cdot q = p\left(1 + \frac{q}{p} \cdot \frac{dp}{dq}\right) = p\left(1 - \frac{1}{-\frac{p}{q} \cdot \frac{dq}{dp}}\right) = p\left(1 - \frac{1}{E}\right).$$

26. Marginal cost equals $C'(q) = k$. On the other hand, by the preceding exercise, marginal revenue equals $p(1 - 1/E)$. Maximum profit will occur when the two are equal

$$k = p(1 - 1/E)$$

Thus

$$k/p = 1 - 1/E$$
$$1/E = 1 - k/p$$

On the other hand,

$$\frac{\text{Profit}}{\text{Revenue}} = \frac{\text{Revenue} - \text{Cost}}{\text{Revenue}} = 1 - \frac{\text{Cost}}{\text{Revenue}} = 1 - \frac{kq}{pq} = 1 - \frac{k}{p} = \frac{1}{E}$$

27. (a) The approximation $E_{C,q} \approx \frac{\Delta C/C}{\Delta q/q}$ shows that $E_{C,q}$ measures the ratio of the fractional change in the cost of production to the fractional change in the quantity produced. Thus, for example, a 1% increase in production will result in an $E_{C,q}$% increase in the cost of production.

(b) Since average cost equals $C/q$ and marginal cost equals $dC/dq$, we have

$$\frac{\text{marginal cost}}{\text{average cost}} = \frac{dC/dq}{C/q} = \frac{q}{C} \cdot \frac{dC}{dq} = E_{C,q}$$

28. (a)

**TABLE 5.5**

| $i$ | Marginal Cost (of $i^{\text{th}}$ filter) | Average Cost (for $i^{\text{th}}$ filter) | Marginal Savings |
|---|---|---|---|
| 0 | $0 | $0 | $0 |
| 1 | $5 | $5 | $64 |
| 2 | $6 | $5.50 | $32 |
| 3 | $7 | $6 | $16 |
| 4 | $8 | $6.50 | $9 |
| 5 | $9 | $7 | $3 |
| 6 | $10 | $7.50 | $3 |
| 7 | $11 | $8 | $0 |

(b) She should install four filters. Up to the fourth filter, the marginal cost is less than the marginal savings. This tells us that for each of the first four filters the developer buys, she will save more on the laundromat than she will have to pay for the filter. From the fifth filter onwards, she pays more for each additional filter than she makes from the laundromat.

(c) If the rack costs $100 she should not buy it. Instead she should let the laundromat protect itself. This is because, if she buys even one filter she has to spend $105 on the filter but her savings as a result of the purchase only amount to $64. If she buys two filters she has to spend $111 on the filters but her total savings as a result of the purchase only amount to $96. If she buys three filters she has to spend $118 on the filters but her total savings as a result of the purchase only amount to $112. If she buys four filters she has to spend $126 on the filters but her total savings as a result of the purchase only amount to $121 etc.

(d) If the rack costs $50 she should buy it and install four filters since she covers the price of the rack with the first purchase of filters i.e., having bought one filter she will have spent $50 + $5 = $55 but she will have saved $64 on the laundromat business. Her marginal cost and marginal savings do not change after this point and so she still saves more on laundromat than she spends on filters if she buys the second, third and fourth filter.

**29.**

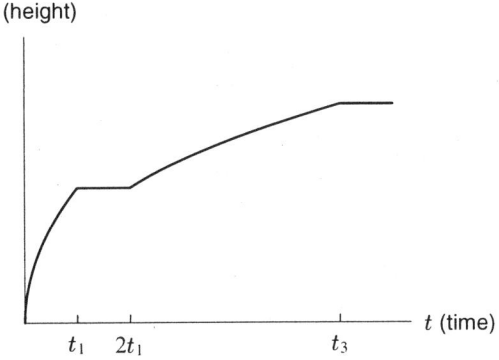

Suppose $t_1$ is the time to fill the left side to the top of the middle ridge. Since the container gets wider as you go up, the rate $dH/dt$ decreases with time. Therefore, for $0 \leq t \leq t_1$, graph is concave down.

At $t = t_1$, water starts to spill over to right side and so depth of left side doesn't change. It takes as long for the right side to fill to the ridge as the left side, namely $t_1$. Thus the graph is horizontal for $t_1 \leq t \leq 2t_1$.

For $t \geq 2t_1$, water level is above the central ridge. The graph is climbing because the depth is increasing, but at a slower rate than for $t \leq t_1$ because the container is wider. The graph is concave down because width is increasing with depth. Time $t_3$ represents the time when container is full.

**30.** (a) The concavity changes at $t_1$ and $t_3$, as shown below.

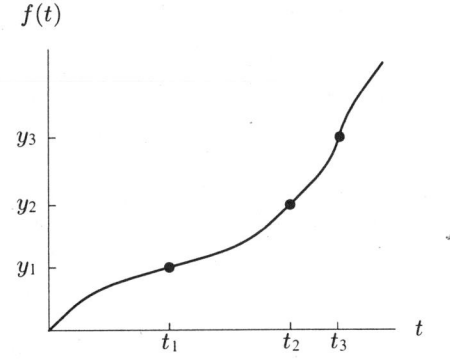

(b) $f(t)$ grows most quickly where the vase is skinniest (at $y_3$) and most slowly where the vase is widest (at $y_1$). The diameter of the widest part of the vase looks to be about 4 times as large as the diameter at the skinniest part. Since the area of a cross section is given by $\pi r^2$, where $r$ is the radius, the ratio between areas of cross sections at these two places is about $4^2$, so the growth rates are in a ratio of about 1 to 16 (the wide part being 16 times slower).

**31.** We see that
$$m_1 = \frac{\text{Rise}}{\text{Run}} = \frac{-p_0}{q_0}$$
We also have
$$m_2 = \frac{dp}{dq}$$
Therefore
$$\frac{m_1}{m_2} = \frac{\frac{-p_0}{q_0}}{\frac{dp}{dq}} = -\frac{p_0}{q_0} \cdot \frac{dq}{dp} = E$$

**32.** It is interesting to note that to draw a graph of $C'(q)$ for this problem, you never have to know what $C(q)$ looks like, although you *could* draw a graph of $C(q)$ if you wanted to. By the definition of average cost, we know that $C(q) = q \cdot a(q)$. Using the product rule we get that $C'(q) = a(q) + q \cdot a'(q)$.

We are given a graph of $a(q)$ which is linear, so $a(q) = b + mq$, where $b = a(0)$ is the $y$-intercept and $m$ is the slope. Therefore

$$C'(q) = a(q) + q \cdot a'(q) = b + mq + q \cdot m$$
$$= b + 2mq.$$

In other words, $C'(q)$ is also linear, and it has twice the slope and the same $y$–intercept as $a(q)$.

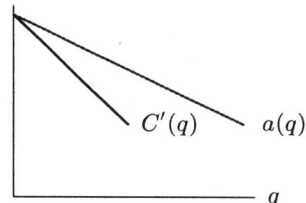

**33.** Recall that the natural logarithm is undefined for $x \leq 0$, so the domain of $f$ is $x > 0$. We see from looking at the graph of $f(x) = x - \ln x$ in the text that this function has one local minimum. We want to assign values to $a$ and $b$ so that this local minimum occurs at $x = 2$. The function must therefore have a critical point at $x = 2$. We find the derivative of $f(x) = a(x - b \ln x)$ and the critical points in terms of $a$ and $b$.

$$f'(x) = a\left(1 - b\left(\frac{1}{x}\right)\right) = 0$$
$$1 - b\left(\frac{1}{x}\right) = 0$$
$$1 = \frac{b}{x}$$
$$x = b$$

We see that $f(x)$ has only one critical point, at $x = b$. Since we want a critical point at $x = 2$, we choose $b = 2$.
Since $b = 2$, we have $f(x) = a(x - 2\ln x)$. We now use the condition that $f(2) = 5$ to find $a$:

$$f(2) = 5$$
$$a(2 - 2\ln 2) = 5$$
$$a = 5/(2 - 2\ln 2)$$
$$a \approx 8.147.$$

We let $a = 8.147$ and $b = 2$, so the function is $f(x) = 8.147(x - 2\ln x)$. If we sketch a graph of this function, we see that this function does indeed have a local minimum approximately at the point $(2, 5)$.

**34.** Here is one possible graph of $g$:

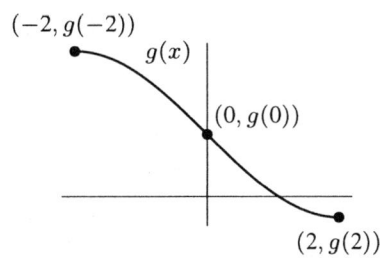

(a) From left to right, the graph of $g(x)$ starts "flat", decreases slowly at first then more rapidly, most rapidly at $x = 0$. The graph then continues to decrease but less and less rapidly until flat again at $x = 2$. The graph should exhibit symmetry about the point $(0, g(0))$.

(b) The graph has an inflection point at $(0, g(0))$ where the slope changes from negative and decreasing to negative and increasing.

(c) The function has a global maximum at $x = -2$ and a global minimum at $x = 2$.

(d) Since the function is decreasing over the interval $-2 \leq x \leq 2$

$$g(-2) = 5 > g(0) > g(2).$$

Since the function appears symmetric about $(0, g(0))$, we have

$$g(-2) - g(0) = g(0) - g(2).$$

**35.** The critical points of $f$ occur where $f'$ is zero. These two points are indicated in the figure below.

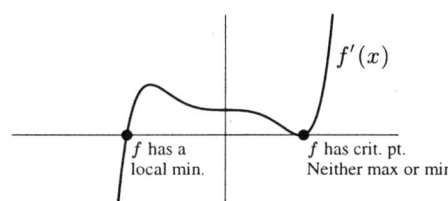

Note that the point labeled as a local minimum of $f$ is not a critical point of $f'$.

**36.** (a) Figure 5.88 shows the rate of sales as a function of time. The point of diminishing returns is reached when the rate of sales is at a maximum; this happens during the 4th month.

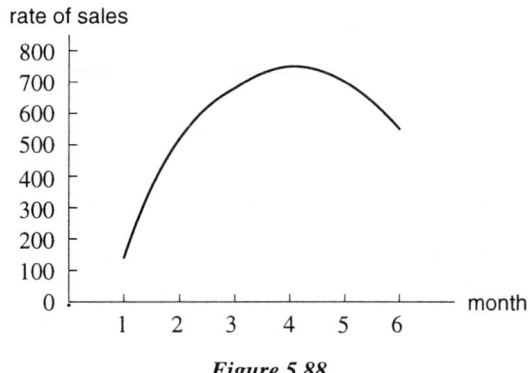

*Figure 5.88*

(b) Total sales for the first four months are

$$140 + 520 + 680 + 750 = 2090 \text{ sales}.$$

(c) Assuming logistic growth, the limiting value should be twice the value at the point of diminishing returns, so total sales should be $2(2090) = 4180$ sales.

**37.** (a) Differentiating using the product rule gives

$$C'(t) = 20e^{-0.03t} + 20t(-0.03)e^{-0.03t}$$
$$= 20(1 - 0.03t)e^{-0.03t}.$$

At the peak concentration, $C'(t) = 0$, so

$$20(1 - 0.03t)e^{-0.03t} = 0$$
$$t = \frac{1}{0.03} = 33.3 \text{ minutes.}$$

When $t = 33.3$, the concentration is

$$C = 20(33.3)e^{-0.03(33.3)} \approx 245 \text{ ng/ml.}$$

See Figure 5.89. The curve peaks after 33.3 minutes with a concentration of 244.9 ng/ml.

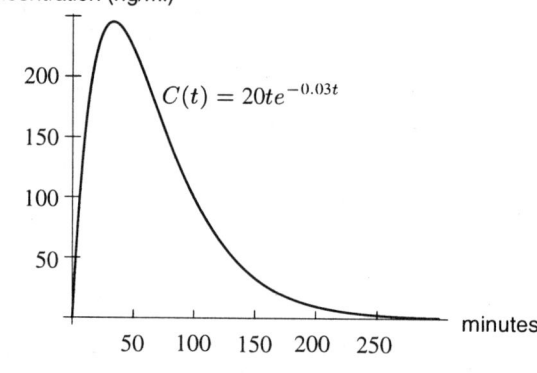

**Figure 5.89**

(b) After 15 minutes, the drug concentration will be $C(15) = 20(15)e^{(-0.03)(15)} \approx 191$ ng/ml. After an hour, the concentration will be $C(60) = 20(60)e^{-0.03(60)} \approx 198$ ng/ml.

(c) We want to know where $C(t) = 10$. We estimate from the graph. It looks like $C(t) = 10$ after 190 minutes or a little over 3 hours.

**38.** (a) Quadratic polynomial (degree 2) with negative leading coefficient.
(b) Exponential.
(c) Logistic.
(d) Logarithmic.
(e) This is a quadratic polynomial (degree 2) and positive leading coefficient.
(f) Exponential.
(g) Surge
(h) Periodic
(i) This looks like a cubic polynomial (degree 3) and negative leading coefficient.

## Solutions to the Projects

1. Since $a(q) = C(q)/q$, we have $C(q) = a(q) \cdot q$. Thus $C'(q) = a'(q)q + a(q)$, and so

$$C'(q_0) = a'(q_0)q_0 + a(q_0).$$

Since $t_1$ is the line tangent to $a(q)$ at $q = q_0$, the slope of $t_1$ is $a'(q_0)$, and the equation of $t_1$ is

$$y = a(q_0) + a'(q_0) \cdot (q - q_0) = a'(q_0) \cdot q + \big(a(q_0) - a'(q_0) \cdot q_0\big).$$

Thus the $y$-intercept of $t_1$ is given by $a(q_0) - a'(q_0)q_0$, and the equation of the line $t_2$ is
$$y = 2 \cdot a'(q_0) \cdot q + \bigl(a(q_0) - a'(q_0) \cdot q_0\bigr)$$
since $t_2$ has twice the slope of $t_1$. Let's compute the $y$-value on $t_2$ when $q = q_0$:
$$y = 2 \cdot a'(q_0) \cdot q_0 + \bigl(a(q_0) - a'(q_0) \cdot q_0\bigr) = a'(q_0)q_0 + a(q_0) = C'(q_0).$$

Hence $C'(q_0)$ is given by the point on $t_2$ where $q = q_0$.

2.

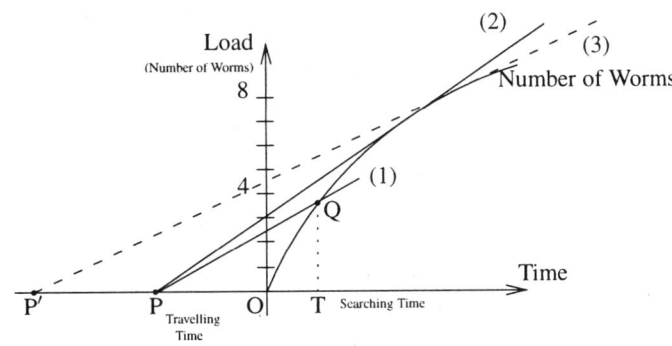

(a) See line (1). For any point $Q$ on the loading curve, the line $PQ$ has slope
$$\frac{QT}{PT} = \frac{QT}{PO + OT} = \frac{\text{load}}{\text{traveling time} + \text{searching time}}.$$

(b) The slope of the line $PQ$ is maximized when the line is tangent to the loading curve, which happens with line (2). The load is then approximately 7 worms.

(c) If the traveling time is increased, the point $P$ moves to the left, to point $P'$, say. If line (3) is tangent to the curve, it will be tangent to the curve further to the right than line (2), so the optimal load is larger. This makes sense: if the bird has to fly further, you'd expect it to bring back more worms each time.

## Solutions for Section 6.1

1. By counting boxes, we find $\int_1^6 f(x)\,dx = 8.5$, so the average value of $f$ is $\dfrac{8.5}{6-1} = \dfrac{8.5}{5} = 1.7$.

2. (a) Counting the squares yields an estimate of 25 squares, each with area $= 1$, so we conclude that
$$\int_0^5 f(x)\,dx \approx 25.$$
   (b) The average height appears to be around 5.
   (c) Using the formula, we get
$$\text{Average value} = \dfrac{\int_0^5 f(x)\,dx}{5-0} \approx \dfrac{25}{5} \approx 5,$$
   which is consistent with (b).

3. Average value $= \dfrac{1}{2-0}\int_0^2 (1+t)\,dt = \dfrac{1}{2}(4) = 2.$

4. The average value is given by the expression
$$\text{Average value} = \dfrac{1}{10-0} \cdot \int_0^{10} 2^t\,dt = \dfrac{1}{10}\cdot(1475.88) = 147.588.$$
   Note: The value of $\int_0^{10} 2^t\,dt$ was obtained using a calculator.

5. Average value $= \dfrac{1}{10-0}\int_0^{10} e^t\,dt \approx 2202.55$

6. (a) Average value $= \int_0^1 \sqrt{1-x^2}\,dx = 0.79.$
   (b) The area between the graph of $y = 1-x$ and the $x$-axis is 0.5. Because the graph of $y = \sqrt{1-x^2}$ is concave down, it lies above the line $y = 1-x$, so its average value is above 0.5. See figure below.

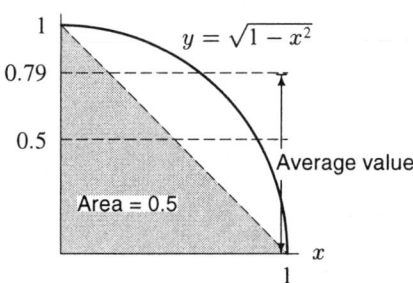

7. By a visual estimate, the average value is $\approx 8$.

8. By a visual estimate, the average value is $\approx -3$.

**260** CHAPTER SIX /SOLUTIONS

9. (a) The average inventory is given by the formula
$$\frac{1}{90-0}\int_0^{90} 5000(0.9)^t\, dt.$$
Using a calculator yields
$$\int_0^{90} 5000(0.9)^t\, dt \approx 47452.5$$
so the average inventory is
$$\frac{47452.5}{90} \approx 527.25.$$

(b) The function is graphed in Figure 6.1. The area of a rectangle of height 527.25 is equal to the area under the curve.

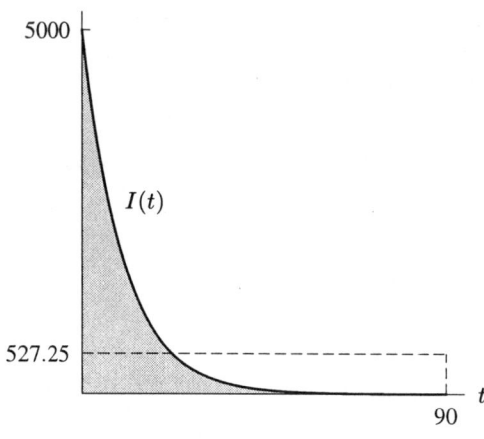

**Figure 6.1**

10. (a) When $t = 0$, $f(0) = 100e^0 = 100$. So there are 100 cases at the start of the six months. When $t = 6$, $f(6) = 100e^{-3} = 4.98$, so there are almost 5 cases left at the end of the half year.
(b) Average number of cases in the inventory can be given by
$$\left(\frac{1}{6-0}\right)\int_0^6 f(t)\, dt \approx \frac{190}{6} \approx 32 \text{ cases.}$$

11. Since $t = 0$ in 1965 and $t = 35$ in 2000, we want:
$$\text{Average Value} = \frac{1}{35-0}\int_0^{35} 225(1.15)^t\, dt$$
$$= \frac{1}{35}(212{,}787) = \$6080.$$

12. (a) $Q(10) = 4(0.96)^{10} \approx 2.7$. $Q(20) = 4(0.96)^{20} \approx 1.8$
(b)
$$\frac{Q(10) + Q(20)}{2} \approx 2.21$$
(c) The average value of $Q$ over the interval is about 2.18.
(d) Because the graph of $Q$ is concave up between $t = 10$ and $t = 20$, the area under the curve is less than what is obtained by connecting the endpoints with a straight line.

13. (a)
$$\text{Average population} = \frac{1}{10}\int_0^{10} 67.38(1.026)^t\, dt$$

Evaluating the integral numerically gives

$$\text{Average population} \approx 76.8 \text{ million}$$

(b) In 1980, $t = 0$, and $P = 67.38(1.026)^0 = 67.38$.
In 1990, $t = 10$, and $P = 67.38(1.026)^{10} = 87.10$.
Average$= \frac{1}{2}(67.38 + 87.10) = 77.24$ million.

(c) If $P$ had been linear, the average value found in (a) would have been the one we found in (b). Since the population graph is concave up, it is below the secant line. Thus, the actual values of $P$ are less than the corresponding values on the secant line, and so the average found in (a) is smaller than that in (b).

14. (a) Average value of $f = \frac{1}{5}\int_0^5 f(x)\, dx$.
(b) Average value of $|f| = \frac{1}{5}\int_0^5 |f(x)|\, dx = \frac{1}{5}(\int_0^2 f(x)\, dx - \int_2^5 f(x)\, dx)$.

15. We'll show that in terms of the average value of $f$,

$$\text{I} > \text{II} = \text{IV} > \text{III}$$

Using Problem ?? (a) on page ??,

$$\frac{\text{Average value}}{\text{of } f \text{ on II}} = \frac{\int_0^2 f(x)dx}{2} = \frac{\frac{1}{2}\int_{-2}^2 f(x)dx}{2}$$

$$= \frac{\int_{-2}^2 f(x)dx}{4}$$

$$= \text{Average value of } f \text{ on IV}.$$

Since $f$ is decreasing on [0,5], the average value of $f$ on the interval $[0, c]$, where $0 \leq c \leq 5$, is decreasing as a function of $c$. The larger the interval the more low values of $f$ are included. Hence

$$\begin{array}{c}\text{Average value of } f \\ \text{on } [0, 1]\end{array} > \begin{array}{c}\text{Average value of } f \\ \text{on } [0, 2]\end{array} > \begin{array}{c}\text{Average value of } f \\ \text{on } [0, 5]\end{array}$$

16. Looking at the graph of $f(x)$, we see that the curve has a horizontal tangent at $x = 1$. Thus $f'(1) = 0$.

Next we must determine if the average value of $f(x)$ is positive or negative on $0 \leq x \leq 4$. The average value of the function from $x = 0$ to $x = 4$ will have the same sign as the integral of the function over the same interval. Because the area between the curve and the $x$-axis that is above the $x$-axis is greater than the area that is below the $x$-axis, the integral is positive. Thus the average value is positive.

The area between the graph of $f(x)$ and the $x$-axis between $x = 0$ and $x = 1$ lies entirely below the $x$-axis. Hence $\int_0^1 f(x)\, dx$ is negative.

We can now arrange the quantities in order from least to greatest: $(c) < (a) < (b)$.

17. In (a), $f'(1)$ is the slope of a tangent line at $x = 1$, which is negative. As for (c), the rate of change in $f(x)$ is given by $f'(x)$, and the average value of this over $0 \leq x \leq a$ is

$$\frac{1}{a-0}\int_0^a f'(x)\, dx = \frac{f(a) - f(0)}{a - 0}.$$

This is the slope of the line through the points $(0, 1)$ and $(a, 0)$, which is less negative that the tangent line at $x = 1$. Therefore, $(a) < (c) < 0$. The quantity (b) is $\left(\int_0^a f(x)\, dx\right)/a$ and (d) is $\int_0^a f(x)\, dx$, which is the net area under the graph of $f$ (counting the area as negative for $f$ below the $x$-axis). Since $a > 1$ and $\int_0^a f(x)\, dx > 0$, we have $0 <$(b)$<$(d). Therefore

$$(a) < (c) < (b) < (d).$$

## Solutions for Section 6.2

1. (a) Looking at the figure in the problem we see that the equilibrium price is roughly $30 giving an equilibrium quantity of 125 units.
   (b) Consumer surplus is the area above $p^*$ and below the demand curve. Graphically this is represented by the shaded area in Figure 6.2. From the graph we can estimate the shaded area to be roughly 14 squares where each square represents ($25/unit)·(10 units). Thus the consumer surplus is approximately

   $$14 \cdot \$250 = \$3500.$$

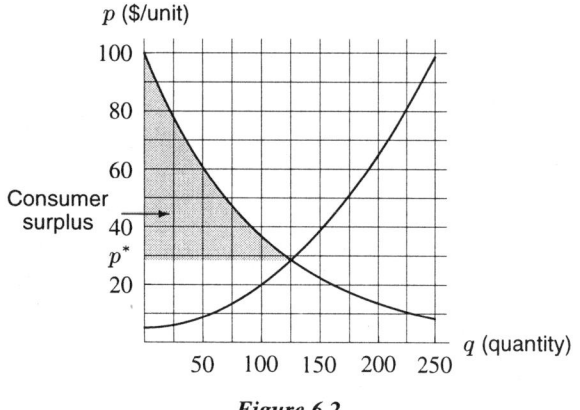
Figure 6.2

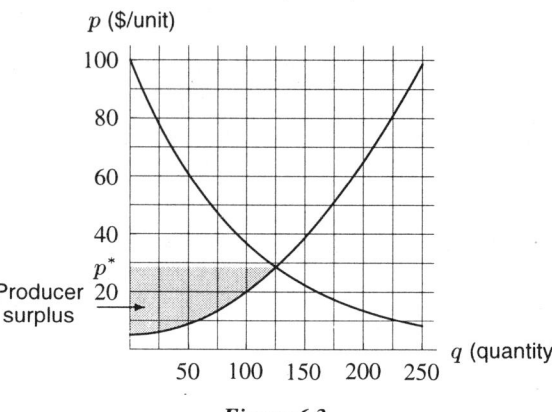
Figure 6.3

   Producer surplus is the area under $p^*$ and above the supply curve. Graphically this is represented by the shaded area in Figure 6.3. From the graph we can estimate the shaded area to be roughly 8 squares where each square represents ($25/unit)·(10 units). Thus the producer surplus is approximately

   $$8 \cdot \$250 = \$2000$$

   (c) We have

   $$\text{Total gains from trade} = \text{Consumer surplus} + \text{producer surplus}$$
   $$= \$3500 + \$2000$$
   $$= \$5500.$$

2. (a) The consumer surplus is the area the between demand curve and the price $40—roughly 9 squares. See Figure 6.4. Since each square represents ($25/unit)·(10 units), the total area is

   $$9 \cdot \$250 = \$2250.$$

   At a price of $40, about 90 units are sold. The producer surplus is the area under $40, above the supply curve, and to the left of $q = 90$. See Figure 6.4. The area is 10.5 squares or

   $$10.5 \cdot \$250 = \$2625.$$

   The total gains from the trade is

   $$\text{Total gain} = \text{Consumer surplus} + \text{Producer surplus} = \$4875.$$

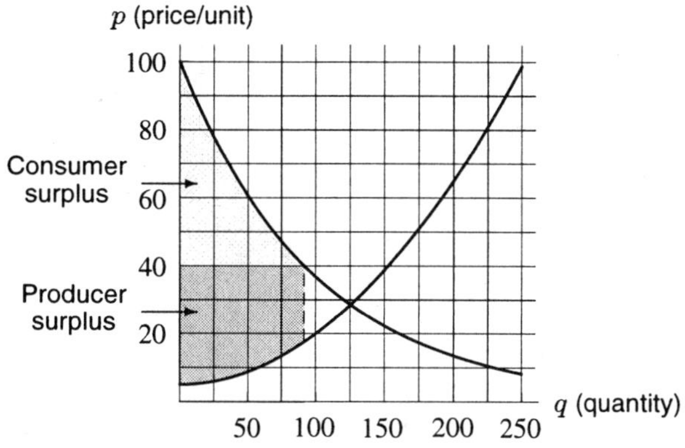

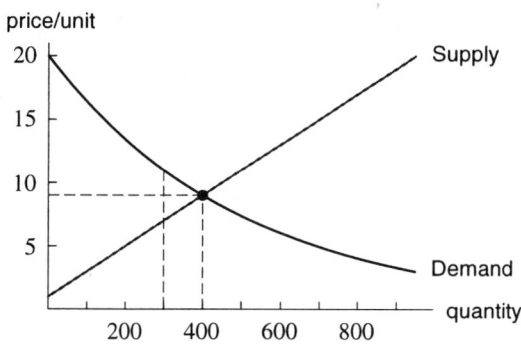

**Figure 6.4**

(b) The consumer surplus is less with the price control.
The producer surplus is greater with the price control.
The total gains from trade are less with the price control.

3. (a) If the quantity is 300 then the demand price is

$$f(300) \approx \$11$$

and the supply price is

$$g(300) = \$7.$$

See Figure 6.5. Thus demand price is higher at 300 units and this will push the quantity supplied higher, towards the equilibrium quantity.

(b)

**Figure 6.5**

Looking at Figure 6.5 we see that

$$p^* \approx \$9 = \text{equilibrium price} \quad q^* \approx 400 = \text{equilibrium quantity}.$$

(c) Consumer surplus is the area under the demand curve and above \$9.

$$\int_0^{400} f(q)\,dq - (400)(\$9) \approx \$1907.$$

See Figure 6.6. Consumers gain \$1907 in buying goods at the equilibrium price instead of at the price they would be willing to pay.

Producer surplus is the area under \$9 and above the supply curve. Thus the producer surplus is

$$(400)(\$9) - \int_0^{400} g(q)\,dq = 1600.$$

See Figure 6.7. Producers gained \$1600 in supplying goods at the equilibrium price instead of the price at which they would have been willing to supply the goods.

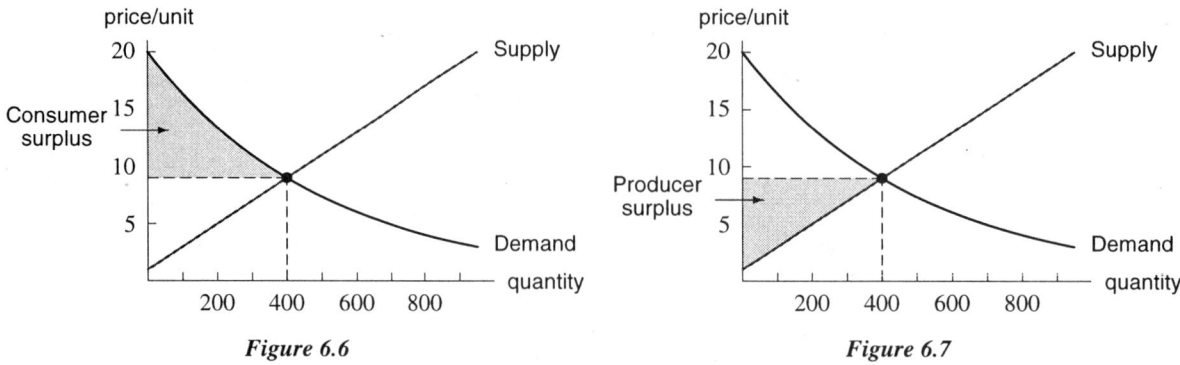

Figure 6.6

Figure 6.7

4. (a) The quantity demanded at a price of $50 is calculated by substituting $p = 50$ into the demand equation $p = 100e^{-0.008q}$. Solving $50 = 100e^{-0.008q}$ for $q$ gives $q \approx 86.6$. In other words, at a price of $50, consumer demand is about 87 units. The quantity supplied at a price of $50 is calculated by substituting by $p = 50$ into the supply equation $p = 4\sqrt{q} + 10$. Solving $50 = 4\sqrt{q} + 10$ for $q$ gives $q = 100$. So at a price of $50, producers supply about 100 units. At a price of $50, the supply is larger than the demand, so some goods remain unsold. We can expect prices to be pushed down.

(b) The supply and demand curves are shown in Figure 6.8. The equilibrium price is about $p^* = \$48$ and the equilibrium quantity is about $q^* = 91$ units. The market will push prices downward from $50 toward the equilibrium price of $48. This agrees with the conclusion to part (a) that prices will drop.

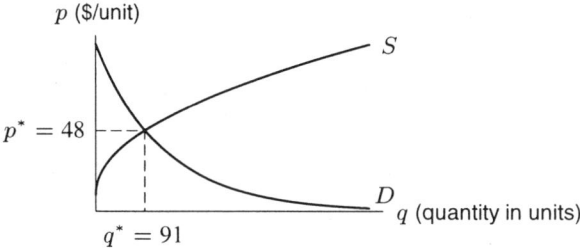

**Figure 6.8**: Demand and supply curves for a product

(c) The consumer surplus is the area under the demand curve and above the line $p = 48$. (See Figure 6.9.) We have

Consumer surplus = Area between demand curve and line $p = p^*$

$$= \left( \int_0^{q^*} f(q)\, dq \right) - p^* q^*$$

$$= \int_0^{91} (100e^{-0.008q})\, dq - (48)(91)$$

$$\approx 6464 - 4368$$

$$= 2096.$$

Consumers gain $2096 by buying goods at the equilibrium price instead of the price they would have been willing to pay.

The producer surplus is the area above the supply curve and below the line $p = 48$. (See Figure 6.10.) We have

Producer surplus = Area between supply curve and line $p = p^*$

$$= p^* q^* - \int_0^{91} (4\sqrt{q} + 10)\, dq$$

$$\approx (48)(91) - 3225$$

$$= 1143.$$

Producers gain $1143 by supplying goods at the equilibrium price instead of the price at which they would have been willing to provide the goods.

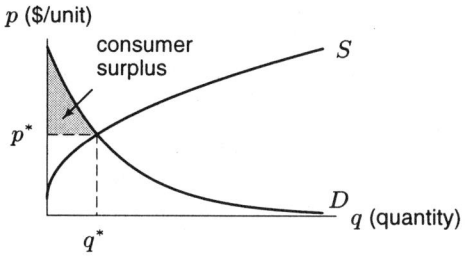

Figure 6.9: Consumer surplus

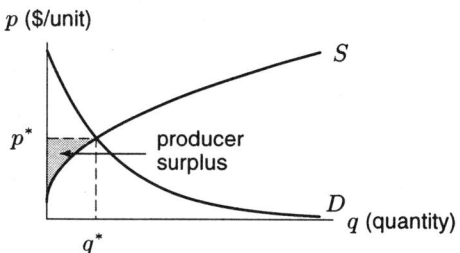

Figure 6.10: Producer surplus

5.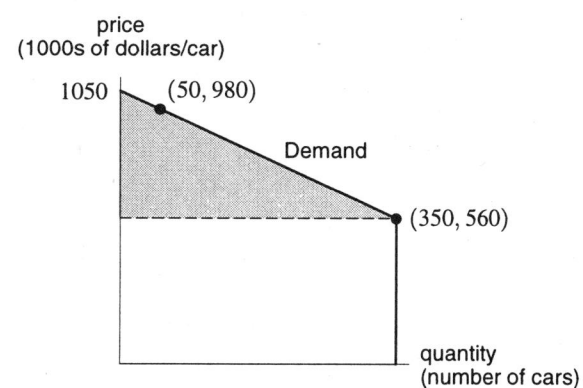

Measuring money in thousands of dollars, the equation of the line representing the demand curve passes through (50, 980) and (350, 560). So the equation is $y - 560 = \frac{420}{-300}(x - 350)$, i.e. $y - 560 = -\frac{7}{5}x + 490$. The consumer surplus is thus

$$\int_0^{350} \left(-\frac{7}{5}x + 1050\right) dx - (350)(560) = 85750.$$

(Note that $85750 = \frac{1}{2} \cdot 490 \cdot 350$, the area of the triangle in the diagram. We thus could have avoided the formula for consumer surplus in solving the problem.)

Recalling that our unit measure for the price axis is $1000/car, the consumer surplus is $85,750,000.

6. (a) Consumer surplus is greater than producer surplus in Figure 6.11.
   (b) Producer surplus is greater than consumer surplus in Figure 6.12.

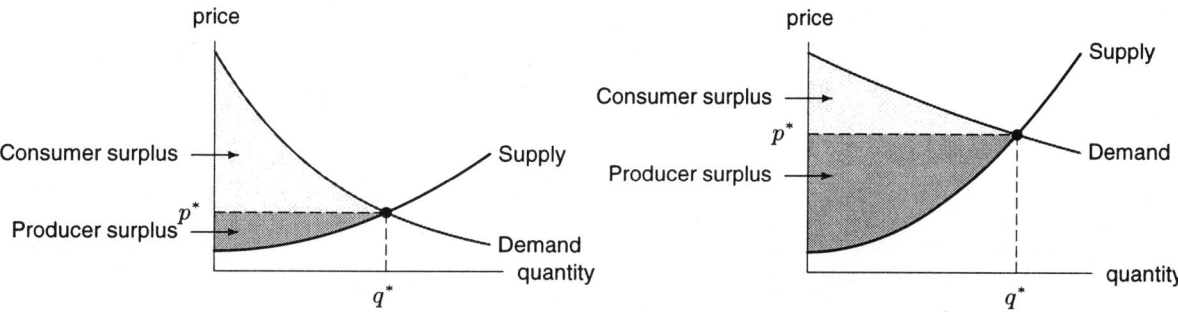

Figure 6.11    Figure 6.12

**266** CHAPTER SIX /SOLUTIONS

7. The total gains from trade at the equilibrium price is shaded in Figure 6.13. We see in Figures 6.14 and 6.15 that if the price is artificially high or low, the quantity sold is less than $q^*$. Thus, the total gains from trade are reduced.

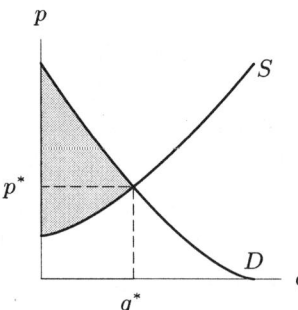

*Figure 6.13*: Shade area: Total gains from trade at equilibrium price

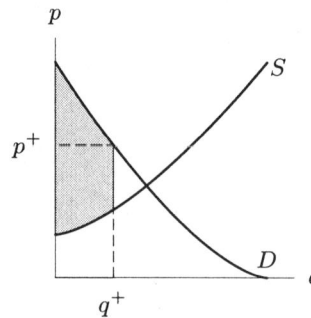

*Figure 6.14*: Shade area: Gains when price is artificially high

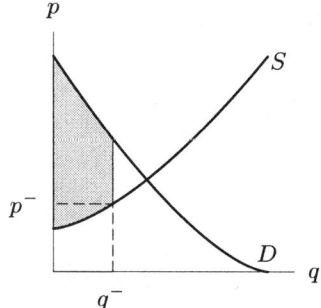

*Figure 6.15*: Shade area: Gains when price is artificially low

8.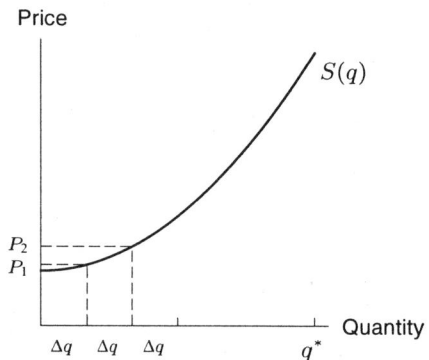

The supply curve, $S(q)$, represents the minimum price $p$ per unit that the suppliers will be willing to supply some quantity $q$ of the good for. If the suppliers have $q^*$ of the good and $q^*$ is divided into subintervals of size $\Delta q$, then if the consumers could offer the suppliers for each $\Delta q$ a price increase just sufficient to induce the suppliers to sell an additional $\Delta q$ of the good, the consumers' total expenditure on $q^*$ goods would be

$$p_1 \Delta q + p_2 \Delta q + \cdots = \sum p_i \Delta q.$$

As $\Delta q \to 0$ the Riemann sum becomes the integral $\int_0^{q^*} S(q)\, dq$. Thus $\int_0^{q^*} S(q)\, dq$ is the amount the consumers would pay if suppliers could be forced to sell at the lowest price they would be willing to accept.

9. 
$$\int_0^{q^*} (p^* - S(q))\, dq = \int_0^{q^*} p^*\, dq - \int_0^{q^*} S(q)\, dq$$
$$= p^* q^* - \int_0^{q^*} S(q)\, dq.$$

Using Problem 8, this integral is the extra amount consumers pay (i.e., suppliers earn over and above the minimum they would be willing to accept for supplying the good). It results from charging the equilibrium price.

10. (a) $p^* q^* =$ the total amount paid for $q^*$ of the good at equilibrium.

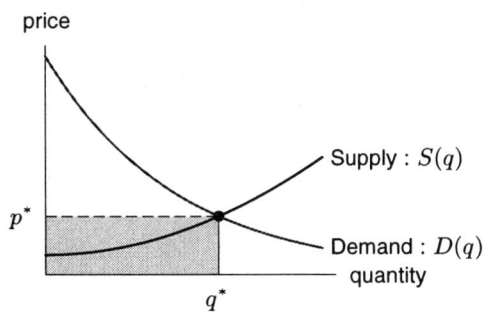

(b) $\int_0^{q^*} D(q)\, dq$ = the maximum consumers would be willing to pay if they had to pay the highest price acceptable to them for each additional unit of the good.

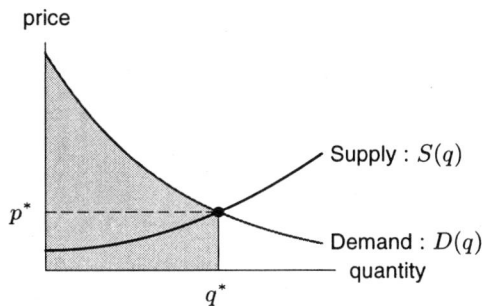

(c) $\int_0^{q^*} S(q)\, dq$ = the minimum suppliers would be willing to accept if they were paid the minimum price acceptable to them for each additional unit of the good.

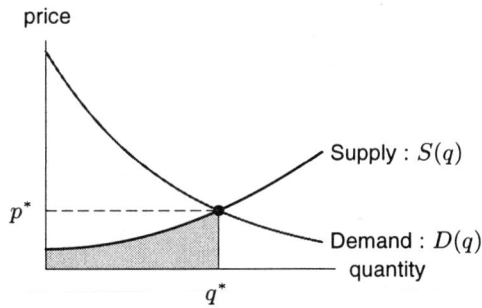

(d) $\int_0^{q^*} D(q)\, dq - p^* q^*$ = consumer surplus.

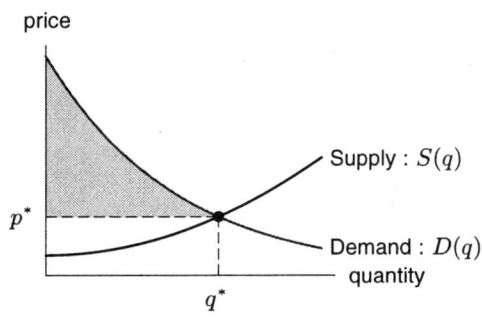

(e) $p^*q^* - \int_0^{q^*} S(q)\,dq$ = producer surplus.

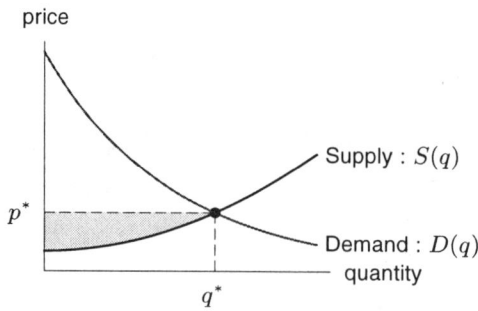

(f) $\int_0^{q^*} (D(q) - S(q))\,dq$ = producer surplus and consumer surplus.

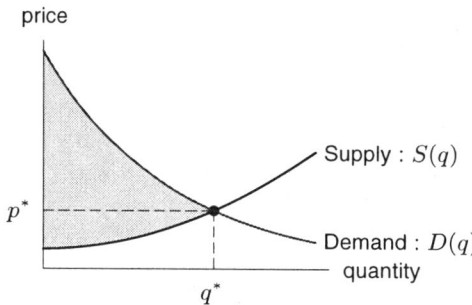

11. Figure 6.16 shows the consumer and producer surplus for the price, $p^-$. For comparsion, Figure 6.17 shows the consumer and producer surplus at the equilibrium price.

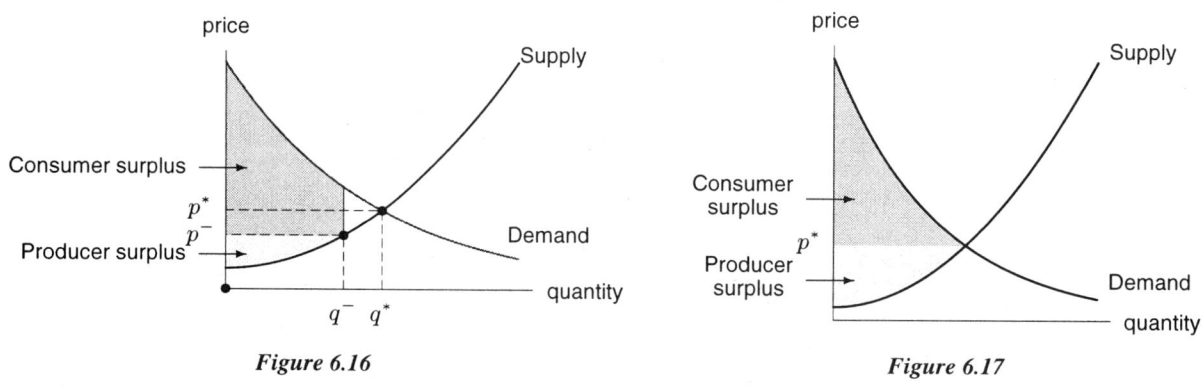

*Figure 6.16*        *Figure 6.17*

(a) The producer surplus is the area on the graph between $p^-$ and the supply curve. Lowering the price also lowers the producer surplus.

(b) The consumer surplus — the area between the supply curve and the line $p^-$ — may increase or decrease depends on the functions describing the supply and demand, and the lowered price. (For example, the consumer surplus seems to be increased in Figure 6.16 but if the price were brought down to $0 then the consumer surplus would be zero, and hence clearly less than the consumer surplus at equilibrium.)

(c) Figure 6.16 shows that the total gains from the trade are decreased.

## Solutions for Section 6.3

1.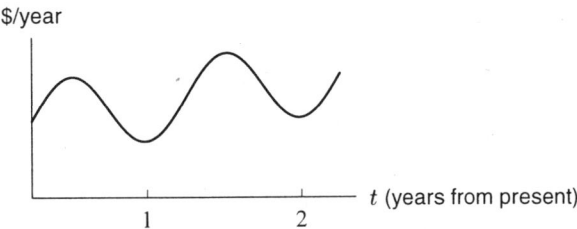

   The graph reaches a peak each summer, and a trough each winter. The graph shows sunscreen sales increasing from cycle to cycle. This gradual increase may be due in part to inflation and to population growth.

2. 
$$\text{Future Value} = \int_0^{15} 3000e^{0.06(15-t)} dt$$
$$\approx \$72{,}980.16$$

$$\text{Present Value} = \int_0^{15} 3000e^{-0.06t} dt$$
$$\approx \$29{,}671.52.$$

   There's a quicker way to calculate the present value of the income stream, since the future value of the income stream is (as we've shown) $72,980.16, the present value of the income stream must be the present value of $72,980.16. Thus,

$$\text{Present Value} = \$72{,}980.16(e^{-.06 \cdot 15})$$
$$\approx \$29{,}671.52,$$

   which is what we got before.

3. 
$$\text{Present Value} = \int_0^{10} (100 + 10t)e^{-.05t} dt$$
$$= \$1{,}147.75.$$

4. (a) (i) If the interest rate is 3%, we have
$$\text{Present value} = \int_0^4 5000e^{-0.03t} dt = \$18{,}846.59.$$

   (ii) If the interest rate is 10%, we have
$$\text{Present value} = \int_0^4 5000e^{-0.10t} dt = \$16{,}484.00.$$

   (b) At the end of the four-year period, if the interest rate is 3%,
$$\text{Value} = 18{,}846.59 e^{0.03(4)} = \$21{,}249.47.$$

   At 10%,
$$\text{Value} = 16{,}484.00 e^{0.10(4)} = \$24{,}591.24.$$

5.  (a) We first find the present value, $P$, of the income stream:

$$P = \int_0^{10} 6000 e^{-0.05t}\, dt = \$47{,}216.32.$$

We use the present value to find the future value, $F$:

$$F = Pe^{rt} = 47126.32 e^{0.05(10)} = \$77{,}846.55.$$

(b) The income stream contributed \$6000 per year for 10 years, or \$60,000. The interest earned was $77{,}846.55 - 60{,}000 = \$17{,}846.55$.

6. The future value is \$20,000 in 5 years. We first find the present value. Since

$$20{,}000 = Pe^{0.06(5)} = Pe^{0.3}.$$

Solving for $P$ gives

$$P = \frac{20{,}000}{e^{0.3}} = \$14{,}816.36.$$

Now we solve for the income stream $S$ which gives a present value of \$14,816.36:

$$14{,}816.36 = \int_0^5 S e^{-0.06t}\, dt.$$

Since $S$ is a constant, we bring it outside the integral sign:

$$14{,}816.36 = S\int_0^5 e^{-0.06t}\, dt = S(4.320)$$

Solving for $S$ gives

$$S = \frac{14816.36}{4.320} \approx 3430.$$

Money must be deposited at a rate of about \$3430 per year, or about \$66 per week.

7.  (a) The future value in 10 years is \$100,000. We first find the present value, $P$:

$$100000 = Pe^{0.10(10)}$$
$$P = \$36{,}787.94$$

We solve for the income stream $S$:

$$36{,}787.94 = \int_0^{10} S e^{-0.10t}\, dt$$
$$36{,}787.94 = S\int_0^{10} e^{-0.10t}\, dt$$
$$36{,}787.94 = S(6.321)$$
$$S = \$5820.00 \text{ per year.}$$

The income stream required is about \$5820 per year (or about \$112 per week).
(b) The present value is \$36,787.94. This is the amount that would have to be deposited now.

8. At any time $t$, the company receives income of $s(t) = 50e^{-t}$ thousands of dollars per year. Thus the present value is

$$\text{Present value} = \int_0^2 s(t) e^{-0.06t}\, dt$$
$$= \int_0^2 (50 e^{-t}) e^{-0.06t}\, dt$$
$$= \$41{,}508.$$

9. (a) Since the income stream is $7035 million per year and the interest rate is 8.5%,

$$\text{Present value} = \int_0^1 7035 e^{-0.085t} \, dt$$
$$= 6744.31 \text{ million dollars}.$$

The present value of Intel's profits over a one-year time period is about 6744 million dollars.

(b) The value at the end of one year is $6744.31 e^{0.085(1)} = 7342.64$, or about 7343, million dollars. This is the value, at the end of one year, of Intel's profits over a one-year time period.

10. January 1, 1996 through January 1, 2003 is a seven-year time period, and $t = 0$ corresponds to January 1, 1996, so the value on January 1, 1996 of the sales over this seven-year time period is

$$\text{Value on Jan. 1, 1996} = \int_0^7 1431 e^{0.134t} e^{-0.075t} \, dt$$
$$= \int_0^7 1431 e^{0.059t} \, dt$$
$$= 12{,}402.28 \quad \text{million dollars}.$$

The value, on January 1, 1996, of Harley-Davidson sales over the time period from January 1, 1996 through January 1, 2003 is about 12,402 million dollars.

11. January 1, 1995 through January 1, 2000 is a five-year time period, and $t = 0$ corresponds to January 1, 1995. The value of the net profit over this five-year time period is given by

$$\text{Present value} = \int_0^5 (28.5t + 265.75) e^{-0.02t} \, dt = 1597.84 \quad \text{million dollars}.$$

Therefore, the value of January 1, 2000 is given by

$$\text{Value 5 years later} = 1597.84 e^{0.02(5)} = 1765.89 \text{ million dollars}.$$

The value, on January 1, 2000, of Hershey's net profit over the time period from January 1, 1995 through January 1, 2000 is 1766 million dollars to the nearest million dollars.

12. (a) Since the rate at which revenue is generated is at least 10,600 million dollars per year, the present value of the revenue over a five-year time period is at least

$$\int_0^5 10{,}600 e^{-0.09t} \, dt = 10{,}600 \int_0^5 e^{-0.09t} \, dt = 42{,}679.35.$$

Since the rate at which revenue is generated is at most 12,600 million dollars per year, the present value of the revenue over a five-year time period is at most

$$\int_0^5 12{,}600 e^{-0.09t} \, dt = 12{,}600 \int_0^5 e^{-0.09t} \, dt = 50{,}732.06.$$

The present value of McDonald's revenue over a five year time period is between 42,679 and 50,732 million dollars.

(b) The present value of the revenue over a twenty-five year time period is at least

$$\int_0^{25} 10{,}600 e^{-0.09t} \, dt = 10{,}600 \int_0^{25} e^{-0.09t} \, dt = 105{,}364.09.$$

The present value of the revenue over a twenty-five year time period is at most

$$\int_0^{25} 12{,}600 e^{-0.09t} \, dt = 12{,}600 \int_0^{25} e^{-0.09t} \, dt = 125{,}244.11$$

The present value of McDonald's revenue over a twenty-five year time period is between 105,364 and 125,244 million dollars.

13. Find the time T at which
$$130{,}000 = \int_0^T 80{,}000 e^{-0.085 t}\, dt.$$
Trying a few values of $T$, we get $T \approx 1.75$. It takes approximately one year and nine months for the present value of the profit generated by the new machinery to equal the cost of the machinery.

14. (a) The present value of the revenue during the first year is the sum of the present value during the first six months (from $t = 0$ to $t = 1/2$) and the present value during the second six months (from $t = 1/2$ to $t = 1$).

$$\text{Present value of revenue during the first year} = \int_0^{1/2} (45{,}000 + 60{,}000t) e^{-0.07t}\, dt + \int_{1/2}^1 75{,}000 e^{-0.07t}\, dt$$
$$= 29{,}438.08 + 35{,}583.85 = 65{,}021.93.$$

The present value of the revenue earned by the machine during the first year of operation is about 65,022 dollars.

(b) The present value over a time interval $[0, T]$ with $T > 1/2$ is

$$\text{Present value during first } T \text{ years} = \int_0^{1/2} (45{,}000 + 60{,}000t) e^{-0.07t}\, dt + \int_{1/2}^T 75{,}000 e^{-0.07t}\, dt$$
$$= 29{,}438.08 + \int_{1/2}^T 75{,}000 e^{-0.07t}\, dt$$

If the present value of revenue equals the cost, then

$$150{,}000 = 29{,}438.08 + \int_{1/2}^T 75{,}000 e^{-0.07t}\, dt.$$

Solving for the integral, we get

$$120{,}561.92 = \int_{1/2}^T 75{,}000 e^{-0.07t}\, dt$$

Trying a few values for $T$ gives
$$T \approx 2.27.$$

It will take approximately 2.27 years for the present value of the revenue to equal the cost of the machine.

15. Price in future $= P(1 + 20\sqrt{t})$.
The present value $V$ of price satisfies $V = P(1 + 20\sqrt{t}) e^{-0.05 t}$.
We want to maximize $V$. To do so, we find the critical points of $V(t)$ for $t \geq 0$. (Recall that $\sqrt{t}$ is nondifferentiable at $t = 0$.)

$$\frac{dV}{dt} = P\left[\frac{20}{2\sqrt{t}} e^{-0.05t} + (1 + 20\sqrt{t})(-0.05 e^{-0.05t})\right]$$
$$= P e^{-0.05t} \left[\frac{10}{\sqrt{t}} - 0.05 - \sqrt{t}\right].$$

Setting $\dfrac{dV}{dt} = 0$ gives $\dfrac{10}{\sqrt{t}} - 0.05 - \sqrt{t} = 0$. Using a calculator, we find $t \approx 10$ years. Since $V'(t) > 0$ for $0 < t < 10$ and $V'(t) < 0$ for $t > 10$, we confirm that this is a maximum. Thus, the best time to sell the wine is in 10 years.

16. (a)

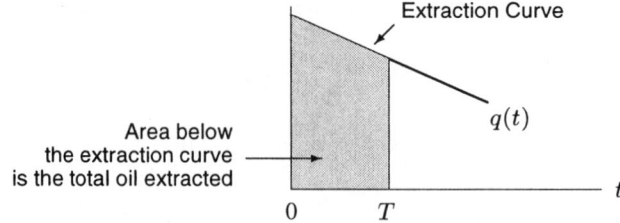

Suppose the oil extracted over the time period $[0, T]$ is $S$. (See above.) Since $q(t)$ is the rate of oil extraction, we have:
$$S = \int_0^T q(t)\, dt = \int_0^T (a - bt)\, dt = \int_0^T (10 - 0.1t)\, dt.$$

To calculate the time at which the oil is exhausted, set $S = 100$ and try different values of $T$. We find $T = 10.6$ gives

$$\int_0^{10.6} (10 - 0.1t)\, dt = 100,$$

so the oil is exhausted in 10.6 years.

(b) Suppose $p$ is the oil price, $C$ is the extraction cost per barrel, and $r$ is the interest rate. We have the present value of the profit as

$$\begin{aligned}
\text{Present value of profit} &= \int_0^T (p - C)q(t)e^{-rt}\, dt \\
&= \int_0^{10.6} (20 - 10)(10 - 0.1t)e^{-0.1t}\, dt \\
&= 624.9 \text{ million dollars.}
\end{aligned}$$

## Solutions for Section 6.4

1. (a) Absolute increase between 1988 and 1989 = $1275 - 813 = 462$ thousand. Between 1992 and 1993, absolute increase = $4820 - 3657 = 1163$ thousand.
   (b) Relative increase between 1988 and 1989 = $\frac{462}{813} \approx 56.8\%$. Relative increase between 1992 and 1993 = $\frac{1163}{3657} \approx 31.8\%$.

2. (a) The 14 is the relative change, as it is the absolute change divided by the initial figure.
   The absolute change is $144.7 - 127 = 17.7$ million.

$$\begin{aligned}
\text{The absolute rate of change} &= \frac{144.7 - 127}{2005 - 1995} \\
&= \frac{17.7}{10} \\
&= 1.77 \text{ million per year.}
\end{aligned}$$

$$\text{The relative rate of change} = \frac{\text{absolute rate of change}}{\text{initial figure}} = \frac{1.77}{127}$$
$$= 1.4\% \text{ per year.}$$

   (b) The 24 is a relative change.
   The 16.2 is an absolute change.
   The 9.1 is an absolute change.
   The 1.3 is an absolute change.
   The 28 is a relative change.

3. (a) The town is growing by 50 people per year. The population in year 1 is thus $1000 + 50 = 1050$. Continuing to add 50 people per year to the population of the town produces the data in Table 6.1.

   **TABLE 6.1**

   | Year | 0 | 1 | 2 | 3 | 4 | ... | 10 |
   |---|---|---|---|---|---|---|---|
   | Population | 1000 | 1050 | 1100 | 1150 | 1200 | ... | 1500 |

   (b) The town is growing at 5% per year. The population in year 1 is thus $1000 + 1000(0.05) = 1000(1.05) = 1050$. Continuing to multiply the town's population by 1.05 each year yields the data in Table 6.2.

   **TABLE 6.2**

   | Year | 0 | 1 | 2 | 3 | 4 | ... | 10 |
   |---|---|---|---|---|---|---|---|
   | Population | 1000 | 1050 | 1103 | 1158 | 1216 | ... | 1629 |

**274** CHAPTER SIX /SOLUTIONS

4. (a) If population is decreasing at 100 per hour,

$$\text{Population} = 4000 - 100(\text{Number of hours}).$$

So

$$P = 4000 - 100t.$$

(b) If population is decreasing at 5% per hour, then it is multiplied by $(1 - 0.05) = 0.95$ for each additional hour. This means

$$P = 4000(1 - 0.05)^t = 4000(0.95)^t.$$

The linear function $(P = 4000 - 100t)$ reaches zero first.

5. The relative birth rate is $\frac{30}{1000} = 0.03$ and the relative death rate is $\frac{20}{1000} = 0.02$. The relative rate of growth is $0.03 - 0.02 = 0.01$

6. (a) The population of Nicaragua is growing exponentially and we use $P = P_0 a^t$. In 1990 (at $t = 0$), we know $P = 3.6$, so $P_0 = 3.6$. Nicaragua is growing at a rate of 3.4% per year, so $a = 1.034$. We have

$$P = 3.6(1.034)^t.$$

(b) We first find the average rate of change between $t = 0$ (1990) and $t = 1$ (1991). At $t = 0$, we know $P = 3.6$ million. At $t = 1$, we find that $P = 3.6(1.034)^t = 3.7224$ million. We have

$$\text{Average rate of change} = \frac{P(1) - P(0)}{1 - 0} = \frac{3.7224 - 3.6}{1} = 0.1224 \text{ people/yr.}$$

Between 1990 and 1991, the population of Nicaragua grew by 0.1224 million people per year, or 122,400 people per year.

When $t = 2$ (1992), we have $P = 3.6(1.034)^2 = 3.8490$. Between 1991 and 1992, we have

$$\text{Average rate of change} = \frac{P(2) - P(1)}{2 - 1} = \frac{3.8490 - 3.7224}{1} = 0.1266 \text{ people/yr.}$$

Between 1990 and 1991, the population of Nicaragua grew by 0.1266 million people per year, or 126,600 people per year. Notice that the population increase was greater between 1991 and 1992. This is because the population was greater in 1991 than in 1990 and so there were more people to have babies.

(c) The relative rate of change between 1990 and 1991 is the absolute rate of change divided by the population in 1990, which is $0.1224/3.6 = 0.034$. The relative rate of change is 3.4% per year, as we expected. Between 1991 and 1992, the relative rate of change is $0.1266/3.7224 = 0.034$. Over small intervals, the relative rate of change will always be approximately 3.4% per year.

7. (a) Since the absolute growth rate for one year periods is constant, the function is linear, and the formula is

$$P = 100 + 10t.$$

(b) Since the relative growth rate for one year periods is constant, the function is exponential, and the formula is

$$P = 100(1.10)^t.$$

(c) See Figure 6.18.

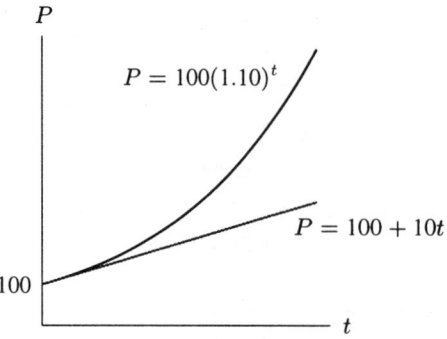

*Figure 6.18*: Two kinds of population growth

8. (a) We compute the left- and right-hand sums:

$$\text{Left sum} = -0.002 - 0.006 - 0.016 \ldots - 0.046 = -0.272$$

and

$$\text{Right sum} = -0.006 - 0.016 \ldots - 0.046 - 0.037 = -0.307,$$

and average these to get our best estimate:

$$\text{Average} = \frac{-0.272 + (-0.307)}{2} = -0.290.$$

We estimate that:

$$\int_{1987}^{1996} \frac{P'(t)}{P(t)} \, dt \approx -0.290.$$

(b) The integral found in part (a) is the change is $\ln P(t)$, so we have

$$\ln P(1996) - \ln P(1987) = \int_{1987}^{1996} \frac{P'(t)}{P(t)} \, dt = -0.290$$

$$\ln \left( \frac{P(1996)}{P(1987)} \right) = -0.290$$

$$\frac{P(1996)}{P(1987)} = e^{-0.290} = 0.749.$$

Burglaries went down by a factor of 0.749 during this time, which means that the number of burglaries in 1996 was 0.749 times the number of burglaries in 1987. This is a decrease of about 25.1% during this period.

9. The relative growth rate is a constant $0.02 = 2\%$. The change in $\ln P(t)$ is the area under the curve which is $10(0.02) = 0.2$. So

$$\ln P(10) - \ln P(0) = \int_0^{10} \frac{P'(t)}{P(t)} \, dt = 0.2$$

$$\ln \left( \frac{P(10)}{P(0)} \right) = 0.2$$

$$\frac{P(10)}{P(0)} = e^{0.2} \approx 1.22.$$

The population has increased by about 22% over the 10-year period.

Another way of looking at this problem is to say that since $P(t)$ is growing at a constant 2% rate, it is growing exponentially, so

$$P(t) = P(0)e^{0.02t}.$$

Substituting $t = 10$ gives the same result as before:

$$\frac{P(10)}{P(0)} = e^{0.02(10)} = e^{0.2} \approx 1.22.$$

10. The change in $\ln P$ is the area under the curve, which is $\frac{1}{2} \cdot 10 \cdot (0.08) = 0.4$. So

$$\ln P(10) - \ln P(0) = \int_0^{10} \frac{P'(t)}{P(t)} \, dt = 0.4$$

$$\ln \left( \frac{P(10)}{P(0)} \right) = 0.4$$

$$\frac{P(10)}{P(0)} = e^{0.4} \approx 1.49.$$

The population has increased by about 49% over the 10-year period.

11. The area between the graph of the relative growth rate and the $t$-axis is 0 (because the areas above and below are equal). Since the change in $\ln P(t)$ is the area under the curve

$$\ln P(10) - \ln P(10) = \int_0^{10} \frac{P'(t)}{P(t)} dt = 0$$

$$\ln\left(\frac{P(10)}{P(0)}\right) = 0$$

$$\frac{P(10)}{P(0)} = e^0 = 1$$

$$P(10) = P(0).$$

The population at $t = 0$ and at $t = 10$ are equal.

12. The area of the triangle is $\frac{1}{2} \cdot 10 \cdot (0.04) = 0.2$. The change in $\ln P(t)$ is $\int_0^{10} \frac{P'(t)}{P(t)} dt$, which is the negative of the area, or $-0.2$. So

$$\ln P(10) - \ln P(0) = \int_0^{10} \frac{P'(t)}{P(t)} dt = -0.2$$

$$\ln\left(\frac{P(10)}{P(0)}\right) = -0.2$$

$$\frac{P(10)}{P(0)} = e^{-0.2} \approx 0.82.$$

The population has decreased by about 18% over the 10-year period.

13. Although the relative growth rate is decreasing, it is everywhere positive, so $f$ is an increasing function for $0 \leq t \leq 10$.

14. Since the relative growth rate is positive for $0 \leq t < 5$ and negative for $5 < t \leq 10$, the function $f$ is increasing for $0 \leq t < 5$ and decreasing for $5 < t \leq 10$.

15. Although the relative growth rate is increasing, it is negative everywhere, so $f$ is a decreasing function for $0 \leq t \leq 10$.

16. (a) Since $f(x) = x^n$, we have $f'(x) = nx^{n-1}$, so

$$\text{Relative growth rate} = \frac{f'(x)}{f(x)} = \frac{nx^{n-1}}{x^n} = \frac{n}{x}.$$

For $x > 0$, the relative growth rate is decreasing to zero.

(b) The relative growth rate of $g(x) = e^{kx}$ is the constant $k$ because

$$\text{Relative growth rate} = \frac{g'(x)}{g(x)} = \frac{ke^{kx}}{e^{kx}} = k.$$

Since $k$ is a positive constant, the relative growth rate of $f(x)$ will, for large $x$, be smaller than $k$. So, for large $x$, the exponential function is growing faster than any power function.

## Solutions for Section 6.5

1. $5x$
2. $\frac{5}{2}x^2$
3. $\frac{1}{3}x^3$
4. $\frac{1}{3}t^3 + \frac{1}{2}t^2$
5. $\frac{x^5}{5}$.
6. $\frac{t^8}{8} + \frac{t^4}{4}$.
7. $\frac{5q^3}{3}$.
8. $6(\frac{x^4}{4}) + 4x = \frac{3x^4}{2} + 4x$.
9. $\frac{2}{3}z^{\frac{3}{2}}$
10. $\ln|z|$

11. $-\dfrac{1}{t}$

12. $-\dfrac{1}{2z^2}$

13. $t^3 + \dfrac{7t^2}{2} + t.$

14. $10x + 8(\dfrac{x^4}{4}) = 10x + 2x^4.$

15. $\dfrac{x^3}{3} - 6(\dfrac{x^2}{2}) + 17x = \dfrac{x^3}{3} - 3x^2 + 17x.$

16. $\dfrac{e^{-3t}}{-3} = \dfrac{-e^{-3t}}{3}.$

17. $x^3 + 5x.$

18. $2x^3 - 4x^2 + 3x.$

19. $\tfrac{2}{3}t^3 + \tfrac{3}{4}t^4 + \tfrac{4}{5}t^5$

20. $\dfrac{t^4}{4} - \dfrac{t^3}{6} - \dfrac{t^2}{2}$

21. $\dfrac{y^5}{5} + \ln|y|$

22. $\dfrac{5}{2}x^2 - \dfrac{2}{3}x^{\frac{3}{2}}$

23. $\sin t$

24. $-\cos t$

25. $f(x) = 3$, so $F(x) = 3x + C$. $F(0) = 0$ implies that $3 \cdot 0 + C = 0$, so $C = 0$. Thus $F(x) = 3x$ is the only possibility.

26. $f(x) = 2x$, so $F(x) = x^2 + C$. $F(0) = 0$ implies that $0^2 + C = 0$, so $C = 0$. Thus $F(x) = x^2$ is the only possibility.

27. $f(x) = -7x$, so $F(x) = \dfrac{-7x^2}{2} + C$. $F(0) = 0$ implies that $-\tfrac{7}{2} \cdot 0^2 + C = 0$, so $C = 0$. Thus $F(x) = -7x^2/2$ is the only possibility.

28. $f(x) = \tfrac{1}{4}x$, so $F(x) = \dfrac{x^2}{8} + C$. $F(0) = 0$ implies that $\tfrac{1}{8} \cdot 0^2 + C = 0$, so $C = 0$. Thus $F(x) = x^2/8$ is the only possibility.

29. $f(x) = x^2$, so $F(x) = \dfrac{x^3}{3} + C$. $F(0) = 0$ implies that $\dfrac{0^3}{3} + C = 0$, so $C = 0$. Thus $F(x) = \dfrac{x^3}{3}$ is the only possibility.

30. $f(x) = x^{1/2}$, so $F(x) = \tfrac{2}{3}x^{3/2} + C$. $F(0) = 0$ implies that $\tfrac{2}{3} \cdot 0^{3/2} + C = 0$, so $C = 0$. Thus $F(x) = \tfrac{2}{3}x^{3/2}$ is the only possibility.

31. $f(x) = 2 + 4x + 5x^2$, so $F(x) = 2x + 2x^2 + \tfrac{5}{3}x^3 + C$. $F(0) = 0$ implies that $C = 0$. Thus $F(x) = 2x + 2x^2 + \tfrac{5}{3}x^3$ is the only possibility.

32. $f(x) = \sin x$, so $F(x) = -\cos x + C$. $F(0) = 0$ implies that $-\cos 0 + C = 0$, so $C = 1$. Thus $F(x) = -\cos x + 1$ is the only possibility.

33. Since $\dfrac{d}{dx}(e^x) = e^x$, we take $F(x) = e^x + C$. Now

$$F(0) = e^0 + C = 1 + C = 0,$$

so

$$C = -1$$

and

$$F(x) = e^x - 1.$$

34. $\dfrac{3x^2}{2} + C$

35. $2t^2 + 7t + C$

36. $4t^2 + 3t + C.$

37. $2x^3 + C.$

38. $\dfrac{t^{13}}{13} + C.$

39. $\dfrac{x^4}{4} - \dfrac{x^2}{2} + C.$

40. $\dfrac{x^3}{3} + x + C.$

41. $\dfrac{x^4}{4} + 2x^2 + 8x + C.$

42. $5e^z + C$

43. $\dfrac{e^{2t}}{2} + C.$

44. $\dfrac{x^2}{2} + 2x^{1/2} + C$

45. $\dfrac{x^3}{3} + \ln|x| + C.$

**278** CHAPTER SIX /SOLUTIONS

46. Since $\frac{d}{dx}(e^{-3t}) = -3e^{-3t}$, we have

$$\int e^{-3t}\, dt = -\frac{1}{3}e^{-3t} + C.$$

47. $\sin\theta + C$

48. $-\cos t + C$

## Solutions for Section 6.6

1. If $F'(t) = t^3$, then $F(t) = \frac{t^4}{4}$. By the Fundamental Theorem, we have

$$\int_0^3 t^3\, dt = F(3) - F(0) = \left.\frac{t^4}{4}\right|_0^3 = \frac{3^4}{4} - \frac{0}{4} = \frac{81}{4}.$$

2. Since $F'(x) = 3x^2$, we take $F(x) = x^3$. Then

$$\int_0^5 3x^2\, dx = F(5) - F(0)$$
$$= 5^3 - 0^3$$
$$= 125.$$

3. If $F'(x) = 6x^2$, then $F(x) = 2x^3$. By the Fundamental Theorem, we have

$$\int_1^3 6x^2\, dx = \left.2x^3\right|_1^3 = 2(27) - 2(1) = 54 - 2 = 52.$$

4. Since $F'(t) = 5t^3$, we take $F(t) = \frac{5}{4}t^4$. Then

$$\int_1^2 5t^3\, dt = F(2) - F(1)$$
$$= \frac{5}{4}(2^4) - \frac{5}{4}(1^4)$$
$$= \frac{5}{4}\cdot 16 - \frac{5}{4}$$
$$= \frac{75}{4}$$

5. Since $F'(y) = y^2 + y^4$, we take $F(y) = \frac{y^3}{3} + \frac{y^5}{5}$. Then

$$\int_0^1 (y^2 + y^4)\, dy = F(3) - F(0)$$
$$= \left(\frac{1^3}{3} + \frac{1^5}{5}\right) - \left(\frac{0^3}{3} + \frac{0^5}{5}\right)$$
$$= \frac{1}{3} + \frac{1}{5} = \frac{8}{15}.$$

6. Since $F'(x) = \sqrt{x}$, we take $F(x) = \dfrac{x^{3/2}}{3/2} = \dfrac{2}{3}x^{3/2}$. Then

$$\int_4^9 \sqrt{x}\, dx = F(9) - F(4)$$
$$= \frac{2}{3} \cdot 9^{3/2} - \frac{2}{3} \cdot 4^{3/2}$$
$$= \frac{2}{3} \cdot 27 - \frac{2}{3} \cdot 8$$
$$= \frac{38}{3}.$$

7. If $f(x) = 1/x$, then $F(x) = \ln|x|$ (since $\dfrac{d}{dx}\ln|x| = \dfrac{1}{x}$). By the Fundamental Theorem, we have

$$\int_1^2 \frac{1}{x}\, dx = \ln|x|\Big|_1^2 = \ln 2 - \ln 1 = \ln 2.$$

8. $\displaystyle\int_0^2 \left(\frac{x^3}{3} + 2x\right) dx = \left(\frac{x^4}{12} + x^2\right)\Big|_0^2 = \frac{4}{3} + 4 = 16/3 \approx 5.333.$

9. $\displaystyle\int_{-3}^{-1} \frac{2}{r^3}\, dr = -r^{-2}\Big|_{-3}^{-1} = -1 + \frac{1}{9} = -8/9 \approx -0.889.$

10. $\displaystyle\int_0^1 2e^x\, dx = 2e^x\Big|_0^1 = 2e - 2 \approx 3.437.$

11. If $f(t) = e^{-0.2t}$, then $F(t) = -5e^{-0.2t}$. (This can be verified by observing that $\dfrac{d}{dt}(-5e^{-0.2t}) = e^{-0.2t}$.) By the Fundamental Theorem, we have

$$\int_0^1 e^{-0.2t}\, dt = (-5e^{-0.2t})\Big|_0^1 = -5(e^{-0.2}) - (-5)(1) = 5 - 5e^{-0.2} \approx 0.906.$$

12. If $f(x) = x^2 + 1$, then $F(x) = \dfrac{x^3}{3} + x$. By the Fundamental Theorem, we have

$$\int_0^2 (x^2 + 1)\, dx = \left(\frac{x^3}{3} + x\right)\Big|_0^2 = \frac{2^3}{3} + 2 - (0 + 0) = \frac{8}{3} + 2 = \frac{14}{3}.$$

13. $\displaystyle\int_0^1 \sin\theta\, d\theta = -\cos\theta\Big|_0^1 = 1 - \cos 1 \approx 0.460.$

14. If $F'(t) = \cos t$, we can take $F(t) = \sin t$, so

$$\int_{-1}^1 \cos t\, dt = \sin t\Big|_{-1}^1 = \sin 1 - \sin(-1).$$

Since $\sin(-1) = -\sin 1$, we can simplify the answer and write

$$\int_{-1}^1 \cos t\, dt = 2\sin 1$$

15. $\displaystyle\int_0^{\pi/4} (\sin t + \cos t)\, dt = (-\cos t + \sin t)\Big|_0^{\pi/4} = \left(-\frac{\sqrt{2}}{2} + \frac{\sqrt{2}}{2}\right) - (-1 + 0) = 1.$

16. We can get a rough estimate for the area by averaging the left and right hand sums, using 3 rectangles. See Figure 6.19. We have

$$\text{LHS} = f(0)\Delta t + f(1)\Delta t + f(2)\Delta t$$
$$= (0)^3(1) + (1)^3(1) + (2)^3(1) = 9$$
$$\text{RHS} = f(1)\Delta t + f(2)\Delta t + f(3)\Delta t$$
$$= (1)^3(1) + (2)^3(1) + (3)^3(1) = 36$$

The average of the two estimates is $\frac{9+36}{2} = 22.5$ which is close to 20.25, the result of Problem 1.

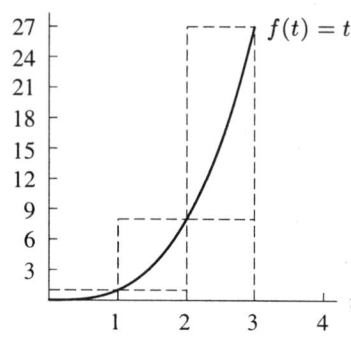
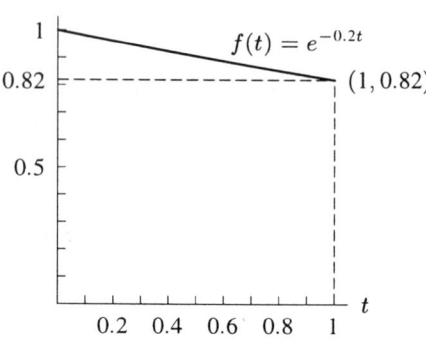

**Figure 6.19**  **Figure 6.20**

17. Figure 6.20 shows the graph of the function $f(t) = e^{-0.2t}$ on the interval $0 \le t \le 1$. To make a rough estimate of the area under the curve we will do two things. First, $f(1) \approx 0.82$, so we look at the rectangle bounded by the $t$- and $y$-axes, the line $y = 0.82$, and the line $t = 1$. The area of this rectangle is $1 \times (0.82) = 0.82$. We can also observe that the graph of the function closely approximates a straight line on the interval $0 \le t \le 1$. Therefore, we have a triangle bounded by the $y$-axis, the line $y = 0.82$, and the function $f(t) = e^{-0.2t}$, which is approximately a straight line. The area of this triangle is $\frac{1}{2}(1 - .82)(1) = \frac{1}{2}(.18) = 0.09$. Our approximation for the total area is $0.82 + 0.09 = 0.91$. (In Problem 11 we obtained a result of $\approx 0.91$ when we evaluated the integral, so our approximation here is quite accurate.)

18. The integral which represents the area under this curve is

$$\text{Area} = \int_0^2 (6x^2 + 1)\,dx.$$

Since $\frac{d}{dx}(2x^3 + x) = 6x^2 + 1$, we can evaluate the definite integral:

$$\int_0^2 (6x^2 + 1)\,dx = (2x^3 + x)\bigg|_0^2 = 2(2^3) + 2 - (2(0) + 0) = 16 + 2 = 18.$$

19. One antiderivative of $f(x) = e^{0.5x}$ is $F(x) = 2e^{0.5x}$. Thus, the definite integral of $f(x)$ on the interval $0 \le x \le 3$ is

$$\int_0^3 e^{0.5x}\,dx = F(3) - F(0) = 2e^{0.5x}\bigg|_0^3.$$

The average value of a function on a given interval is the definite integral over that interval divided by the length of the interval:

$$\text{Average value} = \left(\frac{1}{3-0}\right) \cdot \left(\int_0^3 e^{0.5x}\,dx\right) = \frac{1}{3}\left(2e^{0.5x}\bigg|_0^3\right) = \frac{1}{3}(2e^{1.5} - 2e^0) \approx 2.32.$$

From the graph of $y = e^{0.5x}$ in Figure 6.21 we see that an average value of 2.32 on the interval $0 \le x \le 3$ does make sense.

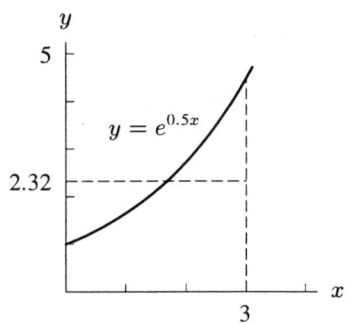

**Figure 6.21**

20. Since $y = x^3 - x = x(x-1)(x+1)$, the graph crosses the axis at the three points shown in Figure 6.22. The two regions have the same area (by symmetry). Since the graph is below the axis for $0 < x < 1$, we have

$$\text{Area} = 2\left(-\int_0^1 (x^3 - x)\, dx\right)$$

$$= -2\left[\frac{x^4}{4} - \frac{x^2}{2}\right]_0^1 = -2\left(\frac{1}{4} - \frac{1}{2}\right) = \frac{1}{2}.$$

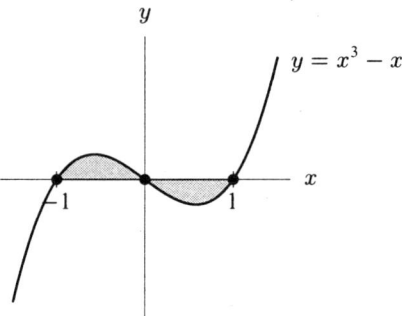

**Figure 6.22**

21. (a) The definite integral which would measure the total quantity of oil consumed would be

$$\int_0^5 (32e^{0.05t})\, dt.$$

(b) The Fundamental Theorem of Calculus states that

$$\int_a^b f(t)\, dt = F(b) - F(a)$$

provided that $F'(t) = f(t)$. To apply this, we need to find $F(t)$ such that $F'(t) = 32e^{0.05t}$. The function $F(t) = \frac{32}{0.05}e^{0.05t} = 640e^{0.05t}$ will satisfy this requirement (since $\frac{d}{dt}\left(\frac{32}{0.05}e^{0.05t}\right) = (0.05)\frac{32}{0.05}e^{0.05t} = 32e^{0.05t}$). Therefore, the total amount of oil consumed equals

$$\int_0^5 (32e^{0.05t})\, dt = F(5) - F(0) = 640e^{0.05t}\Big|_0^5 = 640(e^{0.25} - e^0) \approx 182.$$

Thus, approximately 182 billion barrels of oil were consumed between 1990 and 1995.

22. (a) The graph of $y = e^{-x^2}$ is in Figure 6.23. The integral $\int_{-\infty}^{\infty} e^{-x^2} dx$ represents the entire area under the curve, which is shaded.

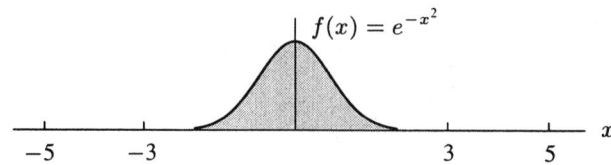

Figure 6.23

(b) Using a calculator or computer, we see that
$$\int_{-1}^{1} e^{-x^2} dx = 1.494, \quad \int_{-2}^{2} e^{-x^2} dx = 1.764, \quad \int_{-3}^{3} e^{-x^2} dx = 1.772, \quad \int_{-5}^{5} e^{-x^2} dx = 1.772$$

(c) From part (b), we see that as we extend the limits of integration, the area appears to get closer and closer to about 1.772. We estimate that
$$\int_{-\infty}^{\infty} e^{-x^2} dx = 1.772$$

23. Figure 6.24 shows the graphs of $y = 1/x^2$ and $y = 1/x^3$. We see that $\int_{1}^{\infty} \frac{1}{x^2} dx$ is larger, since the area under $1/x^2$ is larger than the area under $1/x^3$.

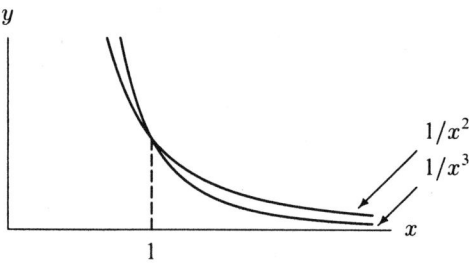

Figure 6.24

24. (a) We sketch $f(x) = xe^{-x}$; see Figure 6.25. The shaded area to the right of the $y$-axis represents the integral $\int_{0}^{\infty} xe^{-x} dx$.

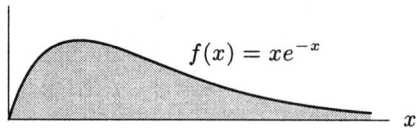

Figure 6.25

(b) Using a calculator or computer, we obtain
$$\int_{0}^{5} xe^{-x} dx = 0.9596 \quad \int_{0}^{10} xe^{-x} dx = 0.9995 \quad \int_{0}^{20} xe^{-x} dx = 0.99999996.$$

(c) The answers to part (b) suggest that the integral converges to 1.

25. (a) Evaluating the integrals with a calculator gives

$$\int_0^{10} xe^{-x/10}\, dx = 26.42$$

$$\int_0^{50} xe^{-x/10}\, dx = 95.96$$

$$\int_0^{100} xe^{-x/10}\, dx = 99.95$$

$$\int_0^{200} xe^{-x/10}\, dx = 100.00$$

(b) The results of part (a) suggest that

$$\int_0^\infty xe^{-x/10}\, dx \approx 100$$

26. (a) An antiderivative of $F'(x) = \frac{1}{x^2}$ is $F(x) = -\frac{1}{x}$ $\left(\text{since } \frac{d}{dx}\left(\frac{-1}{x}\right) = \frac{1}{x^2}\right)$. So by the Fundamental Theorem we have:

$$\int_1^b \frac{1}{x^2}\, dx = -\frac{1}{x}\Big|_1^b = -\frac{1}{b} + 1.$$

(b) Taking a limit, we have

$$\lim_{b\to\infty}\left(-\frac{1}{b} + 1\right) = 0 + 1 = 1.$$

Since the limit is 1, we know that

$$\lim_{b\to\infty}\int_1^b \frac{1}{x^2}\, dx = 1.$$

So the improper integral converges to 1:

$$\int_1^\infty \frac{1}{x^2}\, dx = 1.$$

27. (a) A calculator or computer gives

$$\int_1^{100} \frac{1}{\sqrt{x}}\, dx = 18 \quad \int_1^{1000} \frac{1}{\sqrt{x}}\, dx = 61.2 \quad \int_1^{10000} \frac{1}{\sqrt{x}}\, dx = 198.$$

These values do not seem to be converging.

(b) An antiderivative of $F'(x) = \frac{1}{\sqrt{x}}$ is $F(x) = 2\sqrt{x}$ $\left(\text{since } \frac{d}{dx}(2\sqrt{x}) = \frac{1}{\sqrt{x}}\right)$. So, by the Fundamental Theorem, we have

$$\int_1^b \frac{1}{\sqrt{x}}\, dx = 2\sqrt{x}\Big|_1^b = 2\sqrt{b} - 2\sqrt{1} = 2\sqrt{b} - 2.$$

(c) The limit of $2\sqrt{b} - 2$ as $b \to \infty$ does not exist, as $\sqrt{b}$ grows without bound. Therefore

$$\lim_{b\to\infty}\int_1^b \frac{1}{\sqrt{x}}\, dx = \lim_{b\to\infty}(2\sqrt{b} - 2) \quad \text{does not exist.}$$

So the improper integral $\int_1^\infty \frac{1}{\sqrt{x}}\, dx$ does not converge.

28. (a) The total number of people that get sick is the integral of the rate. The epidemic starts at $t = 0$. Since the rate is positive for all $t$, we use $\infty$ for the upper limit of integration.

$$\text{Total number getting sick} = \int_0^\infty \left(1000te^{-0.5t}\right) dt$$

**284** CHAPTER SIX /SOLUTIONS

(b) The graph of $r = 1000te^{-0.5t}$ is shown in Figure 6.26. The shaded area represents the total number of people who get sick.

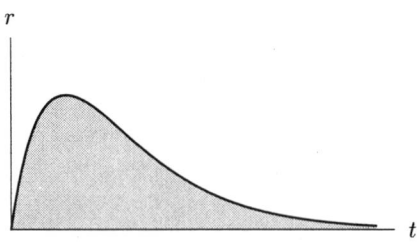

**Figure 6.26**

29. (a) No, it is not reached since
$$\text{Total number of rabbits} = \int_1^\infty \frac{1}{t^2} dt = 1.$$
Thus, the total number of rabbits is 1000.
(b) Yes, since $\int_1^\infty t\, dt$ does not converge to a finite value, which means that infinitely many rabbits could be produced, and therefore 1 million is certainly reached.
(c) Yes, since $\int_1^\infty \frac{1}{\sqrt{t}} dt$ does not converge to a finite value.

## Solutions for Section 6.7

1. Since $F(0) = 0$, $F(b) = \int_0^b f(t)\, dt$. For each $b$ we determine $F(b)$ graphically as follows:
$F(0) = 0$
$F(1) = F(0) + \text{Area of } 1 \times 1 \text{ rectangle} = 0 + 1 = 1$
$F(2) = F(1) + \text{Area of triangle } (\frac{1}{2} \cdot 1 \cdot 1) = 1 + 0.5 = 1.5$
$F(3) = F(2) + \text{Negative of area of triangle} = 1.5 - 0.5 = 1$
$F(4) = F(3) + \text{Negative of area of rectangle} = 1 - 1 = 0$
$F(5) = F(4) + \text{Negative of area of rectangle} = 0 - 1 = -1$
$F(6) = F(5) + \text{Negative of area of triangle} = -1 - 0.5 = -1.5$
The graph of $F(t)$, for $0 \le t \le 6$, is shown in Figure 6.27.

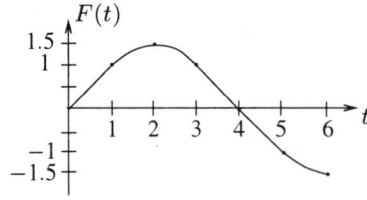

**Figure 6.27**

2. By the Fundamental Theorem,

$$F(1) = F(0) + \int_0^1 F'(t)\,dt$$
$$= 5 - 1.5 = 3.5$$
$$F(2) = F(1) + \int_1^2 F'(t)\,dt$$
$$= 3.5 - 1.5 = 2$$
$$F(3) = F(2) + \int_2^3 F'(t)\,dt$$
$$= 2 - 0.5 = 1.5$$
$$F(4) = F(3) + \int_3^4 F'(t)\,dt$$
$$= 1.5 + 0.5 = 2$$
$$F(5) = F(4) + \int_4^5 F'(t)\,dt$$
$$= 2 + 0.5 = 2.5$$

Thus, our table is as follows:

**TABLE 6.3**

| $t$ | 0 | 1 | 2 | 3 | 4 | 5 |
|---|---|---|---|---|---|---|
| $F(t)$ | 5 | 3.5 | 2 | 1.5 | 2 | 2.5 |

3. First, we observe that
   $g$ is increasing when $g'$ is positive, which is when $0 < x < 4$.
   $g$ is decreasing when $g'$ is negative, which is when $4 < x < 6$.
   Therefore, $x = 4$ is a local maximum. Table 6.4 shows the area between the curve and the $x$-axis for the intervals 0–1, 1–2, etc. It also shows the corresponding change in the value of $g$. These changes are used to compute the values of $g$ using the Fundamental Theorem of Calculus:

$$g(1) - g(0) = \int_0^1 g'(x)\,dx = \frac{1}{2}.$$

Since $g(0) = 0$,

$$g(1) = \frac{1}{2}.$$

Similarly,

$$g(2) - g(1) = \int_1^2 g'(x)\,dx = 1$$
$$g(2) = g(1) + 1 = \frac{3}{2}.$$

Continuing in this way gives the values of $g$ in Table 6.5.

**286** CHAPTER SIX /SOLUTIONS

**TABLE 6.4**

| Interval | Area | Total change in $g = \int_a^b g'(x)dx$ |
|---|---|---|
| 0–1 | 1/2 | 1/2 |
| 1–2 | 1 | 1 |
| 2–3 | 1 | 1 |
| 3–4 | 1/2 | 1/2 |
| 4–5 | 1/2 | −1/2 |
| 5–6 | 1/2 | −1/2 |

**TABLE 6.5**

| $x$ | $g(x)$ |
|---|---|
| 0 | 0 |
| 1 | 1/2 |
| 2 | 3/2 |
| 3 | 5/2 |
| 4 | 3 |
| 5 | 5/2 |
| 6 | 2 |

Notice: the graph of $g$ will be a straight line from 1 to 3 because $g'$ is horizontal there. Furthermore, the tangent line will be horizontal at $x = 4$, $x = 0$ and $x = 6$. The maximum is at $(4, 3)$. See Figure 6.28.

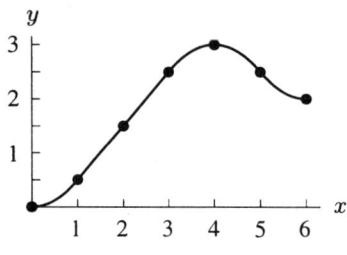

**Figure 6.28**

4. (a) $f(x)$ is increasing when $f'(x)$ is positive. $f'(x)$ is positive when $2 < x < 5$. So $f(x)$ is increasing when $2 < x < 5$.

   $f(x)$ is decreasing when $f'(x)$ is negative. $f'(x)$ is negative when $x < 2$ or $x > 5$. So $f(x)$ is decreasing when $x < 2$ or $x > 5$.

   A function has a local minimum at a point $x$ when its derivative is zero at that point, and when it decreases immediately before $x$ and increases immediately after $x$. $f'(2) = 0$, $f$ decreases to the left of 2, and $f$ increases immediately after 2, therefore $f(x)$ has a local minimum at $x = 2$.

   A function has a local maximum at a point $x$ when its derivative is zero at that point, and when it increases immediately before $x$ and decreases immediately after $x$. $f'(5) = 0$, $f$ increases before 5, and $f$ decreases after 5. Therefore $f(x)$ has a local maximum at $x = 5$.

   (b) Since we do not know any areas or vertical values, we can only sketch a rough graph. We start with the minimum and the maximum, then connect the graph between them. The graph could be more or less steep and further above or below the $x$-axis. See Figure 6.29.

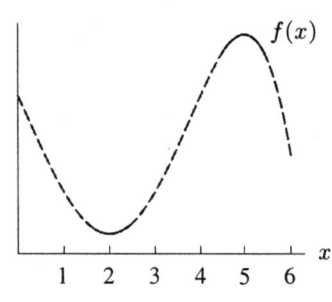

**Figure 6.29**

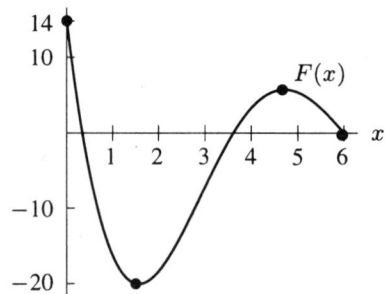

**Figure 6.30**

5. We see that

   $F$ decreases when $x < 1.5$ or $x > 4.67$, because $F'$ is negative there.
   $F$ increases when $1.5 < x < 4.67$, because $F'$ is positive there.

   So

   $F$ has a local minimum at $x = 1.5$.
   $F$ has a local maximum at $x = 4.67$.

   We have $F(0) = 14$. Since $F'$ is negative between 0 and 1.5, the Fundamental Theorem of Calculus gives us

   $$F(1.5) - F(0) = \int_0^{1.5} F'(x)\, dx = -34$$
   $$F(1.5) = 14 - 34 = -20.$$

   Similarly

   $$F(4.67) = F(1.5) + \int_{1.5}^{4.67} F'(x)\, dx = -20 + 25 = 5.$$
   $$F(6) = F(4.67) + \int_{4.67}^{6} F'(x)\, dx = 5 - 5 = 0.$$

   A graph of $F$ is in Figure 6.30. The local maximum is $(4.67, 5)$ and the local minimum is $(1.5, -20)$.

6. The areas given enable us to calculate the changes in the function $F$ as we move along the $t$-axis. Areas above the axis count positively and areas below the axis count negatively. We know that $F(0) = 3$, so

   $$F(2) - F(0) = \int_0^2 F'(t)\, dt = \begin{array}{c}\text{Area under } F' \\ 0 \le t \le 2\end{array} = 5$$

   Thus,
   $$F(2) = F(0) + 5 = 3 + 5 = 8.$$

   Similarly,
   $$F(5) - F(2) = \int_2^5 F'(t)\, dt = -16$$
   $$F(5) = F(2) - 16 = 8 - 16 = -8$$

   and
   $$F(6) = F(5) + \int_5^6 F'(t)\, dt = -8 + 10 = 2.$$

   A graph is shown in Figure 6.31.

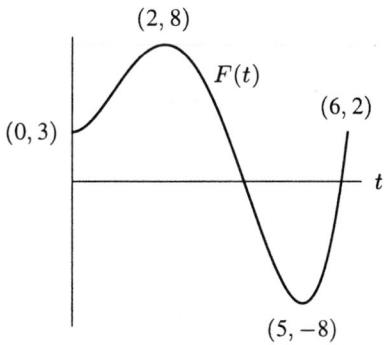

*Figure 6.31*

**288** CHAPTER SIX /SOLUTIONS

7.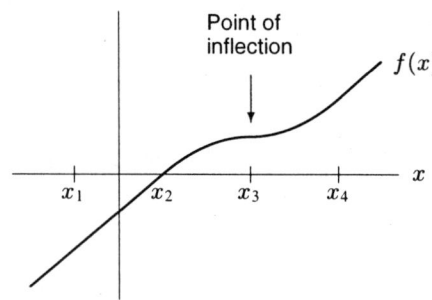

8.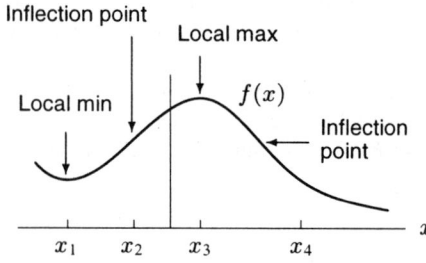

9. By the Fundamental Theorem,
$$f(1) - f(0) = \int_0^1 f'(x)\,dx,$$
Since $f'(x)$ is negative for $0 \leq x \leq 1$, this integral must be negative and so $f(1) < f(0)$.

10. First rewrite each of the quantities in terms of $f'$, since we have the graph of $f'$. If $A_1$ and $A_2$ are the positive areas shown in Figure 6.32:

$$f(3) - f(2) = \int_2^3 f'(t)\,dt = -A_1$$

$$f(4) - f(3) = \int_3^4 f'(t)\,dt = -A_2$$

$$\frac{f(4) - f(2)}{2} = \frac{1}{2}\int_2^4 f'(t)\,dt = -\frac{A_1 + A_2}{2}$$

Since Area $A_1 >$ Area $A_2$,
$$A_2 < \frac{A_1 + A_2}{2} < A_1$$

so
$$-A_1 < -\frac{A_1 + A_2}{2} < -A_2$$

and therefore
$$f(3) - f(2) < \frac{f(4) - f(2)}{2} < f(4) - f(3).$$

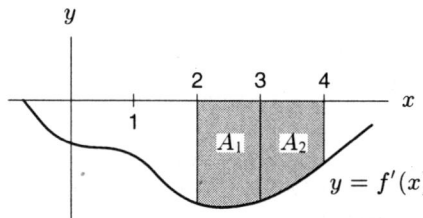

**Figure 6.32**

11.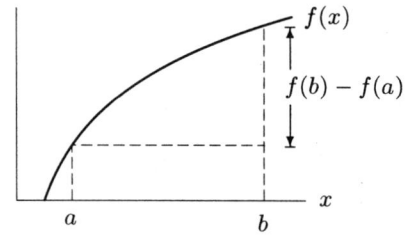

12.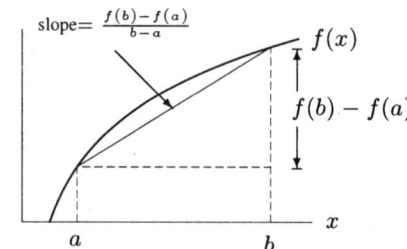

13.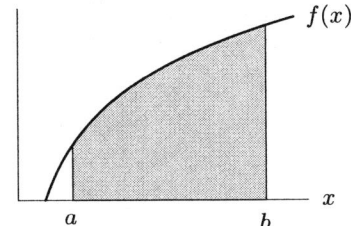

14.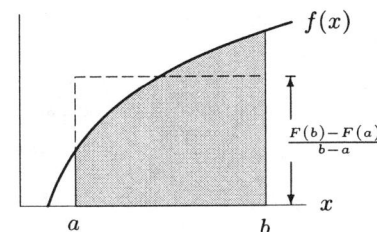

Note that we are using the interpretation of the definite integral as the length of the interval times the average value of the function on that interval, which we developed in Section 6.1.

## Solutions for Section 6.8

1. Since $p(x)$ is a density function, the area under the graph of $p(x)$ is 1, so
$$\text{Area} = \text{Base} \cdot \text{Height} = 15a = 1$$
$$a = \frac{1}{15}.$$

2. Since $p(x)$ is a density function, the area under the graph of $p(x)$ is 1, so
$$\text{Area} = \frac{1}{2}\text{Base} \cdot \text{Height} = \frac{1}{2} \cdot 10 \cdot a = 5a = 1$$
$$a = \frac{1}{5}.$$

3. Since $p(x)$ is a density function, the area under the graph of $p(x)$ is 1, so
$$\text{Area} = \frac{1}{2} \cdot \text{Base} \cdot \text{Height} = \frac{1}{2} \cdot 100 \cdot a = 50a = 1$$
$$a = \frac{1}{50}.$$

4. Since $p(x)$ is a density function,
$$\text{Area under graph} = \frac{1}{2} \cdot 50c = 25c = 1,$$
so $c = 1/25 = 0.04$.

5. Since $p(t)$ is a density function,
$$\text{Area under graph} = \frac{1}{2} \cdot c \cdot 0.01 = 0.005c = 1,$$
so $c = 1/0.005 = 200$.

6. Since $p(t)$ is a density function,
$$\text{Area under graph} = 50 \cdot 2c + 25 \cdot c = 125c = 1,$$
so $c = 1/125 = 0.008$.

7. Since $p(x) = cx$, we know $p(2) = 2c$. Since $p(x)$ is a density function,
$$\text{Area under graph} = \frac{1}{2} \cdot 2 \cdot 2c = 2c = 1,$$
so $c = 1/2 = 0.5$.

8. We use the fact that the area of a rectangle is Base × Height.

   (a) The fraction less than 5 meters high is the area to the left of 5, so
   $$\text{Fraction} = 5 \cdot 0.05 = 0.25.$$

   (b) The fraction above 6 meters high is the area to the right of 6, so
   $$\text{Fraction} = (20 - 6)0.05 = 0.7.$$

   (c) The fraction between 2 and 5 meters high is the area between 2 and 5, so
   $$\text{Fraction} = (5 - 2)0.05 = 0.15.$$

9. We use the fact that the area of a triangle is $\frac{1}{2} \cdot$ Base $\cdot$ Height. Since $p(x)$ is a line with slope $0.1/20 = 0.005$, its equation is
   $$p(x) = 0.005x.$$

   (a) The fraction less than 5 meters high is the area to the left of 5. Since $p(5) = 0.005(5) = 0.025$,
   $$\text{Fraction} = \frac{1}{2} \cdot 5(0.025) = 0.0625.$$

   (b) The fraction more than 6 meters high is the area to the right of 6. Since $p(6) = 0.005(6) = 0.03$,
   $$\text{Fraction} = 1 - (\text{Area to left of 6})$$
   $$= 1 - \frac{1}{2} \cdot 6(0.03) = 0.91.$$

   (c) Fraction between 2 and 5 meters high is area between 2 and 5. Since $p(2) = 0.005(2) = 0.01$,
   $$\text{Fraction} = (\text{Area to left of 5}) - (\text{Area to left of 2})$$
   $$= 0.0625 - \frac{1}{2} \cdot 2 \cdot (0.01) = 0.0525.$$

10. We use the fact that the area of a triangle is $\frac{1}{2} \cdot$ Base $\cdot$ Height. Since $p(x)$ is a line with slope $-0.1/20 = -0.005$ and vertical intercept 0.1, its equation is
    $$p(x) = 0.1 - 0.005x.$$

    (a) The fraction less than 5 meters high is the area to the left of 5. Since $p(5) = 0.1 - 0.005(5) = 0.075$,
    $$\text{Fraction} = 1 - (\text{Area to the right of 5})$$
    $$= 1 - \frac{1}{2} \cdot (20 - 5)0.075 = 0.4375.$$

    (b) The fraction more than 6 meters high is the area to the right of 6. Since $p(6) = 0.1 - 0.005(6) = 0.07$,
    $$\text{Fraction} = \frac{1}{2}(20 - 6)0.07 = 0.49.$$

    (c) The fraction between 2 meters and 5 meters high is the area between 2 and 5. Since $p(2) = 0.1 - 0.005(2) = 0.09$,
    $$\text{Fraction} = (\text{Area to the right of 2}) - (\text{Area to the right of 5})$$
    $$= \left(\frac{1}{2}(20-2)(0.09)\right) - \left(\frac{1}{2}(20-5)(0.075)\right)$$
    $$= 0.81 - 0.5625 = 0.2475$$

11. If $x$ is the yield in kg, the density function is a horizontal line at $p(x) = 1/100$ for $0 \le x \le 100$. See Figure 6.33.

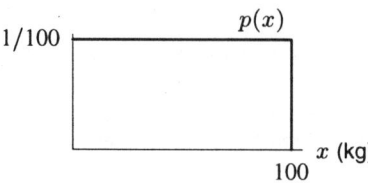

**Figure 6.33**

12. See Figure 6.34. Many other answers are possible.

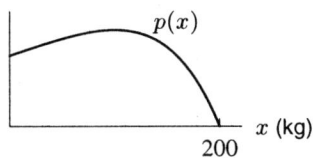

Figure 6.34

13. See Figure 6.35. Many other answers are possible.

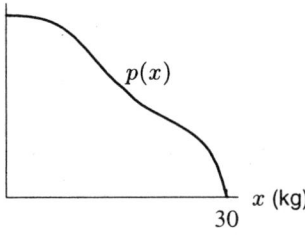

Figure 6.35

14. For a small interval $\Delta x$ around 68, the fraction of the population of American men with heights in this interval is about $(0.2)\Delta x$. For example, taking $\Delta x = 0.1$, we can say that approximately $(0.2)(0.1) = 0.02 = 2\%$ of American men have heights between 68 and 68.1 inches.

15. We can determine the fractions by estimating the area under the curve. Counting the squares for insects in the larval stage between 10 and 12 days we get 4.5 squares, with each square representing $(2) \cdot (3\%)$ giving a total of 27% of the insects in the larval stage between 10 and 12 days.

    Likewise we get 2 squares for the insects in the larval stage for less than 8 days, giving 12% of the insects in the larval stage for less than 8 days.

    Likewise we get 7.5 squares for the insects in the larval stage for more than 12 days, giving 45% of the insects in the larval stage for more than 12 days.

    Since the peak of the graph occurs between 12 and 13 days, the length of the larval stage is most likely to fall in this interval.

16. The fact that most of the area under the graph of the density function is concentrated in two humps, centered at 8 and 12 years, indicates that most of the population belong to one of two groups, those who leave school after finishing approximately 8 years and those who finish about 12 years. There is a smaller group of people who finish approximately 16 years of school.

    The percentage of adults who have completed less than ten years of school is equal to the area under the density function to the left of the vertical line at $t = 10$. (See Figure 6.36.) We know that the total area is 1, so we are estimating the percentage of the total area that lies in this shaded part shown in Figure 6.36. A rough estimate of this area is about 30%.

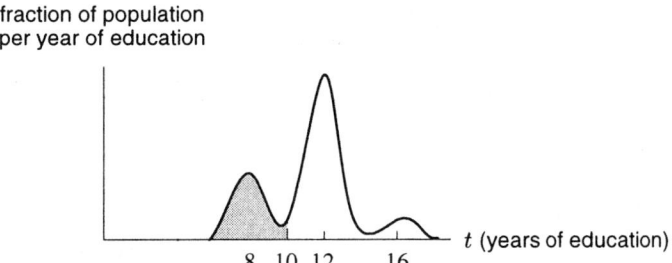

Figure 6.36: What percent has less than 10 years of education?

17. No. Though the density function has its maximum value at 50, this does not mean that a large fraction of the population receives scores near 50. The value $p(50)$ can not be interpreted as a probability. Probability corresponds to *area* under the graph of a density function. Most of the area in this case is in the broad hump covering the range $0 \leq x \leq 40$, very little in the peak around $x = 50$. Most people score in the range $0 \leq x \leq 40$.

18. (a) Most of the earth's surface is below sea level. Much of the earth's surface is either around 3 miles below sea level or exactly at sea level. It appears that essentially all of the surface is between 4 miles below sea level and 2 miles above sea level. Very little of the surface is around 1 mile below sea level.

    (b) The fraction below sea level corresponds to the area under the curve from $-4$ to $0$ divided by the total area under the curve. This appears to be about $\frac{3}{4}$.

19. (a) More insects die in the twelfth month than the first month because $P(t)$ is larger in the twelfth month. This means that the area $\int_{11}^{12} p(t)\, dt$ is larger than the area $\int_0^1 p(t)\, dt$.

    (b) We want to find the fraction dying within the first 6 months. Since $p(12) = 1/6$, we have $p(6) = 1/12$, so

    $$\text{Fraction living up to 6 months} = \int_0^6 p(t)\, dt$$
    $$= \text{Area from } t = 0 \text{ to } t = 6$$
    $$= \frac{1}{2} \cdot 6 \cdot \frac{1}{12} = \frac{1}{4}.$$

    So 1/4 of population lives no more than 6 months.

    (c) We can first find the fraction of insects who die within the first 9 months. Using $p(9) = \frac{1}{6} \cdot \frac{9}{12} = \frac{1}{8}$, we have

    $$\text{Fraction living up to 9 months} = \int_0^9 p(t)\, dt = \text{Area from } t = 0 \text{ to } t = 9$$
    $$= \frac{1}{2} \cdot 9 \cdot \frac{1}{8} = \frac{9}{16}.$$

    So the quantity we want is

    $$\text{Fraction living more than 9 months} = 1 - \frac{9}{16} = \frac{7}{16}.$$

20. (a) The total area under the graph must be 1, so

    $$\text{Area} = 5(0.01) + 5C = 1$$

    So

    $$5C = 1 - 0.05 = 0.95$$
    $$C = 0.19$$

    (b) The machine is more likely to break in its tenth year than first year. It is equally likely to break in its first year and second year.

    (c) Since $p(t)$ is a density function,

    $$\begin{array}{l}\text{Fraction of machines} \\ \text{lasting up to 2 years}\end{array} = \text{Area from 0 to 2} = 2(0.01) = 0.02.$$

    $$\begin{array}{l}\text{Fraction of machines} \\ \text{lasting between 5 and 7 years}\end{array} = \text{Area from 5 to 7} = 2(0.19) = 0.38.$$

    $$\begin{array}{l}\text{Fraction of machines} \\ \text{lasting between 3 and 6 years}\end{array} = \text{Area from 3 to 6}$$
    $$= \text{Area from 3 to 5} + \text{Area from 5 to 6}$$
    $$= 2(0.01) + 1(0.19) = 0.21.$$

## Solutions for Section 6.9

1.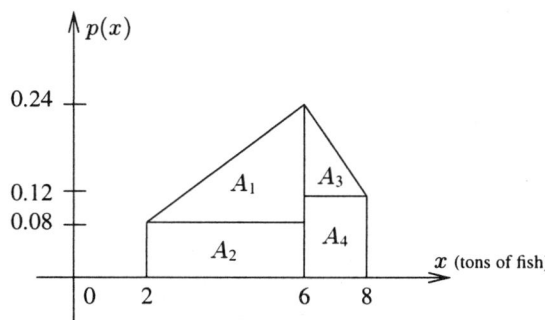

   Splitting the figure into four pieces, we see that

   $$\text{Area under the curve} = A_1 + A_2 + A_3 + A_4$$
   $$= \frac{1}{2}(0.16)4 + 4(0.08) + \frac{1}{2}(0.12)2 + 2(0.12)$$
   $$= 1.$$

   We expect the area to be 1, since $\int_{-\infty}^{\infty} p(x)\,dx = 1$ for any probability density function, and $p(x)$ is 0 except when $2 \leq x \leq 8$.

2. (a) The cumulative distribution function $P(t)$ is defined to be the fraction of patients who get in to see the doctor within $t$ hours. No one gets in to see the doctor in less than 0 minutes, so $P(0) = 0$. We saw in Example 2 part (c) that 60% of patients wait less than 1 hour, so $P(1) = 0.60$. We saw in part (b) of Example 2 that an additional 37.5% of patients get in to see the doctor within the second hour, so 97.5% of patients will see the doctor within 2 hours; $P(2) = 0.975$. Finally, all patients are admitted within 3 hours, so $P(3) = 1$. Notice also that $P(t) = 1$ for all values of $t$ greater than 3. A table of values for $P(t)$ is given in Table 6.6.

   **TABLE 6.6** *Cumulative distribution function for the density function in Example 2*

   | $t$ | 0 | 1 | 2 | 3 | 4 | ... |
   |---|---|---|---|---|---|---|
   | $P(t)$ | 0 | 0.60 | 0.975 | 1 | 1 | ... |

   (b) The graph is in Figure 6.37.

   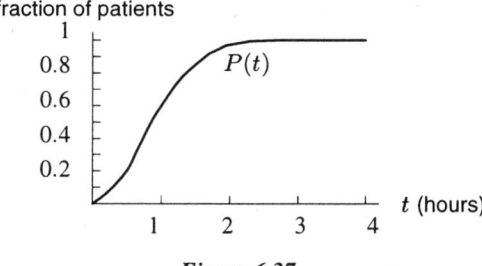

   **Figure 6.37**

3. (a) (i) The probability density function is (III).
       (ii) The cumulative distribution function is (VI).
   (b) (i) The probability density function is (I).
       (ii) The cumulative distribution function is (V).
   (c) (i) The probability density function is (IV).
       (ii) The cumulative distribution function is (II).

4.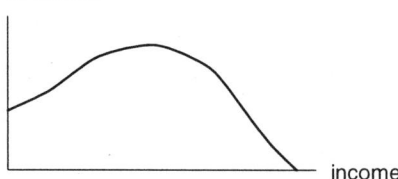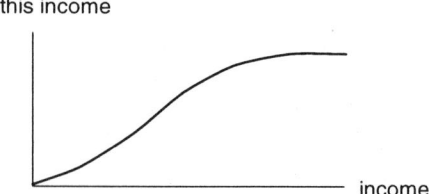

   Figure 6.38: Density function    Figure 6.39: Cumulative distribution function

5.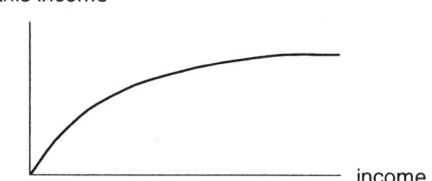

   Figure 6.40: Density function    Figure 6.41: Cumulative distribution function

6.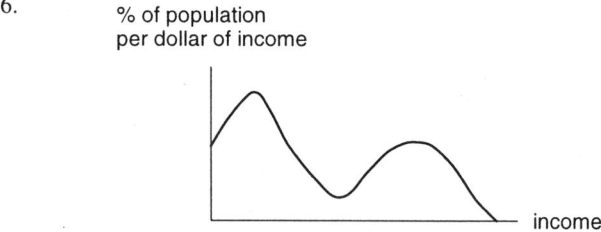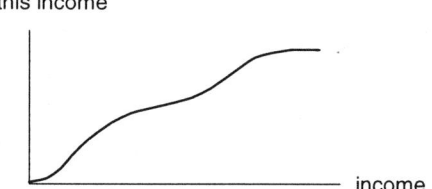

   Figure 6.42: Density function    Figure 6.43: Cumulative distribution function

7. (a) $F(7) = 0.6$ tells us that 60% of the trees in the forest have height 7 meters or less.
   (b) $F(7) > F(6)$. There are more trees of height less than 7 meters than trees of height less than 6 meters because every tree of height $\leq 6$ meters also has height $\leq 7$ meters.

8. The cumulative distribution function

$$P(t) = \int_0^t p(x)dx = \text{Area under graph of density function } p(x) \text{ for } 0 \leq x \leq t$$

$$= \text{Fraction of population who survive } t \text{ years or less after treatment}$$

$$= \text{Fraction of population who survive up to } t \text{ years after treatment.}$$

9. (a) The first item is sold at the point at which the graph is first greater than zero. Thus the first item is sold at $t = 30$ or January 30. The last item is sold at the $t$ value at which the function is first equal to 100%. Thus the last item is sold at $t = 240$ or August 28, unless its a leap year.
   (b) Looking at the graph at $t = 121$ we see that roughly 65% of the inventory has been sold by May 1.

(c) The percent of the inventory sold during May and June is the difference between the percent of the inventory sold on the last day of June and the percent of the inventory sold on the first day of May. Thus, the percent of the inventory sold during May and June is roughly 25%.

(d) The percent of the inventory left after half a year is

$$100 - \text{(percent inventory sold after half year)}.$$

Thus, roughly 10% of the inventory is left after half a year.

(e) The items probably went on sale on day 100 and were on sale until day 120. Roughly from April 10 until April 30.

10.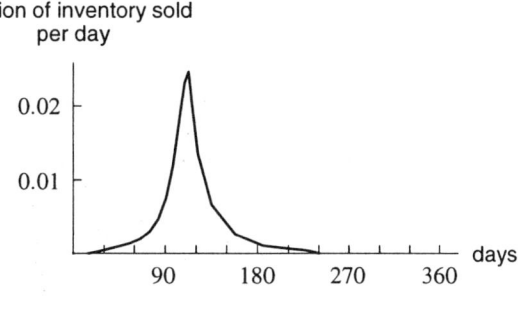

*Figure 6.44*

11. (a) Let $P(x)$ be the cumulative distribution function of the heights of the unfertilized plants. As do all cumulative distribution functions, $P(x)$ rises from 0 to 1 as $x$ increases. The greatest number of plants will have heights in the range where $P(x)$ rises the most. The steepest rise appears to occur at about $x = 1$ m. Reading from the graph we see that $P(0.9) \approx 0.2$ and $P(1.1) \approx 0.8$, so that approximately $P(1.1) - P(0.9) = 0.8 - 0.2 = 0.6 = 60\%$ of the unfertilized plants grow to heights between 0.9 m and 1.1 m. Most of the plants grow to heights in the range 0.9 m to 1.1 m.

(b) Let $P_A(x)$ be the cumulative distribution function of the plants that were fertilized with A. Since $P_A(x)$ rises the most in the range $0.7 \text{ m} \leq x \leq 0.9$ m, many of the plants fertilized with A will have heights in the range 0.7 m to 0.9 m. Reading from the graph of $P_A$, we find that $P_A(0.7) \approx 0.2$ and $P_A(0.9) \approx 0.8$, so $P_A(0.9) - P_A(0.7) \approx 0.8 - 0.2 = 0.6 = 60\%$ of the plants fertilized with A have heights between 0.7 m and 0.9 m. Fertilizer A had the effect of stunting the growth of the plants.

On the other hand, the cumulative distribution function $P_B(x)$ of the heights of the plants fertilized with B rises the most in the range $1.1 \text{ m} \leq x \leq 1.3$ m, so most of these plants have heights in the range 1.1 m to 1.3 m. Fertilizer B caused the plants to grow about 0.2 m taller than they would have with no fertilizer.

12. (a) The two functions are shown below. The choice is based on the fact that the cumulative distribution does not decrease.

(b) The cumulative distribution levels off to 1, so the top mark on the vertical scale must be 1.

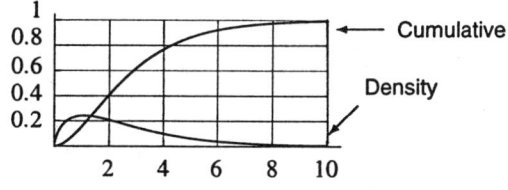

The total area under the density function must be 1. Since the area under the density function is about 2.5 boxes, each box must have area $1/2.5 = 0.4$. Since each box has a height of 0.2, the base must be 2.

13. (a) The fraction of students passing is given by the area under the curve from 2 to 4 divided by the total area under the curve. This appears to be about $\frac{2}{3}$.

(b) The fraction with honor grades corresponds to the area under the curve from 3 to 4 divided by the total area. This is about $\frac{1}{3}$.

(c) The peak around 2 probably exists because many students work to get just a passing grade.

(d)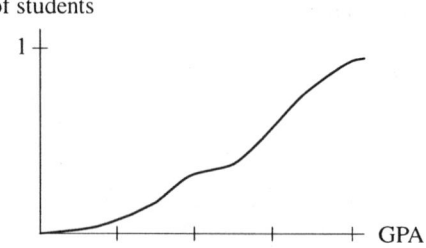

14. (a) The percentage of calls lasting from 1 to 2 minutes is given by the integral

$$\int_1^2 p(x)\,dx \approx 0.221 = 22.1\%.$$

(b) A similar calculation (changing the limits of integration) gives the percentage of calls lasting 1 minute or less as

$$\int_0^1 p(x)\,dx \approx 0.33 = 33.0\%.$$

(c) The percentage of calls lasting 3 minutes or more is given by the improper integral

$$\int_3^\infty p(x)\,dx \approx 0.301 = 30.1\%.$$

(d) The cumulative distribution function is the integral of the probability density; thus,

$$C(h) = \int_0^h p(x)\,dx = \int_0^h 0.4e^{-0.4x}\,dx = 1 - e^{-0.4h}$$

15. (a) The shaded region in Figure 6.45 represents the probability that the bus will be from 2 to 4 minutes late.

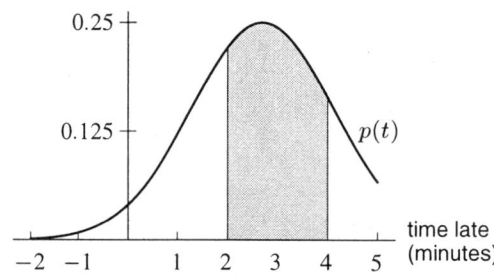

**Figure 6.45**

(b)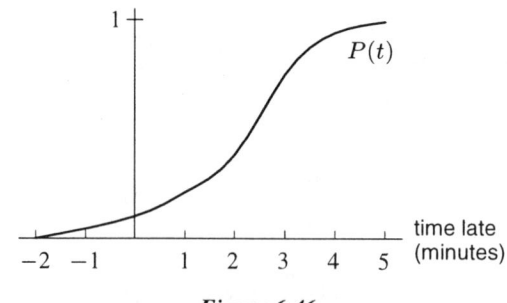

**Figure 6.46**

The probability that the bus will be 2 to 4 minutes late (the area shaded in Figure 6.45) is $P(4) - P(2)$. The inflection point on the graph of $P(t)$ in Figure 6.46 corresponds to where $p(t)$ is a maximum. To the left of the inflection point, $P$ is increasing at an increasing rate, while to the right of the inflection point $P$ is increasing at a decreasing rate. Thus, the inflection point marks where the rate at which $P$ is increasing is a maximum (i.e., where the derivative of $P$, which is $p$, is a maximum).

16. (a) The density function $f(r)$ will be zero outside the range $0 < r < 5$ and equal to a nonzero constant $k$ inside this range. The area of the region under the density curve equals $5k$, which must equal 1, so $k = 0.2$. We have

$$f(r) = \begin{cases} 0 & \text{if } r \leq 0 \\ 0.2 & \text{if } 0 < r < 5 \\ 0 & \text{if } 5 \leq r. \end{cases}$$

The graph of $f(r)$ is given in Figure 6.47.

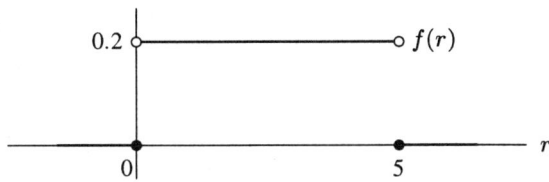

**Figure 6.47**

(b) The cumulative distribution function $F(r)$ equals the area of the region under the density function up to $r$. From the graph in Figure 6.47 we see that the area is zero if $r < 0$; for $0 \leq r \leq 5$ the region is rectangular of height 0.2, width $r$, and area $0.2r$; and for $r > 5$ the area is 1. Thus

$$F(r) = \begin{cases} 0 & \text{if } r < 0 \\ 0.2r & \text{if } 0 \leq r \leq 5 \\ 1 & \text{if } 5 < r. \end{cases}$$

17. Figure 6.48 is a graph of the density function; Figure 6.49 is a graph of the cumulative distribution.

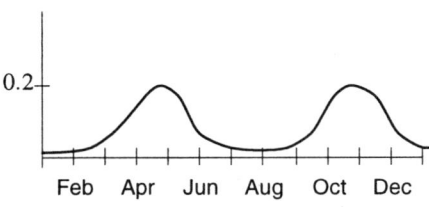

**Figure 6.48**: Density function

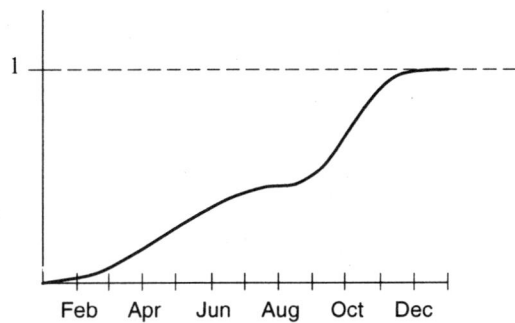

**Figure 6.49**: Cumulative distribution function

18. Since Product $D$ is absorbed most quickly, this is the solution. We see that 80% of the theophylline solution is absorbed within an hour, and all of it within 5 or 6 hours. As we would expect, the timed-release capsules are absorbed more slowly. Product A is absorbed slightly faster than Product B, although both are close to fully absorbed after 24 hours. However, we see very slow absorption with Product C. Even after 28 hours have passed, only about 60% of the drug has been absorbed.

19. (a) The probability that a banana lasts between 1 and 2 weeks is given by

$$\int_1^2 p(t)dt = 0.25$$

Thus there is a 25% probability that the banana will last between one and two weeks.

(b) The formula given for $p(t)$ is valid for up to four weeks; for $t > 4$ we have $p(t) = 0$. So a banana lasting more than 3 weeks must last between 3 and 4 weeks. Thus the probablity is

$$\int_3^4 p(t)dt = 0.325$$

32.5% of the bananas last more than 3 weeks.

(c) Since $p(t) = 0$ for $t > 4$, the probability that a banana lasts more than 4 weeks is 0.

**298** CHAPTER SIX /SOLUTIONS

20. (a) The cumulative distribution, $P(t)$, is the function whose slope is the density function $p(t)$. So $P'(t) = p(t)$. The graph of $P(t)$ starts out with a small slope at $t = 0$; its slope increases as $t$ increases to 3. The graph of $P(t)$ levels off at 1 for $t \geq 4$. See Figure 6.50.

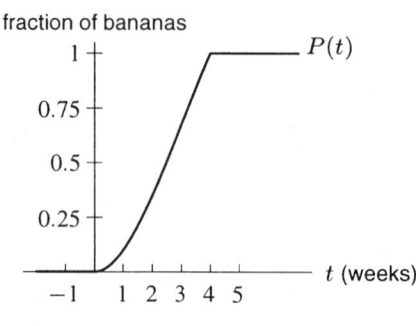

**Figure 6.50**

(b) The probability that a banana will last between 1 and 2 weeks is given by the difference $P(2) - P(1)$ where $P'(t) = p(t)$ and $p(t)$ is the density function. Looking at Figure 6.50 we see that the difference is roughly $0.25 = 25\%$.

## Solutions for Section 6.10

1. The median daily catch is the amount of fish such that half the time a boat will bring back more fish and half the time a boat will bring back less fish. Thus the area under the curve and to the left of the median must be 0.5. There are 25 squares under the curve so the median occurs at 12.5 squares of area. Now

$$\int_2^5 p(x)\,dx = 10.5 \text{ squares}$$

and

$$\int_5^6 p(x)\,dx = 5.5 \text{ squares},$$

so the median occurs at a little over 5 tons. We must find the value $a$ for which

$$\int_5^a p(t)\,dt = 2 \text{ squares},$$

and we note that this occurs at about $a = 0.35$. Hence

$$\int_2^{5.35} p(t)\,dt \approx 12.5 \text{ squares}$$

$$\approx 0.5.$$

The median is about 5.35 tons.

2. Since $p(t) = 0.04 - 0.0008t$, the cumulative distribution which satisfies $P'(t) = p(t)$ is given by

$$P(t) = 0.04t - \frac{0.0008t^2}{2} + C$$

$$= 0.04t - 0.0004t^2 + C.$$

Since $P(0) = 0$, we have $C = 0$, so

$$P(t) = 0.04t - 0.0004t^2.$$

For the median $T$,

$$P(T) = 0.04T - 0.0004T^2 = 0.5.$$

Solving the quadratic equation
$$0.0004T^2 - 0.04T + 0.5 = 0$$
gives
$$T = \frac{0.04 \pm \sqrt{(0.04)^2 - 4(0.0004)(0.5)}}{2(0.0004)}.$$

Evaluating gives $T = 85.4$ and $14.6$. Since $p(t)$ is negative for $t > 50$, the median is $T = 14.6$ days.

3. (a) The median corresponds to the value of $t$ such that $P(t) = 0.5$. Since $P(36) \approx 0.5$, the median $\approx 36$.
   (b) The density function is positive wherever the derivative of $P(t)$ is positive, namely from about $t = 5$ until roughly $t = 65$. The derivative function is increasing everywhere $P(t)$ is concave up. So that the density function is increasing until about $t = 35$ and is decreasing after that. The local maximum is where the function is changing from increasing to decreasing so that $t = 35$ is a local maximum.

4. We know that the median is given by $T$ such that
$$\int_{-\infty}^{T} p(t)dt = 0.5.$$

Trying different values of $T$, we find that $0.5 = \int_0^T p(t)dt$ for $T \approx 2.48$ weeks. Figure 6.51 supports the conclusion that $t = 2.48$ is in fact the median.

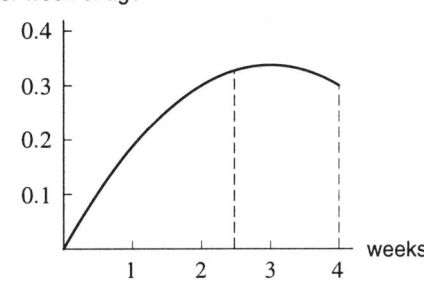

*Figure 6.51*

5. We know that the mean is given by
$$\int_{-\infty}^{\infty} tp(t)dt.$$

Thus we get
$$\text{Mean} = \int_0^4 tp(t)dt$$
$$= \int_0^4 (-0.0375t^3 + 0.225t^2)dt$$
$$\approx 2.4$$

Thus the mean is 2.4 weeks. Figure 6.52 supports the conclusion that $t = 2.4$ is in fact the mean.

**300** CHAPTER SIX /SOLUTIONS

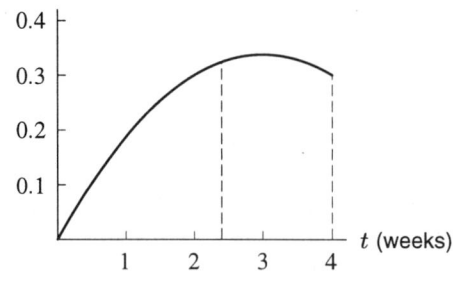

**Figure 6.52**

6. (a) See Figure 6.53. The mean is larger than the median for this distribution; both are less than 15.

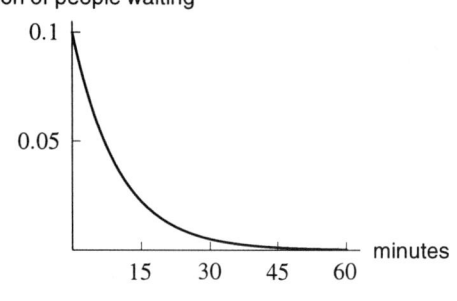

**Figure 6.53**

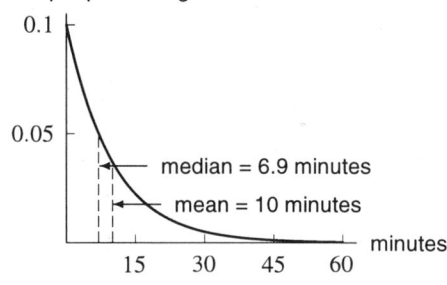

**Figure 6.54**

(b) We know that the median is the value $T$ such that

$$\int_{-\infty}^{T} p(t)dt = 0.5$$

In our case this gives

$$0.5 = \int_{0}^{T} p(t)dt$$
$$= \int_{0}^{T} 0.1e^{-0.1t} dt$$

Substituting different values of $T$ we get

$$\text{Median} = T \approx 6.9.$$

See Figure 6.54. We know that the mean is given by

$$\text{Mean} = \int_{-\infty}^{\infty} tp(t)dt$$
$$= \int_{0}^{60} (0.1te^{-0.1t})dt$$
$$\approx 9.83.$$

(c) The median tells us that exactly half of the people waiting at the stop wait less than 6.9 minutes.
The fact that the mean is 9.83 minutes can be interpreted in the following way: If all the people waiting at the stop were to wait exactly 9.83 minutes, the total time waited would be the same.

7. The median is the value $T$ such that
$$\int_{-\infty}^{T} p(x)dx = 0.5$$

Thus we get
$$0.5 = \int_{-\infty}^{T} p(x)dx$$
$$= \int_{0}^{T} 0.122 e^{-0.122x} dx$$

Substituting different values for $T$ we get $T \approx 5.68$. Thus the median occurs at 5.68 seconds
We know that the mean is
$$\int_{-\infty}^{\infty} xp(x)dx.$$

Thus we get
$$\text{Mean} = \int_{-\infty}^{\infty} xp(x)dx$$
$$= \int_{0}^{40} x(0.122 e^{-0.122x})dx \approx 7.83 \text{ seconds}.$$

The median tells us that fifty percent of the time gaps between cars are less than 5.68 seconds, and fifty percent of the time gaps between cars are more than 5.68 seconds. The mean tells us that over all time gaps, the average time gap between cars is 7.83 seconds.

8. (a) We can find the proportion of students by integrating the density $p(x)$ between $x = 1.5$ and $x = 2$:
$$P(2) - P(1.5) = \int_{1.5}^{2} \frac{x^3}{4} dx$$
$$= \frac{x^4}{16} \bigg|_{1.5}^{2}$$
$$= \frac{(2)^4}{16} - \frac{(1.5)^4}{16} = 0.684,$$

so that the proportion is $0.684 : 1$ or $68.4\%$.

(b) We find the mean by integrating $x$ times the density over the relevant range:
$$\text{Mean} = \int_{0}^{2} x\left(\frac{x^3}{4}\right) dx$$
$$= \int_{0}^{2} \frac{x^4}{4} dx$$
$$= \frac{x^5}{20} \bigg|_{0}^{2}$$
$$= \frac{2^5}{20} = 1.6 \text{ hours}.$$

(c) The median will be the time $T$ such that exactly half of the students are finished by time $T$, or in other words
$$\frac{1}{2} = \int_{0}^{T} \frac{x^3}{4} dx$$
$$\frac{1}{2} = \frac{x^4}{16} \bigg|_{0}^{T}$$
$$\frac{1}{2} = \frac{T^4}{16}$$
$$T = \sqrt[4]{8} = 1.682 \text{ hours}.$$

**302** CHAPTER SIX /SOLUTIONS

9. (a) $P$ is the cumulative distribution function, so the percentage of the population that made between $20,000 and $50,000 is
$$P(50) - P(20) = 99\% - 75\% = 24\%.$$
Therefore $\frac{6}{25}$ of the population made between $20,000 and $50,000.

(b) The median income is the income such that half the people made less than this amount. Looking at the chart, we see that $P(12.6) = 50\%$, so the median must be $12,600.

(c) The cumulative distribution function looks something like this:

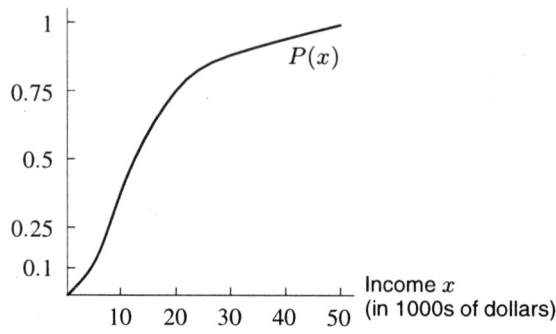

The density function is the derivative of the cumulative distribution. Qualitatively it looks like:

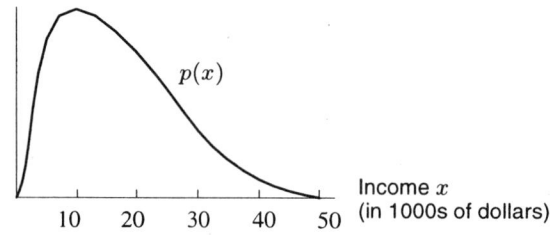

The density function has a maximum at about $8000. This means that more people have incomes around $8000 than around any other amount. On the density function, this is the highest point. On the cumulative distribution, this is the point of steepest slope (because $P' = p$), which is also the point of inflection.

## Solutions for Chapter 6 Review

1. The average value equals
$$\frac{1}{3}\int_0^3 f(x)\,dx = \frac{24}{3} = 8.$$

2. (a) Since $f(x)$ is positive on the interval from 0 to 6, the integral is equal to the area under the curve. By examining the graph, we can measure and see that the area under the curve is 20 square units, so
$$\int_0^6 f(x)\,dx = 20.$$

(b) The average value of $f(x)$ on the interval from 0 to 6 equals the definite integral we calculated in part (a) divided by the size of the interval. Thus
$$\text{Average Value} = \frac{1}{6}\int_0^6 f(x)\,dx = 3\frac{1}{3}.$$

3. The change in $\ln P(t)$ is the area under the curve, which is $\frac{1}{2} \cdot 10 \cdot (0.02) = 0.1$. So

$$\ln P(10) - \ln P(0) = \int_0^{10} \frac{P'(t)}{P(t)} dt = 0.1$$

$$\ln\left(\frac{P(10)}{P(0)}\right) = 0.1$$

$$\frac{P(10)}{P(0)} = e^{0.1} \approx 1.11.$$

The population has increased by about 11% over the 10-year period.

4. (a) See Figure 6.55. More sales were made in the second half of the year, because the area under the second half of the curve is greater.

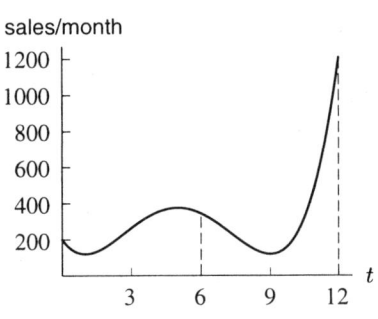

**Figure 6.55**

(b) The total sales for the first six months amount to

$$\int_0^6 r(t)dt = \$1531.20$$

The total sales for the last 6 months of the year amount to

$$\int_6^{12} r(t)dt = \$1963.20$$

(c) Thus the total sales for the year amount to

$$\int_0^{12} r(t)dt = \int_0^6 r(t)dt + \int_6^{12} r(t)dt = \$1531.20 + \$1963.20 = \$3494.40$$

(d) The average sales per month is the quotient of the total sales with 12 months giving

$$\frac{\text{Total sales}}{12 \text{ months}} = \frac{3494.4}{12} = \$291.20/\text{month}.$$

5. Looking at the graph we see that the supply and demand curves intersect at roughly the point $(345, 8)$. Thus the equilibrium price is \$8 per unit and the equilibrium quantity is 345 units. Figures 6.56 and 6.57 show the shaded areas corresponding to the consumer surplus and the producer surplus.

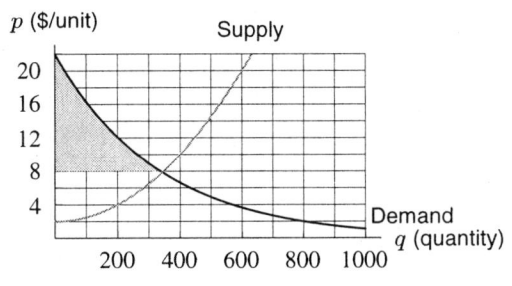
**Figure 6.56**: Consumer surplus

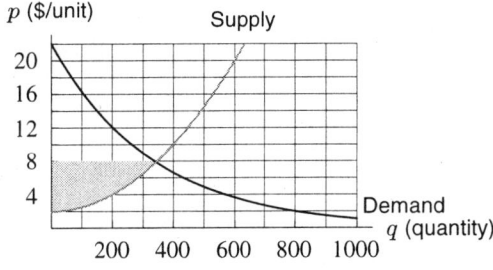
**Figure 6.57**: Producer surplus

Counting boxes we see that the consumer surplus is roughly \$2000 while the producer surplus is roughly \$1400.

**6.** The area under the curve is about 4.5 boxes, and each box has area $(0.05)(2) = 0.1$, so the area under the curve is about 0.45. The change in $\ln P$ is the area under the curve, so

$$\ln P(8) - \ln P(0) = \int_0^8 \frac{P'(t)}{P(t)} \, dt = 0.45$$

$$\ln\left(\frac{P(8)}{P(0)}\right) = 0.45$$

$$\frac{P(8)}{P(0)} = e^{0.45} = 1.57$$

The population grows by 57% over the 8-year period. Since

$$\frac{P(8)}{P(0)} = 1.57,$$

solving for $P(8)$ and substituting $P(0) = 10,000$ gives

$$P(8) = 1.57 \cdot P(0)$$
$$= 1.57(10,000)$$
$$= 15,700.$$

The population is about 15,700 after 8 years.

**7.** We compute the present value of the company:

$$\text{Present value} = \int_0^8 50,000 e^{-0.07t} \, dt = \$306,279.24.$$

You should not buy the company for $350,000, since the present value of the company is less than this amount.

**8.** We first find the present value:

$$\text{Present value} = \int_0^{20} 12000 e^{-0.06t} \, dt = \$139,761.16.$$

We use the present value to find the future value:

$$\text{Future value} = Pe^{rt} = 139761.16 e^{0.06(20)} = \$464,023.39.$$

**9.** (a) If $P$ is the present value, then the value in two years at 9% interest is $Pe^{0.09(2)}$:

$$500,000 = Pe^{0.09(2)}$$

$$P = \frac{500,000}{e^{0.09(2)}} = 417,635.11$$

The present value of the renovations is $417,635.11.

(b) If money is invested at a constant rate of $$S$ per year, then

$$\text{Present value of deposits} = \int_0^2 Se^{-0.09t} \, dt$$

Since $S$ is constant, we can take it out in front of the integral sign:

$$\text{Present value of deposits} = S \int_0^2 e^{-0.09t} \, dt$$

We want the rate $S$ so that the present value is 417,635.11:

$$417,635.11 = S \int_0^2 e^{-0.09t} \, dt$$

$$417,635.11 = S(1.830330984)$$

$$S = \frac{417,635.11}{1.830330984} = 228,174.64.$$

Money deposited at a continuous rate of 228,174.64 dollars per year and earning 9% interest per year has a value of $500,000 after two years.

10. Antiderivative $F(x) = \dfrac{x^2}{2} + \dfrac{x^6}{6} - \dfrac{x^{-4}}{4} + C$

11. $F(z) = e^z + 3z + C$

12. $P(r) = \pi r^2 + C$

13. $\ln|x| - \dfrac{1}{x} - \dfrac{1}{2x^2} + C$

14. $F(x) = \dfrac{x^7}{7} - \dfrac{1}{7}\left(\dfrac{x^{-5}}{-5}\right) + C = \dfrac{x^7}{7} + \dfrac{1}{35}x^{-5} + C$

15. $P(y) = \ln|y| + y^2/2 + y + C$

16. $G(t) = 5t + \sin t + C$

17. $G(\theta) = -\cos\theta - 2\sin\theta + C$

18. Antiderivative $G(x) = \dfrac{x^4}{4} + x^3 + \dfrac{3x^2}{2} + x + C = \dfrac{(x+1)^4}{4} + C$

19. $3x^3 + C$.

20. $\tfrac{5}{2}x^2 + 7x + C$

21. $\dfrac{t^3}{3} - 3t^2 + 5t + C$.

22. $\displaystyle\int (x+1)^2\, dx = \dfrac{(x+1)^3}{3} + C.$

    Another way to work the problem is to expand $(x+1)^2$ to $x^2 + 2x + 1$ as follows:
    $$\int (x+1)^2\, dx = \int (x^2 + 2x + 1)\, dx = \dfrac{x^3}{3} + x^2 + x + C.$$

    These two answers are the same, since $\dfrac{(x+1)^3}{3} = \dfrac{x^3 + 3x^2 + 3x + 1}{3} = \dfrac{x^3}{3} + x^2 + x + \dfrac{1}{3}$, which is $\dfrac{x^3}{3} + x^2 + x$, plus a constant.

23. $e^x + 5x + C$

24. $\dfrac{-1}{0.05}e^{-0.05t} + C = -20e^{-0.05t} + C.$

25. $3\ln|t| + \dfrac{2}{t} + C$

26. Since $f(x) = \dfrac{x+1}{x} = 1 + \dfrac{1}{x}$, the indefinite integral is $x + \ln|x| + C$

27. $2x^4 + \ln|x| + C$.

28. $3\sin x + 7\cos x + C$

29. $\dfrac{x^2}{2} + 2\ln|x| - \pi\cos x + C$

30. $2e^x - 8\sin x + C$

31. If $f(t) = 3t^2 + 4t + 3$, then $F(t) = t^3 + 2t^2 + 3t$. By the Fundamental Theorem, we have
    $$\int_0^2 (3t^2 + 4t + 3)\, dt = (t^3 + 2t^2 + 3t)\Big|_0^2 = 2^3 + 2(2^2) + 3(2) - 0 = 22.$$

32. If $f(q) = 6q^2 + 4$, then $F(q) = 2q^3 + 4q$. By the Fundamental Theorem, we have
    $$\int_0^1 (6q^2 + 4)\, dq = (2q^3 + 4q)\Big|_0^1 = 2(1) + 4(1) - (0 + 0) = 6.$$

33. If $f(t) = e^{0.05t}$, then $F(t) = 20e^{0.05t}$ (you can check this by observing that $\dfrac{d}{dt}(20e^{0.05t}) = e^{0.05t}$). By the Fundamental Theorem, we have
    $$\int e^{0.05t}\, dt = 20e^{0.05t}\Big|_0^3 = 20e^{0.15} - 20e^0 = 20(e^{0.15} - 1).$$

**34.** Since $F'(t) = 1/(2t)$, we take $F(t) = \frac{1}{2}\ln|t|$. Then

$$\int_1^2 \frac{1}{2t}\,dt = F(2) - F(1)$$
$$= \frac{1}{2}\ln|2| - \frac{1}{2}\ln|1|$$
$$= \frac{1}{2}\ln 2.$$

**35.** Since $F'(t) = 1/t^2 = t^{-2}$, we take $F(t) = \dfrac{t^{-1}}{-1} = -1/t$. Then

$$\int_1^2 \frac{1}{t^2}\,dt = F(2) - F(1)$$
$$= -\frac{1}{2} - \left(-\frac{1}{1}\right)$$
$$= \frac{1}{2}.$$

**36.** $\displaystyle\int_2^5 (x^3 - \pi x^2)\,dx = \left(\frac{x^4}{4} - \frac{\pi x^3}{3}\right)\Bigg|_2^5 = \frac{609}{4} - 39\pi \approx 29.728.$

**37.** Suppose $x$ is the age of death; a possible density function is graphed in Figure 6.58.

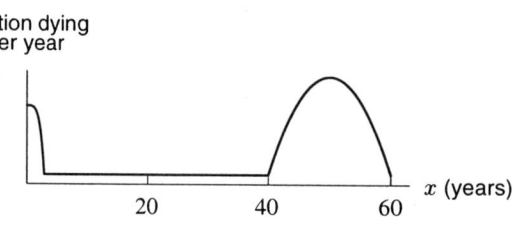

**Figure 6.58**

**38.** If $x$ is the height in feet of each person, a possible density function is graphed in Figure 6.59. The teachers are included.

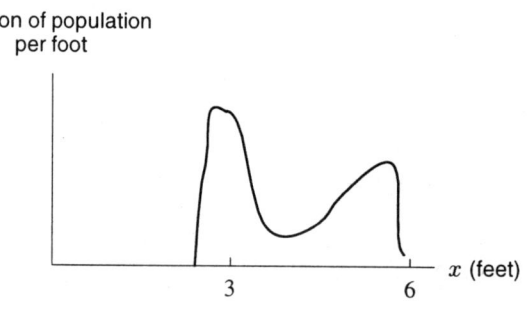

**Figure 6.59**

**39.** Let $C(y)$ be the consumption of petroleum from 1991 through the year $1991 + y$. Let $a = 1.02$ and $K = 1.4 \times 10^{20}$. We are told that in the year $1990 + t$, the annual rate of consumption will be $Ka^t$ joules/year. Thus

$$C(y) = \sum_{t=1}^{y} Ka^t.$$

Since $a > 1$, the function $u = Ka^t$ is increasing and $C(y)$ can be viewed as a right-hand Riemann sum overestimate for a definite integral

$$C(y) \geq \int_0^y Ka^t\, dt.$$

Thus we seek $y$ such that

$$\int_0^y Ka^t\, dt = 10^{22}.$$

Trying different values for $y$, we get $y \approx 45$. So in about 45 years, we will run out of petroleum!

**40.** (a) $E(t) = 1.4e^{0.07t}$

(b)
$$\text{Average Yearly Consumption} = \frac{\text{Total Consumption for the Century}}{100 \text{ years}}$$
$$= \frac{1}{100} \int_0^{100} 1.4e^{0.07t}\, dt$$
$$\approx 219 \text{ million megawatt-hours}.$$

(c) We are looking for $t$ such that $E(t) \approx 219$:
$$1.4e^{0.07t} \approx 219$$
$$e^{0.07t} = 156.4.$$

Taking natural logs,
$$0.07t = \ln 156.4$$
$$t \approx \frac{5.05}{0.07} \approx 72.18.$$

Thus, consumption was closest to the average during 1972.

(d) Between the years 1900 and 2000 the graph of $E(t)$ looks like

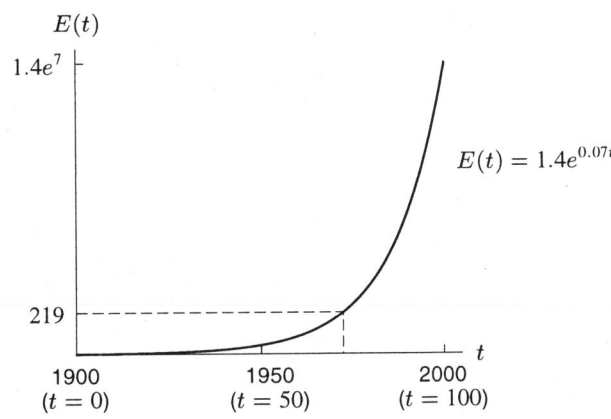

From the graph, we can see the $t$ value such that $E(t) = 219$. It lies to the right of $t = 50$, and is thus in the second half of the century.

**41.** (a) Using a calculator or computer, we get

$$\int_0^3 e^{-2t}\, dt = 0.4988 \qquad \int_0^5 e^{-2t}\, dt = 0.49998$$
$$\int_1^7 e^{-2t}\, dt = 0.4999996 \qquad \int_0^{10} e^{-2t}\, dt = 0.499999999.$$

The values of these integrals are getting closer to 0.5. A reasonable guess is that the improper integral converges to 0.5.

**308** CHAPTER SIX /SOLUTIONS

(b) Since $-\frac{1}{2}e^{-2t}$ is an antiderivative of $e^{-2t}$, we have

$$\int_0^b e^{-2t}\,dt = -\frac{1}{2}e^{-2t}\Big|_0^b = -\frac{1}{2}e^{-2b} - \left(-\frac{1}{2}e^0\right) = -\frac{1}{2}e^{-2b} + \frac{1}{2}.$$

(c) Since $e^{-2b} = 1/e^{2b}$, we have

$$e^{2b} \to \infty \quad \text{as} \quad b \to \infty, \quad \text{so} \quad e^{-2b} = \frac{1}{e^{2b}} \to 0.$$

Therefore,

$$\lim_{b \to \infty} \int_0^b e^{-2t}\,dt = \lim_{b \to \infty} \left(-\frac{1}{2}e^{-2b} + \frac{1}{2}\right) = 0 + \frac{1}{2} = \frac{1}{2}.$$

So the improper integral converges to $1/2 = 0.5$:

$$\int_0^\infty e^{-2t}\,dt = \frac{1}{2}.$$

42. (a) Since $v(t) = 60/50^t$ is never 0, the car never stops.
    (b) For time $t \geq 0$,
    
    $$\text{Distance traveled} = \int_0^\infty \frac{60}{50^t}\,dt.$$
    
    (c) Evaluating $\int_0^b \frac{60}{50^t}\,dt$ for $b = 1, 5, 10$ gives
    
    $$\int_0^1 \frac{60}{50^t}\,dt = 15.0306 \quad \int_0^5 \frac{60}{50^t}\,dt = 15.3373 \quad \int_0^{10} \frac{60}{50^t}\,dt = 15.3373,$$
    
    so the integral appears to converge to 15.3373; so we estimate the distance traveled to be 15.34 miles.

43. (a) See Figure 6.60. This is a cumulative distribution function.

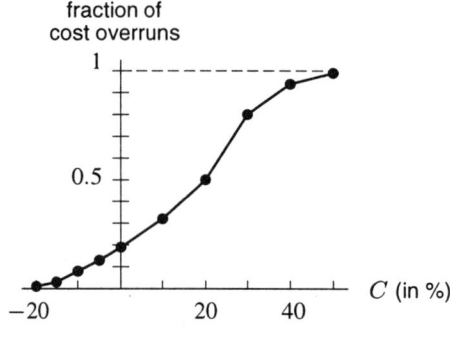

*Figure 6.60*

(b) The density function is the derivative of the cumulative distribution function. See Figure 6.61.

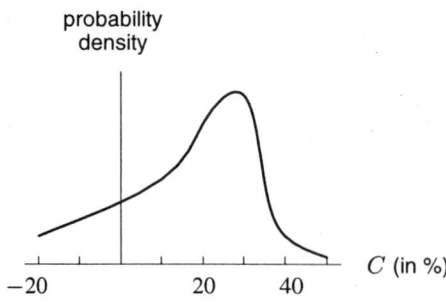

*Figure 6.61*

(c) Let's call the cumulative distribution function $F(C)$. The probability that there will be a cost overrun of more than 50% is $1 - F(50) = 0.01$, a 1% chance. The probability that it will be between 20% and 50% is $F(50) - F(20) = 0.99 - 0.50 = 0.49$, or 49%. The most likely amount of cost overrun occurs when the slope of the tangent line to the cumulative distribution function is a maximum. This occurs at the inflection point of the cumulative distribution graph, at about $C = 28\%$.

44. (a) Since $d(e^{-ct})/dt = ce^{-ct}$, we have

$$c \int_0^6 e^{-ct} dt = -e^{-ct}\Big|_0^6 = 1 - e^{-6c} = 0.1,$$

so

$$c = -\frac{1}{6} \ln 0.9 \approx 0.0176.$$

(b) Similarly, with $c = 0.0176$, we have

$$c \int_6^{12} e^{-ct} dt = -e^{-ct}\Big|_6^{12}$$

$$= e^{-6c} - e^{-12c} = 0.9 - 0.81 = 0.09,$$

so the probability is 9%.

45. (a) The probability you dropped the glove within a kilometer of home is given by

$$\int_0^1 2e^{-2x} dx = -e^{-2x}\Big|_0^1 = -e^{-2} + 1 \approx 0.865.$$

(b) Since the probability that the glove was dropped within $y$ km $= \int_0^y p(x) dx = 1 - e^{-2y}$, we solve

$$1 - e^{-2y} = 0.95$$
$$e^{-2y} = 0.05$$
$$y = \frac{\ln 0.05}{-2} \approx 1.5 \text{ km}.$$

46. It is not (a) since a probability density must be a non-negative function; not (c) since the total integral of a probability density must be 1; (b) and (d) are probability density functions, but (d) is not a good model. According to (d), the probability that the next customer comes after 4 minutes is 0. In real life there should be a positive probability of not having a customer in the next 4 minutes. So (b) is the best answer.

47. One good way to approach the problem is in terms of present values. In 1980, the present value of Germany's loan was 20 billion DM. Now let's figure out the rate that the Soviet Union would have to give money to Germany to pay off 10% interest on the loan by using the formula for the present value of a continuous stream. Since the Soviet Union sends gas at a constant rate, the rate of deposit, $P(t)$, is a constant $c$. Since they don't start sending the gas until after 5 years have passed, the present value of the loan is given by:

$$\text{Present Value} = \int_5^\infty P(t) e^{-rt} dt.$$

We want to find $c$ so that

$$20{,}000{,}000{,}000 = \int_5^\infty ce^{-rt} dt = c \int_5^\infty e^{-rt} dt$$

$$= c \int_5^\infty e^{-0.10t} dt$$

$$\approx 6.065c.$$

Dividing, we see that $c$ should be about 3.3 billion DM per year. At 0.10 DM per m$^3$ of natural gas, the Soviet Union must deliver gas at the constant, continuous rate of about 33 billion m$^3$ per year.

## Solutions to the Projects

1. (a) If the poorest $p\%$ of the population has exactly $p\%$ of the goods, then $F(x) = x$.
   (b) Any such $F$ is increasing. For example, the poorest 50% of the population includes the poorest 40%, and so the poorest 50% must own more than the poorest 40%. Thus $F(0.4) \leq F(0.5)$, and so, in general, $F$ is increasing. In addition, it is clear that $F(0) = 0$ and $F(1) = 1$.

   The graph of $F$ is concave up by the following argument. Consider $F(0.05) - F(0.04)$. This is the fraction of resources the fifth poorest percent of the population has. Similarly, $F(0.20) - F(0.19)$ is the fraction of resources that the twentieth poorest percent of the population has. Since the twentieth poorest percent owns more than the fifth poorest percent, we have
   $$F(0.05) - F(0.04) \leq F(0.20) - F(0.19).$$
   More generally, we can see that
   $$F(x_1 + \Delta x) - F(x_1) \leq F(x_2 + \Delta x) - F(x_2)$$
   for any $x_1$ smaller than $x_2$ and for any increment $\Delta x$. Dividing this inequality by $\Delta x$ and taking the limit as $\Delta x \to 0$, we get
   $$F'(x_1) \leq F'(x_2).$$
   So, the derivative of $F$ is an increasing function, i.e. $F$ is concave up.
   (c) $G$ is twice the shaded area below in the following figure.

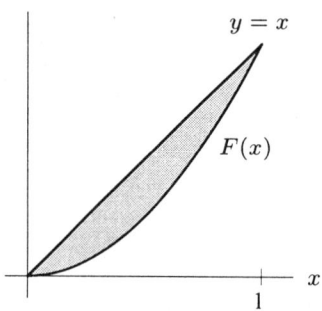

   (d) The most equitable distribution is the one modeled by the equation $F(x) = x$, as this tells us that the wealth is evenly distributed. The less equitable the distribution the further the function is from resembling $F(x) = x$. Thus in our case, Country A has more equitable distribution.
   (e) The function $F$ goes through the points $(0, 0)$ and $(1, 1)$ and is increasing and concave up. Graphical representations of the two extremes for $G$ are shown in Figures 6.62 and 6.63. Since Gini's index is twice the area shown, the maximum possible value is 1 and the minimum possible value is 0. If Gini's index is 1, the distribution of resources is as unequitable as it can get, with one person holding all resources and everyone else having none. (See Figure 6.62.) If Gini's index is 0, the distribution of resources is totally equitable, with $F(x) = x$ as in part (a). (See Figure 6.63.) The closer Gini's index is to 0, the more equitable the distribution.

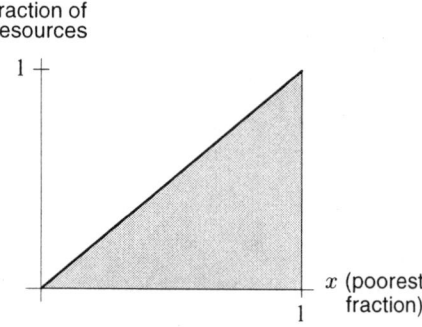

Figure 6.62: Maximum value of Gini's index: Function $F$ with $G = 1$

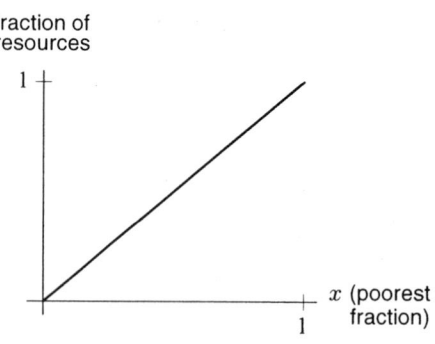

Figure 6.63: Minimum value of Gini's index: Function $F$ with $G = 0$

2. (a)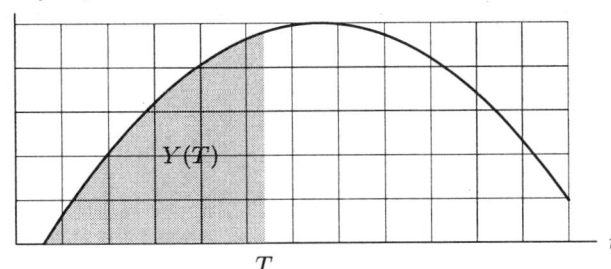

$Y(T)$ is the area of the shaded region in the picture. Thus, $Y(T) = \int_0^T y(t)\,dt$.

(b) Here is a graph of $Y(T)$. Note that the graph of $y$ looked like the graph of a quadratic function. Thus, the graph of $Y$ should look like a cubic, which indeed it does.

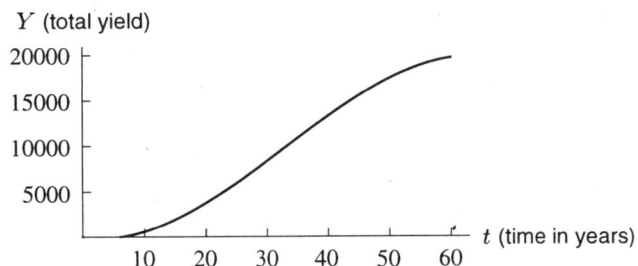

(c) $a(T) = \frac{1}{T}Y(T) = \frac{1}{T}\int_0^T y(t)\,dt$.

(d) (i) If the function $a(T)$ takes on its maximum at some point $T$, then $a'(T) = 0$. Since $a(T) = \frac{1}{T}Y(T)$, we may differentiate using the quotient rule:

$$a'(T) = \frac{TY'(T) - Y(T)}{T^2} = 0;$$

or, equivalently, $TY'(T) = Y(T)$.

(ii) The expression above may be rewritten in terms of $y$, giving us

$$T\frac{d}{dT}\int_0^T y(t)\,dt = \int_0^T y(t)\,dt.$$

Simplifying, we obtain $Ty(T) = \int_0^T y(t)\,dt$, or, equivalently,

$$y(T) = \frac{1}{T}\int_0^T y(t)\,dt = a(T),$$

as a condition on $y(T)$ for maximization of $a(T)$.

To find the value of $T$ which satisfies $Ty(T) = Y(T)$, notice that $Y(T)$ is the area under the curve from 0 to T, and that $Ty(T)$ is the area of a rectangle of height $y(T)$. Thus we want the area under the curve to be equal to the area of the rectangle, or $A = B$ in the figure below. This happens when $T \approx 50$ years. In other words, the orchard should be cut down after about 50 years.

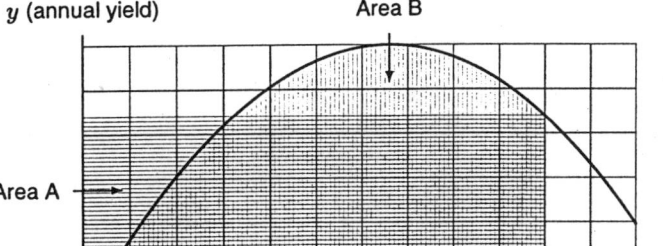

## Solutions to Practice Problems on Integration

1. $\int (t^3 + 6t^2)\, dt = \dfrac{t^4}{4} + 6 \cdot \dfrac{t^3}{3} + C = \dfrac{t^4}{4} + 2t^3 + C$

2. $\int (u^4 + 5)\, du = \dfrac{u^5}{5} + 5u + C$

3. $\int (x^2 + x^{-2})\, dx = \dfrac{x^3}{3} + \dfrac{x^{-1}}{-1} + C = \dfrac{x^3}{3} - \dfrac{1}{x} + C$

4. $\int e^{3r}\, dr = \dfrac{1}{3}e^{3r} + C$

5. $\int 3w^{1/2}\, dw = 3 \cdot \dfrac{w^{3/2}}{3/2} + C = 2w^{3/2} + C$

6. $\int (ax^2 + b)\, dx = a \cdot \dfrac{x^3}{3} + bx + C$

7. $\int (t^2 + 5t + 1)\, dt = \dfrac{t^3}{3} + 5 \cdot \dfrac{t^2}{2} + t + C$

8. $\int 100e^{-0.5t}\, dt = 100\left(\dfrac{1}{-0.5}e^{-0.5t}\right) + C = -200e^{-0.5t} + C$

9. $\int (w^4 - 12w^3 + 6w^2 - 10)\, dw = \dfrac{w^5}{5} - 12 \cdot \dfrac{w^4}{4} + 6 \cdot \dfrac{w^3}{3} - 10 \cdot w + C$
$\phantom{\int (w^4 - 12w^3 + 6w^2 - 10)\, dw} = \dfrac{w^5}{5} - 3w^4 + 2w^3 - 10w + C$

10. $\int \left(p^2 + \dfrac{5}{p}\right) dp = \dfrac{p^3}{3} + 5\ln|p| + C$

11. $\int q^{-1/2}\, dq = \dfrac{q^{1/2}}{1/2} + C = 2q^{1/2} + C$

12. $\int 3\sin\theta\, d\theta = -3\cos\theta + C$

13. $\int \left(\dfrac{4}{x} + 5x^{-2}\right) dx = 4\ln|x| + \dfrac{5x^{-1}}{-1} + C = 4\ln|x| - \dfrac{5}{x} + C$

14. $\int P_0 e^{kt}\, dt = P_0\left(\dfrac{1}{k}e^{kt}\right) + C = \dfrac{P_0}{k}e^{kt} + C$

15. $\int (q^3 + 8q + 15)\, dq = \dfrac{q^4}{4} + 8 \cdot \dfrac{q^2}{2} + 15q + C$
$\phantom{\int (q^3 + 8q + 15)\, dq} = \dfrac{q^4}{4} + 4q^2 + 15q + C$

16. $\int 1000e^{0.075t}\, dt = 1000\left(\dfrac{1}{0.075}e^{0.075t}\right) + C = 13333e^{0.075t} + C$

17. $\int (5\sin x + 3\cos x)\, dx = -5\cos x + 3\sin x + C$

18. $\int (10 + 5\sin x)\, dx = 10x - 5\cos x + C$

19. $\int \pi r^2 h\, dr = \pi h\left(\dfrac{r^3}{3}\right) + C = \dfrac{\pi}{3}hr^3 + C$

20. $\int (q + q^{-3})\, dq = \dfrac{q^2}{2} + \dfrac{q^{-2}}{-2} + C = \dfrac{q^2}{2} - \dfrac{1}{2q^2} + C$

## SOLUTIONS TO PRACTICE PROBLEMS ON INTEGRATION

21. $\int 15p^2 q^4 \, dp = 15 \left(\dfrac{p^3}{3}\right) q^4 + C = 5p^3 q^4 + C$

22. $\int 15p^2 q^4 \, dq = 15p^2 \left(\dfrac{q^5}{5}\right) + C = 3p^2 q^5 + C$

23. $\int (3x^2 + 6e^{2x}) \, dx = 3 \cdot \dfrac{x^3}{3} + 6 \cdot \dfrac{e^{2x}}{2} + C$
    $= x^3 + 3e^{2x} + C$

24. $\int \dfrac{5}{w} \, dw = 5 \ln |w| + C$

25. $\int 5e^{2q} \, dq = 5 \cdot \dfrac{1}{2} e^{2q} + C = 2.5 e^{2q} + C$

26. $\int \left(p^3 + \dfrac{1}{p}\right) dp = \dfrac{p^4}{4} + \ln |p| + C$

27. $\int (Ax^3 + Bx) \, dx = \dfrac{Ax^4}{4} + \dfrac{Bx^2}{2} + C$

28. $\int (6x^{1/2} + 15) \, dx = 6 \cdot \dfrac{x^{3/2}}{3/2} + 15x + C = 4x^{3/2} + 15x + C$

29. $\int (x^2 + 8 + e^x) \, dx = \dfrac{x^3}{3} + 8x + e^x + C$

30. $\int 25 e^{-0.04q} \, dq = 25 \left(\dfrac{1}{-0.04} e^{-0.04q}\right) + C = -625 e^{-0.04q} + C$

# CHAPTER SEVEN

## Solutions for Section 7.1

1. (a) Beef consumption by households making $20,000/year is given by Row 1 of Table 7.2 on page 358 of the text.

    **TABLE 7.1**

    | $p$ | 3.00 | 3.50 | 4.00 | 4.50 |
    |---|---|---|---|---|
    | $f(20, p)$ | 2.65 | 2.59 | 2.51 | 2.43 |

    For households making $20,000/year, beef consumption decreases as price goes up.

    (b) Beef consumption by households making $100,000/year is given by Row 5 of Table 7.2.

    **TABLE 7.2**

    | $p$ | 3.00 | 3.50 | 4.00 | 4.50 |
    |---|---|---|---|---|
    | $f(100, p)$ | 5.79 | 5.77 | 5.60 | 5.53 |

    For households making $100,000/year, beef consumption also decreases as price goes up.

    (c) Beef consumption by households when the price of beef is $3.00/lb is given by Column 1 of Table 7.2.

    **TABLE 7.3**

    | $I$ | 20 | 40 | 60 | 80 | 100 |
    |---|---|---|---|---|---|
    | $f(I, 3.00)$ | 2.65 | 4.14 | 5.11 | 5.35 | 5.79 |

    When the price of beef is $3.00/lb, beef consumption increases as income increases.

    (d) Beef consumption by households when the price of beef is $4.00/lb is given by Column 3 of Table 7.2.

    **TABLE 7.4**

    | $I$ | 20 | 40 | 60 | 80 | 100 |
    |---|---|---|---|---|---|
    | $f(I, 4.00)$ | 2.51 | 3.94 | 4.97 | 5.19 | 5.60 |

    When the price of beef is $4.00/lb, beef consumption increases as income increases.

2. If the price of beef is held constant, beef consumption for households with various incomes can be read from a fixed column in Table 7.2 on page 7.2 of the text. For example, the column corresponding to $p = 3.00$ gives the function $h(I) = f(I, 3.00)$; it tells you how much beef a household with income $I$ will buy at $3.00/lb. Looking at the column from the top down, you can see that it is an increasing function of $I$. This is true in every column. This says that at any fixed price for beef, consumption goes up as household income goes up—which makes sense. Thus, $f$ is an increasing function of $I$ for each value of $p$.

3. The amount of money spent on beef equals the product of the unit price $p$ and the quantity $C$ of beef consumed:
$$M = pC = pf(I, p).$$

Thus, we multiply each entry in Table 7.2 on page 358 of the text by the price at the top of the column. This yields Table 7.5.

**TABLE 7.5** *Amount of money spent on beef ($/household/week)*

|  |  | Price |  |  |  |
|---|---|---|---|---|---|
|  |  | 3.00 | 3.50 | 4.00 | 4.50 |
| Income | 20 | 7.95 | 9.07 | 10.04 | 10.94 |
|  | 40 | 12.42 | 14.18 | 15.76 | 17.46 |
|  | 60 | 15.33 | 17.50 | 19.88 | 21.78 |
|  | 80 | 16.05 | 18.52 | 20.76 | 22.82 |
|  | 100 | 17.37 | 20.20 | 22.40 | 24.89 |

4. Table 7.7 gives the amount $M$ spent on beef per household per week. Thus, the amount the household spent on beef in a year is $52M$. Since the household's annual income is $I$ thousand dollars, the proportion of income spent on beef is

$$P = \frac{52M}{1000I} = 0.052\frac{M}{I}.$$

Thus, we need to take each entry in Table 7.7, divide it by the income at the left, and multiply by 0.052. Table 7.6 shows the results.

**TABLE 7.6** *Proportion of annual income spent on beef*

|  |  | Price of Beef ($) |  |  |  |
|---|---|---|---|---|---|
|  |  | 3.00 | 3.50 | 4.00 | 4.50 |
| Income ($1,000) | 20 | 0.021 | 0.024 | 0.026 | 0.028 |
|  | 40 | 0.016 | 0.018 | 0.020 | 0.023 |
|  | 60 | 0.013 | 0.015 | 0.017 | 0.019 |
|  | 80 | 0.010 | 0.012 | 0.013 | 0.015 |
|  | 100 | 0.009 | 0.011 | 0.012 | 0.013 |

**TABLE 7.7** *Money spent on beef ($/household/week)*

|  |  | Price of Beef ($) |  |  |  |
|---|---|---|---|---|---|
|  |  | 3.00 | 3.50 | 4.00 | 4.50 |
| Income ($1,000) | 20 | 7.95 | 9.07 | 10.04 | 10.94 |
|  | 40 | 12.42 | 14.18 | 15.76 | 17.46 |
|  | 60 | 15.33 | 17.50 | 19.88 | 21.78 |
|  | 80 | 16.05 | 18.52 | 20.76 | 22.82 |
|  | 100 | 17.37 | 20.20 | 22.40 | 24.89 |

5. In the answer to Problem 4 we saw that

$$P = 0.052\frac{M}{I},$$

and in the answer to Problem 3 we saw that

$$M = pf(I, p).$$

Putting the expression for $M$ into the expression for $P$, gives:

$$P = 0.052\frac{pf(I, p)}{I}.$$

6. We have $M = f(B, t) = B(1.05)^t$.

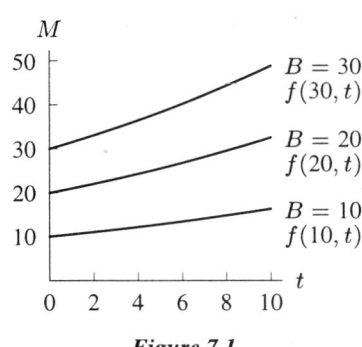

**Figure 7.1**

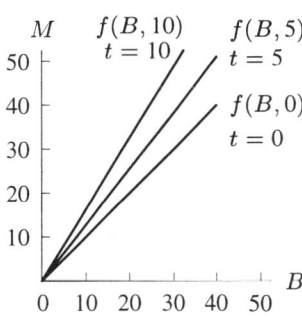

**Figure 7.2**

Figure 7.1 gives the graphs of $f$ as a function of $t$ for $B$ fixed at 10, 20, and 30. For each fixed $B$, the function $f(B, t)$ is an increasing function of $t$. The larger the fixed value of $B$, the larger $f(B, t)$ is.

Figure 7.2 gives the graphs of $f$ as a function of $B$ for $t$ fixed at 0, 5, and 10. For each fixed $t$, $f(B, t)$ is an increasing (and in fact linear) function of $B$. The larger $t$ is, the larger the slope of the line.

7. (a) Decreasing, because, other things being equal, we expect the sales of cars to drop if gas prices increase.
   (b) Decreasing, because, other things being equal, we expect car sales to drop if car prices increase.

8. Asking if $f$ is an increasing or decreasing function of $p$ is the same as asking how does $f$ vary as we vary $p$, when we hold $a$ fixed. Intuitively, we know that as we increase the price $p$, total sales of the product will go down. Thus, $f$ is a decreasing function of $p$. Similarly, if we increase $a$, the amount spent on advertising, we can expect $f$ to increase and therefore $f$ is an increasing function of $a$.

9. To see whether $f$ is an increasing or decreasing function of $x$, we need to see how $f$ varies as we increase $x$ and hold $y$ fixed. We note that each column of the table corresponds to a fixed value of $y$. Scanning down the $y = 2$ column, we can see that as $x$ increases, the value of the function decreases from 114 when $x = 0$ down to 93 when $x = 80$. Thus, $f$ may be decreasing. In order for $f$ to actually be decreasing however, we have to make sure that $f$ decreases for *every* column. In this case, we see that $f$ indeed does decrease for every column. Thus, $f$ is a decreasing function of $x$. Similarly, to see whether $f$ is a decreasing function of $y$ we need to look at the rows of the table. As we can see, $f$ increases for every row as we increase $y$. Thus, $f$ is an increasing function of $y$.

10. We make a table by calculating values for $C = f(d, m)$ for each value of $d$ and $m$. Such a table is shown in Table 7.8

**TABLE 7.8**

|   |   | \multicolumn{4}{c}{d} |   |   |
|---|---|-----|-----|-----|-----|
|   |   | 1   | 2   | 3   | 4   |
| m | 100 | 55  | 95  | 135 | 175 |
|   | 200 | 70  | 110 | 150 | 190 |
|   | 300 | 85  | 125 | 165 | 205 |
|   | 400 | 100 | 140 | 180 | 220 |

11. (a) $f(3, 200) = 40(3) + 0.15(200) = 150$. Renting a car for three days and driving it 200 miles will cost $150.
    (b) $f(3, m)$ is the value of $C$ for various values of $m$ with $d$ fixed at 3. In other words, it is the cost of renting a car for three days and driving it various numbers of miles. A plot of $C = f(3, m)$ against $m$ is shown in Figure 7.3.

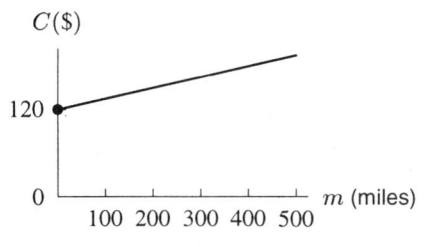

**Figure 7.3**

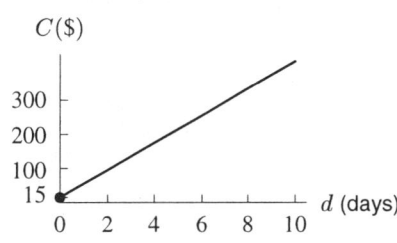

**Figure 7.4**

(c) $f(d, 100)$ is the value of $C$ for various values of $d$ with $m$ fixed at 100. In other words, it is the cost of renting a car for various lengths of time and driving it 100 miles. A graph of $C = f(d, 100)$ against $d$ is shown in Figure 7.4.

12. (a) The daily fuel cost is calculated:

$$\text{Cost} = \text{Price per gallon} \times \text{Number of gallons}.$$

**TABLE 7.9** *Cost vs. Price & Gallons*

|  |  | \multicolumn{7}{c}{Price per gallon (dollars)} |
|---|---|---|---|---|---|---|---|---|
|  |  | 1.00 | 1.05 | 1.10 | 1.15 | 1.20 | 1.25 | 1.30 |
| Number of gallons | 5 | 5.00 | 5.25 | 5.50 | 5.75 | 6.00 | 6.25 | 6.50 |
|  | 6 | 6.00 | 6.30 | 6.60 | 6.90 | 7.20 | 7.50 | 7.80 |
|  | 7 | 7.00 | 7.35 | 7.70 | 8.05 | 8.40 | 8.75 | 9.10 |
|  | 8 | 8.00 | 8.40 | 8.80 | 9.20 | 9.60 | 10.00 | 10.40 |
|  | 9 | 9.00 | 9.45 | 9.90 | 10.35 | 10.80 | 11.25 | 11.70 |
|  | 10 | 10.00 | 10.50 | 11.00 | 11.50 | 12.00 | 12.50 | 13.00 |
|  | 11 | 11.00 | 11.55 | 12.10 | 12.65 | 13.20 | 13.75 | 14.30 |
|  | 12 | 12.00 | 12.60 | 13.20 | 13.80 | 14.40 | 15.00 | 15.60 |

Note: Table entries may vary depending on the price increase interval chosen.

(b) If the daily travel distance and the price of gas are given, then:

$$\text{Number of gallons} = (\text{Distance in miles}) / (30 \text{ miles per gallon})$$
$$\text{Cost} = \text{Price per gallon} \times \text{Number of gallons}$$

**TABLE 7.10** *Cost vs. Price & Distance*

|  |  | \multicolumn{7}{c}{Price per gallon (dollars)} |
|---|---|---|---|---|---|---|---|---|
|  |  | 1.00 | 1.05 | 1.10 | 1.15 | 1.20 | 1.25 | 1.30 |
| Distance (miles) | 100 | 3.33 | 3.50 | 3.67 | 3.83 | 4.00 | 4.17 | 4.33 |
|  | 150 | 5.00 | 5.25 | 5.50 | 5.75 | 6.00 | 6.25 | 6.50 |
|  | 200 | 6.67 | 7.00 | 7.33 | 7.67 | 8.00 | 8.33 | 8.67 |
|  | 250 | 8.33 | 8.75 | 9.17 | 9.58 | 10.00 | 10.42 | 10.83 |
|  | 300 | 10.00 | 10.50 | 11.00 | 11.50 | 12.00 | 12.50 | 13.00 |
|  | 350 | 11.67 | 12.25 | 12.83 | 13.42 | 14.00 | 14.58 | 15.17 |
|  | 400 | 13.33 | 14.00 | 14.67 | 15.33 | 16.00 | 16.67 | 17.33 |
|  | 450 | 15.00 | 15.75 | 16.50 | 17.25 | 18.00 | 18.75 | 19.50 |

Note: Table entries may vary depending on the price increase interval chosen.

13. (a) According to Table 7.4 of the problem, it feels like $-31°F$.
   (b) A wind of 10 mph, according to Table 7.4.
   (c) About 5.5 mph. Since at a temperature of 25°F, when the wind increases from 5 mph to 10 mph, the temperature adjusted for wind-chill decreases from 21°F to 10°F, we can say that a 5 mph increase in wind speed causes an 11°F decrease in the temperature adjusted for wind-chill. Thus, each 0.5 mph increase in wind speed brings *about* a 1°F drop in the temperature adjusted for wind-chill.
   (d) With a wind of 15 mph, approximately 23.5°F would feel like 0°F. With a 15 mph wind speed, when air temperature drops five degrees from 25°F to 20°F, the temperature adjusted for wind-chill drops 7 degrees from 2°F to $-5°F$. We can say that for every 1°F decrease in temperature there is *about* a 1.4°F $(= 7/5)$ drop in the temperature you feel.

14.

**TABLE 7.11** *Temperature adjusted for wind-chill at 20° F*

| Wind speed (mph)       | 5  | 10 | 15 | 20  | 25  |
|------------------------|----|----|----|-----|-----|
| Adjusted temperature (°F) | 16 | 3  | −5 | −10 | −15 |

**TABLE 7.12** *Temperature adjusted for wind-chill at 0° F*

| Wind speed (mph)       | 5  | 10  | 15  | 20  | 25  |
|------------------------|----|-----|-----|-----|-----|
| Adjusted temperature (°F) | −5 | −22 | −31 | −39 | −44 |

15.

**TABLE 7.13** *Temperature adjusted for wind-chill at 5 mph*

| Temperature (°F)          | 35 | 30 | 25 | 20 | 15 | 10 | 5 | 0  |
|---------------------------|----|----|----|----|----|----|---|----|
| Adjusted temperature (°F) | 33 | 27 | 21 | 16 | 12 | 7  | 0 | −5 |

**TABLE 7.14** *Temperature adjusted for wind-chill at 20 mph*

| Temperature (°F)          | 35 | 30 | 25 | 20  | 15  | 10  | 5   | 0   |
|---------------------------|----|----|----|-----|-----|-----|-----|-----|
| Adjusted temperature (°F) | 12 | 4  | −3 | −10 | −17 | −24 | −31 | −39 |

16. (a) It feels like 81°F.
    (b) At 30% relative humidity, 90°F feels like 90°F.
    (c) By finding the temperature which has heat index 105°F for each humidity level, we get Table 7.15:

**TABLE 7.15** *Estimates of danger temperatures*

| Relative humidity(%) | 0   | 10  | 20  | 30  | 40 | 50 | 60 |
|----------------------|-----|-----|-----|-----|----|----|----|
| Temperature(°F)      | 117 | 110 | 105 | 101 | 97 | 94 | 92 |

   (d) With a high humidity your body cannot cool itself as well by sweating. With a low humidity your body is capable of cooling itself to below the actual temperature. Therefore a high humidity feels hotter and a low humidity feels cooler.

17.

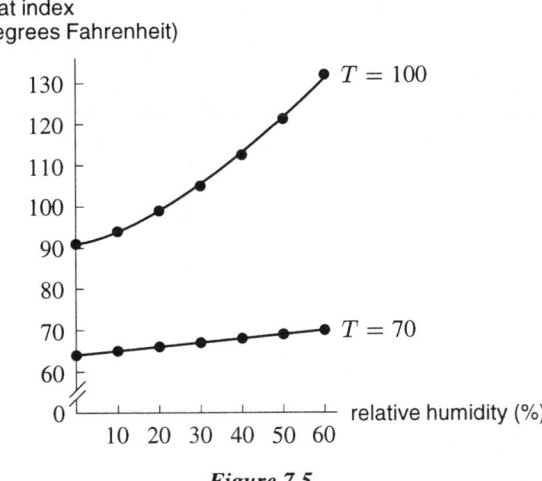

*Figure 7.5*

Both graphs are increasing because at any fixed temperature the air feels hotter as the humidity increases. The fact that the graph for $T = 100$ increases more rapidly with humidity than the graph for $T = 70$ tells us that when it is hot (100°F), high humidity has more effect on how we feel than at lower temperatures (70°F).

18. (a) See Figure 7.6.

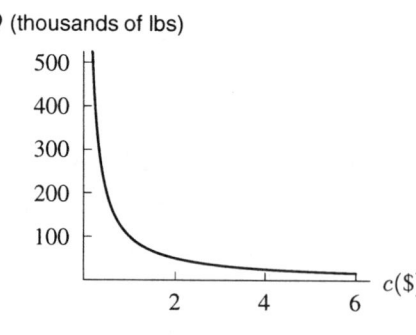
Figure 7.6

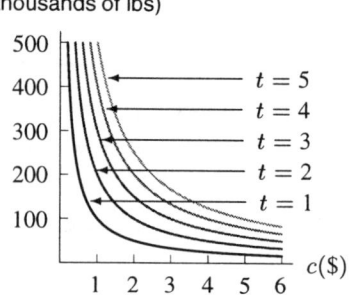
Figure 7.7

(b) See Figure 7.7. The demand curve of $f$ is a hyperbola, reflecting the inversely proportional relation between the demand, $Q$, and the price of coffee, $c$. As $t$ increases, we can see from the graph that the demand curve rises, showing that $Q$ is an increasing function of $t$; in addition $Q$ is a decreasing function of $c$. This makes sense since as the price of tea increases, the demand for tea will decrease and more consumers will opt for coffee over the increasingly expensive tea. As the price of coffee rises, the demand for coffee will fall.

19. $Q$ is a decreasing function of $c$ and an increasing function of $t$. This is because when the price of coffee rises, consumers drink less. When the price of tea rises, some consumers switch from tea to coffee and the demand for coffee increases.

20. Table 7.16 shows, as we concluded in Problem 19, that $Q$ is a decreasing function of $C$ and an increasing function of $t$.

**TABLE 7.16** *Demand Q*

|   |   | \ | t |   |   |
|---|---|---|---|---|---|
|   |   | 1 | 2 | 3 | 4 |
|   | 1 | 100 | 200 | 300 | 400 |
| $c$ | 2 | 50 | 100 | 150 | 200 |
|   | 3 | 33 | 67 | 100 | 133 |
|   | 4 | 25 | 50 | 75 | 100 |

21. It stands to reason that the demand for tea, $D$, will have a similar formula to the demand for coffee. The demand $D$ will be a decreasing function of the price of tea, $t$, and an increasing function of the price of coffee $c$. A possible formula for the demand for tea is
$$D = 100\frac{c}{t}.$$

22. (a) For $t = 0$, we have $y = f(x, 0) = \sin x$, $0 \le x \le \pi$, as in Figure 7.8.

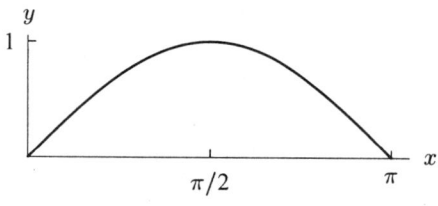
Figure 7.8

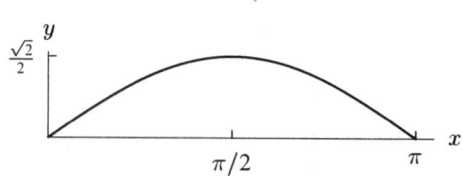
Figure 7.9

For $t = \pi/4$, we have $y = f(x, \pi/4) = \frac{\sqrt{2}}{2}\sin x$, $0 \le x \le \pi$, as in Figure 7.9.
For $t = \pi/2$, we have $y = f(x, \pi/2) = 0$, as in Figure 7.10.

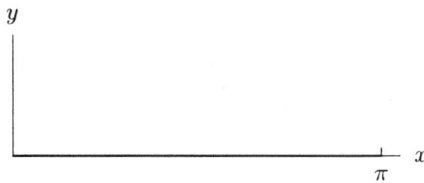

*Figure 7.10*

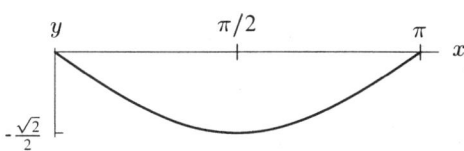

*Figure 7.11*

For $t = 3\pi/4$, we have $y = f(x, 3\pi/4) = \frac{-\sqrt{2}}{2}\sin x$, $0 \leq x \leq \pi$, as in Figure 7.11.
For $t = \pi$, we have $y = f(x, \pi) = -\sin x$, $0 \leq x \leq \pi$, as in Figure 7.12.

(b) The graphs show an arch of a sine wave which is above the $x$-axis, concave down at $t = 0$, is straight along the $x$-axis at $t = \pi/2$, and below the $x$-axis, concave up at $t = \pi$, like a guitar string vibrating up and down.

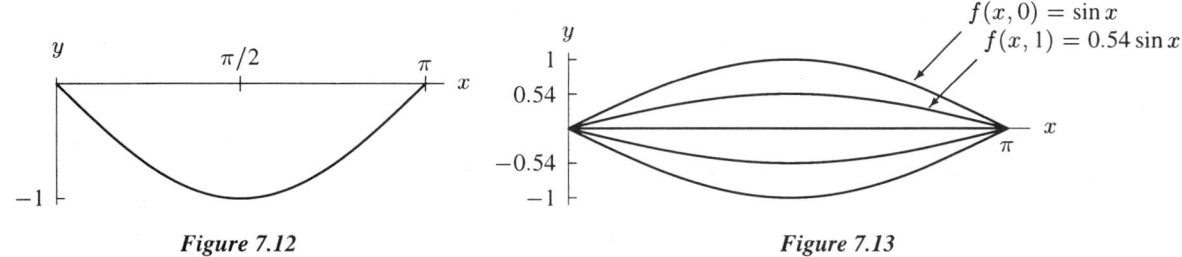

*Figure 7.12*  *Figure 7.13*

23. The function $y = f(x, 0) = \cos 0 \sin x = \sin x$ gives the displacement of each point of the string when time is held fixed at $t = 0$. The function $f(x, 1) = \cos 1 \sin x = 0.54 \sin x$ gives the displacement of each point of the string at time $t = 1$. Graphing $f(x, 0)$ and $f(x, 1)$ gives in each case an arch of the sine curve, the first with amplitude 1 and the second with amplitude 0.54. For each different fixed value of $t$, we get a different snapshot of the string, each one a sine curve with amplitude given by the value of $\cos t$. The result looks like the sequence of snapshots shown in Figure 7.13.

24. The function $f(0, t) = \cos t \sin 0 = 0$ gives the displacement of the left end of the string as time varies. Since that point remains stationary, the displacement is zero. The function $f(1, t) = \cos t \sin 1 = 0.84 \cos t$ gives the displacement of the point at $x = 1$ as time varies. Since $\cos t$ oscillates back and forth between 1 and $-1$, this point moves back and forth with maximum displacement of 0.84 in either direction. Notice the maximum displacements are greatest at $x = \pi/2$ where $\sin x = 1$.

25. (a) For $g(x, t) = \cos 2t \sin x$, our snapshots for fixed values of $t$ are still one arch of the sine curve. The amplitudes, which are governed by the $\cos 2t$ factor, now change twice as fast as before. That is, the string is vibrating twice as fast.

(b) For $y = h(x, t) = \cos t \sin 2x$, the vibration of the string is more complicated. If we hold $t$ fixed at any value, the snapshot now shows one full period, i.e. one crest and one trough, of the sine curve. The magnitude of the sine curve is time dependent, given by $\cos t$. Now the center of the string, $x = \pi/2$, remains stationary just like the end points. This is a vibrating string with the center held fixed, as shown in Figure 7.14.

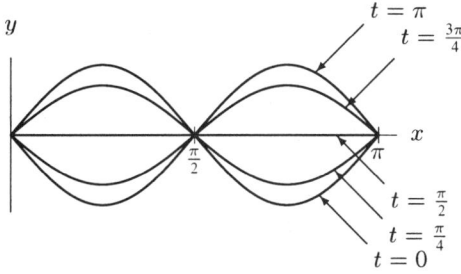

*Figure 7.14*: Another vibrating string: $y = h(x, t) = \cos t \sin 2x$

26. (a) We wish to find $H(20, 0.3)$. The plot of $H(T, 0.3)$ is shown on the given graph. To find $H(20, 0.3)$, we can just find the value of $H(T, 0.3)$ where $T = 20$. Looking at the graph, we see that $H(20, 0.3) \approx 260$ cal/m$^3$. It takes about 260 calories to clear one cubic meter of air at 20° C with 0.3 grams of water in it.

(b) Proceeding in an analogous manner we can estimate $H$ for the other combinations of $T$ and $w$; these are shown in Table 7.17:

**TABLE 7.17**

|  |  | $w$ (gm/m$^3$) | | | |
|---|---|---|---|---|---|
|  |  | 0.1 | 0.2 | 0.3 | 0.4 |
| $T$ (°C) | 0 | 150 | 290 | 425 | 590 |
|  | 10 | 110 | 240 | 330 | 450 |
|  | 20 | 100 | 180 | 260 | 350 |
|  | 30 | 70 | 150 | 220 | 300 |
|  | 40 | 65 | 140 | 200 | 270 |

## Solutions for Section 7.2

1. (a) 80-90°F
   (b) 60-72°F
   (c) 60-100°F

2.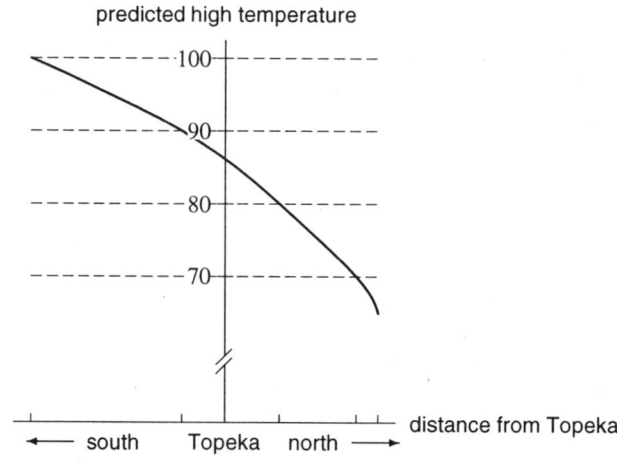

   *Figure 7.15*

3.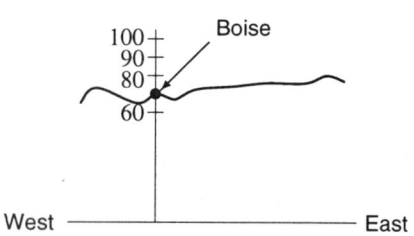

   *Figure 7.16*     *Figure 7.17*

4. The contour for $C = 50$ is given by
$$40d + 0.15m = 50.$$
This is the equation of a line with intercepts $d = 50/40 = 1.25$ and $m = 50/0.15 \approx 333$. (See Figure 7.18.) The contour for $C = 100$ is given by
$$40d + 0.15m = 100.$$
This is the equation of a parallel line with intercepts $d = 100/40 = 2.5$ and $m = 100/0.15 \approx 667$. The contours for $C = 150$ and $C = 200$ are parallel lines drawn similarly. (See Figure 7.18.)

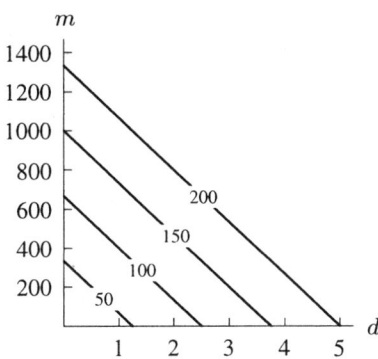

*Figure 7.18:* A contour diagram for
$C = 40d + 0.15m$

5. We can see that as we move horizontally to the right, we are increasing $x$ but not changing $y$. As we take such a path at $y = 2$, we cross decreasing contour lines, starting at the contour line 6 at $x = 1$ to the contour line 1 at around $x = 5.7$. This trend holds true for all of horizontal paths. Thus, $z$ is a decreasing function of $x$. Similarly, as we move up along a vertical line, we cross increasing contour lines and thus $z$ is an increasing function of $y$.

6. Looking at the contour diagram, we can see that $Q(x, y)$ is an increasing function of $x$ and a decreasing function of $y$. It stands to reason that as the price of orange juice goes up, the demand for orange juice will go down. Thus, the demand is a decreasing function of the price of orange juice and thus the $y$-axis corresponds to the price of orange juice. Also, as the price of apple juice goes up, so will the demand for orange juice and therefore the demand for orange juice is an increasing function of the price of apple juice. Thus, the price of apple juice corresponds to the $x$-axis.

7. Using our economic intuition, we know that the total sales of a product should be an increasing function of the amount spent on advertising. From the graph, $Q$ is a decreasing function of $x$ and an increasing function of $y$. Thus, the $y$-axis corresponds to the amount spent on advertising and the $x$-axis corresponds to the price of the product.

8. To find the hills, we look for areas of the map with concentric contour arcs that increase as we move closer to the center of the area. We see three hills: a height 9 hill centered near $(0, 2)$, a height 7 hill with a peak at $(3, 7)$, and a height 7 hill near $(7, 1.5)$. The highest hill is just the one with the highest contour line, in this case the hill near $(0, 2)$.

   We can expect the river to look like a valley, which is represented on a contour diagram as roughly parallel lines. We see such a pattern running from the bottom left at around $(2, 0)$ to the upper right at $(8, 8)$. To see where the river is flowing, we have to see which end of the river is at a higher altitude. At the bottom left of the map, the river has altitude between 3 and 4 and at the upper right, it reaches an altitude of 0. Thus, since rivers flow from areas of high altitude to areas of low altitude, the river flows toward the upper right—or northeast—section of the map.

9. To draw a contour for a wind-chill of $W = 20$, we need a few combinations of temperature and wind velocity $(T, v)$ such that $W(T, v) = 20$. Estimating from the table, some such points are $(24, 5)$ and $(33, 10)$. We can connect these points to get a contour for $W = 20$. Similarly, some points that have wind-chill of about $0°F$ are $(5, 5)$, $(17.5, 10)$, $(23.5, 15)$, $(27, 20)$, and $(29, 25)$. By connecting these points we get the contour for $W = 0$. If we carry out this procedure for more values of $W$, we get a full contour diagram such as is shown in Figure 7.19:

**324** CHAPTER SEVEN /SOLUTIONS

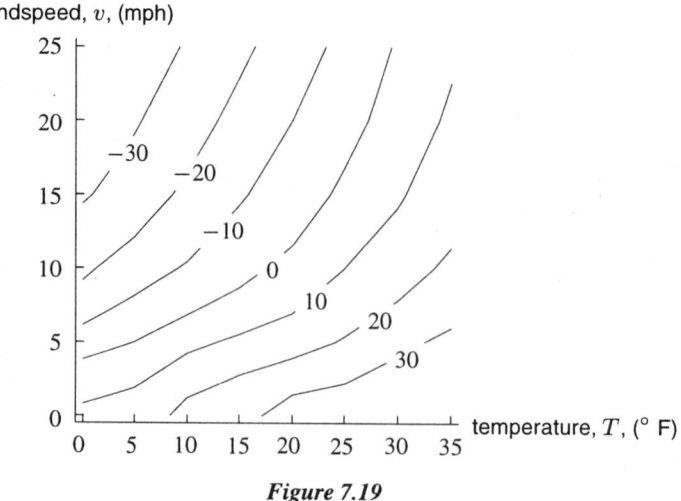

**Figure 7.19**

10. We first investigate the behavior of $f$ with $t$ fixed. We choose a value for $t$ and move horizontally across the diagram looking at how the values on the contours change. For $t = 1$ hour, as we move from the left at $x = 0$ to the right at $x = 5$, we cross contours of 0.1, 0.2, 0.3, 0.4, and 0.5. The concentration of the drug is increasing as the initial dose, $x$, increases. (This is what we saw in Figure 7.2 in Section 7.1.) For each choice of $f$ with $t$ fixed (each horizontal line), we see that the contour values are increasing as we move from left to right, showing that, at any given time, the concentration of the drug increases as the size of the dose increases.

The values of $f$ with $x$ fixed are read vertically. For $x = 4$, as we move up from $t = 0$ to $t = 5$, we see that the contours go up from 0.1 to 0.2 to 0.3, and then back down from 0.3 to 0.2 to 0.1. The maximum value of $C$ is very close to 0.4 and is reached at about $t = 1$ hour. The concentration increases quickly at first (the contour lines are closer together), reaches its maximum, and then decreases slowly (the contour lines are farther apart.) All of the contours of $f$ with $x$ fixed are similar to this, although the maximum value varies. For the contour at $x = 3$, we see that the maximum value is slightly below 0.2. If the dose of the drug is 1 mg, the concentration of the drug in the bloodstream never gets as high as 0.1, since we cross no contours on the cross-section of $f$ for $x = 1$.

11. (a) (i) As money increases, with love fixed, your happiness goes up, reaches a maximum and then goes back down. Evidently, there is such a thing as too much money.
    (ii) On the other hand, as love increases, with money fixed, your happiness keeps going up.
    (b) A cross-section with love fixed will show your happiness as money increases; the curve goes up to a maximum then back down, as in Figure 7.20. The higher cross-section, showing more overall happiness, corresponds to a larger amount of love, because as love increases so does happiness. Figure 7.21 shows two cross-sections with money fixed. Happiness increases as love increases. We cannot say, however, which cross-section corresponds to a larger fixed amount of money, because as money increases happiness can either increase or decrease.

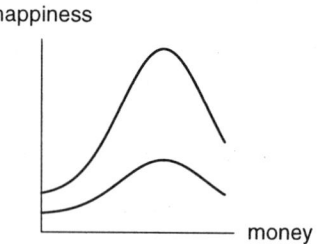

**Figure 7.20**: Cross-sections with love fixed

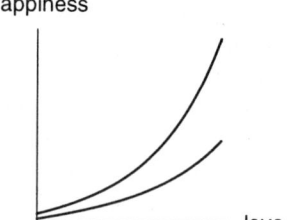

**Figure 7.21**: Cross-sections with money fixed

12. (a) The point representing 13% and $6000 on the graph lies between the 120 and 140 contours. We estimate the monthly payment to be about $137.

(b) Since the interest rate has dropped, we will be able to borrow more money and still make a monthly payment of $137. To find out how much we can afford to borrow, we find where the interest rate of 11% intersects the $137 contour and read off the loan amount to which these values correspond. Since the $137 contour is not shown, we estimate its position from the $120 and $140 contours. We find that we can borrow an amount of money that is more than $6000 but less than $6500. So we can borrow about $250 more without increasing the monthly payment.

(c) The entries in the table will be the amount of loan at which each interest rate intersects the 137 contour. Using the $137 contour from (b) we make table 7.18.

**TABLE 7.18** *Amount borrowed at a monthly payment of $137.*

| Interest Rate (%) | 0 | 1 | 2 | 3 | 4 | 5 | 6 | 7 |
|---|---|---|---|---|---|---|---|---|
| Loan Amount ($) | 8200 | 8000 | 7800 | 7600 | 7400 | 7200 | 7000 | 6800 |
| Interest rate (%) | 8 | 9 | 10 | 11 | 12 | 13 | 14 | 15 |
| Loan Amount ($) | 6650 | 6500 | 6350 | 6250 | 6100 | 6000 | 5900 | 5800 |

13. (a) Graph with kilometers north fixed at 50:

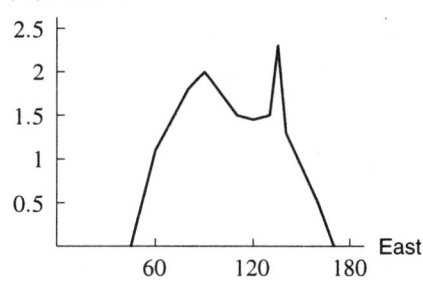

*Figure 7.22*

(b) Graph with kilometers north fixed at 100:

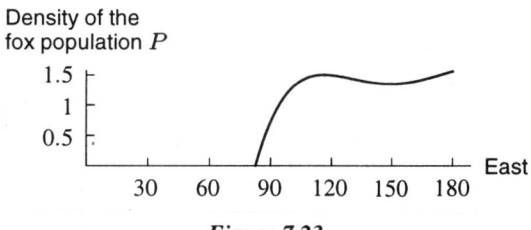

*Figure 7.23*

(c) Graph with kilometers east fixed at 60:

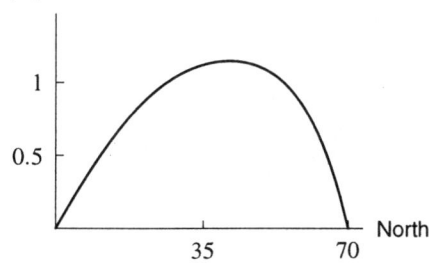

*Figure 7.24*

(d) Graph with kilometers east fixed at 120:

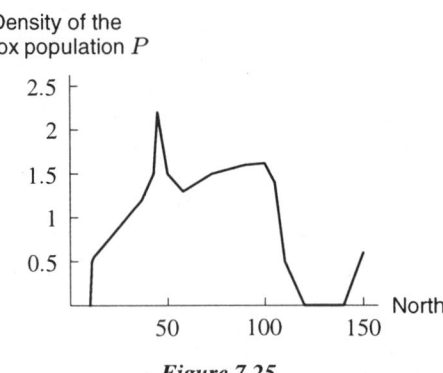

*Figure 7.25*

14. The contour where $f(x,y) = x + y = c$, or $y = -x + c$, is the graph of the straight line with slope $-1$ as shown in Figure 7.26. Note that we have plotted the contours for $c = -3, -2, -1, 0, 1, 2, 3$.

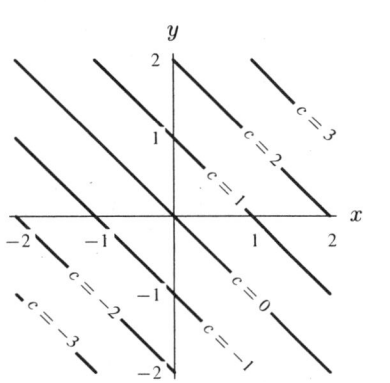

*Figure 7.26*

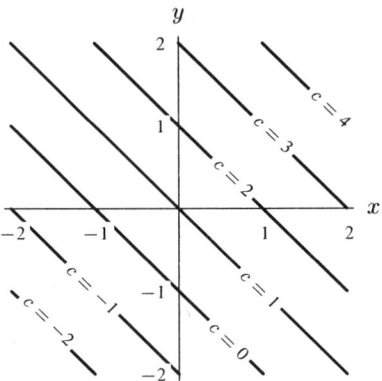

*Figure 7.27*

15. The contour where $f(x,y) = x + y + 1 = c$ or $y = -x + c - 1$ is the graph of the straight line of slope $-1$ as shown in Figure 7.27. Note that we have plotted the contours for $c = -2, -1, 0, 1, 2, 3, 4$.

16. The contour where $f(x,y) = 3x + 3y = c$ or $y = -x + c/3$ is the graph of the straight line of slope $-1$ as shown in Figure 7.28. Note that we have plotted the contours for $c = -9, -6, -3, 0, 3, 6, 9$.

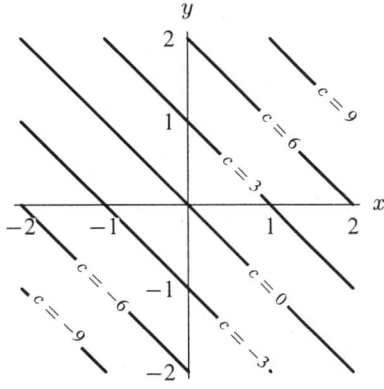

*Figure 7.28*

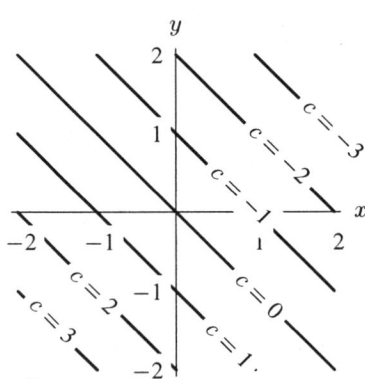

*Figure 7.29*

17. The contour where $f(x,y) = -x - y = c$ or $y = -x - c$ is the graph of the straight line of slope $-1$ as shown in Figure 7.29. Note that we have plotted contours for $c = -3, -2, -1, 0, 1, 2, 3$.

18. The contour where $f(x,y) = 2x - y = c$ is the graph of the straight line $y = 2x - c$ of slope 2 as shown in Figure 7.30. Note that we have plotted the contours for $c = -3, -2, -1, 0, 1, 2, 3$.

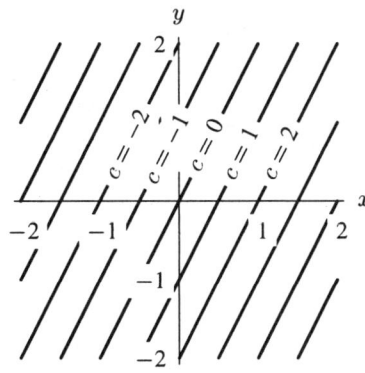

**Figure 7.30**

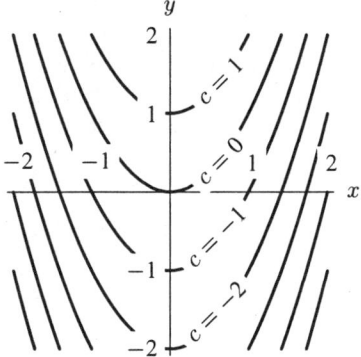

**Figure 7.31**

19. The contour where $f(x,y) = y - x^2 = c$ is the graph of the parabola $y = x^2 + c$, with vertex $(0, c)$ and symmetric about the $y$-axis, shown in Figure 7.31. Note that we have plotted the contours for $c = -2, -1, 0, 1$. The contours become more closely packed as we move farther from the $y$-axis.

20. One possible answer follows.

    (a) If the middle line is a highway in the city, there may be many people living around it since it provides mass transit nearby that nearly everyone needs and uses. Thus, the population density would be greatest near the highway, and less dense farther from the highway, as is shown in diagram (I). In an alternative solution, highways may be considered inherently very noisy and dirty. In this scenario, the people in a certain community may purposely live away from the highway. Then the population density would follow the pattern in diagram (III), or in an extreme case, even diagram (II).

    (b) If the middle line is a sewage canal, there will be no one living within it, and very few people living close to it. This is represented in diagram (II).

    (c) If the middle line is a railroad line in the city, then one scenario is that very few people would enjoy living nearby a railroad yard, with all of the noise and difficulty in crossing it. Then the population density would be smallest near the middle line, and greatest farther from the railroad, as in diagram (III). In an alternative solution, a community might well depend upon the railroad for transit, news or supplies, in which case there would be a denser population nearer the railroad than farther from it, as in diagram (I).

21. A possible contour diagram showing light intensity in the park as a function of position is in Figure 7.32.

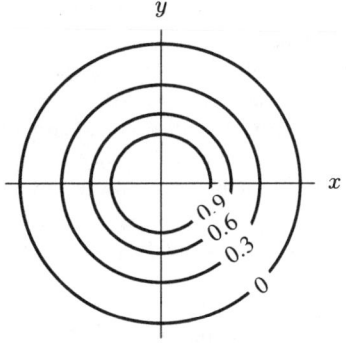

**Figure 7.32**

We can either label the contours as in Figure 7.32, or in reverse order. Figure 7.32 assumes that the park is well lit, and that the coastline is dark. Alternatively, we could assume that the park is dark at night, but that near the coastline are a lot of tourist areas, maybe a carnival, or a clambake and bonfire. Many answers are possible.

**328** CHAPTER SEVEN /SOLUTIONS

22. (a) *False*. The values on the level curves are decreasing as you go northward in Canada.
    (b) *True*. In general, the contour levels are increasing from peninsulas to mainland. This is true for all three examples. For instance, the density on the Baja peninsula is mostly below 180 whereas on the mainland nearby it is 200 and up to 280.
    (c) *False*. It is below 100, since the values on the level curves are decreasing as you go southward in Florida and Miami is south of the 100 level curve.
    (d) *True*. Pick the point $P$ where the level curves are the closest together, and pick the direction in which the values on the level curves are increasing fastest. See the arrow marked in Figure 7.33.

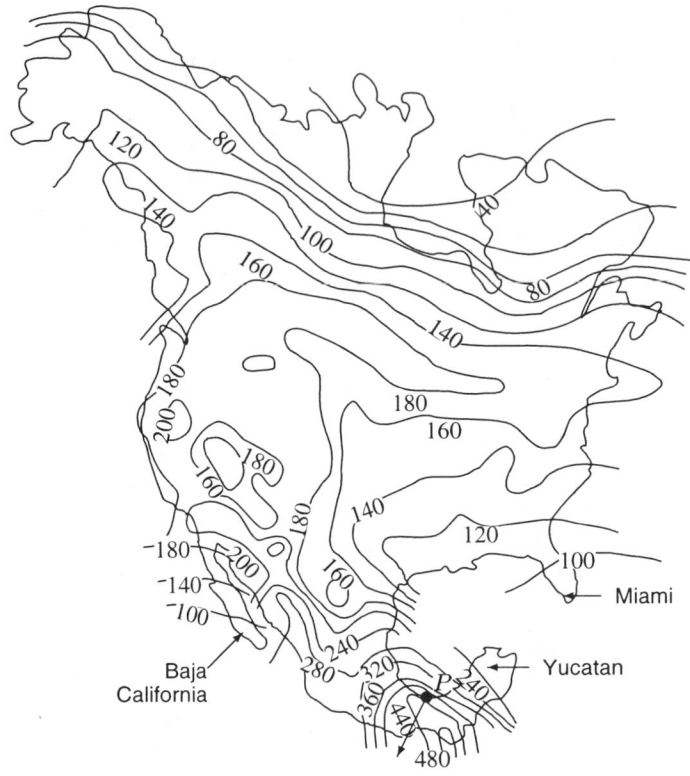

*Figure 7.33*

23. One possible answer follows.
    (a) If there is a city at the center of the diagram, then the population is very dense at the center, but progressively less dense as you move into the suburbs, further from the city center. This scenario corresponds to diagram (I) or (II). We pick (I) because it has the highest density, as we would expect in a city.
    (b) If the center of the diagram is a lake, and is a very busy and thriving center, where lake front property is considered the most desirable, then the most dense area will be at lakeside, and decrease as you move further from the lake in the center, as in diagram (I) or (II). We pick (II) because we expect the population density at lake front to be less than that in the center of a city.
        In an alternative solution, if the lake were in the middle of nowhere, the entire area would be very sparsely populated, and there would be slightly fewer people living on the actual lake shore, as in diagram (III).
    (c) If the center of the diagram is a power plant and if the plant is not in a densely populated area, where people can and will choose not to live anywhere near it, the population density will then be very low nearby, increasing slightly further from the plant, as in diagram (III).

24. Many different answers are possible. Answers are in degrees Celsius.

   (a) Minnesota in winter.

   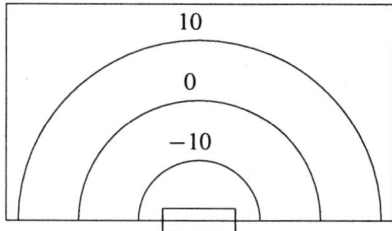

   (b) San Francisco in winter.

   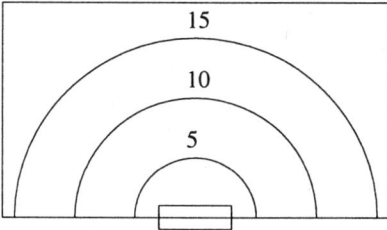

   (c) Houston in summer.

   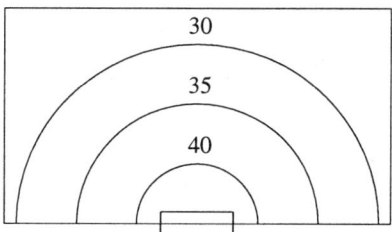

   (d) Oregon in summer.

   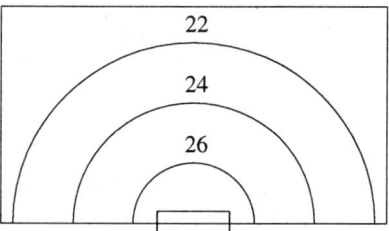

25. (a) The contour lines are much closer together on path A, so path A is steeper.
   (b) If you are on path A and turn around to look at the countryside, you find hills to your left and right, obscuring the view. But the ground falls away on either side of path $B$, so you are likely to get a much better view of the countryside from path $B$.
   (c) There is more likely to be a stream alongside path A, because water follows the direction of steepest descent.

**330** CHAPTER SEVEN /SOLUTIONS

26. (a) In this company success only increases when money increases, so success will remain constant along the work axis. However, as money increases so does success, which is shown in Graph (III).
    (b) As both work and money increase, success never increases, which corresponds to Graph (II).
    (c) If the money doesn't matter, then regardless of how much the money increases success will be constant along the money axis. However, success increases as work increases. This is best represented in Graph (I).
    (d) This company's success increases as both money and work increase, which is demonstrated in Graph (IV).

27. (a) If we have iron stomachs and can consume cola and pizza endlessly without ill effects, then we expect our happiness to increase without bound as we get more cola and pizza. Graph (IV) shows this since it increases along both the pizza and cola axes throughout.
    (b) If we get sick upon eating too many pizzas or drinking too much cola, then we expect our happiness to decrease once either or both of those quantities grows past some optimum value. This is depicted in graph (I) which increases along both axes until a peak is reached, and then decreases along both axes.
    (c) If we do get sick after too much cola, but are always able to eat more pizza, then we expect our happiness to decrease after we drink some optimum amount of cola, but continue to increase as we get more pizza. This is shown by graph (III) which increases continuously along the pizza axis but, after reaching a maximum, begins to decrease along the cola axis.

28. The values in Table (a) are not constant along rows or columns and therefore cannot be the lines shown in (I) or (IV). Also observe that as you move away from the origin, whose contour value is 0, the $z$-values on the contours increase. Thus, this table corresponds to diagram (II).

    The values in Table (b) are also not constant along rows or columns. Since the contour values are decreasing as you move away from the origin, this table corresponds to diagram (III).

    Table (c) shows that for each fixed value of $x$, we have constant contour value, suggesting a straight vertical line at each $x$-value, as in diagram (IV).

    Table (d) also shows lines, however these are horizontal since for each fixed value of $y$ we have constant contour values. Thus, this table matches diagram (I).

29. The point $x = 10$, $t = 5$ is between the contours $H = 70$ and $H = 75$, a little closer to the former. Therefore, we estimate $H(10, 5) \approx 72$, i.e., it is about $72°$F. Five minutes later we are at the point $x = 10$, $t = 10$, which is just above the contour $H = 75$, so we estimate that it has warmed up to $76°$F by then.

30. The line $t = 5$ crosses the contour $H = 80$ at about $x = 4$; this means that $H(4, 5) \approx 80$, and so the point $(4, 80)$ is on the graph of the one-variable function $y = H(x, 5)$. Each time the line crosses a contour, we can plot another point on the graph of $H(x, 5)$, and thus get a sketch of the graph. See Figure 7.34. Each data point obtained from the contour map has been indicated by a dot on the graph. The graph of $H(x, 20)$ was obtained in a similar way.

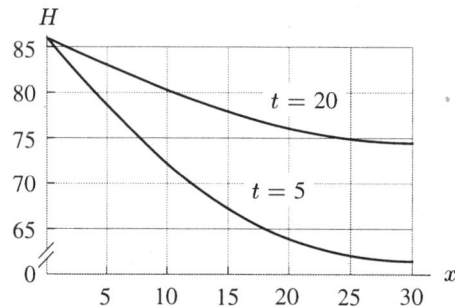

**Figure 7.34**: Graph of $H(x, 5)$ and $H(x, 20)$: heat as a function of distance from the heater at $t = 5$ and $t = 20$ minutes

These two graphs describe the temperature at different positions as a function of $x$ for $t = 5$ and $t = 20$.

Notice that the graph of $H(x, 5)$ descends more steeply than the graph of $H(x, 20)$; this is because the contours are quite close together along the line $t = 5$, whereas they are more spread out along the line $t = 20$. In practical terms the shape of the graph of $H(x, 5)$ tells us that the temperature drops quickly as you move away from the heater, which makes sense, since the heater was turned on just five minutes ago. On the other hand, the graph of $H(x, 20)$ descends more slowly, which makes sense, because the heater has been on for 20 minutes and the heat has had time to diffuse throughout the room.

31. (a) To find cardiac output $o$ when pressure $p = 4$ mm Hg and time $t = 0$ hours (i.e. when the patient goes into shock) we go to the coordinate $(0, 4)$ on the graph; we see that this point falls at where $o = 12$ L/min, so initial cardiac output for $p = 4$ mm Hg is 12 L/min. Three hours later, $t = 3$ and $p$ still $= 4$. Going to $(3, 4)$ on the graph, we find that we are somewhat to the right of where $o = 8$. Since output is decreasing as we move to the right, output at $(3, 4)$ will be somewhat less than 8. Estimating, 3 hours after shock with pressure of 4 mm Hg, $o \approx 7.5$ L/min. Since $o = 12$ at $t = 0$, we wish to find the time corresponding to $o = 12/2 = 6$ and $p$ still $= 4$. Going along the line of $o = 6$, we find that this intersects $p = 4$ at $t \approx 4.2$. So about 4.2 hours have elapsed when cardiac output halves.

(b) Looking at the graph, we see that output increases as we move up a vertical line.

(c) Looking at the graph, we see that output decreases as we move to the right along a horizontal line.

(d) If $p = 3$, the cardiac output decreases rather slowly between $t = 0$ and $t = 2$, the first two hours. We can see this because there is a large gap between initial output, which is approximately 9 L/min, and $o = 8$. In these two hours, output only falls by 1 L/min, a fairly gradual change. Between $t = 2$ and $t = 4$, output falls from 8 L/min to 6 L/min. In the last hour, output falls all the way from 6 L/min to 0 L/min, corresponding to death. So the decrease in output accelerates rapidly. This information might be very useful to physicians treating shock. It tells them that in the first two hours the patient's condition will not worsen very much; in the next two hours, it will deteriorate at a faster rate. After four hours, the patient may well die unless something is done to maintain the patient's cardiac output.

32. (a) We will draw the contour diagram with $c$ on the horizontal axis and $t$ on the vertical axis. First, we will find the equation of the contour with $Q = 25$. We substitute $Q = 25$ into our equation.

$$25 = 100\frac{t}{c}.$$

Solving for $t$ give $t = c/4$. This contour is a line through the origin with a slope of $1/4$. Substituting $Q = 50$ gives $t = c/2$, a line through the origin with slope $1/2$. For $Q = 100$, our contour line is $t = c$, a line through the origin with slope 1, and if $Q = 200$, we have $t = 2c$, a line through the origin with slope 2. These contour lines are drawn in Figure 7.35.

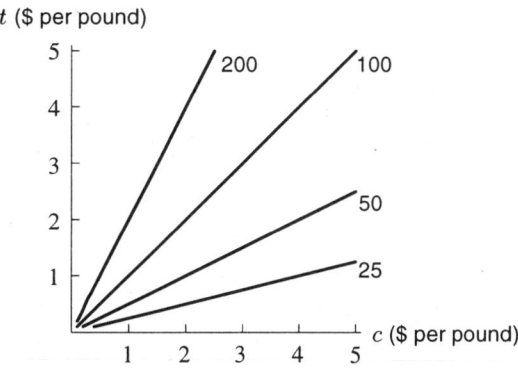

*Figure 7.35:* A contour diagram of demand for coffee

(b) To see if $Q$ is an increasing or a decreasing function of $c$, we read across contour lines from left to right to see how the contour values change. We see that the contour values are decreasing, and $Q$ is a decreasing function of $c$.

To see if $Q$ is an increasing or a decreasing function of $t$, we see how $Q$ changes as $t$ increases. In other words, we see how the contour values change as we read from the bottom to the top of the contour diagram. Since these values are increasing, $Q$ is an increasing function of $t$.

This makes sense. As the price of coffee goes up, the demand for coffee goes down, so $Q$ is a decreasing function of $c$. As the price of tea goes up, people buy less tea and more coffee, so the demand for coffee goes up and $Q$ is an increasing function of $t$.

(c) We fix the price of tea at 1, so we are looking along the horizontal line with $t = 1$. On the left, for a small change in $c$, we cross several contour lines. This corresponds to a large change in demand. On the right, for an equivalent small change in $c$, the demand does not change very much since the contour lines are farther apart. We know that demand for coffee is the most elastic when prices are low. At low prices, a small increase in price will cause a relatively large decrease in demand.

## Solutions for Section 7.3

1. (a) We expect the demand for coffee to decrease as the price of coffee increases (assuming the price of tea is fixed.) Thus we expect $f_c$ to be negative. We expect people to switch to coffee as the price of tea increases (assuming the price of coffee is fixed), so that the demand for coffee will increase. We expect $f_t$ to be positive.
   (b) The statement $f(3,2) = 780$ tells us that if coffee costs \$3 per pound and tea costs \$2 per pound, we can expect 780 pounds of coffee to sell each week. The statement $f_c(3,2) = -60$ tells us that, if the price of coffee then goes up \$1 and the price of tea stays the same, the demand for coffee will go down by about 60 pounds. The statement $20 = f_t(3,2)$ tells us that if the price of tea goes up \$1 and the price of coffee stays the same, the demand for coffee will go up by about 20 pounds.

2. (a) The units of $\partial c/\partial x$ are units of concentration/distance. (For example, (gm/cm$^3$)/cm.) The practical interpretation of $\partial c/\partial x$ is the rate of change of concentration with distance as you move down the blood vessel at a fixed time. We expect $\partial c/\partial x < 0$ because the further away you get from the point of injection, the less of the drug you would expect to find (at a fixed time).
   (b) The units of $\partial c/\partial t$ are units of concentration/time. (For example, (gm/cm$^3$)/sec.) The practical interpretation of $\partial c/\partial t$ is the rate of change of concentration with time, as you look at a particular point in the blood vessel. We would expect the concentration to first increase (as the drug reaches the point) and then decrease as the drug dies away. Thus, we expect $\partial c/\partial t > 0$ for small $t$ and $\partial c/\partial t < 0$ for large $t$.

3. (a) This means you must pay a mortgage payment of \$1090.08/month if you have borrowed a total of \$92,000 at an interest rate of 14%, on a 30-year mortgage.
   (b) This means that the rate of change of the monthly payment with respect to the interest rate is \$72.82; i.e., your monthly payment will go up by approximately \$72.82 for one percentage point increase in the interest rate for the \$92,000 borrowed under a 30-year mortgage.
   (c) It should be *positive*, because the monthly payments will increase if the total amount borrowed is increased.
   (d) It should be *negative*, because as you increase the number of years in which to pay the mortgage, you should have to pay less each month.

4. (a) If you borrow \$8000 at an interest rate of 1% per month and pay it off in 24 months, your monthly payments are \$376.59.
   (b) The increase in your monthly payments for borrowing an extra dollar under the same terms as in (a) is about 4.7 cents.
   (c) If you borrow the same amount of money for the same time period as in (a), but if the interest rate increases by 1%, the increase in your monthly payments is about \$44.83.

5. $\partial P/\partial t$: The unit is dollars per month. This is the rate at which payments change as the number of months it takes to pay off the loan changes. The sign is negative because payments decrease as the pay-off time increases.

   $\partial P/\partial r$: The unit is dollars per percentage point. This is the rate at which payments change as the interest rate changes. The sign is positive because payments increase as the interest rate increases.

6. (a) If we move horizontally by increasing $x$ along any of the rows of this table, we see that the value of $f$ decreases when $x$ increases, so $f_x < 0$. Similarly, if we move vertically by increasing $y$ along any of the columns of this table, we see that the value of $f$ increases, so $f_y > 0$.
   (b) We know that the formula for $f_x$ at a point $(x, y)$ is

   $$f_x(x,y) \approx \frac{f(x+\Delta x, y) - f(x,y)}{\Delta x}.$$

   Approximating with $\Delta x = 10$ at the point $(10, 6)$ gives:

   $$\begin{aligned}f_x(10,6) &\approx \frac{f(10+10, 6) - f(10, 6)}{10} \\ &= \frac{f(20,6) - f(10,6)}{10} \\ &= \frac{92 - 98}{10} = -0.6.\end{aligned}$$

Similarly for $f_y$, we approximate with $\Delta y = 2$.

$$f_y(10, 6) \approx \frac{f(10, 6+2) - f(10, 6)}{2}$$
$$= \frac{f(10, 8) - f(10, 6)}{2}$$
$$= \frac{105 - 98}{2} = 3.5.$$

(c) We use $\Delta f \approx \Delta x f_x + \Delta y f_y$. To calculate $f(30, 9)$ using the value for $f(30, 8)$ shown in the table, we need to calculate the partial derivative $f_y(30, 8)$. Using $\Delta y = -2$, we have

$$f_y(30, 8) \approx \frac{f(30, 8-2) - f(30, 8)}{-2} = 3.$$

Then

$$f(30, 9) \approx f(30, 8) + 0 \cdot f_x(30, 8) + 1 \cdot f_y(30, 8)$$
$$\approx 94 + 0 + 3$$
$$= 97.$$

To calculate $f(34, 8)$ using the value for $f(30, 8)$ shown in the table, we need to calculate the partial derivative $f_x(30, 8)$. Using $\Delta x = -10$, we have

$$f_x(30, 8) \approx \frac{f(30 - 10, 8) - f(30, 8)}{-10} = -0.5.$$

Then

$$f(34, 8) \approx f(30, 8) + 4 \cdot f_x(30, 8) + 0 \cdot f_y(30, 8)$$
$$\approx 94 - 2 + 0$$
$$= 92.$$

7. (a) Positive.
   (b) Negative.
   (c) Positive.
   (d) Zero.

8. $f_x(5, 2)$, because the contour lines in the positive $x$ direction are closer together at $(5, 2)$ than at $(3, 1)$.

9. If $h$ is small, then

$$f_x(3, 2) \approx \frac{f(3+h, 2) - f(3, 2)}{h}.$$

With $h = 0.01$, we find

$$f_x(3, 2) \approx \frac{f(3.01, 2) - f(3, 2)}{0.01} = \frac{\frac{3.01^2}{(2+1)} - \frac{3^2}{(2+1)}}{0.01} = 2.00333.$$

With $h = 0.0001$, we get

$$f_x(3, 2) \approx \frac{f(3.0001, 2) - f(3, 2)}{0.0001} = \frac{\frac{3.0001^2}{(2+1)} - \frac{3^2}{(2+1)}}{0.0001} = 2.0000333.$$

Since the difference quotient seems to be approaching 2 as $h$ gets smaller, we conclude

$$f_x(3, 2) \approx 2.$$

To estimate $f_y(3, 2)$, we use

$$f_y(3, 2) \approx \frac{f(3, 2+h) - f(3, 2)}{h}.$$

With $h = 0.01$, we get

$$f_y(3,2) \approx \frac{f(3, 2.01) - f(3,2)}{0.01} = \frac{\frac{3^2}{(2.01+1)} - \frac{3^2}{(2+1)}}{0.01} = -0.99668.$$

With $h = 0.0001$, we get

$$f_y(3,2) \approx \frac{f(3, 2.0001) - f(3,2)}{0.0001} = \frac{\frac{3^2}{(2.0001+1)} - \frac{3^2}{(2+1)}}{0.0001} = -0.9999667.$$

Thus, it seems that the difference quotient is approaching $-1$, so we estimate

$$f_y(3,2) \approx -1.$$

10. Estimate $\partial P/\partial r$ and $\partial P/\partial L$ by using difference quotients and reading values of $P$ from the graph:

$$\left.\frac{\partial P}{\partial r}\right|_{(8,5000)} \approx \frac{P(15, 5000) - P(8, 5000)}{15 - 8}$$
$$= \frac{120 - 100}{7} = 2.9,$$

and

$$\left.\frac{\partial P}{\partial L}\right|_{(8,5000)} \approx \frac{P(8, 4000) - P(8, 5000)}{4000 - 5000}$$
$$= \frac{80 - 100}{-1000} = 0.02.$$

$P_r(8, 5000) \approx 2.9$ means that at an interest rate of 8% and a loan amount of $5000 the monthly payment increases by approximately $2.90 for every one percent increase of the interest rate. $P_L(8, 5000) \approx 0.02$ means the monthly payment increases by approximately $0.02 for every $1 increase in the loan amount at an 8% rate and a loan amount of $5000.

11. (a) We know we can approximate $f_p$ at a point $(I, p)$ by $f_p(I, p) \approx \frac{f(I, p+\Delta p) - f(I,p)}{\Delta p}$. Approximating with $\Delta p = 0.5$ at the point $(80, 4.0)$ gives:

$$f_p(80, 4.0) \approx \frac{f(80, 4.0 + 0.5) - f(80, 4.0)}{0.5}$$
$$= \frac{f(80, 4.5) - f(80, 4.0)}{0.5}$$
$$= \frac{5.07 - 5.19}{0.5} = -0.24$$

When beef rises in price by a dollar per pound, the average household with an income of $80,000 will buy 0.24 fewer pounds of beef per week.

(b) We estimate $f_I$ similarly; we approximate with $\Delta I = 20$.

$$f_I(80, 4.0) \approx \frac{f(80 + 20, 4.0) - f(80, 4.0)}{20}$$
$$= \frac{f(100, 4.0) - f(80, 4.0)}{20}$$
$$= \frac{5.60 - 5.19}{20} = 0.0205$$

When a household's income rises by $1000, it will buy 0.0205 more pounds of beef per week.

(c) We wish to find $f(110, 4.0)$. To do this, we employ the formula $\Delta f \approx \Delta I f_I + \Delta p f_p$. Using our results for $f_I(80, 4.0)$ and $f_p(110, 4.0)$ from above gives

$$f(110, 4.0) \approx f(80, 4.0) + 30 \cdot f_I(80, 4.0) + 0 \cdot f_p(80, 4.0)$$
$$\approx 5.60 + 0.615 + 0$$
$$= 6.215 \text{ lbs.}$$

12. (a) We know that we can approximate $f_R$ at a point $(R, T)$ by $f_R(R, T) \approx \frac{f(R+\Delta R, T) - f(R, T)}{\Delta R}$. Approximating with $\Delta R = 3$ at the point $(15, 76)$ gives:

$$f_R(15, 76) \approx \frac{f(15+3, 76) - f(15, 76)}{3}$$
$$= \frac{f(18, 76) - f(15, 76)}{3}$$

Looking at the graph, we see that $f(15, 76) \approx 100$ and $f(18, 76) \approx 110$, so

$$f_R(15, 76) \approx \frac{110 - 100}{3} \approx 3.3\% \text{ per in.}$$

Each additional inch of rainfall increases corn yield by about 3.3% from the present production level.

(b) We estimate $f_T$ similarly; we approximate with $\Delta T = 2$.

$$f_T(15, 76) \approx \frac{f(15, 76+2) - f(15, 76)}{2}$$
$$= \frac{f(15, 78) - f(15, 76)}{2}$$

Looking at the graph, we see that $f(15, 76) \approx 100$ and $f(15, 78) \approx 90$, so

$$f_T(15, 76) \approx \frac{90 - 100}{2} = -5\% \text{ per } °F.$$

For every Fahrenheit degree that the temperature increases, corn yield falls by 5% from the current production level.

13. We can use the formula $\Delta f \approx \Delta x f_x + \Delta y f_y$. Applying this formula in order to estimate $f(105, 21)$ from the known value of $f(100, 20)$ gives

$$f(105, 21) \approx f(100, 20) + (105 - 100)f_x(100, 20) + (21 - 20)f_y(100, 20)$$
$$= 2750 + 5 \cdot 4 + 1 \cdot 7$$
$$= 2777.$$

14. We can use the formula $\Delta f \approx \Delta r f_r + \Delta s f_s$. Applying this formula in order to estimate $f(52, 108)$ from the known value of $f(50, 100)$ gives

$$f(52, 108) \approx f(50, 100) + (52 - 50)f_r(50, 100) + (108 - 100)f_s(50, 100)$$
$$= 5.67 + 2(0.60) + 8(-0.15)$$
$$= 5.67.$$

15. For $f_w(10, 25)$ we get

$$f_w(10, 25) \approx \frac{f(10+h, 25) - f(10, 25)}{h}.$$

Choosing $h = 5$ and reading values from Table 7.4 on page 360 of the text, we get

$$f_w(10, 25) \approx \frac{f(15, 25) - f(10, 25)}{5} = \frac{2 - 10}{5} = -1.6$$

This means that when the wind speed is 10 mph and the true temperature is $25°F$, as the wind speed increases from 10 mph by 1 mph we feel a $1.6°F$ drop in temperature. This rate is negative because the temperature you feel drops as the wind speed increases.

**336** CHAPTER SEVEN /SOLUTIONS

16. Using a difference quotient with $h = 5$, we get

$$f_T(5, 20) \approx \frac{f(5, 20 + 5) - f(5, 20)}{5} = \frac{21 - 16}{5} = 1$$

This means that when the wind speed is 5 mph and the true temperature is 20°F, the apparent temperature increases by approximately 1°F for every increase of 1°F in the true temperature. This rate is positive because the true temperature you feel increases as true temperature increases.

17. Since the average rate of change of the temperature adjusted for wind-chill is about $-2.6$ (drops by 2.6°F), with every 1 mph increase in wind speed from 5 mph to 10 mph, when the true temperature stays constant at 20°F, we know that

$$f_w(5, 20) \approx -2.6.$$

18. We estimate $\partial I/\partial H$ and $\partial I/\partial T$ by using difference quotients. We have

$$\frac{\partial I}{\partial H} \approx \frac{f(H + \Delta H, T) - f(H, T)}{\Delta H} \quad \text{and} \quad \frac{\partial I}{\partial T} \approx \frac{f(H, T + \Delta T) - f(H, T)}{\Delta T}$$

Choosing $\Delta H = 10$ and reading the values from Table 7.6 on page 366 in the text we get

$$\left.\frac{\partial I}{\partial H}\right|_{(10,100)} \approx \frac{f(10 + 10, 100) - f(10, 100)}{10} = \frac{99 - 95}{10} = 0.4.$$

Similarly, choosing $\Delta T = 5$ we get

$$\left.\frac{\partial I}{\partial T}\right|_{(10,100)} \approx \frac{f(10, 100 + 5) - f(10, 100)}{5} = \frac{100 - 95}{5} = 1.$$

The fact that $\left.\dfrac{\partial I}{\partial H}\right|_{(10,100)} \approx 0.4$ means that the rate of change of the heat index per unit increase in humidity is about 0.4. This means that the heat index increases by approximately 0.4°F for every percentage point increase in humidity. This rate is positive, because as the humidity increases, the heat index increases.

The partial derivative $\partial I/\partial T$ gives the rate of increase of heat index with respect to temperature. It is positive because the heat index increases as temperature increases. Knowing that $\left.\dfrac{\partial I}{\partial T}\right|_{(10,100)} \approx 1$ tells us that as the temperature increases by 1°F, the temperature you feel increases by 1°F also. In other words, at this humidity and temperature, the changes in temperature that you feel are approximately equal to the actual changes in temperature.

The fact that $\dfrac{\partial I}{\partial T} > \dfrac{\partial I}{\partial H}$ at $(10, 100)$ tells us that a unit increase in temperature has a greater effect on the heat index in Tucson than a unit increase in humidity.

19. We estimate $\partial I/\partial H$ and $\partial I/\partial T$ by using difference quotients. We have

$$\frac{\partial I}{\partial H} \approx \frac{f(H + \Delta H, T) - f(H, T)}{\Delta H} \quad \text{and} \quad \frac{\partial I}{\partial T} \approx \frac{f(H, T + \Delta T) - f(H, T)}{\Delta T}$$

Choosing $\Delta H = 10$ and reading the values from Table 7.6 on page 366 in the text we have

$$\left.\frac{\partial I}{\partial H}\right|_{(50,80)} \approx \frac{f(50 + 10, 80) - f(50, 80)}{10} = \frac{82 - 81}{10} = 0.1.$$

Similarly, choosing $\Delta T = 5$ we get

$$\left.\frac{\partial I}{\partial T}\right|_{(50,80)} \approx \frac{f(50, 80 + 5) - f(50, 80)}{5} = \frac{88 - 81}{5} = 1.4.$$

The fact that $\left.\dfrac{\partial I}{\partial H}\right|_{(50,80)} \approx 0.1$ means that the average rate of change of the heat index per unit increase in humidity is about 0.1°F. This means that the heat index increases by approximately 0.1 for every percentage point increase in humidity. This rate is positive, because as the humidity increases, the heat index increases.

The partial derivative $\partial I/\partial T$ gives the rate of increase of heat index with respect to temperature. It is positive because the heat index increases as temperature increases. Knowing that $\left.\dfrac{\partial I}{\partial T}\right|_{(50,80)} \approx 1.4$ tells us that as the temperature increases by $1°$F, the temperature you feel increases by $1.4°$F. Thus, at this humidity and temperature, the changes in temperature you feel are even larger then the actual change.

The fact that $\partial I/\partial T > \partial I/\partial H$ at (50,80) tells the residents of Boston that a unit increase in temperature has a greater effect on the heat index than a unit increase in humidity.

20. (a) The table gives $f(200, 400) = 150{,}000$. This means that sales of 200 full-price tickets and 400 discount tickets generate $150,000 in revenue.
   (b) The notation $f_x(200, 400)$ represents the rate of change of $f$ as we fix $y$ at 400 and increase $x$ from 200. What happens to the revenue as we look along the row $y = 400$ in the table? Revenue increases, so $f_x(200, 400)$ is positive. The notation $f_y(200, 400)$ represents the rate of change of $f$ as we fix $x$ at 200 and increase $y$ from 400. What happens to the revenue as we look down the column $x = 200$ in the table? Revenue increases, so $f_y(200, 400)$ is positive.
   
   Both partial derivatives are positive. This makes sense since revenue increases if more of either type of ticket is sold.
   (c) To estimate $f_x(200, 400)$, we calculate $\Delta R/\Delta x$ as $x$ increases from 200 to 300 while $y$ is fixed at 400. We have

$$f_x(200, 400) \approx \frac{\Delta R}{\Delta x} = \frac{185{,}000 - 150{,}000}{300 - 200} = 350 \text{ dollars/ticket}.$$

The partial derivative of $f$ with respect to $x$ is 350 dollars per full-price ticket. This means that the price of a full-price ticket is $350.

To estimate $f_y(200, 400)$, we calculate $\Delta R/\Delta y$ as $y$ increases from 400 and 600 while $x$ is fixed at 200. We have

$$f_y(200, 400) \approx \frac{\Delta R}{\Delta y} = \frac{190{,}000 - 150{,}000}{600 - 400} = 200 \text{ dollars/ticket}.$$

The partial derivative of $f$ with respect to $y$ is 200 dollars per discount ticket. This means that the price of a discount ticket is $200.

21. (a) The partial derivative $f_x = 350$ tells us that $R$ increases by $350 as $x$ increases by 1. Thus, $f(201, 400) = f(200, 400) + 350 = 150{,}000 + 350 = 150{,}350$.
   (b) The partial derivative $f_y = 200$ tells us that $R$ increases by $200 as $y$ increases by 1. Since $y$ is increasing by 5, we have $f(200, 405) = f(200, 400) + 5(200) = 150{,}000 + 1{,}000 = 151{,}000$.
   (c) Here $x$ is increasing by 3 and $y$ is increasing by 6. We have $f(203, 406) = f(200, 400) + 3(350) + 6(200) = 150{,}000 + 1{,}050 + 1{,}200 = 152{,}250$.

   In this problem, the partial derivatives gave exact results, but in general they only give an estimate of the changes in the function.

22. We are asked to find $f(26, 15)$. The closest to this point in the data given in the table is $f(24, 15)$, so we will approximate based on this value. First we need to calculate the partial derivative $f_t$. Using $\Delta t = -2$, we have

$$f(24, 15) \approx \frac{f(22, 15) - f(24, 15)}{-2} = \frac{58 - 36}{-2} = -11.$$

Given this, we can use $\Delta f \approx \Delta t f_t + \Delta c f_c$ to obtain

$$f(26, 15) \approx f(24, 15) + 2 \cdot f_t(24, 15) + 0 \cdot f_c(26, 15)$$
$$\approx 36 - 2 \cdot 11 + 0$$
$$= 14\%.$$

23. (a) An increase in the price of a new car will decrease the number of cars bought annually. Thus $\dfrac{\partial q_1}{\partial x} < 0$. Similarly, an increase in the price of gasoline will decrease the amount of gas sold, implying $\dfrac{\partial q_2}{\partial y} < 0$.
   (b) Since the demands for a car and gas complement each other, an increase in the price of gasoline will decrease the total number of cars bought. Thus $\dfrac{\partial q_1}{\partial y} < 0$. Similarly, we may expect $\dfrac{\partial q_2}{\partial x} < 0$.

**338** CHAPTER SEVEN /SOLUTIONS

24. We know $z_x(1,0)$ is the rate of change of $z$ in the $x$-direction at $(1,0)$. Therefore

$$z_x(1,0) \approx \frac{\Delta z}{\Delta x} \approx \frac{1}{0.5} = 2, \quad \text{so } z_x(1,0) \approx 2.$$

We know $z_x(0,1)$ is the rate of change of $z$ in the $x$-direction at the point $(0,1)$. Since we move along the contour, the change in $z$

$$z_x(0,1) \approx \frac{\Delta z}{\Delta x} \approx \frac{0}{\Delta x} = 0.$$

We know $z_y(0,1)$ is the rate of change of $z$ in the $y$-direction at the point $(0,1)$ so

$$z_y(0,1) \approx \frac{\Delta z}{\Delta y} \approx \frac{1}{0.1} = 10.$$

25. Estimating from the contour diagram, using positive increments for $\Delta x$ and $\Delta y$, we have, for point $A$,

$$\left.\frac{\partial n}{\partial x}\right|_{(A)} \approx \frac{1.5 - 1}{67 - 59} = \frac{1/2}{8} = \frac{1}{16} \approx 0.06 \; \frac{\text{foxes/km}^2}{\text{km}}$$

$$\left.\frac{\partial n}{\partial y}\right|_{(A)} \approx \frac{0.5 - 1}{60 - 51} = -\frac{1/2}{9} = -\frac{1}{18} \approx -0.06 \; \frac{\text{foxes/km}^2}{\text{km}}.$$

So, from point $A$ the fox population density increases as we move eastward. The population density decreases as we move north from $A$.

At point $B$,

$$\left.\frac{\partial n}{\partial x}\right|_{(B)} \approx \frac{0.75 - 1}{135 - 115} = -\frac{1/4}{20} = -\frac{1}{80} \approx -0.01 \; \frac{\text{foxes/km}^2}{\text{km}}$$

$$\left.\frac{\partial n}{\partial y}\right|_{(B)} \approx \frac{0.5 - 1}{120 - 110} = -\frac{1/2}{10} = -\frac{1}{20} \approx -0.05 \; \frac{\text{foxes/km}^2}{\text{km}}.$$

So, fox population density decreases as we move both east and north of $B$. However, notice that the partial derivative $\partial n/\partial x$ at $B$ is smaller in magnitude than the others. Indeed if we had taken a negative $\Delta x$ we would have obtained an estimate of the opposite sign. This suggests that better estimates for $B$ are

$$\left.\frac{\partial n}{\partial x}\right|_{(B)} \approx 0 \; \frac{\text{foxes/km}^2}{\text{km}}$$

$$\left.\frac{\partial n}{\partial y}\right|_{(B)} \approx -0.05 \; \frac{\text{foxes/km}^2}{\text{km}}.$$

At point $C$,

$$\left.\frac{\partial n}{\partial x}\right|_{(C)} \approx \frac{2 - 1.5}{135 - 115} = \frac{1/2}{20} = \frac{1}{40} \approx 0.02 \; \frac{\text{foxes/km}^2}{\text{km}}$$

$$\left.\frac{\partial n}{\partial y}\right|_{(C)} \approx \frac{2 - 1.5}{80 - 55} = \frac{1/2}{25} = \frac{1}{50} \approx 0.02 \; \frac{\text{foxes/km}^2}{\text{km}}.$$

So, the fox population density increases as we move east and north of $C$. Again, if these estimates were made using negative values for $\Delta x$ and $\Delta y$ we would have had estimates of the opposite sign. Thus, better estimates are

$$\left.\frac{\partial n}{\partial x}\right|_{(C)} \approx 0 \; \frac{\text{foxes/km}^2}{\text{km}}$$

$$\left.\frac{\partial n}{\partial y}\right|_{(C)} \approx 0 \; \frac{\text{foxes/km}^2}{\text{km}}.$$

26. Reading values of $H$ from the graph gives Table 7.19. In order to compute $H_T(T, w)$ at $T = 30$, it is useful to have values of $H(T, w)$ for $T = 40° C$. The column corresponding to $w = 0.4$ is not used in this problem.

**TABLE 7.19**  *Estimated values of $H(T, w)$ (in calories/meter$^3$)*

|  |  | \multicolumn{4}{c}{$w$ (gm/m$^3$)} |
|---|---|---|---|---|---|
|  |  | 0.1 | 0.2 | 0.3 | 0.4 |
| $T$ (°C) | 10 | 110 | 240 | 330 | 450 |
|  | 20 | 100 | 180 | 260 | 350 |
|  | 30 | 70 | 150 | 220 | 300 |
|  | 40 | 65 | 140 | 200 | 270 |

The estimates for $H_T(T, w)$ in Table 7.20 are now computed using the formula

$$H_T(T, w) \approx \frac{H(T + 10, w) - H(T, w)}{10}.$$

**TABLE 7.20**  *Estimated values of $H_T(T, w)$ (in calories/meter$^3$)*

|  |  | \multicolumn{3}{c}{$w$ (gm/m$^3$)} |
|---|---|---|---|---|
|  |  | 0.1 | 0.2 | 0.3 |
| $T$ (°C) | 10 | $-1.0$ | $-6.0$ | $-7.0$ |
|  | 20 | $-3.0$ | $-3.0$ | $-4.0$ |
|  | 30 | $-0.5$ | $-1.0$ | $-2.0$ |

Practically, the values $H_T(T, W)$ show how much less heat (in calories per cubic meter of fog) is needed to clear the fog if the temperature is increased by 10° C, with $w$ fixed.

27. **TABLE 7.21**  *Estimated values of $H(T, w)$ (in calories/meter$^3$)*

|  |  | \multicolumn{4}{c}{$w$ (gm/m$^3$)} |
|---|---|---|---|---|---|
|  |  | 0.1 | 0.2 | 0.3 | 0.4 |
| $T$ (°C) | 10 | 110 | 240 | 330 | 450 |
|  | 20 | 100 | 180 | 260 | 350 |
|  | 30 | 70 | 150 | 220 | 300 |
|  | 40 | 65 | 140 | 200 | 270 |

Values of $H$ from the graph are given in Table 7.21. In order to compute $H_w(T, w)$ for $w = 0.3$, it is useful to have the column corresponding to $w = 0.4$. The row corresponding to $T = 40$ is not used in this problem. The partial derivative $H_w(T, w)$ can be approximated by

$$H_w(10, 0.1) \approx \frac{H(10, 0.1 + h) - H(10, 0.1)}{h} \quad \text{for small } h.$$

We choose $h = 0.1$ because we can read off a value for $H(10, 0.2)$ from the graph. If we take $H(10, 0.2) = 240$, we get the approximation

$$H_w(10, 0.1) \approx \frac{H(10, 0.2) - H(10, 0.1)}{0.1} = \frac{240 - 110}{0.1} = 1300.$$

In practical terms, we have found that for fog at 10° C containing 0.1 g/m³ of water, an increase in the water content of the fog will increase the heat requirement for dissipating the fog at the rate given by $H_w(10, 0.1)$. Specifically, a 1 g/m³ increase in the water content will increase the heat required to dissipate the fog by about 1300 calories per cubic meter of fog. Wetter fog is harder to dissipate. Other values of $H_w(T, w)$ in Table 7.22 are computed using the formula

$$H_w(T, w) \approx \frac{H(T, w + 0.1) - H(T, w)}{0.1},$$

where we have used Table 7.21 to evaluate $H$.

**TABLE 7.22** *Table of values of $H_w(T, w)$ (in cal/gm)*

|  |  | \multicolumn{3}{c}{$w$ (gm/m³)} | | |
|---|---|---|---|---|
|  |  | 0.1 | 0.2 | 0.3 |
| $T$ (°C) | 10 | 1300 | 900 | 1200 |
|  | 20 | 800 | 800 | 900 |
|  | 30 | 800 | 700 | 800 |

## Solutions for Section 7.4

1.
$$f(1, 2) = (1)^3 + 3(2)^2 = 13$$
$$f_x(x, y) = 3x^2 + 0 \Rightarrow f_x(1, 2) = 3(1)^2 = 3$$
$$f_y(x, y) = 0 + 6y \Rightarrow f_y(1, 2) = 6(2) = 12$$

2.
$$f(3, 1) = 5(3)(1)^2 = 15$$
$$f_u(u, v) = 5v^2 \Rightarrow f_u(3, 1) = 5(1)^2 = 5$$
$$f_v(u, v) = 10uv \Rightarrow f_v(3, 1) = 10(3)(1) = 30$$

3. $f_x(x, y) = 2x + 2y$, $f_y(x, y) = 2x + 3y^2$.

4. $\dfrac{\partial z}{\partial x} = 2xe^y$

5. $f_x(x, y) = 4x + 0 = 4x$
   $f_y(x, y) = 0 + 6y = 6y$

6. $\dfrac{\partial Q}{\partial p} = 5a^2 - 9ap^2$

7. $\dfrac{\partial P}{\partial r} = 100te^{rt}$

8. $f_t(t, a) = 3 \cdot 5a^2t^2 = 15a^2t^2$.

9. $f_x(x, y) = 2 \cdot 100xy = 200xy$
   $f_y(x, y) = 100x^2 \cdot 1 = 100x^2$

10. $f_x(x, y) = 20xe^{3y}$, $f_y(x, y) = 30x^2e^{3y}$.

11. $z_x = 2xy + 10x^4y$

12. $f_u(u, v) = 2u + 5v + 0 = 2u + 5v$
    $f_v(u, v) = 0 + 5u + 2v = 5u + 2v$

13. $\dfrac{\partial A}{\partial h} = \dfrac{1}{2}(a + b)$

14. $\dfrac{\partial}{\partial m}\left(\dfrac{1}{2}mv^2\right) = \dfrac{1}{2}v^2$

15. To calculate $\partial B/\partial t$, we hold $P$ and $r$ constant and differentiate $B$ with respect to $t$:

$$\frac{\partial B}{\partial t} = \frac{\partial}{\partial t}(Pe^{rt}) = Pre^{rt}.$$

In financial terms, $\partial B/\partial t$ represents the change in the amount of money in the bank as one unit of time passes by.

To calculate $\partial B/\partial r$, we hold $t$ and $P$ constant and differentiate with respect to $r$:

$$\frac{\partial B}{\partial r} = \frac{\partial}{\partial r}(Pe^{rt}) = Pte^{rt}.$$

In financial terms, $\partial B/\partial r$ represents the rate of change in the amount of money in the back as the interest rate changes.

To calculate $\partial B/\partial P$, we hold $t$ and $r$ constant and differentiate $B$ with respect to $P$:

$$\frac{\partial B}{\partial P} = \frac{\partial}{\partial P}(Pe^{rt}) = e^{rt}.$$

In financial terms, $\partial B/\partial P$ represents the rate of change in the amount of money in the bank at time $t$ as you change the amount of money that was initially deposited.

16. Differentiating with respect to $d$ gives

$$\frac{\partial C}{\partial d} = 40 + 0 = \$40 \text{ per day,}$$

so the cost of renting a car an additional day is \$40.

Differentiating with respect to $m$ gives

$$\frac{\partial C}{\partial m} = 0 + 0.15 = \$0.15 \text{ per mile,}$$

so the cost of driving the car an additional mile is \$0.15.

17.
$$f(500, 1000) = 16 + 1.2(500) + 1.5(1000) + 0.2(500)(1000)$$
$$= \$102{,}116$$

The cost of producing 500 units of item 1 and 1000 units of item 2 is \$102,116.

$$f_{q_1}(q_1, q_2) = 0 + 1.2 + 0 + 0.2q_2$$

So $f_{q_1}(500, 1000) = 1.2 + 0.2(1000) = \$201.20$ per unit. When the company is producing at $q_1 = 500$, $q_2 = 1000$, the cost of producing one more unit of item 1 is \$201.20.

$$f_{q_2}(q_1, q_2) = 0 + 0 + 1.5 + 0.2q_1$$

So $f_{q_2}(500, 1000) = 1.5 + 0.2(500) = \$101.50$ per unit. When the company is producing at $q_1 = 500$, $q_2 = 1000$, the cost of producing one more unit of item 2 is \$101.50.

18. (a) From contour diagram,

$$f_x(2, 1) \approx \frac{f(2.3, 1) - f(2, 1)}{2.3 - 2}$$
$$= \frac{6 - 5}{0.3} = 3.3,$$
$$f_y(2, 1) \approx \frac{f(2, 1.4) - f(2, 1)}{1.4 - 1}$$
$$= \frac{6 - 5}{0.4} = 2.5.$$

(b) A table of values for $f$ is given in Table 7.23.

**TABLE 7.23**

|   |     | \multicolumn{3}{c}{$y$} |     |
|---|-----|------|------|------|
|   |     | 0.9  | 1.0  | 1.1  |
|   | 1.9 | 4.42 | 4.61 | 4.82 |
| $x$ | 2.0 | 4.81 | 5.00 | 5.21 |
|   | 2.1 | 5.22 | 5.41 | 5.62 |

From Table 7.23 we estimate $f_x(2,1)$ and $f_y(2,1)$ using difference quotients:

$$f_x(2,1) \approx \frac{5.41 - 5.00}{2.1 - 2} = 4.1$$

$$f_y(2,1) \approx \frac{5.21 - 5.00}{1.1 - 1} = 2.1.$$

We obtain better estimates by finer data in the table.

(c) $f_x(x,y) = 2x$, $f_y(x,y) = 2y$. So the true values are $f_x(2,1) = 4$, $f_y(2,1) = 2$.

19. The function in $h(x,t)$ tells us the height of the head of the spectator in seat $x$ at time $t$ seconds. Thus, $h_x(2,5)$ is in feet per seat and $h_t(2,5)$ is in feet per second. So

$$h_x(x,t) = -0.5\sin(0.5x - t)$$
$$h_x(2,5) = -0.5\sin(0.5(2) - 5) \approx -0.38 \text{ ft/seat.}$$

and

$$h_t(x,t) = \sin(0.5x - t)$$
$$h_t(2,5) = \sin(0.5(2) - 5) = 0.76 \text{ ft/sec.}$$

The value of $h_x(2,5)$ is the rate of change of height of heads as you move along the row of seats. The value of $h_t(2,5)$ is the vertical velocity of the head of the person at seat 2 at time $t = 5$.

20. $N$ is the number of workers, so $N = 80$. $V$ is the value of equipment in units of \$25,000, so $V = $ (value of equipment)$/\$25,000 = 30$.

$$f(80, 30) = 5(80)^{0.75}(30)^{0.25} = 313 \text{ tons.}$$

Using 80 workers and \$750,000 worth of equipment produces about 313 tons of output.

$$f_N(N,V) = 0.75 \cdot 5N^{-0.25}V^{0.25}, \quad \text{so } f_N(80,30) = 0.75 \cdot 5(80)^{-0.25}(30)^{0.25} = 2.9 \text{ tons/worker.}$$

When the company is using 80 workers and \$750,000 worth of equipment, each additional worker would add about 2.9 tons to total output.

$$f_V(N,V) = 0.25 \cdot 5N^{0.75}V^{-0.75}, \quad \text{so } f_V(80,30) = 0.25 \cdot 5(80)^{0.75}(30)^{-0.75} = 2.6 \text{ tons/\$25,000.}$$

When the company is using 80 workers and \$750,000 worth of equipment, each additional \$25,000 of equipment would add about 2.6 tons to total output.

21. (a) $Q_K = 18.75K^{-0.25}L^{0.25}$, $Q_L = 6.25K^{0.75}L^{-0.75}$.
    (b) When $K = 60$ and $L = 100$,

$$Q = 25 \cdot 60^{0.75} \cdot 100^{0.25} = 1704.33$$
$$Q_K = 18.75 \cdot 60^{-0.25} 100^{0.25} = 21.3$$
$$Q_L = 6.25 \cdot 60^{0.75} 100^{-0.75} = 4.26$$

(c) $Q$ is actual quantity being produced. $Q_K$ is how much more could be produced if you increased $K$ by one unit. $Q_L$ is how much more could be produced if you increased $L$ by 1.

22. **(a)** We have $N = 300$ and $V = 200$ so production

$$P = 2(300)^{0.6}(200)^{0.4} = 510.17 \text{ thousand pages per day.}$$

**(b)** If the labor force, $N$, is doubled to 600 workers and the value of the equipment, $V$, remains at 200, we have

$$P = 2(600)^{0.6}(200)^{0.4} = 773.27.$$

The production $P$ has increased from 510.17 to 773.27, an increase of 263.10 thousand pages per day.

**(c)** If the value of the equipment, $V$, is doubled to 400 and the labor force, $N$, remains at 300 workers, we have

$$P = 2(300)^{0.6}(400)^{0.4} = 673.17.$$

The production $P$ has increased from 510.17 to 673.17, an increase of 163.00 thousand pages per day. We see that doubling the work force has a greater effect on production than doubling the value of the equipment. We might have expected this, since the exponent for $N$ is larger than the exponent for $V$.

**(d)** What happens if we double both $N$ and $V$? Then

$$P = 2(600)^{0.6}(400)^{0.4} = 1020.34 \text{ thousand pages per day.}$$

This is exactly double our original production level of 510.17. In this case, doubling both $N$ and $V$ doubled $P$.

23. **(a)** Substituting $L = 70$ and $K = 50$, we have

$$\begin{aligned} Q &= 900 L^{1/2} K^{2/3} \\ &= 900 \cdot 70^{1/2} \cdot 50^{2/3} \\ &= 102{,}197 \end{aligned}$$

**(b)** When $L = 140$ and $K = 100$, $Q = 229{,}425$. To understand the general effect of doubling $L$ and $K$, calculate:

$$\begin{aligned} Q(\geq L, \geq K) &= 900(2L)^{1/2}(2K)^{2/3} \\ &= 900 L^{1/2} K^{2/3} \cdot (2)^{1/2}(2)^{2/3} \\ &= 2^{7/6} \cdot Q \end{aligned}$$

so $Q$ increases by a factor of about 2.2. Substituting $L = 140$ and $K = 100$, gives $Q = 229{,}425 = 2^{7/6} \cdot Q$

24. **(A)** In graph I, $L = 1, K = 1$ gives us $F = 1$, and $L = 3, K = 3$ gives us $F = 3$. So tripling all inputs in graph I triples output; graph I corresponds to statement (A).

**(B)** In graph II, $L = 1, K = 1$ gives us $F = 1$, and $L = 2.2, K = 2.2$ gives us $F = 1.5$. Extrapolating from this ratio, $L = 4, K = 4$ should gives us $F = 2$. So, quadrupling all inputs in graph II doubles output; graph II corresponds to statement (B).

**(C)** In graph III, $L = 1, K = 1$ gives us $F = 1$, and $L = 2, K = 2$ gives us $F = 2.8$. So, doubling the inputs in graph III almost tripled output; graph III corresponds to statement (C).

25. $f_x = 2xy$ and $f_y = x^2$, so $f_{xx} = 2y$, $f_{xy} = 2x$, $f_{yy} = 0$ and $f_{yx} = 2x$.

26. $f_x = 2x + 2y$ and $f_y = 2x + 2y$, so $f_{xx} = 2$, $f_{xy} = 2$, $f_{yy} = 2$ and $f_{yx} = 2$.

27. $f_x = e^y$ and $f_y = xe^y$, so $f_{xx} = 0$, $f_{xy} = e^y$, $f_{yy} = xe^y$ and $f_{yx} = e^y$.

28. $f_x = 2/y$ and $f_y = -2x/y^2$, so $f_{xx} = 0$, $f_{xy} = -2/y^2$, $f_{yy} = 4x/y^3$ and $f_{yx} = -2/y^2$.

29. $f_x = 2xy^2$ and $f_y = 2x^2 y$, so $f_{xx} = 2y^2$, $f_{xy} = 4xy$, $f_{yy} = 2x^2$ and $f_{yx} = 4xy$.

30. $f_x = ye^{xy}$ and $f_y = xe^{xy}$, so $f_{xx} = y^2 e^{xy}$, $f_{xy} = xye^{xy} + e^{xy} = (xy+1)e^{xy}$, $f_{yy} = x^2 e^{xy}$ and $f_{yx} = xye^{xy} + e^{xy} = (xy+1)e^{xy}$.

31. $Q_{p_1} = 10 p_1 p_2^{-1}$ and $Q_{p_2} = -5 p_1^2 p_2^{-2}$, so $Q_{p_1 p_1} = 10 p_2^{-1}$, $Q_{p_2 p_2} = 10 p_1^2 p_2^{-3}$ and $Q_{p_1 p_2} = Q_{p_2 p_1} = -10 p_1 p_2^{-2}$.

32. $V_r = 2\pi r h$ and $V_h = \pi r^2$, so $V_{rr} = 2\pi h$, $V_{hh} = 0$ and $V_{rh} = V_{hr} = 2\pi r$.

33. $P_K = 2L^2$ and $P_L = 4KL$, so $P_{KK} = 0$, $P_{LL} = 4K$ and $P_{KL} = P_{LK} = 4L$.

34. $B_x = 5e^{-2t}$ and $B_t = -10xe^{-2t}$, so $B_{xx} = 0$, $B_{tt} = 20xe^{-2t}$ and $B_{xt} = B_{tx} = -10e^{-2t}$.

35. $f_x = -8xt$ and $f_t = 3t^2 - 4x^2$, so $f_{xx} = -8t$, $f_{tt} = 6t$ and $f_{xt} = f_{tx} = -8x$.

36. $f_r = 100te^{rt}$ and $f_t = 100re^{rt}$, so $f_{rr} = 100t^2 e^{rt}$, $f_{rt} = f_{tr} = 100tre^{rt} + 100e^{rt} = 100(rt+1)e^{rt}$ and $f_{tt} = 100r^2 e^{rt}$.

37. Since $f_x(x,y) = 4x^3 y^2 - 3y^4$, we could have
$$f(x,y) = x^4 y^2 - 3xy^4.$$

In that case,
$$f_y(x,y) = \frac{\partial}{\partial y}(x^4 y^2 - 3xy^4) = 2x^4 y - 12xy^3$$

as expected. More generally, we could have $f(x,y) = x^4 y^2 - 3xy^4 + C$, where $C$ is any constant.

## Solutions for Section 7.5

1. Mississippi lies entirely within a region designated as 80s so we expect both the maximum and minimum daily high temperatures within the state to be in the 80s. The southwestern-most corner of the state is close to a region designated as 90s, so we would expect the temperature here to be in the high 80s, say 87-88. The northern-most portion of the state is located near the center of the 80s region. We might expect the high temperature there to be between 83-87.

    Alabama also lies completely within a region designated as 80s so both the high and low daily high temperatures within the state are in the 80s. The southeastern tip of the state is close to a 90s region so we would expect the temperature here to be about 88-89 degrees. The northern-most part of the state is near the center of the 80s region so the temperature there is 83-87 degrees.

    Pennsylvania is also in the 80s region, but it is touched by the boundary line between the 80s and a 70s region. Thus we expect the low daily high temperature to occur there and be about 80 degrees. The state is also touched by a boundary line of a 90s region so the high will occur there and be 89-90 degrees.

    New York is split by a boundary between an 80s and a 70s region, so the northern portion of the state is likely to be about 74-76 while the southern portion is likely to be in the low 80s, maybe 81-84 or so.

    California contains many different zones. The northern coastal areas will probably have the daily high as low as 65-68, although without another contour on that side, it is difficult to judge how quickly the temperature is dropping off to the west. The tip of Southern California is in a 100s region, so there we expect the daily high to be 100-101.

    Arizona will have a low daily high around 85-87 in the northwest corner and a high in the 100s, perhaps 102-107 in its southern regions.

    Massachusetts will probably have a high daily high around 81-84 and a low daily high of 70.

2. Searching over the entire table for the largest value (the highest UV exposure), we see that at latitude $-60°$ in the year 2010, the exposure is rated 11.97.

3. At a critical point $f_x = 2x + 6 = 0$ and $f_y = 2y - 10 = 0$, so $(-3, 5)$ is the only critical point. Since $f_{xx} = 2 > 0$ and $f_{xx}f_{yy} - f_{xy}^2 = 4 > 0$, the point $(-3, 5)$ is a local minimum.

4. At a critical point $f_x = 2x + 4 = 0$ and $f_y = 2y = 0$, so $(-2, 0)$ is the only critical point. Since $f_{xx} = 2 > 0$ and $f_{xx}f_{yy} - f_{xy}^2 = 4 > 0$, the point $(-2, 0)$ is a local minimum.

5. At a critical point $f_x = 2x + y = 0$ and $f_y = x + 3 = 0$, so $(-3, 6)$ is the only critical point. Since $f_{xx}f_{yy} - f_{xy}^2 = -1 < 0$, the point $(-3, 6)$ is neither a local maximum nor a local minimum.

6. At a critical point $f_x = -3y + 6 = 0$ and $f_y = 3y^2 - 3x = 0$, so $(4, 2)$ is the only critical point. Since $f_{xx}f_{yy} - f_{xy}^2 = -9 < 0$, the point $(4, 2)$ is neither a local maximum nor a local minimum.

7. We can identify local extreme points on a contour diagram because these points will either be the centers of a series of concentric circles that close around them, or will lie on the edges of the diagram. Looking at the graph, we see that $(2, 10)$, $(6, 4)$, $(6.5, 16)$ and $(9, 10)$ appear to be such points. Since the points near $(2, 10)$ decrease in functional value as they close around $(2, 10)$, $f(2, 10)$ will be somewhat less than its nearest contour. So $f(2, 10) \approx 0.5$. Similarly, since the contours near $(2, 10)$ are greater in functional value than $f(2, 10)$, $f(2, 10)$ is a local minimum. Applying analogous arguments to the point $(6, 4)$, we see that $f(6, 4) \approx 9.5$ and is a local maximum. The contour values are increasing as we approach $(6.5, 16)$ along any path, so $f(6.5, 16) \approx 10$ is a local maximum and $(9, 10)$ is a local minimum.

    Since none of the local minima are less in value than $f(2, 10) \approx 0.5$, $f(2, 10)$ is a global minimum. Since none of the local maxima are greater in value than $f(6.5, 16) \approx 10$, $f(6.5, 16)$ is a global maximum.

8. We can identify local extreme points on contour plots because these points will be the centers of a series of concentric circles that close around them or will lie on the edges of the plot. Looking at the graph, we see that $(0, 0)$, $(13, 30)$, $(37, 18)$, and $(32, 34)$ appear to be such points. Since the points near $(0, 0)$ increase in functional value as they close around $(0, 0)$, $f(0, 0)$ will be somewhat more than its nearest contour. So $f(0, 0) \approx 12.5$. Similarly, since the contours near $(0, 0)$ are less in functional value than $f(0, 0)$, $f(0, 0)$ is a local maximum. Applying analogous arguments to the other extreme points, we see that $f(13, 30) \approx 4.5$ and is a local minimum; $f(37, 18) \approx 2.5$ and is a local minimum; and $f(32, 34) \approx 10.5$ and is a local maximum. Since none of the local maxima are greater in value than $f(0, 0) \approx 12.5$, $f(0, 0)$ is a global maximum. Since none of the local minima are lower in value than $f(37, 18) \approx 2.5$, $f(37, 18)$ is a global minimum.

9. At a local maximum value of $f$,
$$\frac{\partial f}{\partial x} = -2x - B = 0.$$
We are told that this is satisfied by $x = -2$. So $-2(-2) - B = 0$ and $B = 4$. In addition,
$$\frac{\partial f}{\partial y} = -2y - C = 0$$
and we know this holds for $y = 1$, so $-2(1) - C = 0$, giving $C = -2$. We are also told that the value of $f$ is 15 at the point $(-2, 1)$, so
$$15 = f(-2, 1) = A - ((-2)^2 + 4(-2) + 1^2 - 2(1)) = A - (-5), \text{ so } A = 10.$$

Now we check that these values of $A$, $B$, and $C$ give $f(x, y)$ a local maximum at the point $(-2, 1)$. Since
$$f_{xx}(-2, 1) = -2,$$
$$f_{yy}(-2, 1) = -2$$
and
$$f_{xy}(-2, 1) = 0,$$
we have that $f_{xx}(-2, 1)f_{yy}(-2, 1) - f_{xy}^2(-2, 1) = (-2)(-2) - 0 > 0$ and $f_{xx}(-2, 1) < 0$. Thus, $f$ has a local maximum value 15 at $(-2, 1)$.

10. (a) We first express the revenue $R$ in terms of the prices $p_1$ and $p_2$:
$$\begin{aligned} R(p_1, p_2) &= p_1 q_1 + p_2 q_2 \\ &= p_1(517 - 3.5p_1 + 0.8p_2) + p_2(770 - 4.4p_2 + 1.4p_1) \\ &= 517p_1 - 3.5p_1^2 + 770p_2 - 4.4p_2^2 + 2.2p_1 p_2. \end{aligned}$$

(b) We compute the partial derivatives and set them to zero:
$$\frac{\partial R}{\partial p_1} = 517 - 7p_1 + 2.2p_2 = 0,$$
$$\frac{\partial R}{\partial p_2} = 770 - 8.8p_2 + 2.2p_1 = 0.$$

Solving these equations, we find that
$$p_1 = 110 \quad \text{and} \quad p_2 = 115.$$
To see whether or not we have a found a local maximum, we compute the second-order partial derivatives:
$$\frac{\partial^2 R}{\partial p_1^2} = -7, \quad \frac{\partial^2 R}{\partial p_2^2} = -8.8, \quad \frac{\partial^2 R}{\partial p_1 \partial p_2} = 2.2.$$

Therefore,
$$D = \frac{\partial^2 R}{\partial p_1^2} \frac{\partial^2 R}{\partial p_2^2} - \left(\frac{\partial^2 R}{\partial p_1 \partial p_2}\right)^2 = (-7)(-8.8) - (2.2)^2 = 56.76,$$
and so we have found a local maximum point. The graph of $R(p_1, p_2)$ has the shape of an upside down bowl. Therefore, $(110, 115)$ is a global maximum point.

11. The total revenue is
$$R = pq = (60 - 0.04q)q = 60q - 0.04q^2,$$
and as $q = q_1 + q_2$, this gives
$$R = 60q_1 + 60q_2 - 0.04q_1^2 - 0.08q_1q_2 - 0.04q_2^2.$$
Therefore, the profit is
$$P(q_1, q_2) = R - C_1 - C_2$$
$$= -13.7 + 60q_1 + 60q_2 - 0.07q_1^2 - 0.08q_2^2 - 0.08q_1q_2.$$
At a local maximum point, we have:
$$\frac{\partial P}{\partial q_1} = 60 - 0.14q_1 - 0.08q_2 = 0,$$
$$\frac{\partial P}{\partial q_2} = 60 - 0.16q_2 - 0.08q_1 = 0.$$
Solving these equations, we find that
$$q_1 = 300 \quad \text{and} \quad q_2 = 225.$$
To see whether or not we have found a local maximum, we compute the second-order partial derivatives:
$$\frac{\partial^2 P}{\partial q_1^2} = -0.14, \quad \frac{\partial^2 P}{\partial q_2^2} = -0.16, \quad \frac{\partial^2 P}{\partial q_1 \partial q_2} = -0.08.$$
Therefore,
$$D = \frac{\partial^2 P}{\partial q_1^2} \frac{\partial^2 P}{\partial q_2^2} - \frac{\partial^2 P}{\partial q_1 \partial q_2} = (-0.14)(-0.16) - (-0.08)^2 = 0.016,$$
and so we have found a local maximum point. The graph of $P(q_1, q_2)$ has the shape of an upside down paraboloid since $P$ is quadratic in $q_1$ and $q_2$, hence (300, 225) is a global maximum point.

12. (a) The revenue $R = p_1q_1 + p_2q_2$. Profit $= P = R - C = p_1q_1 + p_2q_2 - 2q_1^2 - 2q_2^2 - 10$.
$$\frac{\partial P}{\partial q_1} = p_1 - 4q_1 = 0 \quad \text{gives } q_1 = \frac{p_1}{4}$$
$$\frac{\partial P}{\partial q_2} = p_2 - 4q_2 = 0 \quad \text{gives } q_2 = \frac{p_2}{4}$$
Since $\frac{\partial^2 P}{\partial q_1^2} = -4, \frac{\partial^2 P}{\partial q_2^2} = -4$ and $\frac{\partial^2 P}{\partial q_1 \partial q_2} = 0$, at $(p_1/4, p_2/4)$ we have that the discriminant, $D = (-4)(-4) > 0$ and $\frac{\partial^2 P}{\partial q_1^2} < 0$, thus $P$ has a local maximum value at $(q_1, q_2) = (p_1/4, p_2/4)$. Since $P$ is quadratic in $q_1$ and $q_2$, this is a global maximum. So $P = \frac{p_1^2}{4} + \frac{p_2^2}{4} - 2\frac{p_1^2}{16} - 2\frac{p_2^2}{16} - 10 = \frac{p_1^2}{8} + \frac{p_2^2}{8} - 10$ is the maximum profit.

(b) The rate of change of the maximum profit as $p_1$ increases is
$$\frac{\partial}{\partial p_1}(\max P) = \frac{2p_1}{8} = \frac{p_1}{4}.$$

## Solutions for Section 7.6

1. We wish to optimize $f(x, y) = xy$ subject to the constraint $g(x, y) = 5x + 2y = 100$. To do this we must solve the following system of equations:
$$f_x(x, y) = \lambda g_x(x, y), \quad \text{so } y = 5\lambda$$
$$f_y(x, y) = \lambda g_y(x, y), \quad \text{so } x = 2\lambda$$
$$g(x, y) = 100, \quad \text{so } 5x + 2y = 100$$
We substitute in the third equation to obtain $5(2\lambda) + 2(5\lambda) = 100$, so $\lambda = 5$. Thus,
$$x = 10 \quad y = 25 \quad \lambda = 5$$
corresponding to optimal $f(x, y) = (10)(25) = 250$.

2. We wish to optimize $f(x,y) = x^2 + 3y^2 + 100$ subject to the constraint $g(x,y) = 8x + 6y = 88$. To do this we must solve the following system of equations:

$$f_x(x,y) = \lambda g_x(x,y), \quad \text{so } x = 4\lambda$$
$$f_y(x,y) = \lambda g_y(x,y), \quad \text{so } y = \lambda$$
$$g(x,y) = 88, \quad \text{so } 8x + 6y = 88$$

Solving these equations produces:
$$x = 9.26 \quad y = 2.32 \quad \lambda = 2.32$$
corresponding to optimal $f(x,y) = (9.26)^2 + 3(2.32)^2 + 100 = 201.9$.

3. We wish to optimize $f(x,y) = x^2 + 4xy$ subject to the constraint $g(x,y) = x + y = 100$. To do this we must solve the following system of equations:

$$f_x(x,y) = \lambda g_x(x,y), \quad \text{so } 2x + 4y = \lambda$$
$$f_y(x,y) = \lambda g_y(x,y), \quad \text{so } 4x = \lambda$$
$$g(x,y) = 100, \quad \text{so } x + y = 100$$

Solving these equations produces:
$$x \approx 66.7 \quad y \approx 33.3 \quad \lambda \approx 266.8$$
corresponding to optimal $f(x,y) \approx (66.7)^2 + 4(66.7)(33.3) \approx 13{,}333$.

4. We wish to optimize $f(x,y) = 5xy$ subject to the constraint $g(x,y) = x + 3y = 24$. To do this we must solve the following system of equations:

$$f_x(x,y) = \lambda g_x(x,y), \text{ so } 5y = \lambda$$
$$f_y(x,y) = \lambda g_y(x,y), \text{ so } 5x = 3\lambda$$
$$g(x,y) = 24, \text{ so } x + 3y = 24$$

Solving these equations produces:
$$x = 12 \quad y = 4 \quad \lambda = 20$$
corresponding to optimal $f(x,y) = 5(12)(4) = 240$.

5. Our objective function is $f(x,y) = x + y$ and our equation of constraint is $g(x,y) = x^2 + y^2 = 1$. To optimize $f(x,y)$ with Lagrange multipliers, we solve the following system of equations

$$f_x(x,y) = \lambda g_x(x,y), \quad \text{so } 1 = 2\lambda x$$
$$f_y(x,y) = \lambda g_y(x,y), \quad \text{so } 1 = 2\lambda y$$
$$g(x,y) = 1, \quad \text{so } x^2 + y^2 = 1$$

Solving for $\lambda$ gives
$$\lambda = \frac{1}{2x} = \frac{1}{2y},$$
which tells us that $x = y$. Going back to our equation of constraint, we use the substitution $x = y$ to solve for $y$:

$$g(y,y) = y^2 + y^2 = 1$$
$$2y^2 = 1$$
$$y^2 = \frac{1}{2}$$
$$y = \pm\sqrt{\frac{1}{2}} = \pm\frac{\sqrt{2}}{2}.$$

Since $x = y$, our critical points are $(\frac{\sqrt{2}}{2}, \frac{\sqrt{2}}{2})$ and $(-\frac{\sqrt{2}}{2}, -\frac{\sqrt{2}}{2})$. Evaluating $f$ at these points we find that the maximum value is $f(\frac{\sqrt{2}}{2}, \frac{\sqrt{2}}{2}) = \sqrt{2}$ and the minimum value is $f(-\frac{\sqrt{2}}{2}, -\frac{\sqrt{2}}{2}) = -\sqrt{2}$.

6. Our objective function is $f(x,y) = 3x - 2y$ and our equation of constraint is $g(x,y) = x^2 + 2y^2 = 44$. To optimize $f(x,y)$ with Lagrange multipliers we solve the following system of equations

$$f_x(x,y) = \lambda g_x(x,y), \quad \text{so } 3 = 2\lambda x$$
$$f_y(x,y) = \lambda g_y(x,y), \quad \text{so } -2 = 4\lambda y$$
$$g(x,y) = 44, \quad \text{so } x^2 + y^2 = 44$$

Solving for $\lambda$ gives us

$$\lambda = \frac{3}{2x} = \frac{-2}{4y},$$

which we can use to find $x$ in terms of $y$:

$$\frac{3}{2x} = \frac{-2}{4y}$$
$$-4x = 12y$$
$$x = -3y.$$

Using this relation in our equation of constraint, we can solve for $y$:

$$x^2 + 2y^2 = 44$$
$$(-3y)^2 + 2y^2 = 44$$
$$9y^2 + 2y^2 = 44$$
$$11y^2 = 44$$
$$y^2 = 4$$
$$y = \pm 2.$$

Thus, the critical points are $(-6, 2)$ and $(6, -2)$. Evaluating $f$ at these points, we find that the maximum is $f(6, -2) = 18 + 4 = 22$ and the minimum value is $f(-6, 2) = -18 - 4 = -22$.

7. To maximize $f(x,y)$ subject to $g(x,y) = c$, we wish to find the highest possible value of $f(x,y)$ where $(x,y)$ is on the line corresponding to our constraint. From our constraint, we see that $f = 400$ is the highest curve that intersects the constraint line. We can verify this either by noting that all higher contours lie above the constraint $g(x,y) = c$ or by noting that the contour $f = 400$ is tangent to the constraint. The point of intersection, from looking at the graph, occurs at approximately $x = 6, y = 6$, giving us maximal $f(x,y) = 400$.

8. We want to minimize cost $C = 100L + 200K$ subject to $Q = 900L^{1/2}K^{2/3} = 36000$. We solve the system of equations:

$$C_L = \lambda Q_L, \quad \text{so } 100 = \lambda 450 L^{-1/2} K^{2/3}$$
$$C_K = \lambda Q_K, \quad \text{so } 200 = \lambda 600 L^{1/2} K^{-1/3}$$
$$Q = 36000, \quad \text{so } 900 L^{1/2} K^{2/3} = 36000$$

Since $\lambda \neq 0$ this gives

$$450 L^{-1/2} K^{2/3} = 300 L^{1/2} K^{-1/3}.$$

Solving, we get $L = (3/2)K$. Substituting into $Q = 36{,}000$ gives

$$900 \left(\frac{3}{2}K\right)^{1/2} K^{2/3} = 36{,}000.$$

Solving yields $K = \left[40 \cdot \left(\frac{2}{3}\right)^{1/2}\right]^{6/7} \approx 19.85$, so $L \approx \frac{3}{2}(19.85) = 29.78$. We can thus calculate cost using $K = 20$ and $L = 30$ which gives $C = \$7{,}000$.

9. (a) To be producing the maximum quantity $Q$ under the cost constraint given, the firm should be using $K$ and $L$ values given by

$$\frac{\partial Q}{\partial K} = 0.6aK^{-0.4}L^{0.4} = 20\lambda$$
$$\frac{\partial Q}{\partial L} = 0.4aK^{0.6}L^{-0.6} = 10\lambda$$
$$20K + 10L = 150.$$

Hence $\dfrac{0.6aK^{-0.4}L^{0.4}}{0.4aK^{0.6}L^{-0.6}} = 1.5\dfrac{L}{K} = \dfrac{20\lambda}{10\lambda} = 2$, so $L = \dfrac{4}{3}K$. Substituting in $20K + 10L = 150$, we obtain $20K + 10\left(\dfrac{4}{3}\right)K = 150$. Then $K = \dfrac{9}{2}$ and $L = 6$, so capital should be reduced by $\dfrac{1}{2}$ unit, and labor should be increased by 1 unit.

(b) $\dfrac{\text{New production}}{\text{Old production}} = \dfrac{a4.5^{0.6}6^{0.4}}{a5^{0.6}5^{0.4}} \approx 1.01$, so tell the board of directors, "Reducing the quantity of capital by 1/2 unit and increasing the quantity of labor by 1 unit will increase production by 1% while holding costs to \$150."

10. We want to minimize
$$C = f(q_1, q_2) = 2q_1^2 + q_1q_2 + q_2^2 + 500$$
subject to the constraint $q_1 + q_2 = 200$ or $g(q_1, q_2) = q_1 + q_2 = 200$.

We solve the system equations:
$$\begin{aligned} C_{q_1} &= \lambda g_{q_1}, & \text{so } 4q_1 + q_2 &= \lambda \\ C_{q_2} &= \lambda g_{q_2}, & \text{so } 2q_2 + q_1 &= \lambda \\ g &= 200, & \text{so } q_1 + q_2 &= 200. \end{aligned}$$

Solving we get
$$4q_1 + q_2 = 2q_2 + q_1$$
so
$$3q_1 = q_2.$$
We want
$$q_1 + q_2 = 200$$
$$q_1 + 3q_1 = 4q_1 = 200.$$
Therefore
$$q_1 = 50 \text{ units}, \quad q_2 = 150 \text{ units}.$$

11. (a) We wish to maximize $P$ subject to $g = $ total budget $= L + K \leq 1000$. Since additional $L$ and $K$ will always increase production, we will in fact use the entire budget allotted, so $g = L + K = 1000$. We can thus set up the system of equations that will maximize $P$:

$$\begin{aligned} P_L &= \lambda g_L, & \text{so } 9.6L^{-0.6}K^{0.6} &= \lambda \\ P_K &= \lambda g_K, & \text{so } 14.4L^{0.4}K^{-0.4} &= \lambda \\ g &= 1000, & \text{so } L + K &= 1000 \end{aligned}$$

(b) The first two equations tell us
$$9.6K^{0.6}L^{-0.6} = 14.4K^{-0.4}L^{0.4}$$
$$9.6K = 14.4L$$
$$K = 1.5L$$

Thus $L + 1.5L = 1000$ so $L = 400$ and $K = 600$.

(c) At these input levels, the output is
$$P = 24(600)^{0.6}(400)^{0.4} = 12{,}244 \text{ units.}$$

(d) From our equations above we have
$$\lambda = 9.6(600)^{0.6}(400)^{-0.6} \approx 12.24 \text{ units.}$$

This means that if the budget rose by \$1 to \$1001, the maximum possible production would increase by 12 units.

12. (a) We wish to maximize
$$V = 1000D^{0.6}N^{0.3}$$
subject to the budget constraint in dollars
$$40{,}000D + 10{,}000N \leq 600{,}000$$
or (in thousand dollars)
$$\text{Cost } C = 40D + 10N \leq 600.$$

(b) Since additional doctors and nurses will always increase visits, we will use the entire budget allotted to us, so $C = 40D + 10N = 600$. To optimize $V$ subject to this constraint, we must solve the following system of equations:
$$V_D = \lambda C_D, \text{ so } 600D^{-0.4}N^{0.3} = 40\lambda$$
$$V_N = \lambda C_N, \text{ so } 300D^{0.6}N^{-0.7} = 10\lambda$$
$$C = 600, \text{ so } 40D + 10N = 600$$

Thus, we get
$$\frac{600D^{-0.4}N^{0.3}}{40} = \lambda = \frac{300D^{0.6}N^{-0.7}}{10}$$
So
$$N = 2D.$$
To solve for $D$ and $N$, substitute in the budget constraint:
$$40D + 10(2D) = 600$$
$$60D = 600$$
So $D = 10$ and $N = 20$.
$$\lambda = \frac{600(10^{-0.4})(20^{0.3})}{40} \approx 14.67$$
Thus the clinic should hire 10 doctors and 20 nurses. With that staff, the clinic can provide
$$V = 1000(10^{0.6})(20^{0.3}) \approx 9{,}779 \text{ visits per year.}$$

(c) From part (b), the Lagrange multiplier is $\lambda = 14.67$. At the optimum, the Lagrange multiplier tells us that about 14.67 extra visits can be generated through an increase of \$1,000 in the budget. (If we had written out the constraint in dollars instead of thousands of dollars, the Lagrange multiplier would tell us the number of extra visits per dollar.)

13. (a) We wish to maximize $q$ subject to the constraint that cost $C = 10W + 20K = 3000$. To optimize $q$ according to this, we must solve the following system of equations:
$$q_W = \lambda C_W, \quad \text{so } \frac{9}{2}W^{-\frac{1}{4}}K^{\frac{1}{4}} = \lambda 10$$
$$q_K = \lambda C_K, \quad \text{so } \frac{3}{2}W^{\frac{3}{4}}K^{-\frac{3}{4}} = \lambda 20$$
$$C = 3000, \quad \text{so } 10W + 20K = 3000$$

Dividing yields $K = \frac{1}{6}W$, so substituting into $C$ gives
$$10W + 20\left(\frac{1}{6}W\right) = \frac{40}{3}W = 3000.$$
Thus $W = 225$ and $K = 37.5$. Substituting both answers to find $\lambda$ gives
$$\lambda = \frac{\frac{9}{2}(225)^{-\frac{1}{4}}(37.5)^{\frac{1}{4}}}{10} = 0.2875.$$
We also find the optimum quantity produced, $q = 6(225)^{\frac{3}{4}}(37.5)^{\frac{1}{4}} = 862.57$.

(b) When the budget is increased by one dollar, we substitute the relation $K_1 = \frac{1}{6}W_1$ into $10W_1 + 20K_1 = 3001$ which gives $10W_1 + 20(\frac{1}{6}W_1) = \frac{40}{3}W_1 = 3001$. Solving yields $W_1 = 225.075$ and $K_1 = 37.513$, so $q_1 = 862.86 = q + 0.29$. Thus production has increased by $0.29 \approx \lambda$, the Lagrange Multiplier.

14. (a) The objective function is the one we are trying to optimize, so $P(x,y)$ is the objective function. The constraint equations tells us what condition must be satisfied; here, we must have costs equal to $50,000, so $C(x,y) = 50,000$ is the constraint equation. The meaning of $\lambda$ is how much the objective would increase if the budget available increased by 1 and we optimized again with the new budget. Here, this means how much optimal production would increase if the budget increased to $50,001.

   (b) Here, we are trying to minimize cost, so $C(x,y)$ is the objective function. Since we must satisfy $P(x,y) = 2000$, our constraint equation becomes $P(x,y) = 2000$. In this situation, $\lambda$ represents the change in the minimal cost when production is increased by one unit to 2001.

15. The value of $\lambda$ tells us how the optimum value of $f$ changes when we change the constraint. In particular, if the constraint increases by 1, the optimum value of $f$ increases by $\lambda$.

    (a) So, if we raise the quota by 1 product, cost rises by $\lambda = 15$. So $C = 1200 + 15 = 1215$.
    (b) Similarly, if we lower the quota by 1, cost falls by $\lambda = 15$. So $C = 1200 - 15 = 1185$.

16. (a) We want to minimize $C$ subject to $g = x + y = 39$. We solve the system of equations

$$C_x = \lambda g_x, \quad \text{so } 10x + 2y = \lambda$$
$$C_y = \lambda g_y, \quad \text{so } 2x + 6y = \lambda$$
$$g = 39, \quad \text{so } x + y = 39.$$

   The first two equations give $y = 2x$. Solving with $x + y = 39$ gives $x = 13, y = 26, \lambda = 182$. Therefore $C = \$4349$.

   (b) Since $\lambda = 182$, increasing production by 1 will cause costs to increase by approximately \$182. Similarly, decreasing production by 1 will save approximately \$182.

17. (a) The curves are shown in Figure 7.36.

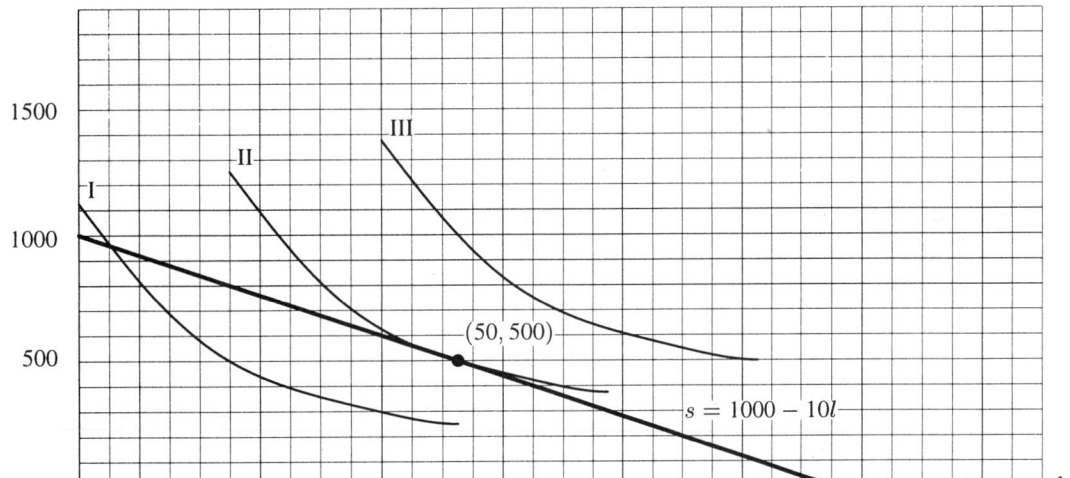

**Figure 7.36**

   (b) The income equals $10/hour times the number of hours of work:
   $$s = 10(100 - l) = 1000 - 10l.$$

   (c) The graph of this constraint is the straight line in Figure 7.36.

   (d) For any given salary, curve III allows for the most leisure time, curve I the least. Similarly, for any amount of leisure time, curve III also has the greatest salary, and curve I the least. Thus, any point on curve III is preferable to any point on curve II, which is preferable to any point on curve I. We prefer to be on the outermost curve that our constraint allows. We want to choose the point on $s = 1000 - 10l$ which is on the most preferable curve. Since all the curves are concave up, this occurs at the point where $s = 1000 - 10l$ is *tangent* to curve II. So we choose $l = 50, s = 500$, and work 50 hours a week.

18. (a) We have 1500 workers and $4,000,000 per month of capital, so $x = 1500, y = 4,000,000/1000 = 4000$. Substituting into the equation for $Q$ gives us $Q = (1500)^{0.4}(4000)^{0.6} = 2702$ cars per month.

(b) Now we are only producing 2000 cars per month. We wish to minimize cost subject to the constraint that the monthly production is 2000 cars. So, our objective function is cost $C = 2100x + 1000y$ and our constraint is that $Q = x^{0.4}y^{0.6} = 2000$. To minimize $C$ according to this, we solve the following system of equations:

$$C_x = \lambda Q_x, \quad \text{so } 0.4x^{-0.6}y^{0.6}\lambda = 2100$$
$$C_y = \lambda Q_x, \quad \text{so } 0.6x^{0.4}y^{-0.4}\lambda = 1000$$
$$Q = 2000, \quad \text{so } x^{0.4}y^{0.6} = 2000$$

Dividing the first two equations gives

$$\frac{0.4x^{-0.6}y^{0.6}\lambda}{0.6x^{0.4}y^{-0.4}\lambda} = \frac{0.4}{0.6}\frac{y}{x} = \frac{2100}{1000} \Rightarrow y = 3.15x.$$

Substituting this into the constraint gives us $x \approx 1004.72$ and $y \approx 3164.858$. So our new level of production uses 1005 workers and $3,164,858 of equipment. So $1500 - 1005 = 495$ workers will be laid off, and monthly investment in capital will fall by $4,000,000 - \$3,164,858 = \$835,142$.

(c) Solving for the Lagrange multiplier $\lambda$ from the above equations gives us

$$\lambda = \frac{2100}{0.4x^{-0.6}y^{0.6}} = \frac{2100}{0.4(1004.72)^{-0.6}(3164.86)^{0.6}} \approx \$2637.4 \text{ per car}$$

This means that to produce on additional car per month would cost about $2637.40 with the lowest-cost use of capital and labor.

## Solutions for Chapter 7 Review

1. (a)

Figure 7.37 — UV exposure in 1970

Figure 7.38 — UV exposure in 1990

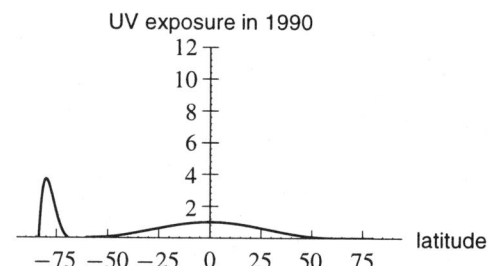

Figure 7.39 — UV exposure in 2000

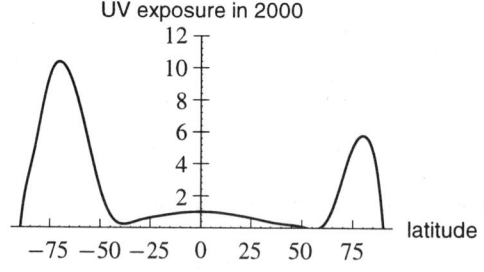

(b) **TABLE 7.24** *Latitude of most severe exposure*

| Year | 1970 | 1975 | 1980 | 1985 | 1990 | 1995 | 2000 | 2005 | 2010 |
|---|---|---|---|---|---|---|---|---|---|
| Latitude | 0 | 0 | 0 | 0 | -80 | -70 | -70 | -70 | -60 |

(c) The latitude of most severe exposure was $-80°$ in 1990 (near the south pole) and moves to $-70°$ and then $-60°$. The reason of this is that a large hole in the ozone is forming above the south pole, and a smaller hole is opening above the north pole. These holes in the ozone allow more UV light to pass through the atmosphere to the surface of the planet.

2. (a) The profit is given by the following:

$$\pi = (\text{Revenue from } q_1) + (\text{ Revenue from } q_2) - \text{Cost}.$$

Measuring $\pi$ in thousands, we obtain:
$$\pi = 3q_1 + 12q_2 - 4.$$

(b) A contour diagram of $\pi$ is in Figure 7.40. Note that the units of $\pi$ are in thousands.

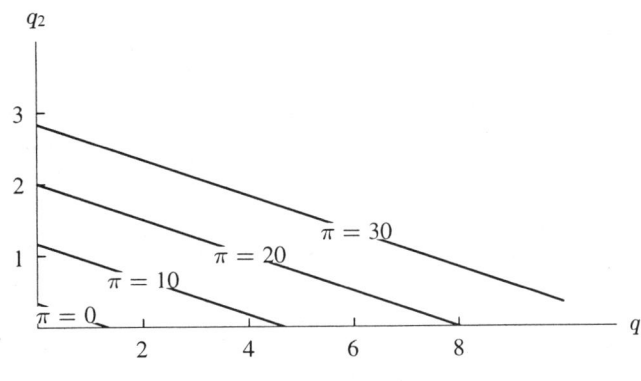

**Figure 7.40**

3. The temperature is decreasing away from the window, suggesting that heat is flowing in from the window. As time goes by the temperature at each point in the room increases. This could be caused by opening the window of an air conditioned room at $t = 0$ thus letting heat from the hot summer day outside raise the temperature inside.

4. At any fixed time $t$, as you get farther from the volcano, the fallout decreases exponentially. At the start of the eruption, there is no fallout; as time passes, the fallout increases. This answer models the behavior of an actual volcano, as you would expect the fallout to increase with time but diminish with distance.

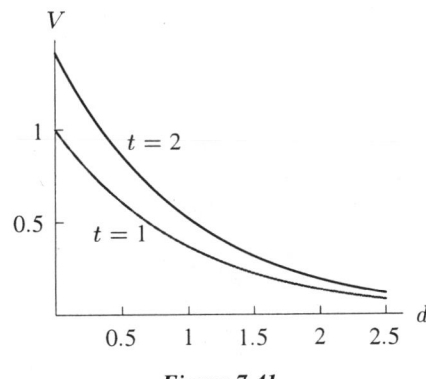

**Figure 7.41**

**354** CHAPTER SEVEN /SOLUTIONS

5. At any fixed position, as time passes, the fallout increases, although increasing more slowly as time passes. Also, the fallout decreases with distance. This answer is as we would expect: the fallout increases with time but decreases with distance from the volcano. See Figure 7.42.

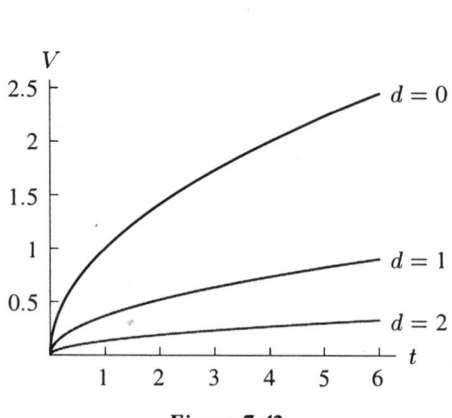

Figure 7.42

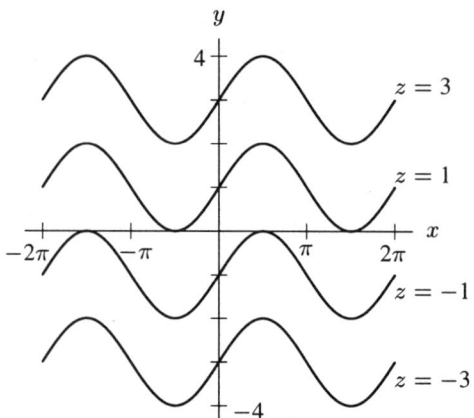

Figure 7.43

6. The contours of $z = y - \sin x$ are of the form $y = \sin x + c$ for a constant $c$. They are sinusoidal graphs shifted vertically by the value of $z$ on the contour line. The contours are equally spaced vertically for equally spaced $z$ values. See Figure 7.43.

7. Contours are lines of the form $3x - 5y + 1 = c$ as shown in Figure 7.44. Note that for the regions of $x$ and $y$ given, the $c$ values range from $-12 < c < 12$ and are equally spaced by 4.

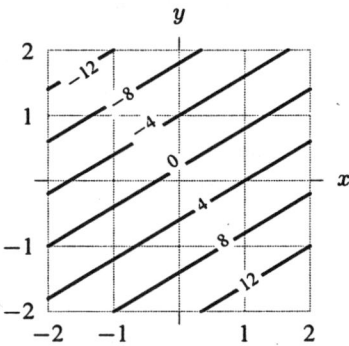

Figure 7.44

8. Contours are ellipses of the form $2x^2 + y^2 = c$ as shown in Figure 7.45. Note that for the ranges of $x$ and $y$ given, the range of $c$ value is $1 \leq c < 9$ and are equally spaced by 2.

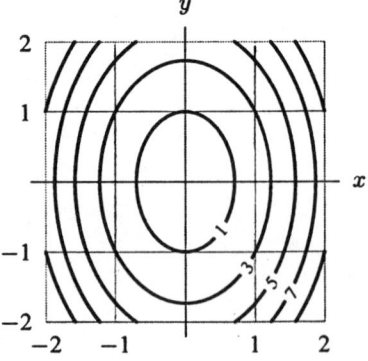

Figure 7.45

9. (a)

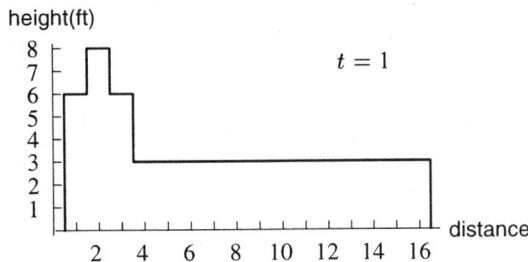

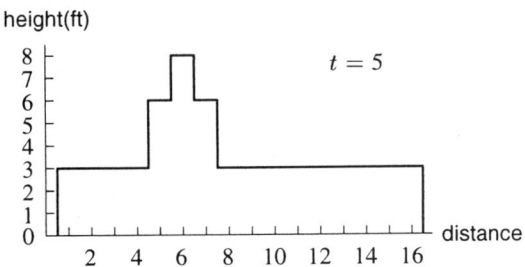

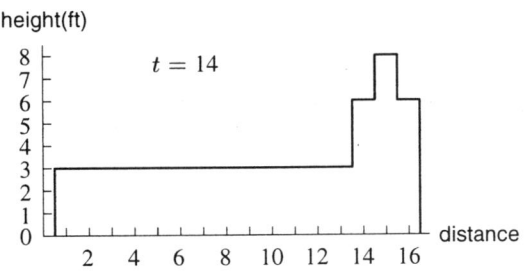

(b)

**Figure 7.46**

(c) The "wave" at a sports arena.

10. (a) The TMS map of an eye of constant curvature will have only one color, with no contour lines dividing the map.
    (b) The contour lines are circles, because the cross-section is the same in every direction. The largest curvature is in the center. See picture below.

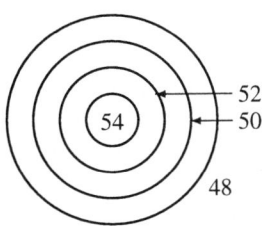

**11.** $f_x = 2x + y$, $f_y = 2y + x$.

**12.** $P_a = 2a - 2b^2$, $P_b = -4ab$.

**13.** $\dfrac{\partial Q}{\partial p_1} = 50p_2$, $\dfrac{\partial Q}{\partial p_2} = 50p_1 - 2p_2$.

**14.** $\dfrac{\partial f}{\partial x} = 5e^{-2t}$, $\dfrac{\partial f}{\partial t} = -10xe^{-2t}$.

**15.** $\dfrac{\partial P}{\partial K} = 7K^{-0.3}L^{0.3}$, $\dfrac{\partial P}{\partial L} = 3K^{0.7}L^{-0.7}$.

**16.** $f_x = \dfrac{x}{\sqrt{x^2+y^2}}$, $f_y = \dfrac{y}{\sqrt{x^2+y^2}}$.

**17.** To read off the cross-sections of $f$ with $t$ fixed, we choose a $t$ value and move horizontally across the diagram looking at the values on the contours. For $t = 0$, as we move from the left at $x = 0$ to the right at $x = \pi$, we cross contours of 0.25, 0.50, 0.75 and reach a maximum at $x = \pi/2$, and then decrease back to 0. That is because if time is fixed at $t = 0$, then $f(x, 0)$ is the displacement of the string at that time: no displacement at $x = 0$ and $x = \pi$ and greatest displacement at $x = \pi/2$. For cross-sections with $t$ fixed at larger values, as we move along a horizontal line, we cross fewer contours and reach a smaller maximum value: the string is becoming less curved. At time $t = \pi/2$, the string is straight so we see a value of 0 all the way across the diagram, namely a contour with value 0. For $t = \pi$, the string has vibrated to the other side and the displacements are negative as we read across the diagram reaching a minimum at $x = \pi/2$.

The cross-sections of $f$ with $x$ fixed are read vertically. At $x = 0$ and $x = \pi$, we see vertical contours of value 0 because the end points of the string have 0 displacement no matter what time it is. The cross-section for $x = \pi/2$ is found by moving vertically up the diagram at $x = \pi/2$. As we expect, the contour values are largest at $t = 0$, zero at $t = \pi/2$, and a minimum at $t = \pi$.

Notice that the spacing of the contours is also important. For example, for the $t = 0$ cross-section, contours are most closely spaced at the end points at $x = 0$ and $x = \pi$ and most spread out at $x = \pi/2$. That is because the shape of the string at time $t = 0$ is a sine curve, which is steepest at the end points and relatively flat in the middle. Thus, the contour diagram shows the steepest terrain at the end points and flattest terrain in the middle.

**18.** (a) Estimating $T(x, t)$ from the figure in the text at $x = 15$, $t = 20$ gives

$$\left.\dfrac{\partial T}{\partial x}\right|_{(15,20)} \approx \dfrac{T(23, 20) - T(15, 20)}{23 - 15} = \dfrac{20 - 23}{8} = -\dfrac{3}{8} \,°\text{C per m},$$

$$\left.\dfrac{\partial T}{\partial t}\right|_{(15,20)} \approx \dfrac{T(15, 25) - T(15, 20)}{25 - 20} = \dfrac{25 - 23}{5} = \dfrac{2}{5} \,°\text{C per min}.$$

At 15 m from heater at time $t = 20$ min, the room temperature decreases by approximately $3/8°$C per meter and increases by approximately $2/5°$C per minute.

(b) We have the estimates,

$$\left.\dfrac{\partial T}{\partial x}\right|_{(5,12)} \approx \dfrac{T(7, 12) - T(5, 12)}{7 - 5} = \dfrac{25 - 27}{2} = -1 \,°\text{C per m},$$

$$\left.\dfrac{\partial T}{\partial t}\right|_{(5,12)} \approx \dfrac{T(5, 40) - T(5, 12)}{40 - 12} = \dfrac{30 - 27}{28} = \dfrac{3}{28} \,°\text{C per min}.$$

At $x = 5$, $t = 12$ the temperature decreases by approximately $1°$C per meter and increases by approximately $3/28°$C per minute.

**19.** The partial derivative, $\partial Q/\partial b$ is the rate of change of the quantity of beef purchased with respect to the price of beef, when the price of chicken stays constant. If the price of beef increases and the price of chicken stays the same, we expect consumers to buy less beef and more chicken. Thus when $b$ increases, we expect $Q$ to decrease, so $\partial Q/\partial b < 0$.

On the other hand, $\partial Q/\partial c$ is the rate of change of the quantity of beef purchased with respect to the price of chicken, when the price of beef stays constant. An increase in the price of chicken is likely to cause consumers to buy less chicken and more beef. Thus when $c$ increases, we expect $Q$ to increase, so $\partial Q/\partial c > 0$.

**20.** The derivative $\partial c/\partial x = b$ is the rate of change of the cost of producing one unit of the product with respect to the amount of labor used (in man hours) when the amount of raw material used stays the same. Thus $\partial c/\partial x = b$ represents the hourly wage.

**21.** The sign of $\partial f/\partial P_1$ tells you whether $f$ (the number of people who ride the bus) increases or decreases when $P_1$ is increased. Since $P_1$ is the price of taking the bus, as it increases, $f$ should decrease. This is because fewer people will be willing to pay the higher price, and more people will choose to ride the train.

On the other hand, the sign of $\frac{\partial f}{\partial P_2}$ tells you the change in $f$ as $P_2$ increases. Since $P_2$ is the cost of riding the train, as it increases, $f$ should increase. This is because fewer people will be willing to pay the higher fares for the train, and more people will choose to ride the bus.

Therefore, $\frac{\partial f}{\partial P_1} < 0$ and $\frac{\partial f}{\partial P_2} > 0$.

22. (a) $C$ represents the quantity of its drinks demanded if $p_1 = p_2 = p_3 = 0$. This represents the quantity of its drinks which would be given away for free when the competing drinks are also free and no money is spent on advertising.

(b) Marginal demand with respect to changes in the company's soft-drink prices is given by

$$\frac{\partial q}{\partial p_1} = -8 \cdot 10^6.$$

Marginal demand with respect to changes in the soft drink prices of other companies is given by

$$\frac{\partial q}{\partial p_2} = 4 \cdot 10^6.$$

Marginal demand with respect to changes in the money spent on advertising is given by

$$\frac{\partial q}{\partial p_3} = 2.$$

The answers are reasonable because the largest effect on quantity sold of the company's soft drinks is caused by changes in their prices and as their prices increase, quantity falls (hence a negative marginal demand). The second largest effect on the quantity sold is caused by the changes in the prices of competing soft drinks. As other companies raise their prices, quantity sold of the first company's drinks increases (hence a positive marginal demand). The effect of changes in money spent on advertising is the smallest and the more money the company spends advertising its drinks, the more drinks it will sell, so there is a positive marginal demand.

23. (a) $\frac{\partial q_1}{\partial x}$ is the rate of change of the quantity of the first brand sold as its price increases. Since this brand competes with another brand, an increase in the price of the first brand should result in a decrease in the quantity sold of this same brand. Thus $\frac{\partial q_1}{\partial x} < 0$. Similarly, $\frac{\partial q_2}{\partial y} < 0$.

(b) Again we take into consideration the competition between the two brands. If the second brand were to increase its price, then more of the first brand should sell. Thus $\frac{\partial q_1}{\partial y} > 0$. Similarly, $\frac{\partial q_2}{\partial x} > 0$.

24. $\frac{\partial q}{\partial I} > 0$ because, other things being constant, as people get richer, more beer will be bought.

$\frac{\partial q}{\partial p_1} < 0$ because, other things being constant, if the price of beer rises, less beer will be bought.

$\frac{\partial q}{\partial p_2} > 0$ because, other things being constant, if the price of other goods rises, but the price of beer does not, more beer will be bought.

25. Since $\frac{\partial f}{\partial y}$ and $\frac{\partial f}{\partial y}$ are defined everywhere, a critical point will occur where $\frac{\partial f}{\partial x} = 0$ and $\frac{\partial f}{\partial y} = 0$. So:

$$\frac{\partial f}{\partial x} = 2x - 4 = 0 \Rightarrow x = 2$$
$$\frac{\partial f}{\partial y} = 6y + 6 = 0 \Rightarrow y = -1$$

$(2, -1)$ is the critical point of $f(x, y)$.

26. Since $\frac{\partial f}{\partial x}$ and $\frac{\partial f}{\partial y}$ are defined everywhere, a critical point will occur where $\frac{\partial f}{\partial x} = 0$ and $\frac{\partial f}{\partial y} = 0$. So:

$$\frac{\partial f}{\partial x} = 3x^2 - 3 = 0, \quad \text{so } x^2 = 1 \text{ and } x = \pm 1$$
$$\frac{\partial f}{\partial y} = 2y = 0, \quad \text{so } y = 0$$

So $(1, 0)$ and $(-1, 0)$ are the critical points. To determine whether these are local extrema, we can examine values of $f$ for $(x, y)$ near the critical points. Functional values for points near $(1, 0)$ are shown in Table 7.25:

**TABLE 7.25**

|   |   | \multicolumn{3}{c}{$x$} | | |
|---|---|---|---|---|
|   |   | 0.99 | 1.00 | 1.01 |
|   | $-0.01$ | $-1.9996$ | $-1.9999$ | $-1.9996$ |
| $y$ | 0.00 | $-1.9997$ | $-2.0000$ | $-1.9997$ |
|   | 0.01 | $-1.9996$ | $-1.9999$ | $-1.9996$ |

As we can see from the table, all the points close to $(1, 0)$ have greater functional value, so $f(1, 0)$ is a local minimum. A similar display of points near $(-1, 0)$ is shown in Table 7.26:

**TABLE 7.26**

|   |   | \multicolumn{3}{c}{$x$} | | |
|---|---|---|---|---|
|   |   | $-1.01$ | $-1.00$ | $-0.99$ |
|   | $-0.01$ | 1.9998 | 2.0001 | 1.9998 |
| $y$ | 0.00 | 1.9997 | 2.0000 | 1.9997 |
|   | 0.01 | 1.9998 | 2.0001 | 1.9998 |

As we can see, some points near $(-1, 0)$ have greater functional value than $f(-1, 0)$ and others have less. So $f(-1, 0)$ is neither a local maximum nor a local minimum.

27. (a) This tells us that an increase in the price of either product causes a decrease in the quantity demanded of both products. An example of products with this relationship is tennis rackets and tennis balls. An increase in the price of either product is likely to lead to a decrease in the quantity demanded of both products as they are used together. In economics, it is rare for the quantity demanded of a product to increase if its price increases, so for $q_1$, the coefficient of $p_1$ is negative as expected. The coefficient of $p_2$ in the expression could be either negative or positive. In this case, it is negative showing that the two products are complementary in use. If it were positive, however, it would indicate that the two products are competitive in use, for example Coke and Pepsi.

(b) The revenue from the first product would be $q_1 p_1 = 150 p_1 - 2 p_1^2 - p_1 p_2$, and the revenue from the second product would be $q_2 p_2 = 200 p_2 - p_1 p_2 - 3 p_2^2$. The total sales revenue of both products, $R$, would be

$$R(p_1, p_2) = 150 p_1 + 200 p_2 - 2 p_1 p_2 - 2 p_1^2 - 3 p_2^2.$$

Note that $R$ is a function of $p_1$ and $p_2$. To find the critical points of $R$, we solve

$$\frac{\partial R}{\partial p_1} = \frac{\partial R}{\partial p_2} = 0.$$

This gives

$$\frac{\partial R}{\partial p_1} = 150 - 2 p_2 - 4 p_1 = 0$$

and

$$\frac{\partial R}{\partial p_2} = 200 - 2 p_1 - 6 p_2 = 0$$

Solving simultaneously, we have $p_1 = 25$ and $p_2 = 25$. Therefore the point $(25, 25)$ is a critical point for $R$. Further,

$$\frac{\partial^2 R}{\partial p_1^2} = -4, \quad \frac{\partial^2 R}{\partial p_2^2} = -6, \quad \frac{\partial^2 R}{\partial p_1 \partial p_2} = -2,$$

so

$$\frac{\partial^2 R}{\partial^2 P_1^2} \frac{\partial^2 R}{\partial^2 P_2^2} - \left( \frac{\partial^2 R}{\partial P_1 \partial P_2} \right)^2 = (-4)(-6) - (-2)^2 = 20.$$

Since $D > 0$ and $\partial^2 R/\partial p_1^2 < 0$, this critical point is a local maximum. Since $R$ is quadratic in $p_1$ and $p_2$, this is a global maximum. Therefore the maximum possible revenue is

$$R = 150(25) + 200(25) - 2(25)(25) - 2(25)^2 - 3(25)^2$$
$$= (6)(25)^2 + 8(25)^2 - 7(25)^2$$
$$= 4375.$$

This is obtained when $p_1 = p_2 = 25$. Note that at these prices, $q_1 = 75$ units, and $q_2 = 100$ units.

28. We compute the partial derivatives:

$$\frac{\partial Q}{\partial K} = b\alpha K^{\alpha-1}L^{1-\alpha} \quad \text{so} \quad K\frac{\partial Q}{\partial K} = b\alpha K^{\alpha}L^{1-\alpha}$$

$$\frac{\partial Q}{\partial L} = b(1-\alpha)K^{\alpha}L^{-\alpha} \quad \text{so} \quad L\frac{\partial Q}{\partial L} = b(1-\alpha)K^{\alpha}L^{1-\alpha}$$

Adding these two results, we have:

$$K\frac{\partial Q}{\partial K} + L\frac{\partial Q}{\partial L} = b(\alpha + 1 - \alpha)K^{\alpha}L^{1-\alpha} = Q.$$

29. (a) About 15 feet along the wall, because that's where there are regions of cold air (55°F and 65°F).
   (b) Roughly between 10 am and 12 noon, and between 4 pm and 6 pm.
   (c) Roughly between midnight and 2 am, between 10 am and 1 pm, and between 4 pm and 9 pm, since that is when the temperature near the heater is greater than 80°F.
   (d)

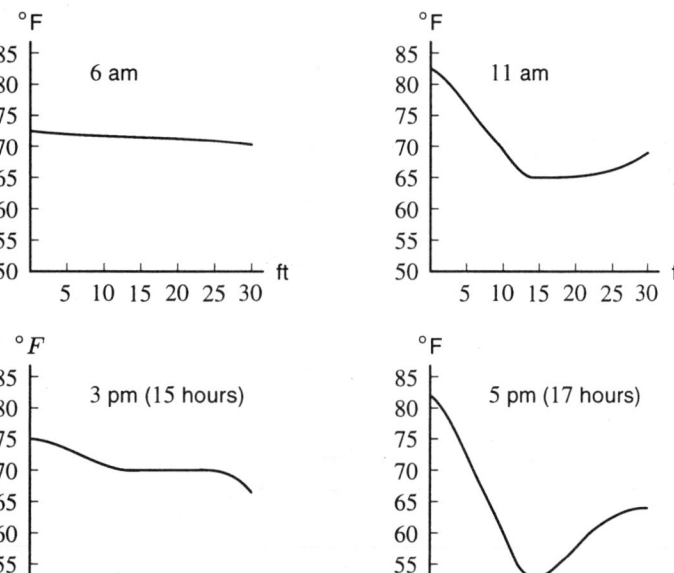

*Figure 7.47*

(e)

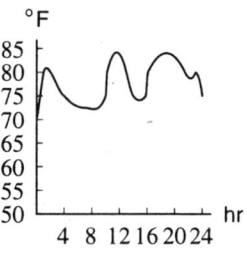

*Figure 7.48*: Temp. vs. Time at heater

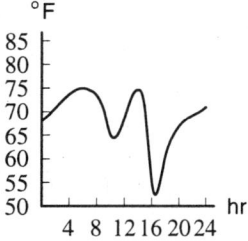

*Figure 7.49*: Temp. vs. Time at window

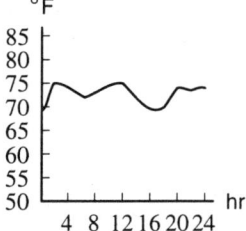

*Figure 7.50*: Temp. vs. Time midway between heater and window

**360** CHAPTER SEVEN /SOLUTIONS

(f) The temperature at the window is colder at 5 pm than at 11 am because the outside temperature is colder at 5 pm than at 11 am.

(g) The thermostat is set to roughly 70°F. We know this because the temperature in the room stays close to 70°F until we get close (a couple of feet) to the window.

(h) We are told that the thermostat is about 2 feet from the window. Thus, the thermostat is either about 13 feet or about 17 feet from the wall. If the thermostat is set to 70°F, every time the temperature at the thermostat goes over or under 70°F, the heater turns off or on. Look at the point at which the vertical lines at 13 feet or about 17 feet cross the 70°F contours. We need to decide which of these crossings correspond best with the times that the heater turns on and off. (These times can be seen along the wall.) Notice that the 17 foot line does not cross the 70°F contour after 16 hours (4 pm). Thus, if the thermostat were 17 feet from the wall, the heater would not turn off after 4 pm. However, the heater does turn off at about 21 hours (9 pm). Since this is the time that the 13 foot line crosses the 70°F contour, we estimate that the thermostat is about 13 feet away from the wall.

30. (a) Because of budget constraints, we are limited in the size of the labor force and the amount of total equipment. This constraint is described in the formula
$$300L + 100K = 15{,}000$$
We can let $L$ range from 0 to about 50. Each choice of $L$ determines a choice of $K$. We have, as a result, the Table 7.27.

**TABLE 7.27**

| $L$ | 5 | 10 | 15 | 20 | 25 | 30 | 35 | 40 | 45 | 50 |
|---|---|---|---|---|---|---|---|---|---|---|
| $K$ | 135 | 120 | 105 | 90 | 75 | 60 | 45 | 30 | 15 | 0 |
| $P$ | 48 | 82 | 111 | 135 | 156 | 172 | 184 | 189 | 181 | 0 |

(b) We wish to maximize $P$ subject to that cost $C$ satisfies $C = 300L + 100K = \$15{,}000$. This is accomplished by solving the following system of equations:

$$P_L = \lambda C_L, \qquad \text{so } 4L^{-0.2}K^{0.2} = 300\lambda$$
$$P_K = \lambda C_K, \qquad \text{so } L^{0.8}K^{-0.8} = 100\lambda$$
$$C = 15{,}000, \qquad \text{so } 300L + 100K = 15{,}000$$

We can solve these equations by dividing the first by the second and then substituting into the budget constraint. Doing so produces the solution $K = 30$, $L = 40$. So the optimal choices of labor and capital are $L = 40$, $K = 30$.

## Solutions to the Projects

1. (a) $\dfrac{\partial p}{\partial c} = f_c(c, s) =$ rate of change in blood pressure as cardiac output increases while systemic vascular resistance remains constant.

   (b) Suppose that $p = kcs$. Note that $c$ (cardiac output), a volume, $s$ (SVR), a resistance, and $p$, a pressure, must all be positive. Thus $k$ must be positive, and our level curves should be confined to the first quadrant. Several level curves are shown in Figure 7.51. Each level curve represents a different blood pressure level. Each point on a given curve is a combination of cardiac output and SVR that results in the blood pressure associated with that curve.

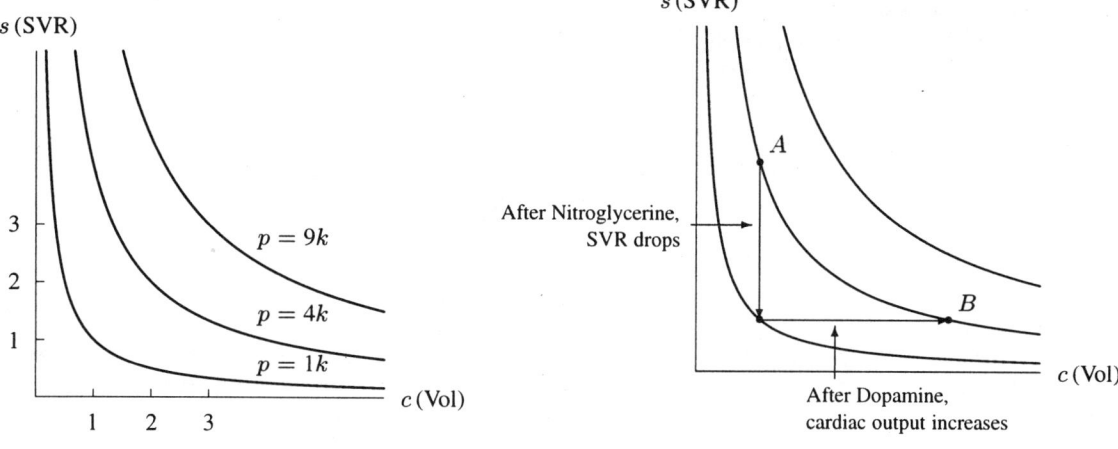

**Figure 7.51**

**Figure 7.52**

(c) Point $B$ in Figure 7.52 shows that if the two doses are correct, the changes in pressure will cancel. The patient's cardiac output will have increased and his SVR will have decreased, but his blood pressure won't have changed.

(d) At point $F$ in Figure 7.53, the patient's blood pressure is normalized, but his/her cardiac output has dropped and his SVR is up.

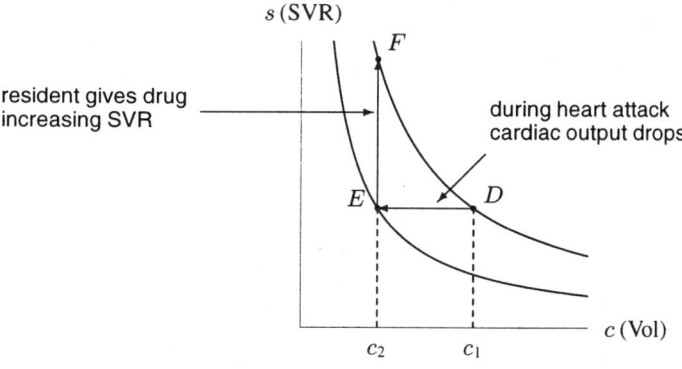

**Figure 7.53**

Note: $c_1$ and $c_2$ are the cardiac outputs before and after the heart attack, respectively.

2. We want to maximize the theater's profit, $P$, as a function of the two variables (prices) $p_c$ and $p_a$. As always, $P = R - C$, where $R$ is the revenue, $R = q_c p_c + q_a p_a$, and $C$ is the cost, which is of the form $C = k(q_c + q_a)$ for some constant $k$. Thus,

$$P(p_c, p_a) = q_c p_c + q_a p_a - k(q_c + q_a)$$
$$= r p_c^{-3} - k r p_c^{-4} + s p_a^{-1} - k s p_a^{-2}$$

To find the critical points, solve

$$\frac{\partial P}{\partial p_c} = -3r p_c^{-4} + 4kr p_c^{-5} = 0$$

$$\frac{\partial P}{\partial p_a} = -s p_a^{-2} + 2ks p_a^{-3} = 0.$$

We get $p_c = 4k/3$ and $p_a = 2k$.

This critical point is a global maximum by the following useful, general argument. Suppose that $F(x, y) = f(x) + g(y)$, where $f$ has a global maximum at $x = b$ and $g$ has a global maximum at $y = d$. Then for all $x, y$:

$$F(x, y) = f(x) + g(y) \leq f(b) + g(d) = F(b, d),$$

so $F$ has global maximum at $x = b$, $y = d$.

The profit function in this problem has the form

$$P(p_c, p_a) = f(p_c) + g(p_a),$$

and the usual single-variable calculus argument using $f'$ and $g'$ shows that $p_c = 4k/3$ and $p_a = 2k$ are global maxima for $f$ and $g$, respectively. Thus the maximum profit occurs when $p_c = 4k/3$ and $p_a = 2k$. Thus,

$$\frac{p_c}{p_a} = \frac{4k/3}{2k} = \frac{2}{3}.$$

## Solutions to Problems on Deriving the Formula for Regression Lines

1. Let the line be in the form $y = b + mx$. When $x$ equals $-1$, $0$ and $1$, then $y$ equals $b - m$, $b$, and $b + m$, respectively. The sum of the squares of the vertical distances, which is what we want to minimize, is

$$f(m, b) = (2 - (b - m))^2 + (-1 - b)^2 + (1 - (b + m))^2.$$

To find the critical points, we compute the partial derivatives with respect to $m$ and $b$,

$$\begin{aligned} f_m &= 2(2 - b + m) + 0 + 2(1 - b - m)(-1) \\ &= 4 - 2b + 2m - 2 + 2b + 2m \\ &= 2 + 4m, \\ f_b &= 2(2 - b + m)(-1) + 2(-1 - b)(-1) + 2(1 - b - m)(-1) \\ &= -4 + 2b - 2m + 2 + 2b - 2 + 2b + 2m \\ &= -4 + 6b. \end{aligned}$$

Setting both partial derivatives equal to zero, we get a system of equations:

$$2 + 4m = 0,$$
$$-4 + 6b = 0.$$

The solution is $m = -1/2$ and $b = 2/3$. You can check that it is a minimum. Hence, the regression line is $y = \frac{2}{3} - \frac{1}{2}x$.

2. Let the line be in the form $y = b + mx$. When $x$ equals $0$, $1$, and $2$, then $y$ equals $b$, $b + m$, and $b + 2m$, respectively. The sum of the squares of the vertical distances, which is what we want to minimize, is

$$f(m, b) = (2 - b)^2 + (4 - (b + m))^2 + (5 - (b + 2m))^2.$$

To find the critical points, we compute the partial derivatives with respect to $m$ and $b$,

$$\begin{aligned} f_m &= 0 + 2(4 - b - m)(-1) + 2(5 - b - 2m)(-2) \\ &= -8 + 2b + 2m - 20 + 4b + 8m \\ &= -28 + 6b + 10m \\ f_b &= 2(2 - b)(-1) + 2(4 - b - m)(-1) + 2(5 - b - 2m)(-1) \\ &= -4 + 2b - 8 + 2b + 2m - 10 + 2b + 4m \\ &= -22 + 6b + 6m. \end{aligned}$$

Setting both partial derivatives equal to zero, we get a system of equations:

$$-28 + 6b + 10m = 0,$$
$$-22 + 6b + 6m = 0.$$

The solution is $m = 1.5$ and $b = 13/6 \approx 2.17$. You can check that it is a minimum. Hence, the regression line is $y = 2.17 + 1.5x$.

3. We have $\sum x_i = 0$, $\sum y_i = 2$, $\sum x_i^2 = 2$, and $\sum y_i x_i = -1$. Thus

$$b = \big((2)(2) - (0)(-1)\big) / \big((3)(2) - (0^2)\big)$$
$$= 4/6 = 2/3$$
$$m = \big((3)(-1) - (0)(2)\big) / \big((3)(2) - (0^2)\big)$$
$$= -3/6 = -1/2$$

The line is $y = \dfrac{2}{3} - \dfrac{1}{2}x$, which agrees with the answer to Problem 1.

4. We have $\sum x_i = 3$, $\sum y_i = 11$, $\sum x_i^2 = 5$, and $\sum y_i x_i = 14$. Thus

$$b = \big((5)(11) - (3)(14)\big) / \big((3)(5) - (3^2)\big)$$
$$= 13/6 \approx 2.17$$
$$m = \big((3)(14) - (3)(11)\big) / \big((3)(5) - (3^2)\big)$$
$$= 9/6 = 1.5$$

The line is $y = 2.17 + 1.5x$, which agrees with the answer to Problem 2.

5. We have $\sum x_i = 6$, $\sum y_i = 5$, $\sum x_i^2 = 14$, and $\sum y_i x_i = 12$. Thus

$$b = \big((14)(5) - (6)(12)\big) / \big((3)(14) - (6^2)\big)$$
$$= -2/6 = -1/3$$
$$m = \big((3)(12) - (6)(5)\big) / \big((3)(14) - (6^2)\big)$$
$$= 6/6 = 1.$$

The line is $y = x - \dfrac{1}{3}$, which agrees with the answer to Example 1.

6. Let $t$ be the number of years since 1960 and let $P(t)$ be the population in the year $1960 + t$. We assume that $P = Ce^{at}$, and therefore

$$\ln P = at + \ln C.$$

So, we plot $\ln P$ against $t$ and find the line of best fit. Our data points are $(0, \ln 180)$, $(10, \ln 206)$, and $(20, \ln 226)$. Applying the method of least squares to find the best-fitting line, we find that

$$a = \frac{\ln 226 - \ln 180}{20} \approx 0.0114,$$
$$\ln C = \frac{\ln 206}{3} - \frac{\ln 206}{6} + \frac{5 \ln 180}{6} \approx 5.20$$

Then, $C = e^{5.20} = 181.3$ and so

$$P(t) = 181.277 e^{0.011378t}.$$

To estimate the population in 1990, we set $t = 30$ and find that $P(30) = 225.026$ million. This is a very good estimate, according to the 1990 census.

7. (a) Let the line take the form of $y = mx + b$, where $x$ equals the number of years since 1920, which is the original year, and $y$ equals the postage corresponding to the year. When the year is 1920, we have $x = 0$, so the postage equals $b$. When the year is 1932, the postage equals $b + 12m$, because the difference in years is 12. And, when the year is 1995, the postage equals $b + 75m$. The sum of the squares of the vertical distances, which is what we wish to minimize, is

$$\begin{aligned}f(m,b) = &(0.02 - b)^2 + (0.03 - (b + 12m))^2 + (0.04 - (b + 38m))^2 \\&+ (0.05 - (b + 43m))^2 + (0.06 - (b + 48m))^2 + (0.08 - (b + 51m))^2 \\&+ (0.1 - (b + 54m))^2 + (0.13 - (b + 55m))^2 + (0.15 - (b + 58m))^2 \\&+ (0.2 - (b + 61m))^2 + (0.22 - (b + 65m))^2 + (0.25 - (b + 68m))^2 \\&+ (0.29 - (b + 71m))^2 + (0.32 - (b + 75m))^2\end{aligned}$$

To find the critical points, calculate the partial derivatives $f_m$ and $f_b$.

$$f_m = 0 + 2(0.03 - (b + 12m))(-12) + 2(0.04 - (b + 38m))(-38)$$
$$+2(0.05 - (b + 43m))(-43) + 2(0.06 - (b + 48m))(-48) + 2(0.08 - (b + 51m))(-51)$$
$$+2(0.1 - (b + 54m))(-54) + 2(0.13 - (b + 55m))(-55) + 2(0.15 - (b + 58m))(-58)$$
$$+2(0.2 - (b + 61m))(-61) + 2(0.22 - (b + 65m))(-65) + 2(0.25 - (b + 68m))(-68)$$
$$+2(0.29 - (b + 71m))(-71) + 2(0.32 - (b + 75m))(-75)$$
$$= 1398b + 81766m - 240.66$$

$$f_b = 2(0.02 - b)(-1) + 2(0.03 - (b + 12m))(-1) + 2(0.04 - (b + 38m))(-1)$$
$$+2(0.05 - (b + 43m))(-1) + 2(0.06 - (b + 48m))(-1) + 2(0.08 - (b + 51m))(-1)$$
$$+2(0.1 - (b + 54m))(-1) + 2(0.13 - (b + 55m))(-1) + 2(0.15 - (b + 58m))(-1)$$
$$+2(0.2 - (b + 61m))(-1) + 2(0.22 - (b + 65m))(-1) + 2(0.25 - (b + 68m))(-1)$$
$$+2(0.29 - (b + 71m))(-1) + 2(0.32 - (b + 75m)(-1)$$
$$= 28b + 1398m - 3.88$$

Setting both partial derivatives equal to zero, we get a system of two equations:

$$1398b + 81766m = 240.66$$
$$28b + 1398m = 3.88$$

with solutions $m \approx 0.0066$ and $b \approx -0.2135$. Thus the equation of the line is $y = 0.0066x - 0.2135$

To predict the cost of a postage stamp in the year 2010, we substitute $x = 2010 - 1920$ into the equation we just created and obtain:

$$y = 0.0066(2010 - 1920) - 0.2135$$
$$y = 0.3805$$

Therefore, the cost of a postage stamp in the year 2010 would be $0.38.

(b) Looking at the data in Figure 7.54 you can see that it does not appear linear over the whole graph, but does look linear after about 1972.

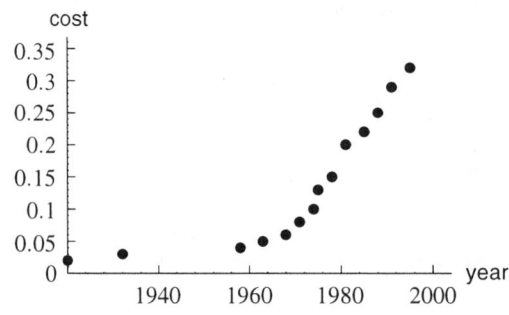

Figure 7.54

(c)

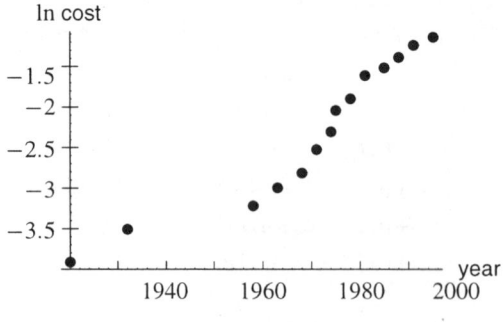

Figure 7.55

Looking at the data in Figure 7.55 you can see that it appears linear after 1960. If it were linear over the entire range, say $\ln y = mx + b$, then $y = e^{mx+b}$ so the price of a stamp is exponential, and is increasing rapidly as time goes on.

To find the line that best fits this data, we use the same method as before. This time

$$f(m,b) = (\ln 0.02 - b)^2 + (\ln 0.03 - (b+12m))^2 + (\ln 0.04 - (b+38m))^2$$
$$+ (\ln 0.05 - (b+43m))^2 + (\ln 0.06 - (b+48m))^2 + (\ln 0.08 - (b+51m))^2$$
$$+ (\ln 0.10 - (b+54m))^2 + (\ln 0.13 - (b+55m))^2 + (\ln 0.15 - (b+58m))^2$$
$$+ (\ln 0.20 - (b+61m))^2 + (\ln 0.22 - (b+65m))^2 + (\ln 0.25 - (b+68m))^2$$
$$+ (\ln 0.29 - (b+71m))^2 + (\ln 0.32 - (b+75m))^2$$

Setting the partial derivatives equal to zero, we get:

$$f_m = 0 + 2(\ln 0.03 - (b+12m))(-12) + 2(\ln 0.04 - (b+38m))(-38)$$
$$+ 2(\ln 0.05 - (b+43m))(-43) + 2(\ln 0.06 - (b+48m))(-48)$$
$$+ 2(\ln 0.08 - (b+51m))(-51) + 2(\ln 0.10 - (b+54m))(-54)$$
$$+ 2(\ln 0.13 - (b+55m))(-55) + 2(\ln 0.15 - (b+58m))(-58)$$
$$+ 2(\ln 0.20 - (b+61m))(-61) + 2(\ln 0.22 - (b+65m))(-65)$$
$$+ 2(\ln 0.25 - (b+68m))(-68) + 2(\ln 0.29 - (b+71m))(-71)$$
$$+ 2(\ln 0.32 - (b+75m))(-75)$$
$$= 1398b + 81766m + 2726.51 = 0,$$

$$f_b = 2(\ln 0.02 - b)(-1) + 2(\ln 0.03 - (b+12m))(-1) + 2(\ln 0.04 - (b+38m))(-1)$$
$$+ 2(\ln 0.05 - (b+43m))(-1) + 2(\ln 0.06 - (b+48m))(-1)$$
$$+ 2(\ln 0.08 - (b+51m))(-1) + 2(\ln 0.10 - (b+54m))(-1)$$
$$+ 2(\ln 0.13 - (b+55m))(-1) + 2(\ln 0.15 - (b+58m))(-1)$$
$$+ 2(\ln 0.20 - (b+61m))(-1) + 2(\ln 0.22 - (b+65m))(-1)$$
$$+ 2(\ln 0.25 - (b+68m))(-1) + 2(\ln 0.29 - (b+71m))(-1)$$
$$+ 2(\ln 0.32 - (b+75m))(-1)$$
$$= 28b + 1398m + 64.20 = 0.$$

These equations have solutions $m \approx 0.04$ and $b \approx -4.2911$. Thus, the equation of this line is

$$\ln y = 0.04x - 4.2911,$$

where $y$ is the price of a stamp.

To predict the cost of the stamp in 2010, we substitute $x = 2010 - 1920$ into the equation and get

$$\ln y = 0.04(2010 - 1920) - 4.2911$$
$$\ln y = -0.6911$$

Therefore, $y = e^{-0.6911} \approx 0.5010$, and so the cost would be \$0.50.

8. (a) Suppose $N = kA^p$. Then the rule of thumb tells us that if $A$ is multiplied by 10, the value of $N$ doubles. Thus

$$2N = k(10A)^p = k10^p A^p.$$

Thus, dividing by $N = kA^p$, we have

$$2 = 10^p$$

so taking logs to base 10 we have

$$p = \log 2 = 0.3010.$$

(where $\log 2$ means $\log_{10} 2$). Thus,

$$N = kA^{0.3010}.$$

(b) Taking natural logs gives

$$\ln N = \ln(kA^p)$$
$$\ln N = \ln k + p \ln A$$
$$\ln N \approx \ln k + 0.301 \ln A$$

Thus, $\ln N$ is a linear function of $\ln A$.

(c) Table 7.28 contains the natural logarithms of the data:

**TABLE 7.28** $\ln N$ and $\ln A$

| Island | $\ln A$ | $\ln N$ |
|---|---|---|
| Redonda | 1.1 | 1.6 |
| Saba | 3.0 | 2.2 |
| Montserrat | 2.3 | 2.7 |
| Puerto Rico | 9.1 | 4.3 |
| Jamaica | 9.3 | 4.2 |
| Hispaniola (Haiti & Dominican Rep.) | 11.2 | 4.8 |
| Cuba | 11.6 | 4.8 |

Using a least squares fit we find the line:

$$\ln N = 1.20 + 0.32 \ln A$$

This yields the power function:

$$N = e^{1.20} A^{0.32} = 3.32 A^{0.32}$$

Since 0.32 is pretty close to $\log 2 \approx 0.301$, the answer does agree with the biological rule.

# CHAPTER EIGHT

## Solutions for Section 8.1

1. Since $y = t^4$, the derivative is $dy/dt = 4t^3$. We have

$$\text{Left-side} = t\frac{dy}{dt} = t(4t^3) = 4t^4.$$
$$\text{Right-side} = 4y = 4t^4.$$

   Since the substitution $y = t^4$ makes the differential equation true, $y = t^4$ is in fact a solution.

2. Since $y = x^3$, we know that $y' = 3x^2$. Substituting $y = x^3$ and $y' = 3x^2$ into the differential equation we get

$$\text{Left-side} = xy' - 3y = x(3x^2) - 3(x^3) = 3x^3 - 3x^3 = 0.$$

   Since the left and right sides are equal for all $x$, we see that $y = x^3$ is a solution.

3. (a) Since $y = x^2$, we have $y' = 2x$. Substituting these functions into our differential equation, we have

$$xy' - 2y = x(2x) - 2(x^2) = 2x^2 - 2x^2 = 0.$$

   Therefore, $y = x^2$ is a solution to the differential equation $xy' - 2y = 0$.

   (b) For $y = x^3$, we have $y' = 3x^2$. Substituting gives:

$$xy' - 2y = x(3x^2) - 2(x^3) = 3x^3 - 2x^3 = x^3.$$

   Since $x^3$ does not equal 0 for all $x$, we see that $y = x^3$ is not a solution to the differential equation.

4. At $t = 0$, we know $P = 70$ and we can compute the value of $dP/dt$:

$$\text{At } t = 0, \quad \text{we have} \quad \frac{dP}{dt} = 0.2P - 10 = 0.2(70) - 10 = 4.$$

   The population is increasing at a rate of 4 million fish per year. At the end of the first year, the fish population will have grown by about 4 million fish, and so we have:

$$\text{At } t = 1, \quad \text{we estimate} \quad P = 70 + 4 = 74.$$

   We can now use this new value of $P$ to calculate $dP/dt$ at $t = 1$:

$$\text{At } t = 1, \quad \text{we have} \quad \frac{dP}{dt} = 0.2P - 10 = 0.2(74) - 10 = 4.8,$$

   and so:

$$\text{At } t = 2, \quad \text{we estimate} \quad P = 74 + 4.8 = 78.8.$$

   Continuing in this way, we obtain the values in Table 8.1.

**TABLE 8.1**

| $t$ | 0 | 1 | 2 | 3 |
|---|---|---|---|---|
| $P$ | 70 | 74 | 78.8 | 84.56 |

5. We know that at time $t = 0$, the value of $y$ is 8. Since we are told that $dy/dt = 0.5y$, we know that at time $t = 0$
$$\frac{dy}{dt} = 0.5(8) = 4.$$
As $t$ goes from 0 to 1, $y$ will increase by 4, so at $t = 1$,
$$y = 8 + 4 = 12.$$
Likewise, we get that at $t = 1$,
$$\frac{dy}{dt} = .5(12) = 6$$
so that at $t = 2$,
$$y = 12 + 6 = 18.$$
At $t = 2$, $\frac{dy}{dt} = .5(18) = 9$ so that at $t = 3$, $y = 18 + 9 = 27$.
At $t = 3$, $\frac{dy}{dt} = .5(27) = 13.5$ so that at $t = 4$, $y = 27 + 13.5 = 40.5$.
Thus we get the following table

| $t$ | 0 | 1 | 2 | 3 | 4 |
|---|---|---|---|---|---|
| $y$ | 8 | 12 | 18 | 27 | 40.5 |

6. We know that at time $t = 0$, the value of $y$ is 8. Since we are told that $dy/dt = 0.5t$, we know that at time $t = 0$
$$\frac{dy}{dt} = 0.5(0) = 0.$$
As $t$ goes from 0 to 1, $y$ will increase by 0, so at $t = 1$,
$$y = 8 + 0(1) = 8.$$
Likewise, we get that at $t = 1$,
$$\frac{dy}{dt} = 0.5(1) = 0.5$$
and so at $t = 2$
$$y = 8 + 0.5(1) = 8.5.$$
At $t = 2$,
$$\frac{dy}{dt} = 0.5(2) = 1$$
then at $t = 3$
$$y = 8.5 + 1(1) = 9.5.$$
At $t = 3$, $\frac{dy}{dt} = 0.5(3) = 1.5$ so that at $t = 4$, $y = 9.5 + 1.5(1) = 11$.
Thus we get the following table

| $t$ | 0 | 1 | 2 | 3 | 4 |
|---|---|---|---|---|---|
| $y$ | 8 | 8 | 8.5 | 9.5 | 11 |

7. When $y = 125$, the rate of change of $y$ is
$$\frac{dy}{dt} = -0.20y = -0.20(125) = -25.$$
The value of $y$ goes down by 25 as $t$ goes up by 1, so when $t = 1$, we have
$$y = \text{Old value of } y + \text{Change in } y$$
$$= 125 + (-25)$$
$$= 100.$$
Continuing in this way, we fill in the table as shown:

| $t$ | 0 | 1 | 2 | 3 | 4 |
|---|---|---|---|---|---|
| $y$ | 125 | 100 | 80 | 64 | 51.2 |

8. We know that at time $t = 0$, the value of $y$ is 8. Since we are told that $dy/dt = 4 - y$, we know that at time $t = 0$

$$\frac{dy}{dt} = 4 - 8 = -4.$$

As $t$ goes from 0 to 1, $y$ will decrease by 4, so at $t = 1$,

$$y = 8 - 4 = 4$$

Likewise, we get that at $t = 1$,

$$\frac{dy}{dt} = 4 - 4 = 0$$

so that at $t = 2$,

$$y = 4 + 0(1) = 4.$$

At $t = 2$, $\frac{dy}{dt} = 4 - 4 = 0$ so that at $t = 3$, $y = 4 + 0 = 4$.

At $t = 3$, $\frac{dy}{dt} = 4 - 4 = 0$ so that at $t = 4$, $y = 4 + 0 = 4$.

Thus we get the following table

| $t$ | 0 | 1 | 2 | 3 | 4 |
|---|---|---|---|---|---|
| $y$ | 8 | 4 | 4 | 4 | 4 |

9. (a) = (III), (b) = (IV), (c) = (I), (d) = (II).

10. (a) = (I), (b) = (IV), (c) = (III). Graph (II) represents an egg originally at 0° C which is moved to the kitchen table (20° C) two minutes after the egg in part (a) is moved.

11. If $P = P_0 e^t$, then

$$\frac{dP}{dt} = \frac{d}{dt}(P_0 e^t) = P_0 e^t = P.$$

12. If $Q = Ce^{kt}$, then

$$\frac{dQ}{dt} = Cke^{kt} = k(Ce^{kt}) = kQ.$$

We are given that $\frac{dQ}{dt} = -0.03Q$, so we know that $kQ = -0.03Q$. Thus we either have $Q = 0$ (in which case $C = 0$ and $k$ is anything) or $k = -0.03$. Notice that if $k = -0.03$, then $C$ can be any number.

13. Since $y = x^2 + k$ we know that

$$y' = 2x.$$

Substituting $y = x^2 + k$ and $y' = 2x$ into the differential equation we get

$$\begin{aligned} 10 &= 2y - xy' \\ &= 2(x^2 + k) - x(2x) \\ &= 2x^2 + 2k - 2x^2 \\ &= 2k \end{aligned}$$

Thus, $k = 5$ is the only solution.

14. Yes. To see why, we substitute $y = x^n$ into the equation $13x\frac{dy}{dx} = y$. We first calculate $\frac{dy}{dx} = \frac{d}{dx}(x^n) = nx^{n-1}$. The differential equation becomes

$$13x(nx^{n-1}) = x^n$$

But $13x(nx^{n-1}) = 13n(x \cdot x^{n-1}) = 13nx^n$, so we have

$$13n(x^n) = x^n$$

This equality must hold for all $x$, so we get $13n = 1$, so $n = 1/13$. Thus, $y = x^{1/13}$ is a solution.

15. We are told that $y$ is a function of $t$ (since the derivative is $dy/dt$) with derivative $2t$. We need to think of a function with derivative $2t$. Since $y = t^2$ has derivative $2t$, we see that $y = t^2$ is a solution to this differential equation. Since the function $y = t^2 + 1$ also has derivative $2t$, we see that $y = t^2 + 1$ is also a solution. In fact, $y = t^2 + C$ is a solution for any constant $C$. The general solution is
$$y = t^2 + C.$$

16.

|     |       |                       |                           |
| --- | ----- | --------------------- | ------------------------- |
| (I) | $y = 2\sin x$, | $dy/dx = 2\cos x$, | $d^2y/dx^2 = -2\sin x$ |
| (II) | $y = \sin 2x$, | $dy/dx = 2\cos 2x$, | $d^2y/dx^2 = -4\sin 2x$ |
| (III) | $y = e^{2x}$, | $dy/dx = 2e^{2x}$, | $d^2y/dx^2 = 4e^{2x}$ |
| (IV) | $y = e^{-2x}$, | $dy/dx = -2e^{-2x}$, | $d^2y/dx^2 = 4e^{-2x}$ |

and so:

(a) (IV)
(b) (III)
(c) (III), (IV)
(d) (II)

17. We first compute $dy/dx$ for each of the functions on the right.

If $y = x^3$ then
$$\frac{dy}{dx} = 3x^2$$
$$= 3\frac{y}{x}.$$

If $y = 3x$ then
$$\frac{dy}{dx} = 3$$
$$= \frac{y}{x}.$$

If $y = e^{3x}$ then
$$\frac{dy}{dx} = 3e^{3x}$$
$$= 3y.$$

If $y = 3e^x$ then
$$\frac{dy}{dx} = 3e^x$$
$$= y.$$

Finally, if $y = x$ then
$$\frac{dy}{dx} = 1$$
$$= \frac{y}{x}.$$

Comparing our calculated derivatives with the right-hand sides of the differential equations we see that (a) is solved by (II) and (V), (b) is solved by (I), (c) is not solved by any of our functions, (d) is solved by (IV) and (e) is solved by (III).

## Solutions for Section 8.2

1. (a)

    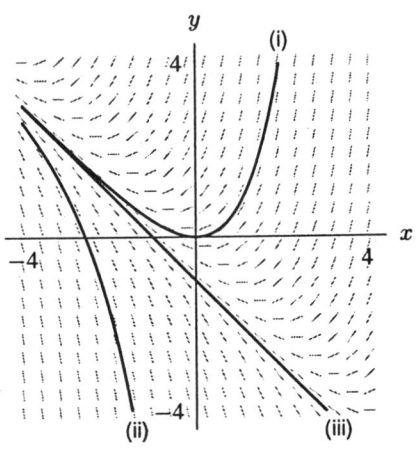

   (b) The solution through $(-1, 0)$ appears to be linear with equation $y = -x - 1$.
   (c) If $y = -x - 1$, then $y' = -1$ and $x + y = x + (-x - 1) = -1$, so this checks as a solution.

2. (a)

    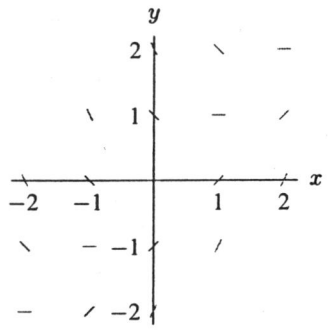

    Figure 8.1

   (b) The point $(1, 0)$ satisfies the equation $y = x - 1$. If $y = x - 1$, then $y' = 1$ and $x - y = x - (x - 1) = 1$, so $y = x - 1$ is the solution to the differential equation through $(1, 0)$.

3. (a) (i), (ii), (iii)

    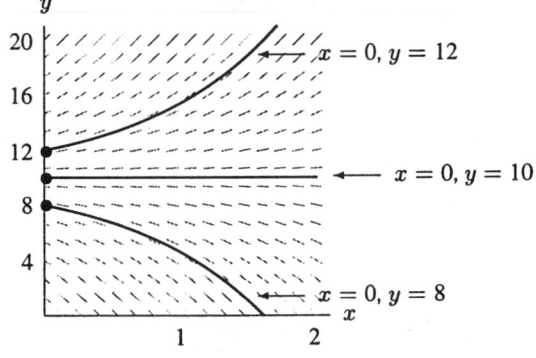

    Figure 8.2

   (b) When $y = 10$, we have $dy/dx = 0$ and so the solution curve is horizontal. This is why the solution curve through $y = 10$ is a horizontal line.

4. (a) The slope at any point $(x, y)$ is equal to $dy/dx$, which is $xy$. Then:

$$\text{At point } (2, 1), \text{ slope} = 2 \cdot 1 = 2,$$
$$\text{At point } (0, 2), \text{ slope} = 0 \cdot 2 = 0,$$
$$\text{At point } (-1, 1), \text{ slope} = -1 \cdot 1 = -1,$$
$$\text{At point } (2, -2), \text{ slope} = 2(-2) = -4.$$

(b)

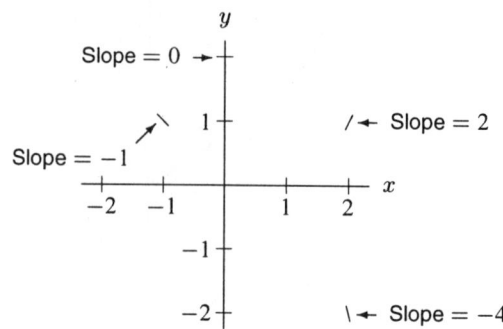

*Figure 8.3*

5.

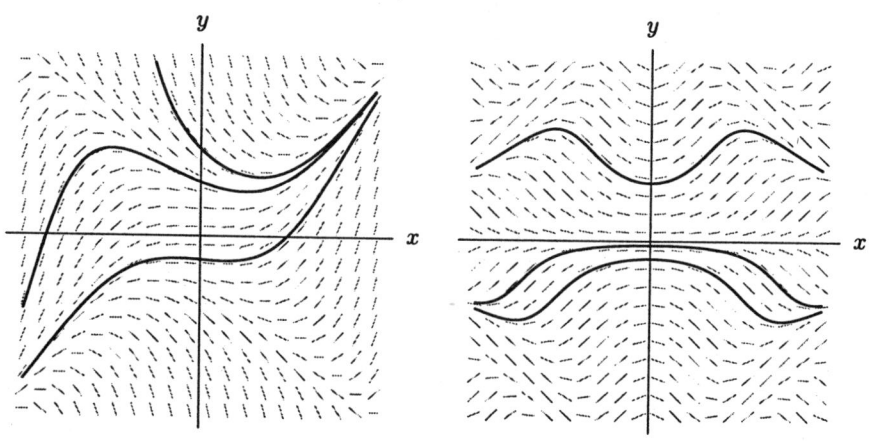

*Figure 8.4*

Other choices of solution curves are, of course, possible.

6. (a) II  (b) VI  (c) IV  (d) I  (e) III  (f) V

7. If the starting point has $y > 0$, then $y \to \infty$ as $x \to \infty$. If the starting point has $y = 0$, then the solution is constant; $y = 0$. If the starting point has $y < 0$, then $y \to -\infty$ as $x \to \infty$.

8. As $x$ increases, $y \to \infty$.

9. As $x \to \infty$, $y \to \infty$, no matter what the starting point is.

10. As $x \to \infty$, $y$ seems to oscillate within a certain range. The range will depend on the starting point, but the *size* of the range appears independent of the starting point.

11. If $y = 4$ for the starting point, then $y = 4$ always, so $y \to 4$ as $x \to \infty$. If $y \neq 4$ for the starting point, then $y \to 4$ as $x \to \infty$.

12. From the slope field, the function looks like a parabola of the form $y = x^2 + C$, where $C$ depends on the starting point. In any case, $y \to \infty$ as $x \to \infty$.

13. When $a = 1$ and $b = 2$, the Gompertz equation is $y' = -y \ln(y/2) = y \ln(2/y) = y(\ln 2 - \ln y)$. This differential equation is similar to the differential equation $y' = y(2 - y)$ in certain ways. For example, in both equations $y'$ is positive for $0 < y < 2$ and negative for $y > 2$. Also, for $y$-values close to 2, $(\ln 2 - \ln y)$ and $(2 - y)$ are both close to 0, so $y(\ln 2 - \ln y)$ and $y(2 - y)$ are approximately equal to zero. Thus around $y = 2$ the slope fields look almost the same. This happens again around $y = 0$, since around $y = 0$ both $y(2 - y)$ and $y(\ln 2 - \ln y)$ go to 0. Finally, for $y > 2$, $\ln y$ grows much slower than $y$, so the slope field for $y' = y(\ln 2 - \ln y)$ is less steep, negatively, than for $y' = y(2 - y)$.

14. (a)

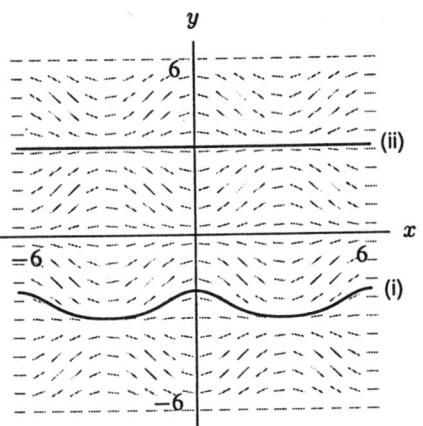

**Figure 8.5**

(b) We can see that the slope lines are horizontal when $y$ is an integer multiple of $\pi$. We conclude from Figure 8.5 that the solution is $y = n\pi$ in this case.

To check this, we note that if $y = n\pi$, then $(\sin x)(\sin y) = (\sin x)(\sin n\pi) = 0 = y'$. Thus $y = n\pi$ is a solution to $y' = (\sin x)(\sin y)$, and it passes through $(0, n\pi)$.

## Solutions for Section 8.3

1. The equation given is in the form
$$\frac{dP}{dt} = kP.$$
Thus we know that the general solution to this equation will be
$$P = Ce^{kt}.$$
And in our case, with $k = 0.02$ and $C = 20$ we get
$$P = 20e^{0.02t}.$$

2. The equation given is in the form
$$\frac{dQ}{dt} = kQ.$$
Thus we know that the general solution to this equation will be
$$Q = Ce^{kt}.$$
And in our case, with $k = \frac{1}{5}$ we get
$$Q = Ce^{\frac{1}{5}t}.$$
We know that $Q = 50$ when $t = 0$. Thus we get
$$Q(t) = Ce^{\frac{1}{5}t}$$
$$Q(0) = 50 = Ce^{0}$$
$$50 = C$$

Thus we get
$$Q = 50e^{\frac{1}{5}t}.$$

3. The equation given is in the form
$$\frac{dm}{dt} = km.$$
Thus we know that the general solution to this equation will be
$$m = Ce^{kt}.$$
And in our case, with $k = 3$ we get
$$m = Ce^{3t}.$$
We know that $m = 5$ when $t = 1$. Thus we get
$$m(t) = Ce^{3t}$$
$$m(1) = 5 = Ce^3$$
$$C = 5e^{-3}$$
Thus we get
$$m = 5e^{-3}e^{3t} = 5e^{3t-3}.$$

4. The equation given is in the form
$$\frac{dI}{dx} = kI.$$
Thus we know that the general solution to this equation will be
$$I = Ce^{kx}.$$
And in our case, with $k = 0.2$ we get
$$I = Ce^{0.2x}.$$
We know that $I = 6$ when $x = -1$. Thus we get
$$I(x) = Ce^{0.2x}$$
$$I(-1) = 6 = Ce^{-0.2}$$
$$C = 6e^{0.2}$$
Thus we get
$$I = 6e^{0.2}e^{0.2x} = 6e^{0.2x+0.2} = 6e^{0.2(x+1)}.$$

5. Rewriting we get
$$\frac{dy}{dx} = -\frac{1}{3}y.$$
We know that the general solution to an equation in the form
$$\frac{dy}{dx} = ky$$
is
$$y = Ce^{kx}.$$
Thus in our case we get
$$y = Ce^{-\frac{1}{3}x}.$$
We are told that $y(0) = 10$ so we get
$$y(x) = Ce^{-\frac{1}{3}x}$$
$$y(0) = 10 = Ce^0$$
$$C = 10$$
Thus we get
$$y = 10e^{-\frac{1}{3}x}.$$

6. Rewriting we get
$$\frac{dz}{dt} = 5z.$$
We know that the general solution to an equation in the form
$$\frac{dz}{dt} = kz$$
is
$$z = Ce^{kt}.$$
Thus in our case we get
$$z = Ce^{5t}.$$
We are told that $z(1) = 5$ so we get
$$z(t) = Ce^{5t}$$
$$z(1) = 5 = Ce^{5}$$
$$C = 5e^{-5}$$
Thus we get
$$z = 5e^{-5}e^{5t} = 5e^{5t-5}.$$

7. Since the right hand side is a function of $t$, we are looking for a function whose derivative is $t^2$. One such function is $t^3/3$, so the general solution of the equation $dy/dt = t^2$ is
$$y = \frac{1}{3}t^3 + C$$
which is neither exponential growth nor exponential decay.

8. The equation given is in the form
$$\frac{dW}{dx} = kW.$$
Thus we know that the general solution to this equation will be
$$W = Ce^{kx}.$$
And in our case, with $k = 5$, we get
$$W = Ce^{5x}.$$
This function is exponential if $C \neq 0$.

9. Since the right hand side is a function of $x$, we are looking for a function whose derivative is $5x$. One such function is $5x^2/2$, so the general solution of the equation $dB/dx = 5x$ is
$$B = \frac{5}{2}x^2 + C$$
which is neither an exponentially growing nor an exponentially decaying function.

10. The equation given is in the form
$$\frac{dR}{dt} = kR.$$
Thus we know that the solution to this equation will be
$$R = Ae^{kt}.$$
This is an exponential function if $A \neq 0$ and $k \neq 0$.

11. Since the right hand side is a function of $t$, we are looking for a function whose derivative is $kt$. One such function is $kt^2/2$, so the general solution of $dy/dt = kt$ is
$$y = \frac{k}{2}t^2 + C$$
which is neither an exponentially growing nor an exponentially decaying function.

**376** CHAPTER EIGHT /SOLUTIONS

12. Rewriting the equation we get
$$\frac{dQ}{dt} = \frac{Q}{k}.$$
The general solution to the equation
$$\frac{dQ}{dt} = \frac{Q}{k}$$
is
$$Q = Ae^{t/k}.$$
This is an exponential function if $A \neq 0$.

13. (a) We know that the balance, $B$, increases at a rate proportional to the current balance. Since interest is being earned at a rate of 7% compounded continuously we have

$$\text{Rate at which interest is earned} = 7\% \text{ (Current balance)}$$

or in other words, if $t$ is time in years,
$$\frac{dB}{dt} = 7\%(B) = 0.07B.$$

(b) The equation is in the form
$$\frac{dB}{dt} = kB$$
so we know that the general solution will be
$$B = B_0 e^{kt}$$
where $B_0$ is the value of $B$ when $t = 0$, i.e., the initial balance. In our case we have $k = 0.07$ so we get
$$B = B_0 e^{0.07t}.$$

(c) We are told that the initial balance, $B_0$, is $5000 so we get
$$B = 5000e^{0.07t}.$$

(d) Substituting the value $t = 10$ into our formula for $B$ we get
$$B = 5000e^{0.07t}$$
$$B(10) = 5000e^{0.07(10)}$$
$$= 5000e^{0.7}$$
$$B(10) \approx \$10,068.75$$

14. (a) The rate of growth of the money in the account is proportional to the amount of money in the account. Thus
$$\frac{dM}{dt} = rM.$$

(b) We know that the equation
$$\frac{dM}{dt} = rM$$
has the general solution
$$M = Ae^{rt}.$$
We know that in 1970 (i.e. $t = 0$) we have $M = 1000$. Thus we get
$$M = Ae^{rt}$$
$$M(0) = 1000 = Ae^{0r}$$
$$1000 = Ae^0$$
$$A = 1000.$$
Thus we get
$$M = 1000e^{rt}.$$

(c)

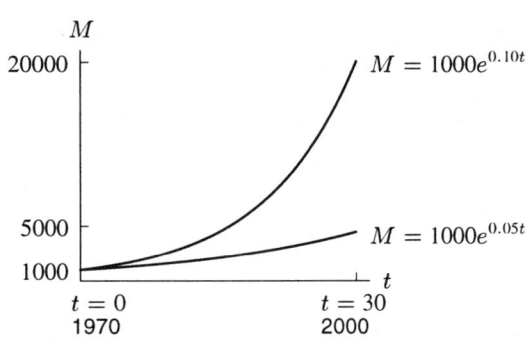

15. (a) $\frac{dB}{dt} = \frac{r}{100}B$. The constant of proportionality is $\frac{r}{100}$.
    (b) We know that the general solution to the equation
    $$\frac{dB}{dt} = kB$$
    is
    $$B = Ae^{kt}.$$
    Thus in our case the solution is
    $$B = Ae^{(r/100)t}.$$
    (c)

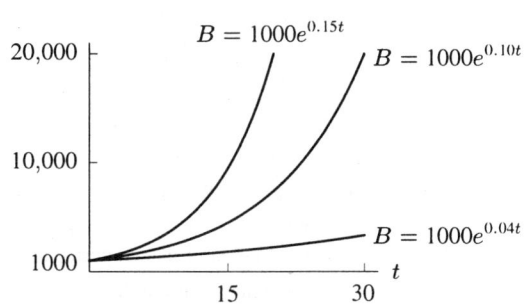

16. Michigan:
    $$\frac{dQ}{dt} = -\frac{r}{V}Q = -\frac{158}{4.9 \times 10^3}Q \approx -0.032Q$$
    so
    $$Q = Q_0 e^{-0.032t}.$$
    We want to find $t$ such that
    $$0.1Q_0 = Q_0 e^{-0.032t}$$
    so
    $$t = \frac{-\ln(0.1)}{0.032} \approx 72 \text{ years}.$$

    Ontario:
    $$\frac{dQ}{dt} = -\frac{r}{V}Q = \frac{-209}{1.6 \times 10^3}Q = -0.131Q$$
    so
    $$Q = Q_0 e^{-0.131t}.$$
    We want to find $t$ such that
    $$0.1Q_0 = Q_0 e^{-0.131t}$$
    so
    $$t = \frac{-\ln(0.1)}{0.131} \approx 18 \text{ years}.$$

    Lake Michigan will take longer because it is larger (4900 km$^3$ compared to 1600 km$^3$) and water is flowing through it at a slower rate (158 km$^3$/year compared to 209 km$^3$/year).

**378** CHAPTER EIGHT /SOLUTIONS

17. Lake Superior will take the longest, because the lake is largest ($V$ is largest) and water is moving through it most slowly ($r$ is smallest). Lake Erie looks as though it will take the least time because $V$ is smallest and $r$ is close to the largest. For Erie, $k = r/V = 175/460 \approx 0.38$. The lake with the largest value of $r$ is Ontario, where $k = r/V = 209/1600 \approx 0.13$. Since $e^{-kt}$ decreases faster for larger $k$, Lake Erie will take the shortest time for any fixed fraction of the pollution to be removed.

For Lake Superior,
$$\frac{dQ}{dt} = -\frac{r}{V}Q = -\frac{65.2}{12{,}200}Q \approx -0.0053Q$$

so
$$Q = Q_0 e^{-0.0053t}.$$

When 80% of the pollution has been removed, 20% remains so $Q = 0.2Q_0$. Substituting gives us
$$0.2Q_0 = Q_0 e^{-0.0053t}$$

so
$$t = -\frac{\ln(0.2)}{0.0053} \approx 301 \text{ years.}$$

(Note: The 301 is obtained by using the exact value of $\frac{r}{V} = \frac{65.2}{12,200}$, rather than 0.0053. Using 0.0053 gives 304 years.)
For Lake Erie, as in the text
$$\frac{dQ}{dt} = -\frac{r}{V}Q = -\frac{175}{460}Q \approx -0.38Q$$

so
$$Q = Q_0 e^{-0.38t}.$$

When 80% of the pollution has been removed
$$0.2Q_0 = Q_0 e^{-0.38t}$$
$$t = -\frac{\ln(0.2)}{0.38} \approx 4 \text{ years.}$$

So the ratio is
$$\frac{\text{Time for Lake Superior}}{\text{Time for Lake Erie}} \approx \frac{301}{4} \approx 75.$$

In other words it will take about 75 times as long to clean Lake Superior as Lake Erie.

18. (a) Since we are told that the rate at which the quantity of the drug decreases is proportional to the amount of the drug left in the body, we know the differential equation modeling this situation is
$$\frac{dQ}{dt} = -kQ.$$

Since we are told that the quantity of the drug is decreasing, we include the negative sign.

(b) We know that the general solution to the differential equation
$$\frac{dQ}{dt} = -kQ$$
is
$$Q = Ce^{-kt}.$$

(c) We are told that the half life of the drug is 3.8 hours. This means that at $t = 3.8$ the amount of the drug in the body is half the amount that was in the body at $t = 0$, or in other words
$$0.5Q(0) = Q(3.8).$$

Solving this equation gives
$$0.5Ce^{-k(0)} = Ce^{-k(3.8)}$$
$$0.5C = Ce^{-k(3.8)}$$
$$0.5 = e^{-k(3.8)}$$
$$\ln(0.5) = -k(3.8)$$
$$k = \frac{-\ln(0.5)}{3.8}$$
$$\approx 0.182.$$

(d) From part (c) we know that the formula for $Q$ is

$$Q = Ce^{-0.182t}.$$

We are told that initially there are 10 mg of the drug in the body. Thus at $t = 0$ we get

$$10 = Ce^{-0.182(0)}$$

$$C = 10.$$

Thus the formula is

$$Q(t) = 10e^{-0.182t}.$$

Substituting in $t = 12$ gives

$$Q(12) = 10e^{-0.182(12)}$$
$$= 10e^{-2.184}$$
$$Q(12) \approx 1.126 \text{ mg}$$

19. (a)

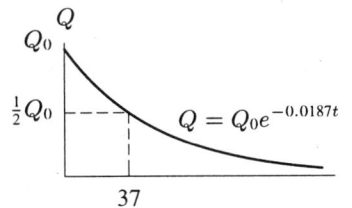

(b) $\dfrac{dQ}{dt} = -kQ$

(c) Since 25% = 1/4, it takes two half-lives = 74 hours for the drug level to be reduced to 25%. Alternatively, $Q = Q_0 e^{-kt}$ and $\frac{1}{2} = e^{-k(37)}$, we have

$$k = -\frac{\ln(1/2)}{37} \approx 0.0187.$$

Therefore $Q = Q_0 e^{-0.0187t}$. We know that when the drug level is 25% of the original level that $Q = 0.25 Q_0$. Setting these equal, we get

$$0.25 = e^{-0.0187t}.$$

giving

$$t = -\frac{\ln(0.25)}{0.0187} \approx 74 \text{ hours} \approx 3 \text{ days}.$$

20. (a) If the world's population grows exponentially, satisfying $\frac{dP}{dt} = kP$, and if the arable land used is proportional to the population, then we'd expect $A$ to satisfy $\frac{dA}{dt} = kA$. One is, of course, also assuming that the total amount of arable land is large compared to the amount that is now being used.

(b) We have $A(t) = A_0 e^{kt} = (1 \times 10^9) e^{kt}$, where $t$ is the number of years after 1950. Since $2 \times 10^9 = (1 \times 10^9) e^{k(30)}$, we have $e^{30k} = 2$, so $k = \frac{\ln 2}{30} \approx 0.023$. Thus, $A \approx (1 \times 10^9) e^{0.023t}$. We want to find $t$ such that $3.2 \times 10^9 = A(t) = (1 \times 10^9) e^{0.023t}$. Taking logarithms yields

$$t = \frac{\ln(3.2)}{0.023} \approx 50.6 \text{ years}.$$

Thus the arable land will have run out by the year 2001.

21. (a) Since the rate of change is proportional to the amount present, $dy/dt = ky$ for some constant $k$.

(b) Solving the differential equation, we have $y = Ae^{kt}$, where $A$ is the initial amount. Since 100 grams become 54.9 grams in one hour, $54.9 = 100e^k$, so $k = \ln(54.9/100) \approx -0.5997$.
Thus, after 10 hours, there remains $100e^{(-0.5997)10} \approx 0.2486$ grams.

22. We know that

$$\text{Rate at which quantity of carbon-14 is changing} = -k(\text{current quantity}).$$

If $Q$ is the quantity of carbon-14 at time $t$ (in years),

$$\text{Rate at which quantity is changing} = \frac{dQ}{dt} = -kQ.$$

This differential equation has solution

$$Q = Q_0 e^{-kt}$$

where $Q_0$ is the initial quantity. Since at the end of one year 9999 parts are left out of 10,000, we know that

$$9999 = 10{,}000 e^{-k(1)}.$$

Solving for $k$ gives

$$k = \ln 0.9999 \approx 0.0001.$$

Thus $Q = Q_0 e^{-0.0001t}$. See Figure 8.6.

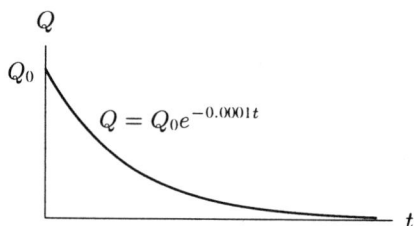

**Figure 8.6**: Exponential decay

23. (a) If $C' = -kC$, and then $C = C_0 e^{-kt}$. Since the half-life is 5730 years, $\frac{1}{2}C_0 = C_0 e^{-5730k}$. Solving for $k$, we have $-5730k = \ln(1/2)$ so $k = \frac{-\ln(1/2)}{5730} \approx 0.000121$.
   (b) From the given information, we have $0.91 = e^{-kt}$, where $t$ is the age of the shroud. Solving for $t$, we have $t = \frac{-\ln 0.91}{k} \approx 779.4$ years.

## Solutions for Section 8.4

1. We know that the general solution to a differential equation of the form

$$\frac{dH}{dt} = k(H - A)$$

is

$$H = A + Ce^{kt}.$$

Thus in our case we get

$$H = 75 + Ce^{3t}.$$

We know that at $t = 0$ we have $H = 0$, so solving for $C$ we get

$$H = 75 + Ce^{3t}$$
$$0 = 75 + Ce^{3(0)}$$
$$-75 = Ce^0$$
$$C = -75.$$

Thus we get

$$H = 75 - 75e^{3t}.$$

2. We know that the general solution to a differential equation of the form
$$\frac{dy}{dt} = k(y - A)$$
is
$$y = A + Ce^{kt}.$$
Thus in our case we get
$$y = 200 + Ce^{0.5t}.$$
We know that at $t = 0$ we have $y = 50$, so solving for $C$ we get
$$y = 200 + Ce^{0.5t}$$
$$50 = 200 + Ce^{0.5(0)}$$
$$-150 = Ce^0$$
$$C = -150.$$
Thus we get
$$y = 200 - 150e^{0.5t}.$$

3. We know that the general solution to a differential equation of the form
$$\frac{dP}{dt} = k(P - A)$$
is
$$P = Ce^{kt} + A.$$
Thus in our case we have $k = 1$, so we get
$$P = Ce^t - 4.$$
We know that at $t = 0$ we have $P = 100$ so solving for $C$ we get
$$P = Ce^t - 4$$
$$100 = Ce^0 - 4$$
$$104 = Ce^0$$
$$C = 104.$$
Thus we get
$$P = 104e^t - 4.$$

4. We know that the general solution to a differential equation of the form
$$\frac{dB}{dt} = k(B - A)$$
is
$$B = A + Ce^{kt}.$$
To get our equation in this form we factor out a 4 to get
$$\frac{dB}{dt} = 4\left(B - \frac{100}{4}\right) = 4(B - 25).$$
Thus in our case we get
$$B = Ce^{4t} + 25.$$
We know that at $t = 0$ we have $B = 20$, so solving for $C$ we get
$$B = 25 + Ce^{4t}$$
$$20 = 25 + Ce^{4(0)}$$
$$-5 = Ce^0$$
$$C = -5.$$
Thus we get
$$B = 25 - 5e^{4t}.$$

5. We know that the general solution to a differential equation of the form

$$\frac{dQ}{dt} = k(Q - A)$$

is

$$H = A + Ce^{kt}.$$

To get our equation in this form we factor out a 0.3 to get

$$\frac{dQ}{dt} = 0.3\left(Q - \frac{120}{0.3}\right) = 0.3(Q - 400).$$

Thus in our case we get

$$Q = 400 + Ce^{0.3t}.$$

We know that at $t = 0$ we have $Q = 50$, so solving for $C$ we get

$$Q = 400 + Ce^{0.3t}$$
$$50 = 400 + Ce^{0.3(0)}$$
$$-350 = Ce^0$$
$$C = -350.$$

Thus we get

$$Q = 400 - 350e^{0.3t}.$$

6. We know that the general solution to a differential equation of the form

$$\frac{dm}{dt} = k(m - A)$$

is

$$m = Ce^{kt} + A.$$

Factoring out a 0.1 on the left side we get

$$\frac{dm}{dt} = 0.1\left(m - \frac{-200}{0.1}\right) = 0.1(m - (-2000)).$$

Thus in our case we get

$$m = Ce^{0.1t} - 2000.$$

We know that at $t = 0$ we have $m = 1000$ so solving for $C$ we get

$$m = Ce^{0.1t} - 2000$$
$$1000 = Ce^0 - 2000$$
$$3000 = Ce^0$$
$$C = 3000.$$

Thus we get

$$m = 3000e^{0.1t} - 2000.$$

7. We know that the general solution to a differential equation of the form

$$\frac{dB}{dt} = k(B - A)$$

is

$$B = Ce^{kt} + A.$$

Rewriting we get

$$\frac{dB}{dt} = -2B + 50.$$

Factoring out a $-2$ on the right side we get
$$\frac{dB}{dt} = -2\left(B - \frac{-50}{-2}\right) = -2(B - 25).$$

Thus in our case we get
$$B = Ce^{-2t} + 25.$$

We know that at $t = 1$ we have $B = 100$ so solving for $C$ we get
$$B = Ce^{-2t} + 25$$
$$100 = Ce^{-2} + 25$$
$$75 = Ce^{-2}$$
$$C = 75e^2.$$

Thus we get
$$B = 75e^2 e^{-2t} + 25 = 75e^{2-2t} + 25.$$

8. Rewrite the differential equation as
$$\frac{dB}{dt} = -0.1B + 10$$

Factoring out $-0.1$ gives
$$\frac{dB}{dt} = -0.1(B - 100),$$

which has solution
$$B = 100 + Ce^{-0.1t}.$$

Substituting $B = 3$ and $t = 2$ gives
$$3 = 100 + Ce^{-0.1(2)}.$$

Solving for $C$ we get
$$C = -\frac{97}{e^{-0.02}} \approx -99$$

So the solution is $B = 100 - 99e^{-0.1t}$.

9. In order to check that $y = A + Ce^{kt}$ is a solution to the differential equation
$$\frac{dy}{dt} = k(y - A),$$
we must show that the derivative of $y$ with respect to $t$ is equal to $k(y - A)$:
$$y = A + Ce^{kt}$$
$$\frac{dy}{dt} = 0 + (Ce^{kt})(k)$$
$$= kCe^{kt}$$
$$= k(Ce^{kt} + A - A)$$
$$= k\left((Ce^{kt} + A) - A\right)$$
$$= k(y - A)$$

10. (a) We know that the general solution to a differential equation of the form
$$\frac{dy}{dt} = k(y - A)$$
is
$$y = Ce^{kt} + A.$$

Factoring out a $-1$ on the left side we get
$$\frac{dy}{dt} = -(y - 100)$$

Thus in our case we get
$$y = Ce^{-t} + 100.$$

This is meaningful if $C \leq 0$, since one cannot know more than 100%.

(b)

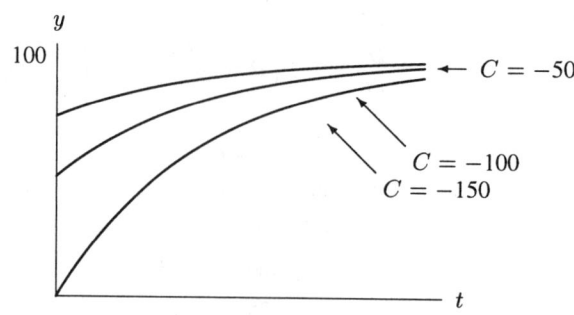

**Figure 8.7**

(c) Substituting $y = 0$ when $t = 0$ gives
$$0 = 100 - Ce^{-0}$$
so $C = 100$. The solution is
$$y = 100 - 100e^{-t}$$

11. Graphically, a function is an equilibrium solution if its graph is a horizontal line. From the slope field we see that the equilibrium solutions are $y = 1$ and $y = 3$. An equilibrium solution is stable if a small change in the initial value conditions gives a solution which tends toward the equilibrium as $t$ tends to positive infinity. Thus by looking at Figure 8.34 we see that $y = 3$ is a stable solution while $y = 1$ is an unstable solution.

12. (a) We know that the equilibrium solution is the solution satisfying the differential equation whose derivative is everywhere 0. Thus we must solve
$$\frac{dy}{dt} = 0.$$
Solving this gives
$$\frac{dy}{dt} = 0$$
$$0.5y - 250 = 0$$
$$y = 500$$

(b) We know that the general solution to a differential equation of the form
$$\frac{dy}{dt} = k(y - A)$$
is
$$y = A + Ce^{kt}.$$
To get our equation in this form we factor out a 0.5 to get
$$\frac{dy}{dt} = 0.5 \left( y - \frac{250}{0.5} \right) = 0.5(y - 500).$$
Thus in our case we get
$$y = 500 + Ce^{0.5t}.$$

(c) The graphs of several solutions is shown in Figure 8.8.

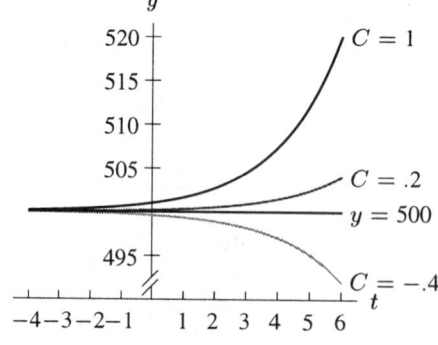

**Figure 8.8**

(d) Looking at Figure 8.8 we see that as $t \to \infty$, the value of $y$ gets further and further away from the line $y = 500$. The equilibrium solution $y = 500$ is unstable.

13. (a) We know that the equilibrium solution are the functions satisfying the differential equation whose derivative everywhere is 0. Thus we must solve the equation
$$\frac{dy}{dt} = 0.$$

Solving we get
$$\frac{dy}{dt} = 0$$
$$0.2(y-3)(y+2) = 0$$
$$(y-3)(y+2) = 0$$

Thus the solutions are $y = 3$ and $y = -2$.

(b) Looking at Figure 8.9 we see that the line $y = 3$ is an unstable solution while the line $y = -2$ is a stable solution.

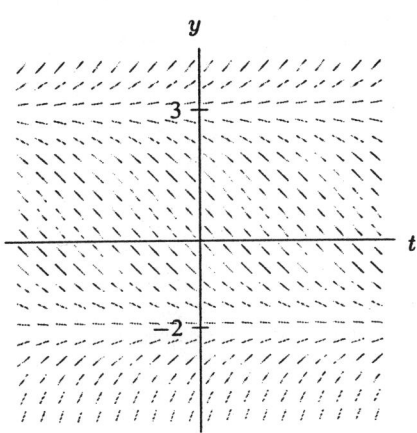

**Figure 8.9**

14. (a) We know that the general solution to a differential equation of the form
$$\frac{dH}{dt} = -k(H - 200)$$
is
$$H = Ce^{-kt} + 200.$$

We know that at $t = 0$ we have $H = 20$ so solving for $C$ we get
$$H = Ce^{-kt} + 200$$
$$20 = Ce^0 + 200$$
$$C = -180.$$

Thus we get
$$H = -180e^{-kt} + 200.$$

(b) Using part (a), we have $120 = 200 - 180e^{-k(30)}$. Solving for $k$, we have $e^{-30k} = \frac{-80}{-180}$, giving
$$k = \frac{\ln \frac{4}{9}}{-30} \approx 0.027.$$

Note that this $k$ is correct if $t$ is given in *minutes*. (If it is given in hours, $k = \frac{\ln \frac{4}{9}}{-\frac{1}{2}} \approx 1.62$.)

**386** CHAPTER EIGHT /SOLUTIONS

15. We find the temperature of the orange juice as a function of time. Newton's Law of Heating says that the rate of change of the temperature is proportional to the temperature difference. If $S$ is the temperature of the juice, this gives us the equation
$$\frac{dS}{dt} = -k(S - 65) \qquad \text{for some constant } k.$$
Notice that the temperature of the juice is increasing, so the quantity $dS/dt$ is positive. In addition, $S = 40$ initially, making the quantity $(S - 65)$ negative.

We know that the general solution to a differential equation of the form
$$\frac{dS}{dt} = -k(S - 65)$$
is
$$S = Ce^{-kt} + 65.$$

Since at $t = 0$, $S = 40$, we have $40 = 65 + C$, so $C = -25$. Thus, $S = 65 - 25e^{-kt}$ for some positive constant $k$. See Figure 8.10 for the graph.

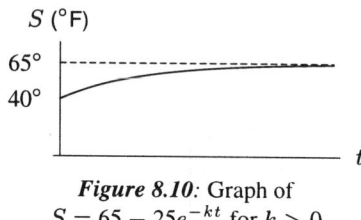

**Figure 8.10**: Graph of $S = 65 - 25e^{-kt}$ for $k > 0$

16. (a) If $B = f(t)$ (where $t$ is in years)
$$\frac{dB}{dt} = \text{Rate at which interest is earned} + \text{Rate at which money is deposited}$$
$$= 0.10B + 1000.$$

(b)
$$\frac{dB}{dt} = 0.1(B + 10{,}000)$$
We know that a differential equation of the form
$$\frac{dB}{dt} = k(B - A)$$
has general solution:
$$B = Ce^{kt} + A.$$
Thus, in our case
$$B = Ce^{0.1t} - 10{,}000.$$
For $t = 0$, $B = 0$, hence $C = 10{,}000$. Therefore, $B = 10{,}000e^{0.1t} - 10{,}000$.

17. (a) $\dfrac{dT}{dt} = -k(T - A)$, where $A = 68°F$ is the temperature of the room.

(b) We know that the general solution to a differential equation of the form
$$\frac{dT}{dt} = -k(T - 68)$$
is
$$T = Ce^{-kt} + 68.$$
We know that the temperature of the body is 90.3°F at 9 am. Thus, letting $t = 0$ correspond to 9 am, we get
$$T = Ce^{-kt} + 68$$
$$T(0) = 90.3 = Ce^{-k(0)} + 68$$
$$90.3 = Ce^0 + 68$$
$$C = 90.3 - 68 = 22.3$$

Thus
$$T = 68 + 22.3e^{-kt}.$$
At $t = 1$, we have
$$89.0 = 68 + 22.3e^{-k}$$
$$21 = 22.3e^{-k}$$
$$k = -\ln\frac{21}{22.3} \approx 0.06.$$

Thus $T = 68 + 22.3e^{-0.06t}$.

We want to know when $T$ was equal to $98.6°$F, the temperature of a live body, so
$$98.6 = 68 + 22.3e^{-0.06t}$$
$$\ln\frac{30.6}{22.3} = -0.06t$$
$$t = \left(-\frac{1}{0.06}\right)\ln\frac{30.6}{22.3}$$
$$t \approx -5.27.$$

The victim was killed approximately $5\frac{1}{4}$ hours prior to 9 am, at 3:45 am.

18. (a) The differential equation is
$$\frac{dT}{dt} = -k(T - A),$$
where $A = 10°$F is the outside temperature.

(b) We know that the general solution to a differential equation of the form
$$\frac{dT}{dt} = -k(T - 10)$$
is
$$T = Ce^{-kt} + 10.$$
We know that initially $T = 68°$F. Thus, letting $t = 0$ correspond to 1 pm, we get
$$T = Ce^{-kt} + 10$$
$$68 = Ce^0 + 10$$
$$C = 58.$$

Thus
$$T = 10 + 58e^{-kt}.$$
Since 10:00 pm corresponds to $t = 9$,
$$57 = 10 + 58e^{-9k}$$
$$\frac{47}{58} = e^{-9k}$$
$$\ln\frac{47}{58} = -9k$$
$$k = -\frac{1}{9}\ln\frac{47}{58} \approx 0.0234.$$

At 7:00 the next morning ($t = 18$) we have
$$T \approx 10 + 58e^{18(-0.0234)}$$
$$= 10 + 58(0.66)$$
$$\approx 48°\text{F},$$
so the pipes won't freeze.

(c) We assumed that the temperature outside the house stayed constant at $10°$F. This is probably incorrect because the temperature was most likely warmer during the day (between 1 pm and 10 pm) and colder after (between 10 pm and 7 am). Thus, when the temperature in the house dropped from $68°$F to $57°$F between 1 pm and 10 pm, the outside temperature was probably higher than $10°$F, which changes our calculation of the value of the constant $k$. The house temperature will most certainly be lower than $48°$F at 7 am, but not by much—not enough to freeze.

19. (a) The smoker smokes 5 cigarettes per hour, and each cigarette contributes 0.4 mg of nicotine, so, every hour, the amount of nicotine is increasing by $5(0.4) = 2.0$ mg. At the same time, the nicotine is being eliminated at a rate of $-0.346$ times the amount of nicotine. Thus, we have

$$\frac{dN}{dt} = \text{Rate in} - \text{Rate out} = 2.0 - 0.346N.$$

(b) We have

$$\frac{dN}{dt} = 2.0 - 0.346N = -0.346(N - 5.78),$$

so the solution is

$$N = 5.78 - 5.78e^{-0.346t}.$$

(c) At $t = 16$, we have $N = 5.78 - 5.78e^{-0.346(16)} = 5.76$ mg.

20. (a) We know that the rate by which the account changes every year is

$$\text{Rate of change of balance} = \text{Rate of increase} - \text{Rate of decrease}.$$

Since $1000 will be withdrawn every year, we know that the account decreases by $1000 every year. We also know that the account accumulates interest at 7% compounded continuously. Thus the amount by which the account increases each year is

$$\text{Rate balance increases per year} = 7\%(\text{Account balance}) = 0.07(\text{Account balance}).$$

Denoting the account balance by $B$ we get

$$\text{Rate balance increases per year} = 0.07B.$$

Thus we get

$$\text{Rate of change of balance} = 0.07B - 1000.$$

or

$$\frac{dB}{dt} = 0.07B - 1000,$$

with $t$ measured in years.

(b) The equilibrium solution makes the derivative 0, so

$$\frac{dB}{dt} = 0$$
$$0.07B - 1000 = 0$$
$$B = \frac{1000}{0.07} \approx \$14{,}285.71.$$

(c) We know that the general solution to a differential equation of the form

$$\frac{dB}{dt} = k(B - A)$$

is

$$B = Ce^{kt} + A.$$

To get our equation in this form we factor out a 0.07 to get

$$\frac{dB}{dt} = 0.07\left(B - \frac{1000}{0.07}\right) \approx 0.07(B - 14{,}285.71).$$

Thus in our case we get

$$B = Ce^{0.07t} + 14{,}285.71.$$

We know that at $t = 0$ we have $B = \$10{,}000$ so solving for $C$ we get

$$B = Ce^{0.07t} + 14{,}285.71$$
$$10{,}000 = Ce^{4(0)} + 14{,}285.71$$
$$-4285.71 = Ce^0$$
$$C = -4285.71.$$

Thus we get

$$B = 14{,}285.71 - (4285.71)e^{0.07t}.$$

(d) Substituting the value $t = 5$ into our function for $B$ we get
$$B(t) = 14{,}285.71 - (4285.71)e^{0.07t}$$
$$B(5) = 14{,}285.71 - (4285.71)e^{0.07(5)}$$
$$= 14{,}285.71 - (4285.71)e^{0.35}$$
$$B(5) \approx \$8204$$

(e) From Figure 8.11 we see that in the long run there is no money left in the account.

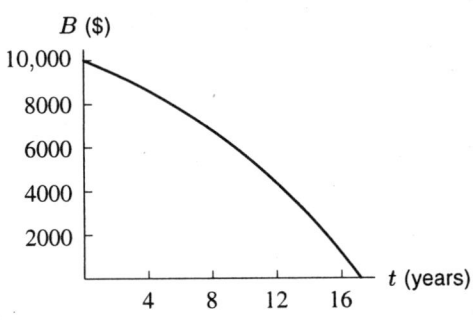

**Figure 8.11**

21. (a) $\dfrac{dy}{dt} = -k(y - a)$, where $k > 0$ and $a$ are constants.

   (b) We know that the general solution to a differential equation of the form
$$\frac{dy}{dt} = -k(y - a)$$
   is
$$y = Ce^{-kt} + a.$$
   We can assume that right after the course is over (at $t = 0$) 100% of the material is remembered. Thus we get
$$y = Ce^{-kt} + a$$
$$1 = Ce^{0} + a$$
$$C = 1 - a.$$
   Thus
$$y = (1 - a)e^{-kt} + a.$$

   (c) As $t \to \infty$, $e^{-kt} \to 0$, so $y \to a$.
   Thus, $a$ represents the fraction of material which is remembered in the long run. The constant $k$ tells us about the rate at which material is forgotten.

22. We know that the amount of morphine, $Q$, in the body can be modeled by an exponential decay function:
$$Q = Q_0 e^{-kt}$$
where $Q_0$ is the initial amount of morphine in the body and $t$ is measured in hours. We are told that the half life of morphine in the body is 2 hours. Thus we know that at $t = 2$ we have $Q(2) = \frac{1}{2}Q_0$. Solving gives
$$Q(2) = \frac{1}{2}Q_0$$
$$Q_0 e^{-2k} = \frac{1}{2}Q_0$$
$$e^{-2k} = \frac{1}{2}$$
$$-2k = \ln\left(\frac{1}{2}\right)$$
$$k = -\frac{\ln(1/2)}{2}$$
$$k \approx 0.347.$$

23. (a) We know that the rate at which the quantity of morphine in the body changes is given by

$$\text{Rate of change of quantity of morphine} = \text{Rate of morphine entering the body} - \text{Rate of morphine leaving the body}$$

Since 2.5 mg are administered every hour, we know that the rate of morphine enters the body is 2.5 mg/hour. We also know that the amount of morphine in the body decays according to

$$\text{Rate of morphine decay} = -0.347 \left( \text{Quantity of morphine in the body} \right) = -0.347Q.$$

Thus we get

$$\text{Rate of change of quantity of morphine} = 2.5 - 0.347Q$$

or

$$\frac{dQ}{dt} = -0.347Q + 2.5$$

where $t$ is measured in hours.

(b) The the equilibrium solution is given by

$$0 = \frac{dQ}{dt}$$
$$= -0.347Q + 2.5$$
$$Q = \frac{2.5}{0.347}$$
$$Q \approx 7.205$$

Thus in the long-term, the system stabilizes with 7.205 mg of morphine in the body.

24. (a) The quantity increases with time. As the quantity increases, the rate at which the drug is excreted also increases, and so the rate at which the drug builds up in the blood decreases; thus the graph of quantity against time is concave down. The quantity rises until the rate of excretion exactly balances the rate at which the drug is entering; at this quantity there is a horizontal asymptote.

(b) Theophylline enters at a constant rate of 43.2mg/hour and leaves at a rate of $0.082Q$, so we have

$$\frac{dQ}{dt} = 43.2 - 0.082Q$$

(c) We know that the general solution to a differential equation of the form

$$\frac{dy}{dt} = k(y - A)$$

is

$$y = Ce^{kt} + A.$$

Thus in our case, since

$$\frac{dQ}{dt} = 43.2 - 0.082Q \approx -0.082(Q - 526.8),$$

we have

$$Q = 526.8 + Ce^{-0.082t}.$$

Since $Q = 0$ when $t = 0$, we can solve for $C$:

$$Q = 526.8 + Ce^{-0.082t}$$
$$0 = 526.8 + Ce^0$$
$$C = -526.8$$

The solution is

$$Q = 526.8 - 526.8e^{-0.082t}.$$

In the long run, the quantity in the body approaches 526.8 mg. See Figure 8.12.

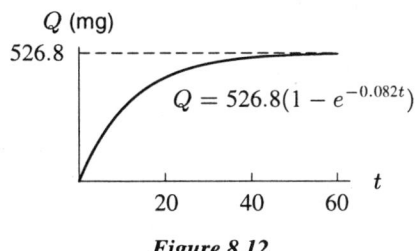

**Figure 8.12**

25. (a) The equilibrium solutions are the solutions where
$$\frac{dP}{dt} = 0$$
$$0.25P(1 - 0.0004P) = 0$$
$$P = 0, P = \frac{1}{0.0004} = 2500.$$

The equilibrium solutions are $P = 0$ and $P = 2500$.

(b)
$$\frac{dP}{dt} = 0.25P(1 - 0.0004P)$$
$$= 0.25P \left(1 - \frac{P}{\frac{1}{0.0004}}\right)$$
$$= 0.25P \left(1 - \frac{P}{2500}\right).$$

So $k = 0.25$ and $L = 2500$.

(c) The solution is of the form $P = \frac{L}{1 + Ce^{-kt}}$, so in this case
$$P = \frac{2500}{1 + Ce^{-0.25t}},$$
where $C$ is an arbitrary constant. The long-term equilibrium population of carp in the lake is 2500.

## Solutions for Section 8.5

1. Here $x$ and $y$ both increase at about the same rate.

2. Initially $x = 0$, so we start with only $y$. Then $y$ decreases while $x$ increases. Then $x$ continues to increase while $y$ starts to increase as well. Finally $y$ continues to increase while $x$ decreases.

3. $x$ decreases quickly while $y$ increases more slowly.

4. The closed trajectory represents populations which oscillate repeatedly.

5. This is an example of a predator-prey relationship. Normally, we would expect the worm population, in the absence of predators, to increase without bound. As the number of worms $w$ increases, so would the rate of increase $dw/dt$; in other words, the relation $dw/dt = w$ might be a reasonable model for the worm population in the absence of predators.

   However, since there are predators (robins), $dw/dt$ won't be that big. We must lessen $dw/dt$. It makes sense that the more interaction there is between robins and worms, the more slowly the worms are able to increase their numbers. Hence we lessen $dw/dt$ by the amount $wr$ to get $dw/dt = w - wr$. The term $-wr$ reflects the fact that more interactions between the species means slower reproduction for the worms.

   Similarly, we would expect the robin population to decrease in the absence of worms. We'd expect the population decrease at a rate related to the current population, making $dr/dt = -r$ a reasonable model for the robin population in absence of worms. The negative term reflects the fact that the greater the population of robins, the more quickly they are dying off. The $wr$ term in $dr/dt = -r + wr$ reflects the fact that the more interactions between robins and worms, the greater the tendency for the robins to increase in population.

**392** CHAPTER EIGHT /SOLUTIONS

6. If there are no worms, then $w = 0$, and $\frac{dr}{dt} = -r$ giving $r = r_0 e^{-t}$, where $r_0$ is the initial robin population. If there are no robins, then $r = 0$, and $\frac{dw}{dt} = w$ giving $w = w_0 e^t$, where $w_0$ is the initial worm population.

7. There is symmetry across the line $r = w$. Indeed, since $\frac{dr}{dw} = \frac{r(w-1)}{w(1-r)}$, if we switch $w$ and $r$ we get $\frac{dw}{dr} = \frac{w(r-1)}{r(1-w)}$, so $\frac{dr}{dw} = \frac{r(1-w)}{w(r-1)}$. Since switching $w$ and $r$ changes nothing, the slope field must be symmetric across the line $r = w$. The slope field shows that the solution curves are either spirals or closed curves. Since there is symmetry about the line $r = w$, the solutions must in fact be closed curves.

8. If $w = 2$ and $r = 2$, then $\frac{dw}{dt} = -2$ and $\frac{dr}{dt} = 2$, so initially the number of worms decreases and the number of robins increases. In the long run, however, the populations will oscillate; they will even go back to $w = 2$ and $r = 2$.

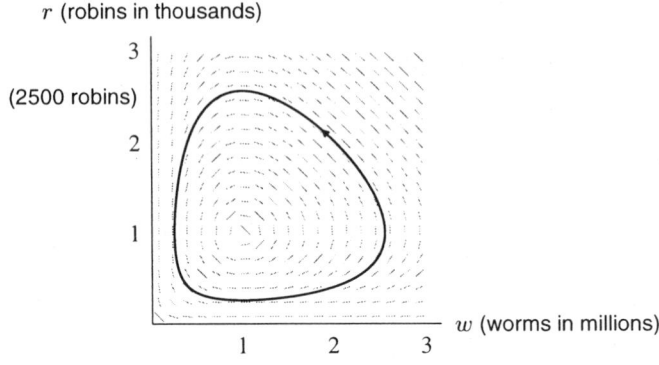

9. Sketching the trajectory through the point $(2, 2)$ on the slope field given shows that the maximum robin population is about 2500, and the minimum robin population is about 500. When the robin population is at its maximum, the worm population is about 1,000,000.

10.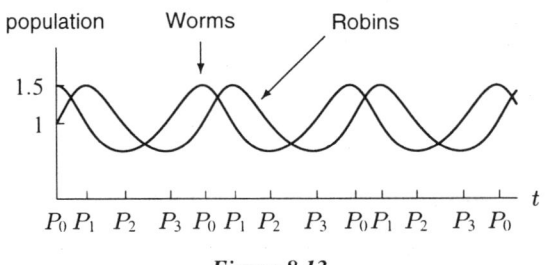

*Figure 8.13*

11. It will work somewhat; the maximum number the robins reach will increase. However, the minimum number the robins reach will decrease as well. (See graph of slope field.) In the long term, the robin-worm populations will again fall into a cycle. Notice, however, if the extra robins are added during the part of the cycle where there are the fewest robins, the new cycle will have smaller variation.

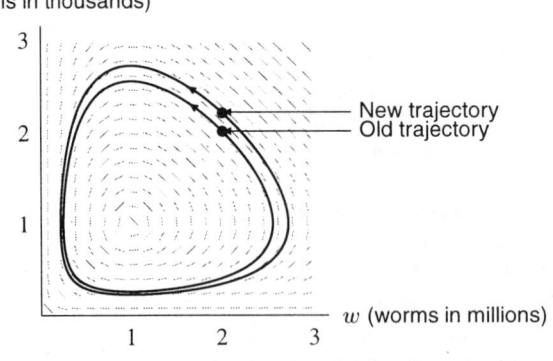

Note that if too many robins are added, the minimum number may get so small the model may fail, since a small number of robins are more susceptible to disaster.

12. The numbers of robins begins to increase while the number of worms remains approximately constant.

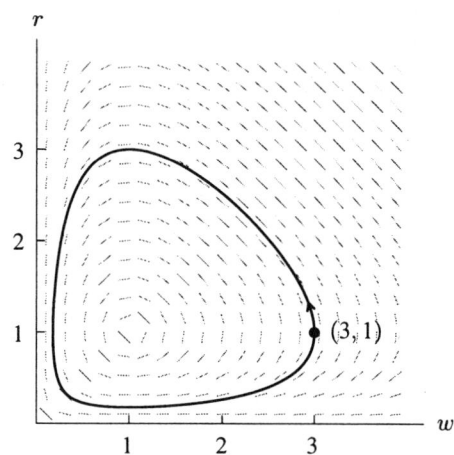

The numbers of robins and worms oscillate periodically between 0.2 and 3, with the robin population lagging behind the worm population.

13. (a) Substituting $w = 2.2$ and $r = 1$ into the differential equations gives

$$\frac{dw}{dt} = 2.2 - (2.2)(1) = 0$$
$$\frac{dr}{dt} = -1 + 1(2.2) = 1.2.$$

(b) Since the rate of change of $w$ with time is 0,

At $t = 0.1$, we estimate $w = 2.2$

Since the rate of change of $r$ is 1.2 thousand robins per unit time,

At $t = 0.1$, we estimate $r = 1 + 1.2(0.1) = 1.12 \approx 1.1$.

(c) We must recompute the derivatives. At $t = 0.1$, we have

$$\frac{dw}{dt} = 2.2 - 2.2(1.12) = -0.264$$
$$\frac{dr}{dt} = -1.12 + 1.12(2.2) = 1.344.$$

Then at $t = 0.2$, we estimate

$$w = 2.2 - 0.264(0.1) = 2.1736 \approx 2.2$$
$$r = 1.12 + 1.344(0.1) = 1.2544 \approx 1.3$$

Recomputing the derivatives at $t = 0.2$ gives

$$\frac{dw}{dt} = 2.1736 - 2.1736(1.2544) = -0.553$$
$$\frac{dr}{dt} = -1.2544 + 1.2544(2.1736) = 1.472$$

Then at $t = 0.3$, we estimate

$$w = 2.1736 - 0.553(0.1) = 2.1183 \approx 2.1$$
$$r = 1.2544 + 1.472(0.1) = 1.4016 \approx 1.4.$$

14. (a) See Figure 8.14.

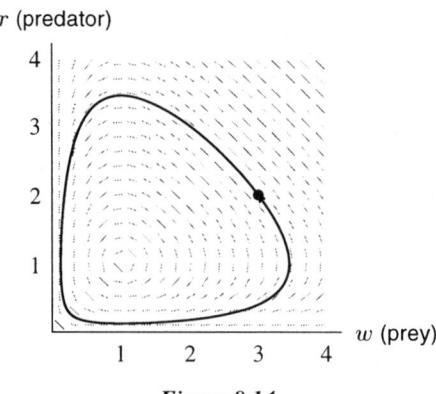

Figure 8.14

(b) The point moves in a counterclockwise direction. If $w = 3$ and $r = 2$, we have

$$\frac{dw}{dt} = w - wr = 3 - (3)(2) = 3 - 6 < 0$$

$$\frac{dr}{dt} = -r + wr = -2 + (3)(2) = -2 + 6 > 0.$$

So $w$ is decreasing and $r$ is increasing. The point moves up and to the left (counterclockwise).

(c) We see in the trajectory that $r$ achieves a maximum value of about 3.5, so the population of robins goes as high as 3.5 thousand robins. At this time, the worm population is at about 1 million.

(d) The worm population goes as high as 3.5 million worms. At this time, the robin population is about 1 thousand.

15. (a) See Figure 8.15.

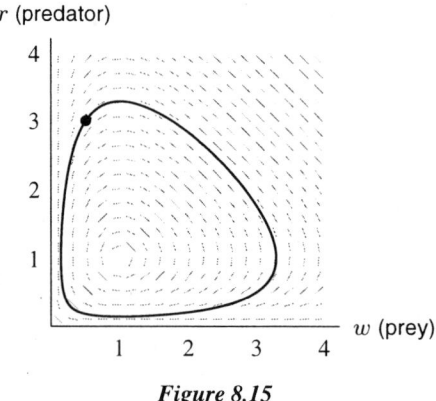

Figure 8.15

(b) If $w = 0.5$ and $r = 3$, we have

$$\frac{dw}{dt} = w - wr = 0.5 - (0.5)(3) < 0$$

$$\frac{dr}{dt} = -r + wr = -3 + (0.5)(3) < 0$$

Both $w$ and $r$ are decreasing, so the point is moving down and to the left (counterclockwise).

(c) The robin population goes up to about 3.3 thousand robins. At this time, the worm population is about 1 million.

(d) The worm population goes up to about 3.3 million worm. At this point, the robin population is about 1 thousand.

16. $\dfrac{dx}{dt} = x - xy, \quad \dfrac{dy}{dt} = y - xy$

17. $\dfrac{dx}{dt} = -x + xy, \quad \dfrac{dy}{dt} = y$

18. $\dfrac{dx}{dt} = -x - xy, \quad \dfrac{dy}{dt} = -y - xy$

19. (a) If alone, the $x$ population grows exponentially, since if $y = 0$ we have $dx/dt = 0.01x$. If alone, the $y$ population decreases to 0 exponentially, since if $x = 0$ we have $dy/dt = -0.2y$.

    (b) This is a predator-prey relationship: interaction between populations $x$ and $y$ decreases the $x$ population and increases the $y$ population. The interaction of lions and gazelles might be modeled by these equations.

20. (a) If alone, the $x$ and $y$ populations each grow exponentially, because the equations become $dx/dt = 0.01x$ and $dy/dt = 0.2y$.

    (b) For each population, the presence of the other decreases their growth rate. The two populations are therefore competitors—they may be eating each other's food, for instance. The interaction of gazelles and zebras might be modeled by these equations.

21. (a) The $x$ population is unaffected by the $y$ population—it grows exponentially no matter what the $y$ population is, even if $y = 0$. If alone, the $y$ population decreases to zero exponentially, because its equation becomes $dy/dt = -0.1y$.

    (b) Here, interaction between the two populations helps the $y$ population but doesn't effect the $x$ population. This is not a predator-prey relationship; instead, this is a one-way relationship, where the $y$ population is helped by the existence of $x$'s. These equations could, for instance, model the interaction of rhinocerouses ($x$) and dung beetles ($y$).

22. (a) Both $x$ and $y$ decrease, since

$$\dfrac{dx}{dt} = 0.2x - 0.5xy = 0.2(2) - 0.5(2)(2) < 0,$$
$$\dfrac{dy}{dt} = 0.6y - 0.8xy = 0.6(2) - 0.8(2)(2) < 0.$$

(b) Population $x$ increases and population $y$ decreases, since

$$\dfrac{dx}{dt} = -2x + 5xy = -2(2) + 5(2)(2) > 0,$$
$$\dfrac{dy}{dt} = -y + 0.2xy = -2 + 0.2(2)(2) < 0.$$

(c) Both $x$ and $y$ increase, since

$$\dfrac{dx}{dt} = 0.5x = 0.5(2) > 0,$$
$$\dfrac{dy}{dt} = -1.6y + 2xy = -1.6(2) + 2(2)(2) > 0$$

(d) Population $x$ decreases and population $y$ increases, since

$$\dfrac{dx}{dt} = 0.3x - 1.2xy = 0.3(2) - 1.2(2)(2) < 0$$
$$\dfrac{dy}{dt} = -0.7x + 2.5xy = 0.7(2) + 2.5(2)(2) > 0$$

23. (a) $\dfrac{dy}{dt} = \dfrac{0.6y - 0.8xy}{0.2x - 0.5xy}$; See Figure 8.16.

    (b) $\dfrac{dy}{dx} = \dfrac{-y + 0.2xy}{-2x + 5xy}$; See Figure 8.17.

    (c) $\dfrac{dy}{dx} = \dfrac{-1.6y + 2xy}{0.5x}$; See Figure 8.18.

    (d) $\dfrac{dy}{dx} = \dfrac{-0.7y + 2.5xy}{0.3x - 1.2xy}$; See Figure 8.19.

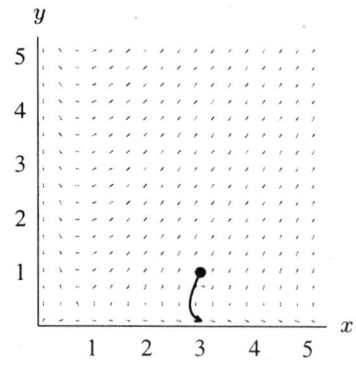

**Figure 8.16:** $\frac{dy}{dx} = \frac{0.6y - 0.8xy}{0.2x - 0.5xy}$

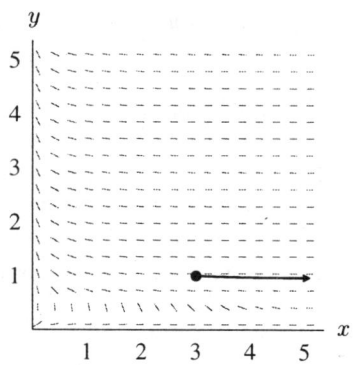

**Figure 8.17:** $\frac{dy}{dx} = \frac{-y + 0.2xy}{-2x + 5xy}$

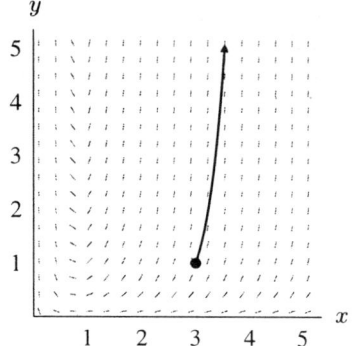

**Figure 8.18:** $\frac{dy}{dx} = \frac{-1.6y + 2xy}{0.5x}$

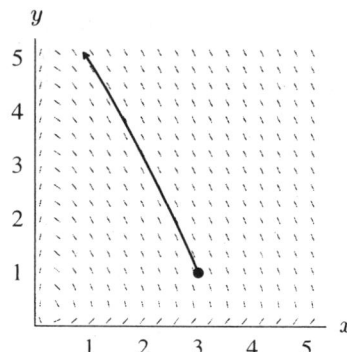

**Figure 8.19:** $\frac{dy}{dx} = \frac{-0.7y + 2.5xy}{0.3x - 1.2xy}$

## Solutions for Section 8.6

1. Since

$$\frac{dS}{dt} = -aSI,$$
$$\frac{dI}{dt} = aSI - bI,$$
$$\frac{dR}{dt} = bI$$

we have

$$\frac{dS}{dt} + \frac{dI}{dt} + \frac{dR}{dt} = -aSI + aSI - bI + bI = 0.$$

Thus $\frac{d}{dt}(S + I + R) = 0$, so $S + I + R = $ constant.

2. Susceptible people are infected at a rate proportional to the product of $S$ and $I$. As susceptible people become infected, $S$ decreases at a rate of $aSI$ and (since these same people are now infected) $I$ increases at the same rate. At the same time, infected people are recovering at a rate proportional to the number infected, so $I$ is decreasing at a rate of $bI$.

3. The epidemic is over when the number of infected people is zero, that is, at the horizontal intercept of the trajectory. This intercept is a very small $S$ value. When the epidemic is over, there are almost no susceptibles left— that is, almost everyone has become infected.

4. (a) The initial values are $I_0 = 1, S_0 = 149$.
   (b) Substituting for $I_0$ and $S_0$, initially we have
   $$\frac{dI}{dt} = 0.0026SI - 0.5I = 0.0026(149)(1) - 0.5(1) < 0.$$
   So, $I$ is decreasing. The number of infected people goes down from 1 to 0. The disease does not spread.

5. (a) $I_0 = 1, S_0 = 349$
   (b) Since $\frac{dI}{dt} = 0.0026SI - 0.5I = 0.0026(349)(1) - 0.5(1) > 0$, so $I$ is increasing. The number of infected people will increase, and the disease will spread. This is an epidemic.

6. (a) See Figure 8.20.

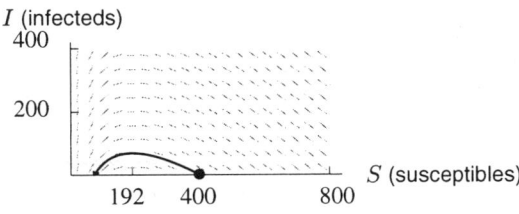

**Figure 8.20**

   (b) $I$ is at a maximum when $S = 192$.

7. The maximum value of $I$ is approximately 300 boys. This represents the maximum number of infected boys who have not (yet) been removed from circulation. It occurs at about $t \approx 6$ days.

8. In the system
$$\frac{dS}{dt} = -aSI$$
$$\frac{dI}{dt} = aSI - bI$$

the constant $a$ represents how infectious the disease is; the larger $a$, the more infectious. The constant $b$ represents $1/$(number of days before removal). Thus, the larger $b$ is, the quicker the infecteds are removed. For the flu example, $a = 0.0026$ and $b = 0.5$.

   (I) Since $0.0026 < 0.04$, this is more infectious. Since $0.2 < 0.5$, infecteds are being removed more slowly. So system (I) corresponds to (a).
   (II) Since $0.002 < 0.0026$, this is less infectious. Since $0.3 < 0.5$, infecteds are being removed more slowly. This corresponds to (c).
   (III) Since the second equation has no $-bI$ term, we have $b = 0$. The infecteds are never removed. This corresponds to (e).

A system of equations corresponding to (b) is
$$\frac{dS}{dt} = -0.04SI$$
$$\frac{dI}{dt} = 0.04SI - 0.7I.$$

A system of equations corresponding to (d) is
$$\frac{dS}{dt} = -0.002SI$$
$$\frac{dI}{dt} = 0.002SI - 0.7I.$$

**398** CHAPTER EIGHT /SOLUTIONS

9. The threshold value of $S$ is the value at which $I$ is a maximum. When $I$ is a maximum,

$$\frac{dI}{dt} = 0.04SI - 0.2I = 0,$$

so

$$S = 0.2/0.04 = 5.$$

10. Since the threshold value of $S$ is given by

$$\frac{dI}{dt} = 0.002SI - 0.3I = 0,$$

we have

$$S = \frac{0.3}{0.002} = 150.$$

So, if $S_0 = 100$, the disease does not spread initially. If $S_0 = 200$, the disease does spread initially.

## Solutions for Chapter 8 Review

1. (a) (i) If $y = Cx^2$, then $\frac{dy}{dx} = C(2x) = 2Cx$. We have

$$x\frac{dy}{dx} = x(2Cx) = 2Cx^2$$

and

$$3y = 3(Cx^2) = 3Cx^2$$

Since $x\frac{dy}{dx} \neq 3y$, this is not a solution.

(ii) If $y = Cx^3$, then $\frac{dy}{dx} = C(3x^2) = 3Cx^2$. We have

$$x\frac{dy}{dx} = x(3Cx^2) = 3Cx^3,$$

and

$$3y = 3Cx^3.$$

Thus $x\frac{dy}{dx} = 3y$, and $y = Cx^3$ is a solution.

(iii) If $y = x^3 + C$, then $\frac{dy}{dx} = 3x^2$. We have

$$x\frac{dy}{dx} = x(3x^2) = 3x^3$$

and

$$3y = 3(x^3 + C) = 3x^3 + 3C.$$

Since $x\frac{dy}{dx} \neq 3y$, this is not a solution.

(b) The solution is $y = Cx^3$. If $y = 40$ when $x = 2$, we have

$$40 = C(2^3)$$
$$40 = C \cdot 8$$
$$C = 5.$$

2. When $y = 100$, the rate of change of $y$ is
$$\frac{dy}{dt} = \sqrt{y} = \sqrt{100} = 10$$
The value of $y$ goes up by 10 units as $t$ goes up 1 unit. When $t = 1$, we have
$$y = \text{Old value of y} + \text{Change in y} = 100 + 10 = 110.$$
Continuing in this way, we obtain the table:

| $t$ | 0 | 1 | 2 | 3 | 4 |
|---|---|---|---|---|---|
| $y$ | 100 | 110 | 120.5 | 131.5 | 143.0 |

3. (a) Slope field $I$ corresponds to $\frac{dy}{dx} = 1 + y$ and slope field $II$ corresponds to $\frac{dy}{dx} = 1 + x$.

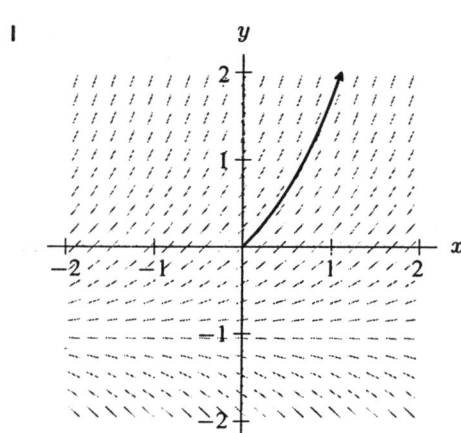 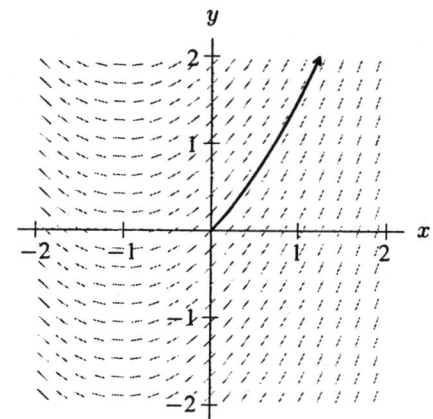

**Figure 8.21**: $\frac{dy}{dx} = 1 + y$  **Figure 8.22**: $\frac{dy}{dx} = 1 + x$

(b) See Figures ?? and ??.

(c) Slope field $I$ has an equilibrium solution at $y = -1$, since $\frac{dy}{dx} = 0$ at $y = -1$. We see in the slope field that this equilibrium solution is unstable. Slope field $II$ does not have any equilibrium solutions.

4. Integrating both sides we get
$$y = \frac{5}{2}t^2 + C,$$
where $C$ is a constant.

5. The general solution is
$$y = Ce^{5t}.$$

6. We know that the general solution to an equation of the form
$$\frac{dP}{dt} = kP$$
is
$$P = Ce^{kt}.$$
Thus in our case the solution is
$$P = Ce^{0.03t},$$
where $C$ is some constant.

**7.** Integrating both sides gives
$$P = \frac{1}{2}t^2 + C,$$
where $C$ is some constant.

**8.** Multiplying both sides by $Q$ gives
$$\frac{dQ}{dt} = 2Q, \quad \text{where } Q \neq 0.$$
We know that the general solution to an equation of the form
$$\frac{dQ}{dt} = kQ$$
is
$$Q = Ce^{kt}.$$
Thus in our case the solution is
$$Q = Ce^{2t},$$
where $C$ is some constant, $C \neq 0$.

**9.** We know that the general solution to the differential equation
$$\frac{dy}{dx} = k(y - A)$$
is
$$y = Ce^{kx} + A.$$
Thus in our case we factor out 0.2 to get
$$\frac{dy}{dx} = 0.2\left(y - \frac{8}{0.2}\right) = 0.2(y - 40).$$
Thus the general solution to our differential equation is
$$y = Ce^{0.2x} + 40,$$
where $C$ is some constant.

**10.** We know that the general solution to the differential equation
$$\frac{dP}{dt} = k(P - A)$$
is
$$P = Ce^{kt} + A.$$
Thus in our case we factor out $-2$ to get
$$\frac{dP}{dt} = -2\left(P + \frac{10}{-2}\right) = -2(P - 5).$$
Thus the general solution to our differential equation is
$$P = Ce^{-2t} + 5,$$
where $C$ is some constant.

**11.** We know that the general solution to the differential equation
$$\frac{dH}{dt} = k(H - A)$$
is
$$H = Ce^{kt} + A.$$
Thus in our case we factor out 0.5 to get
$$\frac{dH}{dt} = 0.5\left(H + \frac{10}{0.5}\right) = 0.5(H - (-20)).$$
Thus the general solution to our differential equation is
$$H = Ce^{0.5t} - 20,$$
where $C$ is some constant.

12. Since $\frac{dy}{dt} = -(y - 100)$, the general solution is $y = 100 + Ce^{-t}$.

13. We know that the general solution to an equation of the form

$$\frac{dP}{dt} = kP$$

is

$$P = Ce^{kt}.$$

Thus in our case the solution is

$$P = Ce^{0.08t}.$$

We know that $P = 5000$ when $t = 0$ so solving for $C$ we get

$$P = Ce^{0.08t}$$
$$5000 = Ce^0$$
$$C = 5000.$$

Thus the solution is

$$P = 5000e^{0.08t}.$$

A graph of this function is shown in Figure 8.23.

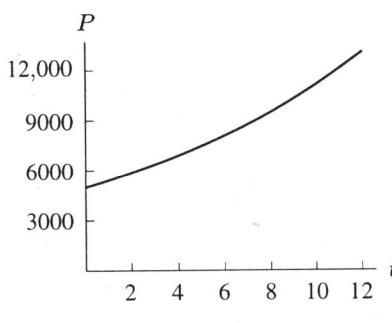

Figure 8.23

14. The general solution is $y = Ce^{-0.2t}$. Since $y = 25$ when $t = 0$, we have $25 = Ce^0$ and so $C = 25$. The solution is

$$y = 25e^{-0.2t}.$$

See Figure 8.24 for a graph of function.

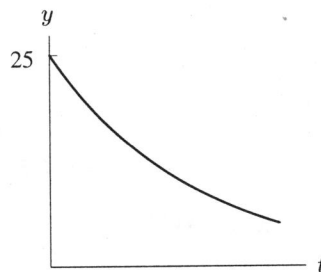

Figure 8.24

**15.** We know that the general solution to the differential equation

$$\frac{dP}{dt} = k(P - A)$$

is

$$P = Ce^{kt} + A.$$

Thus in our case we factor out 0.08 to get

$$\frac{dP}{dt} = 0.08\left(P - \frac{50}{0.08}\right) = 0.08(P - 625).$$

Thus the general solution to our differential equation is

$$P = Ce^{0.08t} + 625.$$

Solving for $C$ with $P(0) = 10$ we get

$$P(t) = Ce^{0.08t} + 625$$
$$10 = Ce^0 + 625$$
$$C = -615.$$

Thus the solution is

$$P = 625 - 615e^{0.08t}.$$

The graph of this function is shown in Figure 8.25

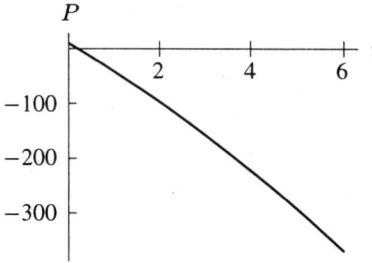

**Figure 8.25**

**16.** We know that the general solution to the differential equation

$$\frac{dH}{dt} = k(H - A)$$

is

$$H = Ce^{kt} + A.$$

Thus in our case we factor out $-0.5$ to get

$$\frac{dH}{dt} = -0.5\left(H - \frac{100}{-0.5}\right) = -0.5(H - (-200)).$$

Thus the general solution to our differential equation is

$$H = Ce^{-0.5t} - 200.$$

Solving for $C$ with $H(0) = 40$ we get

$$H(t) = Ce^{-0.5t} - 200$$
$$40 = Ce^0 - 200$$
$$C = 240.$$

Thus the solution is
$$H = 240e^{-0.5t} - 200.$$
The graph of this function is shown in Figure 8.26

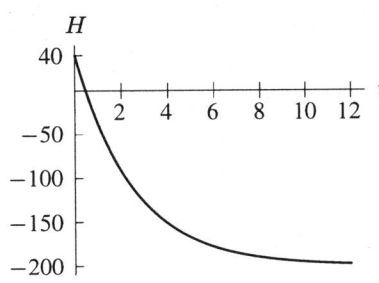

**Figure 8.26**

17. (a) We know that the equilibrium solutions are those functions which satisfy the differential equation and whose derivative is everywhere 0. Thus we must solve
$$0 = \frac{dy}{dx}$$
$$= 0.5y(y-4)(2+y)$$
Thus the equilibrium solutions are $y = 0$, $y = 4$, and $y = -2$.

(b) The slope field of the differential equation is shown in Figure 8.27. An equilibrium solution is stable if a small change in the initial conditions gives a solution which tends toward the equilibrium as the independent variable tends to positive infinity. Looking at Figure 8.27 we see that the only stable solution is $y = 0$.

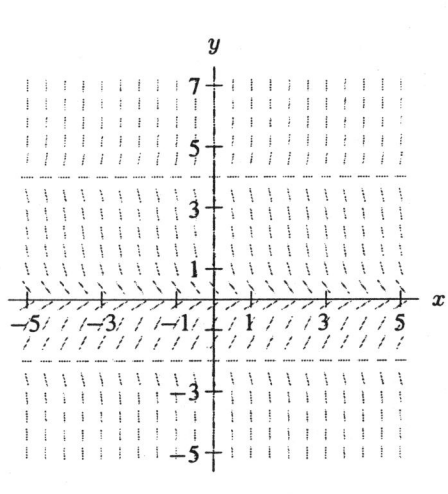

**Figure 8.27**

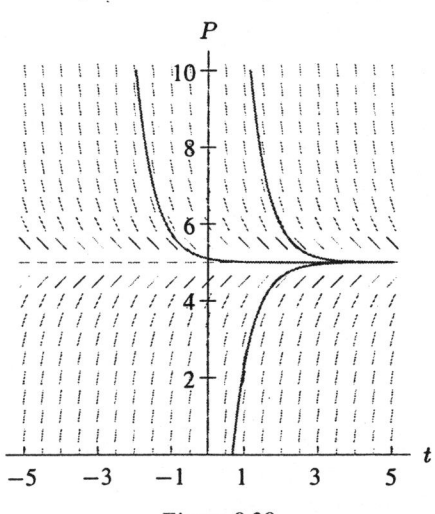

**Figure 8.28**

18. The graph of the slope field of
$$\frac{dP}{dt} = 10 - 2P$$
and three solutions are shown in Figure 8.28.

19. (a) Since interest is earned continuously,
$$\text{Rate of change of balance} = 1.5\%(\text{Balance})$$
so
$$\frac{dB}{dt} = 0.015B.$$

(b) $B = Ce^{0.015t}$ is the general solution. Since $B = 5000$ when $t = 0$, we have $C = 5000$. The solution is $B = 5000e^{0.015t}$

(c) When $t = 10$, $B = 5000e^{0.015(10)} = \$5809.17$.

**404** CHAPTER EIGHT /SOLUTIONS

20. (a) $k$ is positive because if $T > A$, then the body will lose heat so that its temperature falls to $A$. Thus, $dT/dt$ should be negative, so $k$ should be positive. Similarly, if $T < A$ then $dT/dt$ should be positive, so $k$ again should be positive.

    (b) The units for $dT/dt$ is degrees/time. Since the units for $T - A$ are degrees, the units for $k$ are (time)$^{-1}$ or 1/time. Thus, if we change from days to hours, $k$ would be different. For example,
    $$\frac{1}{1 \text{ day}} = \frac{1}{24}\left(\frac{1}{1 \text{ hour}}\right).$$
    In words, Daily $k = 24$(Hourly $k$). Similarly, the numerical value of $k$ for hours is 60 times the value of $k$ for minutes.

    (c) Everything else being equal, coffee will cool faster in a thin china cup than in styrofoam. Thus, for a given temperature difference, $dT/dt$ should be larger in magnitude for china, and therefore $k$ should be larger.

    (d) We have
    $$\frac{dT}{dt} = -0.14(T - 70),$$
    and $T = 170$ when $t = 0$. Solving this, we have $T = 70 + Ae^{-0.14t}$. The initial condition gives us that $170 = 70 + A$, so $A = 100$. Therefore the solution to the differential equation is $T = 70 + 100e^{-0.14t}$. To find the amount of time we must wait for the coffee to cool to 120 degrees or less, we put $T = 120$ into the solution and solve for $t$.
    $$120 = 70 + 100e^{-0.14t} \text{ giving } t \approx 5 \text{ minutes}.$$
    To find how soon we must drink the coffee before it cools to less than 90 degrees, we put $T = 90$ into the solution and solve for $t$.
    $$90 = 70 + 100e^{-0.14t} \text{ giving } t \approx 11.5 \text{ minutes}.$$
    Therefore our fussiness requires that we not drink our coffee until it has cooled for 5 minutes and then we must finish it before it has cooled for more than 11.5 minutes.

21. (a) The value of the company satisfies

    Rate of change of value = Rate interest earned − Rate expenses paid

    so
    $$\frac{dV}{dt} = 0.02V - 80{,}000.$$

    (b) We find $V$ when
    $$\frac{dV}{dt} = 0$$
    $$0.02V - 80{,}000 = 0$$
    $$0.02V = 80{,}000$$
    $$V = 4{,}000{,}000$$
    There is an equilibrium solution at $V = \$4{,}000{,}000$. If the company has \$4,000,000 in assets, its earnings will exactly equal its expenses.

    (c) The general solution is
    $$V = 4{,}000{,}000 + Ce^{0.02t}.$$

    (d) If $V = 3{,}000{,}000$ when $t = 0$, we have $C = -1{,}000{,}000$. The solution is
    $$V = 4{,}000{,}000 - 1{,}000{,}000e^{0.02t}.$$
    When $t = 12$, we have
    $$V = 4{,}000{,}000 - 1{,}000{,}000e^{0.02(12)}$$
    $$= 4{,}000{,}000 - 1{,}271{,}249$$
    $$= \$2{,}728{,}751.$$

    The company is losing money.

**22.** (a) Since the amount leaving the blood is proportional to the quantity in the blood,

$$\frac{dQ}{dt} = -kQ \quad \text{for some positive constant } k.$$

Thus $Q = Q_0 e^{-kt}$, where $Q_0$ is the initial quantity in the bloodstream. Only 20% is left in the blood after 3 hours. Thus $0.20 = e^{-3k}$, so $k = \frac{\ln 0.20}{-3} \approx 0.5365$. Therefore $Q = Q_0 e^{-0.5365t}$.

(b) Since 20% is left after 3 hours, after 6 hours only 20% of that 20% will be left. Thus after 6 hours only 4% will be left, so if the patient is given 100 mg, only 4 mg will be left 6 hours later.

**23.** (a) We have

$$\frac{dQ}{dt} = r - \alpha Q.$$

We know that the general solution to a differential equation of the form

$$\frac{dQ}{dt} = k(Q - A)$$

is

$$Q = Ce^{kt} + A.$$

Factoring out a $-\alpha$ on the left side we get

$$\frac{dQ}{dt} = -\alpha \left( Q - \frac{r}{\alpha} \right).$$

Thus in our case we get

$$Q = Ce^{-\alpha t} + \frac{r}{\alpha}.$$

We know that at $t = 0$ we have $Q = 0$ so solving for $C$ we get

$$Q = Ce^{-\alpha t} + \frac{r}{\alpha}$$
$$0 = Ce^0 + \frac{r}{\alpha}$$
$$C = -\frac{r}{\alpha}.$$

Thus we get

$$Q = -\frac{r}{\alpha} e^{-\alpha t} + \frac{r}{\alpha}.$$

So,

$$Q_\infty = \lim_{t \to \infty} Q = \frac{r}{\alpha}.$$

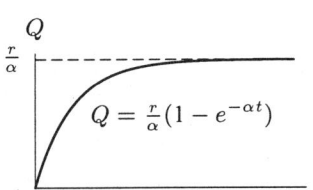

(b) Doubling $r$ doubles $Q_\infty$. Since $Q_\infty = r/\alpha$, the time to reach $\frac{1}{2} Q_\infty$ is obtained by solving

$$\frac{r}{2\alpha} = \frac{r}{\alpha}(1 - e^{-\alpha t})$$
$$\frac{1}{2} = 1 - e^{-\alpha t}$$
$$e^{-\alpha t} = \frac{1}{2}$$
$$t = -\frac{\ln(1/2)}{\alpha} = \frac{\ln 2}{\alpha}.$$

**406** CHAPTER EIGHT /SOLUTIONS

So altering $r$ doesn't alter the time it takes to reach $\frac{1}{2}Q_\infty$. See Figure 8.29.

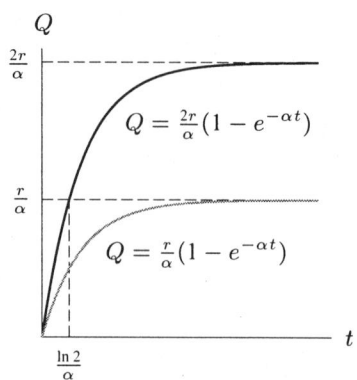

*Figure 8.29*

(c) $Q_\infty$ is halved by doubling $\alpha$, and so is the time, $t = \frac{\ln 2}{\alpha}$, to reach $\frac{1}{2}Q_\infty$.

**24.** (a) Use the fact that

$$\begin{pmatrix} \text{Rate at which} \\ \text{balance is changing} \end{pmatrix} = \begin{pmatrix} \text{Rate interest} \\ \text{is accrued} \end{pmatrix} - \begin{pmatrix} \text{Rate payments} \\ \text{are made} \end{pmatrix}.$$

Thus
$$\frac{dB}{dt} = 0.05B - 12{,}000.$$

(b) We know that the general solution to a differential equation of the form
$$\frac{dB}{dt} = k(B - A)$$
is
$$B = Ce^{kt} + A.$$

Factoring out a 0.05 on the left side we get
$$\frac{dB}{dt} = 0.05\left(B - \frac{12{,}000}{0.05}\right) = 0.05(B - 240{,}000).$$

Thus in our case we get
$$B = Ce^{0.05t} + 240{,}000.$$

We know that the initial balance is $B_0$, thus we get
$$B_0 = Ce^0 + 240{,}000$$
$$C = B_0 - 240{,}000.$$

Thus we get
$$B = (B_0 - 240{,}000)e^{0.05t} + 240{,}000.$$

(c) To find the initial balance such that the account has a 0 balance after 20 years, we solve
$$0 = (B_0 - 240{,}000)e^{(0.05)20} + 240{,}000 = (B_0 - 240{,}000)e^1 + 240{,}000,$$
$$B_0 = 240{,}000 - \frac{240{,}000}{e} \approx \$151{,}708.93.$$

**25.** Using (Rate balance changes) = (Rate interest is added) − (Rate payments are made), when the interest rate is $i$, we have
$$\frac{dB}{dt} = iB - 100.$$
We know that the general solution to a differential equation of the form
$$\frac{dB}{dt} = k(B - A)$$
is
$$B = Ce^{kt} + A.$$
Factoring out an $i$ on the left side we get
$$\frac{dB}{dt} = i\left(B - \frac{100}{i}\right).$$
Thus in our case we get
$$B = Ce^{it} + \frac{100}{i}.$$
We know that at $t = 0$ we have $B = 1000$ so solving for $C$ we get
$$1000 = Ce^0 + \frac{100}{i}$$
$$C = 1000 - \frac{100}{i}.$$
Thus $B = \frac{100}{i} + (1000 - \frac{100}{i})e^{it}$.
When $i = 0.05$, $B = 2000 - 1000e^{0.05t}$.
When $i = 0.1$, $B = 1000$.
When $i = 0.15$, $B = 666.67 + 333.33e^{0.15t}$.
We now look at the graph when $i = 0.05$, $i = 0.1$, and $i = 0.15$.

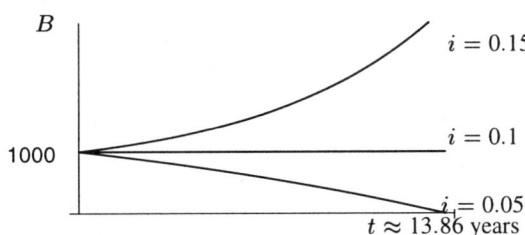

**26.** (a) For this situation,
$$\begin{pmatrix}\text{Rate money added}\\ \text{to account}\end{pmatrix} = \begin{pmatrix}\text{Rate money added}\\ \text{via interest}\end{pmatrix} + \begin{pmatrix}\text{Rate money}\\ \text{deposited}\end{pmatrix}$$
Translating this into an equation yields
$$\frac{dB}{dt} = 0.1B + 1200.$$
(b) We know that the general solution to the differential equation
$$\frac{dB}{dt} = k(B + A)$$
is
$$B = Ce^{kt} - A$$
We factor out 0.1 to put our equation in the form
$$\frac{dB}{dt} = 0.1\left(B + \frac{1200}{0.1}\right) = 0.1(B + 12{,}000).$$
This equation has the solution
$$B = Ce^{0.1t} - 12{,}000.$$
Solving for $C$ with $B(0) = 0$ we get $C = 12{,}000$ and so
$$B = f(t) = 12{,}000(e^{0.1t} - 1).$$

(c) After 5 years, the balance is
$$B = f(5) = 12{,}000(e^{(0.1)(5)} - 1)$$
$$\approx 7784.66 \text{ dollars.}$$

27. Let $D(t)$ be the quantity of dead leaves, in grams per square centimeter. Then $\frac{dD}{dt} = 3 - 0.75D$, where $t$ is in years. We know that the general solution to a differential equation of the form
$$\frac{dD}{dt} = k(D - B)$$
is
$$D = Ae^{kt} + B.$$

Factoring out a $-0.75$ on the left side we get
$$\frac{dD}{dt} = -0.75\left(D - \frac{-3}{-0.75}\right) = -0.75(D - 4).$$

Thus in our case we get
$$D = Ae^{-0.75t} + 4.$$

If initially the ground is clear, the solution looks like the following graph:

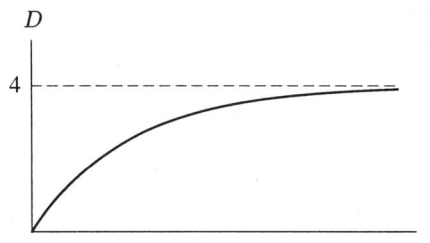

The equilibrium level is 4 grams per square centimeter, regardless of the initial condition.

28. (a) Since the rate of change of the weight is equal to
$$\frac{1}{3500}(\text{Intake} - \text{Amount to maintain weight})$$
we have
$$\frac{dW}{dt} = \frac{1}{3500}(I - 20W).$$

(b) We know that the general solution to a differential equation of the form
$$\frac{dW}{dt} = k(W - A)$$
is
$$W = Ce^{kt} + A.$$

Factoring out a $-20$ on the left side we get
$$\frac{dW}{dt} = \frac{-20}{3500}\left(W - \frac{-I}{-20}\right) = -\frac{2}{350}\left(W - \frac{I}{20}\right).$$

Thus in our case we get
$$W = Ce^{-\frac{2}{350}t} + \frac{I}{20}.$$

Let us call the person's initial weight $W_0$ at $t = 0$. Then $W_0 = \frac{I}{20} + Ce^0$, so $C = W_0 - \frac{I}{20}$. Thus
$$W = \frac{I}{20} + \left(W_0 - \frac{I}{20}\right)e^{-\frac{2}{350}t}.$$

(c) Using part (b), we have $W = 150 + 10e^{-\frac{2}{350}t}$. This means that $W \to 150$ as $t \to \infty$. See the following figure.

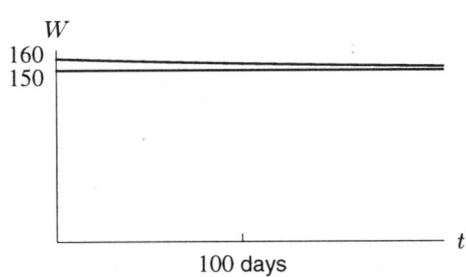

**29.** (a) $dp/dt = -k(p - p_0)$, where $k$ is constant. Notice that $k > 0$, since if $p > p_0$ then $dp/dt$ should be negative, and if $p < p_0$ then $dp/dt$ should be positive.

(b) We know that the general solution to a differential equation of the form

$$\frac{dp}{dt} = -k(p - A)$$

is

$$p = Ce^{-kt} + A.$$

In our case $A = p_0$, so we get

$$p = Ce^{-kt} + p_0.$$

If $I$ is the initial price, we get

$$p = Ce^{-kt} + p_0$$
$$I = Ce^0 + p_0$$
$$C = I - p_0.$$

Thus we get

$$p = p_0 + (I - p_0)e^{-kt}.$$

(c)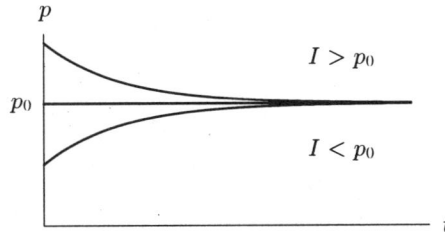

(d) As $t \to \infty$, $p \to p_0$. We see this in the solution in part (b), since as $t \to \infty$, $e^{-kt} \to 0$. (Remember $k > 0$.) In other words, as $t \to \infty$, $p$ approaches the equilibrium price $p_0$.

**30.** (a) The species need each other to survive. Both would die out without the other, and they help each other.

(b) If $x = 2$ and $y = 1$,

$$\frac{dx}{dt} = -3x + 2xy = -3(2) + 2(2)(1) < 0,$$

and so population $x$ decreases. If $x = 2$ and $y = 1$,

$$\frac{dy}{dx} = -y + 5xy = -1 + 5(2)(1) > 0,$$

and so population $y$ increases.

(c) $\dfrac{dy}{dx} = \dfrac{-y + 5xy}{-3x + 2xy}.$

(d) See Figure 8.30.

(e) See Figure 8.30.

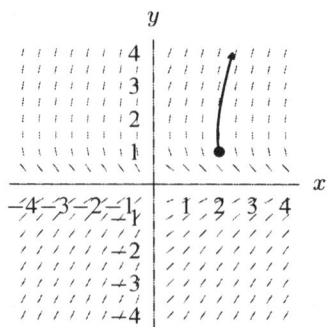

**Figure 8.30**

31. (I) Both companies start with about 4 million dollars, and both initially lose money. In the long run, however, Company $A$ makes money and Company $B$ looks like it goes out of business.

(II) Initially, Company $A$ has 2 million dollars and Company $B$ has 4 million dollars. Both companies lose money in the beginning. Company $A$ continues to lose money and probably goes out of business, but Company $B$ eventually makes money and does well.

(III) Company $A$ starts with well under 1 million dollars and Company $B$ starts with 1 million dollars. Company $B$ makes money the whole time and does well. Company $A$ shows a small profit for a while but then loses money and probably goes out of business.

(IV) Both companies start with well under 1 million dollars. Company $A$ makes money and does well, Company $B$ holds steady for a time but then loses money and probably goes out of business.

32. (a)
$$p(x) = \text{the number of people with incomes} \geq x.$$
$$p(x + \Delta x) = \text{the number of people with incomes} \geq x + \Delta x.$$

So the number of people with incomes between $x$ and $x + \Delta x$ is
$$p(x) - p(x + \Delta x) = -\Delta p.$$

Since all the people with incomes between $x$ and $x + \Delta x$ have incomes of about $x$ (if $\Delta x$ is small), the total amount of money earned by people in this income bracket is approximately $x(-\Delta p) = -x\Delta p$.

(b) Pareto's law claims that the average income of all the people with incomes $\geq x$ is $kx$. Since there are $p(x)$ people with income $\geq x$, the total amount of money earned by people in this group is $kxp(x)$.

The total amount of money earned by people with incomes $\geq (x + \Delta x)$ is therefore $k(x + \Delta x)p(x + \Delta x)$. Then the total amount of money earned by people with incomes between $x$ and $x + \Delta x$ is
$$kxp(x) - k(x + \Delta x)p(x + \Delta x).$$

Since $\Delta p = p(x + \Delta x) - p(x)$, we can substitute $p(x + \Delta x) = p(x) + \Delta p$. Thus the total amount of money earned by people with incomes between $x$ and $x + \Delta x$ is
$$kxp(x) - k(x + \Delta x)(p(x) + \Delta p).$$

Multiplying out, we have
$$kxp(x) - kxp(x) - k(\Delta x)p(x) - kx\Delta p - k\Delta x\Delta p$$

Simplifying and dropping the second order term $\Delta x \Delta p$ gives the total amount of money earned by people with incomes between $x$ and $x + \Delta x$ as
$$-kp\Delta x - kx\Delta p.$$

(c) Setting the answers to parts (a) and (b) equal gives

$$-x\Delta p = -kp\Delta x - kx\Delta p.$$

Dividing by $\Delta x$, and letting $\Delta x \to 0$ so that
$$\frac{\Delta p}{\Delta x} = \frac{p(x + \Delta x) - p(x)}{\Delta x} \to p',$$

$$x\frac{\Delta p}{\Delta x} = kp + kx\frac{\Delta p}{\Delta x}$$
$$xp' = kp + kxp'$$

so

$$(1-k)xp' = kp.$$

(d) Looking at Figure 8.31 we note that the larger the value of $k$, the slower $p(x) \to 0$ as $x \to \infty$. Also, a larger value of $k$ implies a more widespread wealth distribution.

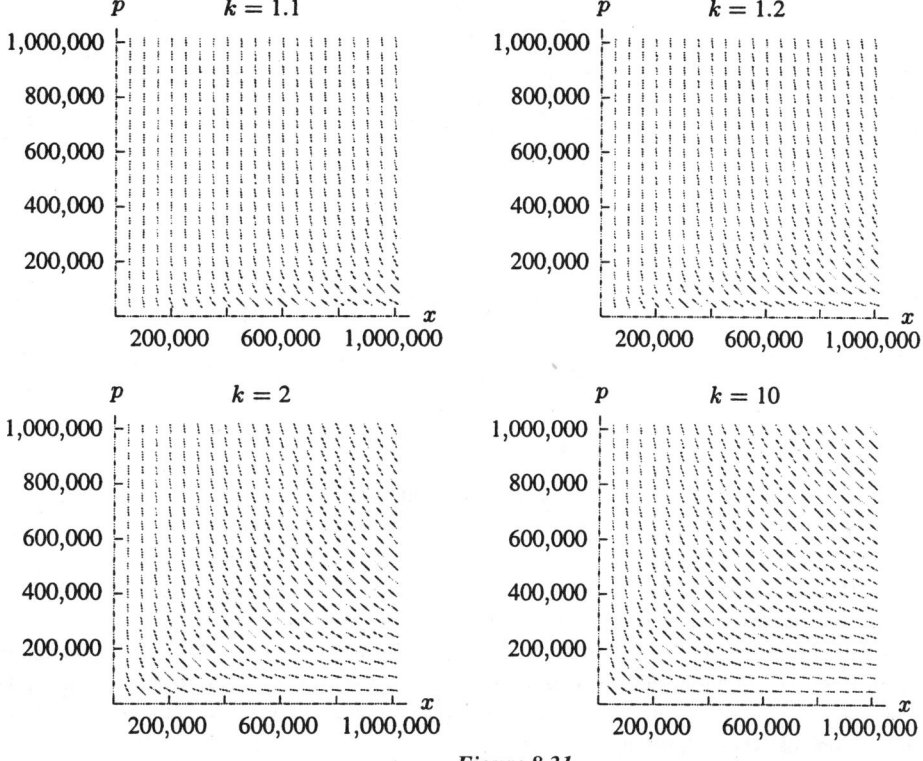

**Figure 8.31**

## Solutions to the Projects

1. (a) Equilibrium values are $N = 0$ (unstable) and $N = 200$ (stable). The graphs are shown in Figures 8.32 and 8.33.

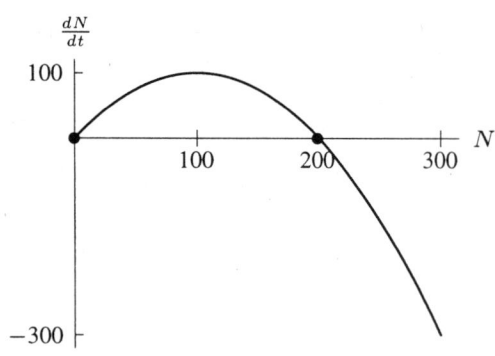

**Figure 8.32:** $dN/dt = 2N - 0.01N^2$

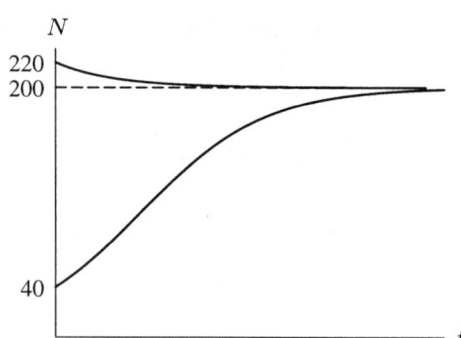

**Figure 8.33:** Solutions to $dN/dt = 2N - 0.01N^2$

(b) When there is no fishing the rate of population change is given by $\frac{dN}{dt} = 2N - 0.01N^2$. If fishermen remove fish at a rate of 75 fish/year, then this results in a decrease in the growth rate, $\frac{dP}{dt}$, by 75 fish/year. This is reflected in the differential equation by including the $-75$.

(c) 

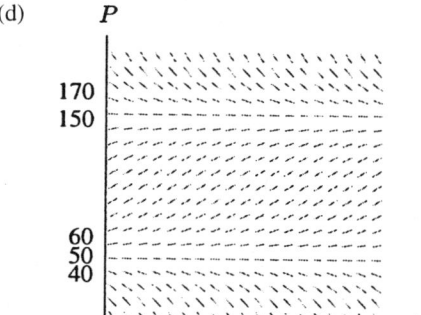

**Figure 8.34:** $dP/dt = 2P - 0.01P^2 - 75$

(d)

**Figure 8.35**

(e) In Figure 8.34, we see that $dP/dt = 0$ when $P = 50$ and when $P = 150$, that $dP/dt$ is positive when $P$ is between 50 and 150, and that $dP/dt$ is negative when $P$ is less than 50 or greater than 150.

 (i) Since $dP/dt = 0$ at $P = 50$ and at $P = 150$, these are the two equilibrium values.
 (ii) Since $dP/dt$ is positive when $P$ is between 50 and 150, we know that $P$ increases for initial values in this interval. It increases toward the equilibrium value of $P = 150$.
 (iii) Since $dP/dt$ is negative for $P$ less than 50 or $P$ greater than 150, we know $P$ decreases for starting values in these intervals. If the initial value of $P$ is less than 50, then $P$ decreases to zero and the fish all die out. If the initial value of $P$ is greater than 150, then the fish population decreases toward the equilibrium value of 150.

The solutions look like those shown in Figure 8.36.

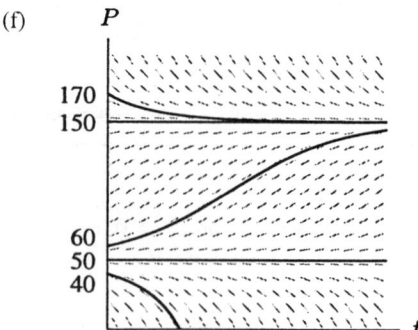

**Figure 8.36:** Solutions to $dP/dt = 2P - 0.01P^2 - 75$

(f)

**Figure 8.37**

(g) The two equilibrium populations are $P = 50, 150$. The stable equilibrium is $P = 150$, while $P = 50$ is unstable. Notice that $P = 50$ and $P = 150$ are solutions of $dP/dt = 0$:

$$\frac{dP}{dt} = 2P - 0.01P^2 - 75 = -0.01(P^2 - 200P + 7500) = -0.01(P-50)(P-150).$$

(h) (i)

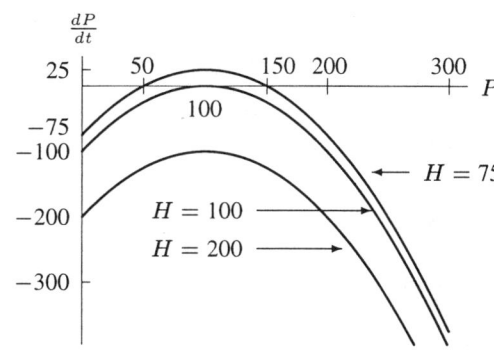

(ii) For $H = 75$, the equilibrium populations (where $dP/dt = 0$) are $P = 50$ and $P = 150$. If the population is between 50 and 150, $dP/dt$ is positive. This means that when the initial population is between 50 and 150, the population will increase until it reaches 150, when $dP/dt = 0$ and the population no longer increases or decreases. If the initial population is greater than 150, then $dP/dt$ is negative, and the population decreases until it reaches 150. Thus 150 is a stable equilibrium. However, 50 is unstable.

For $H = 100$, the equilibrium population (where $dP/dt = 0$) is $P = 100$. For all other populations, $dP/dt$ is negative and so the population decreases. If the initial population is greater than 100, it will decrease to the equilibrium value, $P = 100$. However, for populations less than 100, the population decreases until the species dies out.

For $H = 200$, there are no equilibrium points where $dP/dt = 0$, and $dP/dt$ is always negative. Thus, no matter what the initial population, the population always dies out eventually.

(iii) If the population is not to die out, looking at the three cases above, there must be an equilibrium value where $dP/dt = 0$, i.e. where the graph of $dP/dt$ crosses the $P$ axis. This happens if $H \leq 100$. Thus provided fishing is not more than 100 fish/year, there are initial values of the population for which the population will not be depleted.

(iv) Fishing should be kept below the level of 100 per year.

2. (a) In each generation, mutation causes the fraction of $b$ genes to decrease $k_1$ times the fraction of $b$ genes (as $b$ genes mutate to $B$ genes). Likewise, in every generation, mutation causes the fraction of $b$ genes to increase by $k_2$ times the fraction of $B$ genes (as $B$ genes mutate to $b$ genes). Therefore, $q$ decreases by $k_1 q$ and increases by $k_2(1-q)$, and we have:

$$\frac{dq}{dt} = -k_1 q + k_2(1-q).$$

(b) We have

$$\begin{aligned}\frac{dq}{dt} &= -0.0001q + 0.0004(1-q)\\ &= -0.0001q + 0.0004 - 0.0004q\\ &= -0.0005q + 0.0004\\ &= -0.0005(q - 0.8).\end{aligned}$$

The solution to this differential equation is

$$q = 0.8 + Ce^{-0.0005t}.$$

If $q_0 = 0.1$, then $C = -0.7$ and the solution is $q = 0.8 - 0.7e^{-0.0005t}$. If $q_0 = 0.9$, then $C = 0.1$ and the solution is $q = 0.8 + 0.1e^{-0.0005t}$. These solution are in Figure 8.38.

The equilibrium value is $q = 0.8$. From Figure 8.38, we see that as generations pass, the fraction of genes responsible for the recessive trait gets closer to 0.8.

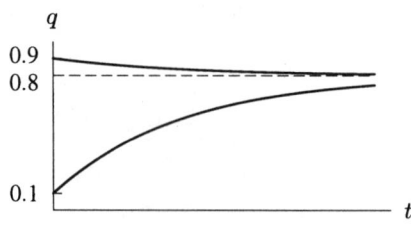

**Figure 8.38**

The equilibrium is given by the solution to the equation

$$\frac{dq}{dt} = -0.0005q + 0.0004 = 0.0005(q - 0.8) = 0.$$

Therefore the equilibrium is given by $0.0004/0.0005 = 0.8$ and so is completely determined by the values of $k_1$ and $k_2$.

(c) We have

$$\begin{aligned}\frac{dq}{dt} &= -0.0003q + 0.0001(1 - q) \\ &= -0.0003q + 0.0001 - 0.0001q \\ &= -0.0004q + 0.0001 \\ &= -0.0004(q - 0.25).\end{aligned}$$

The solution to this differential equation is

$$q = 0.25 + Ce^{-0.0004t}.$$

If $q_0 = 0.1$, then $C = -0.15$ and the solution is $q = 0.25 - 0.15e^{-0.0004t}$. If $q_0 = 0.9$, then $C = 0.65$ and the solution is $q = 0.25 + 0.65e^{-0.0004t}$. These solutions are shown in Figure 8.39.

The equilibrium value is $q = 0.25$. As more generations pass, the fraction of genes responsible for the recessive trait gets and closer to 0.25.

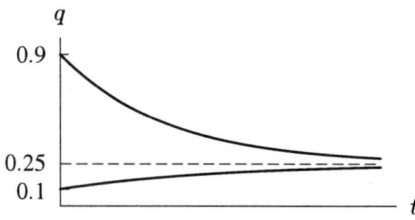

**Figure 8.39**

# APPENDIX

## Solutions to the Malthus Project

1. Answer depends on spreadsheet program used.
2. (a) The ratio is a maximum in the year 2013.
   (b) The ratios is 1 in the year 2076.
3. (a) If the population growth rate is 2.5%, the ratio is about 1 in the year 2100.
   (b) If the food supply grows at a rate of 175,000 per year, the ratio will be about 1 in the year 2100.
4. (a) See Figure A.1.

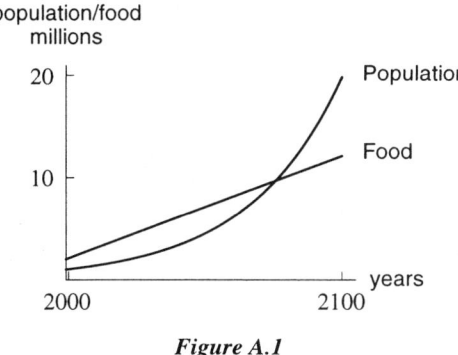

*Figure A.1*

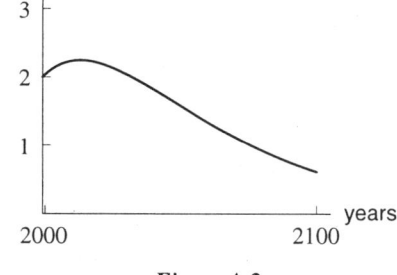

*Figure A.2*

(b) See Figure A.2.

## Solutions to the Credit Card Debt Project

1. The effective annual rate is about 19.56%, since $(1.015)^{12} \approx 1.1956$.
2. If the minimum required payment didn't at least match the interest charges, then the amount you owe would always be increasing, and the credit card company would never get back all of the money that you owe them. So, the company makes sure that, even if you never pay more than the minimum amount required, they will eventually always get all of their money (plus interest). However, it's to their advantage to have customers drag out their payments for long periods; in fact, most card companies consider people who conscientiously pay off their entire balance every month to be "problem customers" in the sense that they aren't generating any revenue in the form of interest payments.
3. Your guess should be *at least* 40 months (3 years, 4 months). The reason is that it would take 40 equal payments of $50 each to pay off $2000 assuming that there was no interest. Thus, since of course there is interest, it must take longer than 40 months; as it happens, it takes quite a bit longer.
4. As you pay off your debt, your balance decreases. Since your monthly payment is calculated as a percentage of your balance, it decreases when your balance does. The fact that your monthly payments decrease means that your balance goes down more and more slowly over time. This should affect your answer to Question 3: as your monthly payments become very small, it takes longer and longer to make a dent in your remaining balance. (Another way to look at this is to see that every month your balance drops by 1%, because 2.5% − 1.5% = 1%. This 1% drop is of a large number at first—$2000—but later, the 1% is of smaller numbers, seeing as you've paid off much of what you owe. Thus, the rate of decline gradually gets slower and slower.)
5. This is a tough guess to make, since the amount of money you owe is always decreasing; thus, your interest payments will also be decreasing. For this reason, people are almost always surprised by how much they'd actually end up paying.
6. After 368 months have elapsed, and you have made your 368$^{th}$ payment, (30 years and 8 months) your balance is down to $49.52. This might startle you, especially in light of the fact that it would take just over three years to pay off an interest-free loan with payments of $50. Charging interest really does make a difference.

7. You end up paying $4876.20 to bring your balance down to $49.52. After 368 months have elapsed and you've made your 368$^{th}$ payment, your balance is $49.52, and you have paid a total of $4876.20, so the total cost is $4876.20 + $49.52 = $4925.72. This means that to borrow $2000, you have to spend $2925.72 (assuming that you pay off the remaining $49.52 all at once), which is a lot of money. (However, it's important to remember that the $2876.20 we spend is spread out over a long period of time—over thirty years—which in a real sense can make it less alarming than a lump sum payment of the same amount.)

8. You *can't* bring your balance down to $0. The problem is that you never quite manage to pay off the entire debt; each month, you're only paying off 1% of the debt (the 2.5% payment less the 1.5% interest). This leads to the unrealistic situation of making extremely small monthly payments, which are eventually only fractions of a penny (far less than the price of postage). Most credit card companies avoid this kind of situation by requiring a minimum monthly payment of (for example) 2.5% or twenty dollars—whichever is higher. (You might want to figure out how long it would take for you to pay off the $2000 debt under this scheme, and how much it would cost you.)

9. (a) It requires 263 monthly payments (with the first payment at the end of the first month) to bring your balance down to $49.37. After 263 months have elapsed and you've made your 263$^{rd}$ payment, your balance is $49.37, and you have paid a total of $4482.06, so the total cost is $4482.06 + $49.37 = $4531.43. Thus, the total cost of paying off your debt is $4531.43. That's eight years, nine months faster than the original method, and you save $4925.72 − $4531.43 = $394.29. That's a big difference for just $1 per month.
    (b) It requires 61 monthly payments (with the first payment at the end of the first month) to bring your balance down to $26.84. This would cost $3050. Since you still owe $26.84, the total cost would be $3076.84. Thus, you can pay off your loan in just over five years (instead of over thirty), with a whopping savings of $1848.88, for only $50 per month.
    (c) It requires 245 monthly payments (with the first payment at the end of the first month) to bring the balance down to $49.31. The total cost of paying off your debt is $3300.47, resulting in a savings of $1625.25. That's a truly substantial savings in time and money.

10. You can realize enormous savings by doing a few simple things. First, you can reduce the amount of time that you owe money, as was the case in Question 9(b) where you paid off the loan in 5 years instead of 30. Or, you can find a credit card that charges lower interest, as was the case in Question 9(c) where a decrease in 0.5% per month resulted in huge savings. In Question 9(a), you can see that even modest contributions, if made consistently, can significantly reduce the amount of money you end up paying to a creditor.

## *Solutions to the Bank Loan Project*

1. The cheapest loan is the 9.25% biweekly loan: by making 52 biweekly payments of $42.95 each, you'll pay off your loan at a total cost of $2233.40. The next cheapest loan is the 9% two-year loan: by making 24 monthly payments of $95.00, you'll pay off the loan at a total cost of $2280.00. The most expensive loan is the three-year loan at 10%; this isn't surprising, since it has the longest term and the highest interest rate. By making 36 monthly payments of $65.24 each, you'll pay off the loan at a total cost of $2348.64. Note that in each case, we figured out the monthly or biweekly payment using the method of guess-and-check.

2. The three-year loan is the easiest on your budget—each monthly payment is more than $20 cheaper than each biweekly payment and almost $30 less than each monthly payment for the 9% loan. (You might want to try and figure out why the biweekly loan is less expensive overall than the 9% loan. After all, this is somewhat surprising, insofar as both loans are for two years, and the cheaper biweekly loan actually has the higher interest rate.)

## *Solutions to the Mortgage Project*

1. The answer depends on the specific figures that you used. However, the following rule of thumb will probably apply: the longer the period of the loan, the lower the monthly payment, but the higher the total payoff.

2. The answer depends on the specific figures that you used. However, the following rule of thumb will probably apply: the longer the period of the loan, the lower the monthly payment, but the higher the total payoff.

3. The answer depends on the specific figures that you used. However, the following rule of thumb will probably apply: the longer the period of the loan, the lower the monthly payment, but the higher the total payoff.

4. Let $x$ be the monthly payment. Initially the debt is $P$. Since $r$ is defined as $1 + i/12$, where $i$ is the annual interest rate (expressed as a decimal, not a percentage), then after one month,

$$\text{Remaining debt} = Pr - x.$$

After the second month,
$$\text{Remaining debt} = (Pr - x)r - x = Pr^2 - xr - x.$$

After the third month,
$$\text{Remaining debt} = (Pr^2 - xr - x)r - x = Pr^3 - xr^2 - xr - x.$$

The pattern continues, so that after $n$ months,
$$\text{Remaining debt} = Pr^n - xr^{n-1} - \cdots - xr - x.$$

But this expression equals zero, since after $n$ months we've paid off all our debt. Thus, solving for $x$, we find
$$x = \frac{Pr^n}{r^{n-1} + r^{n-2} + \cdots + 1}.$$

By summing the geometric series in the denominator, we arrive at the formula used above.

## Solutions to the Lottery Project

1. At 5%, the present value is about $17.5 million, or 65.4% of the total prize. At 10%, the present value is about $12.5 million, or 46.8% of the total prize. At 15%, the present value is about $9.6 million, or about 36% of the total prize.

2. An interest rate between 8.5% and 9%. (See the graph for Question 3.)

3. See Figure E.3. The graph drops quickly at first, and then starts to level off. Since the present value is always less than the face value of the prize, it is to the Lottery Commission's advantage to delay payments into the future.

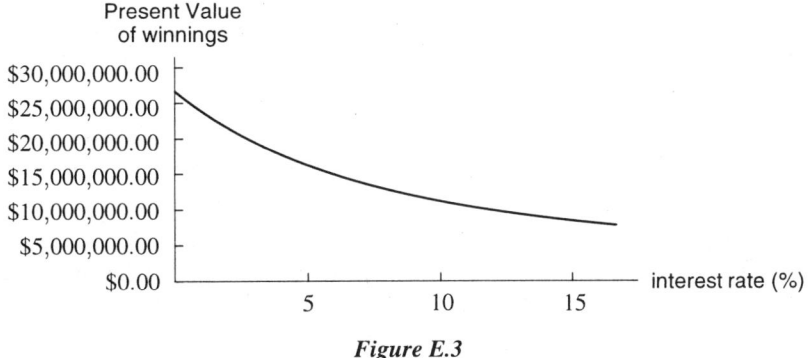

**Figure E.3**

## Solutions to the Investment Project

1. At 4%, the present value of $A$ is $3704.44, and the present value of $B$ is $6134.01. Thus, $B$ seems preferable.

2. At 16%, the present value of $A$ is $785.66, and the present value of $B$ is $163.00. Thus, $A$ seems preferable.

3. Notice that in project $A$, most of the income is in the near future, whereas in project $B$, there is a longer wait. If the discount rate is low, the penalty for waiting is small, so $B$ is favored despite the initial loss. If the discount rate is high, the penalty for waiting is great, and so $A$ is favored because of the low initial loss.

4. The internal rate of return for $A$ is just over 20%. The internal rate of return for $B$ is just over 16%.

5. The graph is found in Figure F.4. The points where the two curves cross the horizontal axis (the $x$-intercepts of the curves) correspond to the internal rates of return.

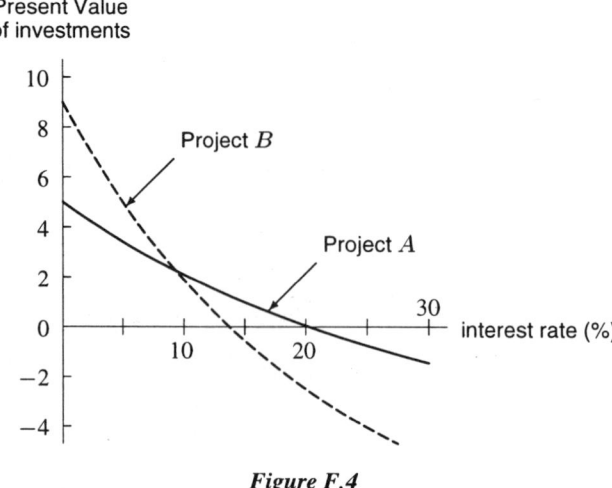

*Figure F.4*

## Solutions to the Tuition Project

1. The exact amount required is $59,712.99.
2. The present value is $59,712.99.
3. The answers are the same. It is not a coincidence. In Question 1, you figured out the exact amount of money that if deposited today would generate a given future payment stream. But that means that, by definition, this deposit is the present value of that payment stream, which is what you calculated in Question 2.
4. The account would need to earn about 11.81%, since this is the discount rate that gives the eight annual $10,000 payments a present value of $50,000.

## Solutions to the Verhulst Project

1. The equation $r = 0.0001(10,000 - P)$ is linear. Furthermore, it states that $r = 1 = 100\%$ when $P$ is zero, and that $r = 0$ when $P$ is at 10,000 (the carrying capacity). This is what Verhulst wanted: a decreasing linear relationship between $P$ and $r$.
2. The graph is found in Figure H.5. The population doesn't seem to change much for the first six to eight months. It then rises rapidly and seems to be increasing the fastest after about thirteen months. After sixteen months, though, it has leveled off, and hardly seems to be changing at all.

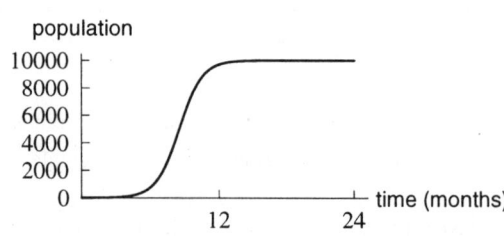

*Figure H.5*: Rabbit population over the first two years

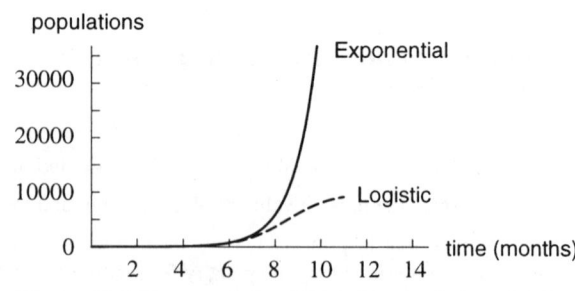

*Figure H.6*: Exponential and logistic growth over the first 14 months

3. The graph is found in Figure H.6. During the first 7 months the two populations seem identical. But then the exponentially growing population really takes off, and after sixteen months, when the logistic population has reached its carrying capacity of 10,000, the exponential population is over ten times as large. Since in real life populations always level off eventually, the logistic model offers a true advantage over the exponential model.

4. See Figure H.7. At first, the growth rate is almost steady at 100%. After about 6 months, though, it begins to drop – gradually at first, then rapidly, until it finally levels off at about 0% after 16 months or so. Thus, the relative growth rate is not at all linear with respect to time. However, it is linear with respect to the size of the growing population.

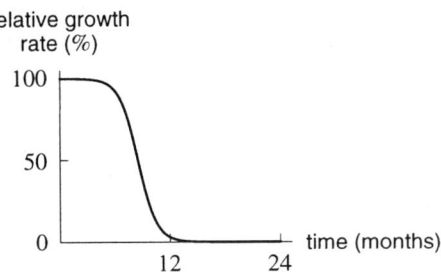

*Figure H.7:* Relative growth rate versus time over the first two years

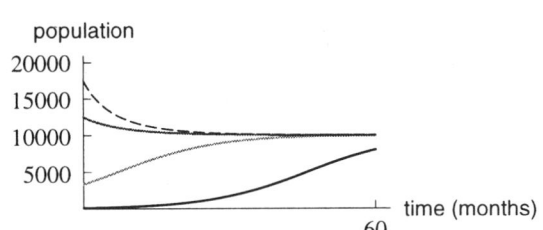

*Figure H.8:* Populations with various initial conditions over first five years

5. (a) The formula for $r$ in terms of $P$ is $r = 0.00001(10000 - P)$.
   (b) See Figure H.8. When the initial population is greater than the carrying capacity of the island, the population falls instead of rises. This is because the island cannot sustain more than 10,000 rabbits; consequently, they die off—rapidly at first, and later on more gradually. (Notice that our formula for $r$ automatically gives negative values for populations exceeding the carrying capacity.)

## *Solutions to the Spread of Information Project*

1.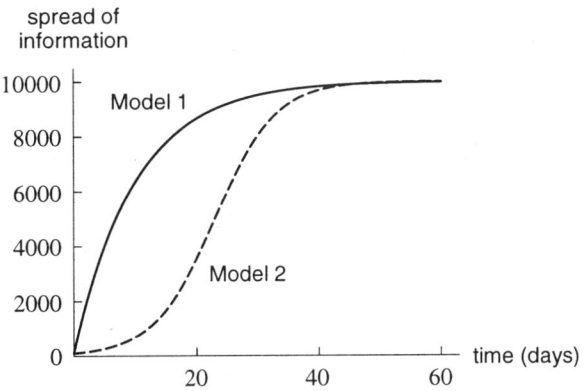

*Figure I.9:* Spread of information over first 60 days

In Model 1, the information spreads very rapidly at first, and later levels off gradually until the point where almost everybody has the information. This makes sense given the presence of mass media, because one would expect that it would not take long for most people to gain the information if it were broadcast widely. In Model 2, the information hardly seems to spread at all at first. However, the number of people having the information eventually seems to reach a "critical mass", and after that the information spreads extremely rapidly. The spread slows down as the information saturates the population. This makes sense assuming that the information is spread by word of mouth alone: at first, very few people learn the information, but after awhile, enough people know that word starts to get around pretty quickly.

2. The second model is logistic; the first is not. One way to tell is by the appearance of the graphs you made: the second one simply looks logistic and has the expected "sigmoid" shape, while the first one does not look at all logistic. Another way to tell is to plot the relative growth rate against the population. If you do this using the numbers for the second model, you get a descending line, as expected. If you do this using the numbers for the first model, you don't get a straight line. The first model is actually exhibiting a type of exponential growth. The number of people who don't have the information is decreasing by 10% every week. Thus, the number of people without the information is decreasing exponentially. This means that the number of people with the information is the difference between a constant and a decreasing exponential function.

3. Let $r$ be the absolute rate of change, and $P$ be the current population. From the description of the second model, we have $r$ equaling 0.002% times the product of the number who know and the number who don't know. This equation can be written as
$$r = 0.00002(P)(10000 - P).$$
We divide both sides of this equation by $P$:
$$\frac{r}{P} = 0.00002(10000 - P).$$
If $r$ is the rate of change of the population, then for any unit of time, the amount of change will be $r$. (For example, if the rate of change in population is 40 people per day, then in one day, the change in population will be 40.) Thus, the quantity $r/P$ is just the fractional (or percent) change, that is, the relative rate of change. From the equation, we see that the relative rate of change, $r/P$, is a decreasing linear function of $P$—the salient feature of logistic growth.

4. We've been assuming that everything is constant in between days, months, years, or whatever unit of time we're using. Take as an example the case of the spread of information: on the first day, the initial rate of change (using model one) is 990 people per day. But surely not every one of these people gain this information simultaneously. The problem is that, after the first few people out of the 990 gain the information, the rate of change changes. In model one, once just one new person gains the information, there would be 101 people in possession of it; this means that 9899 are without the information, and so the rate of change is now 989.9 people per day, slightly slower than the initial 990 per day. When the next person gains the information, the rate will drop again, to 989.8; and so on. Thus, by assuming that the rate is constant over the course of each day, we are only finding approximate solutions. How valid do you suppose these approximations are? Do you see a way that we could improve these approximations?